Molecular Simulations and Visualization

University of Nottingham, Nottingham, UK

7-9 May 2014

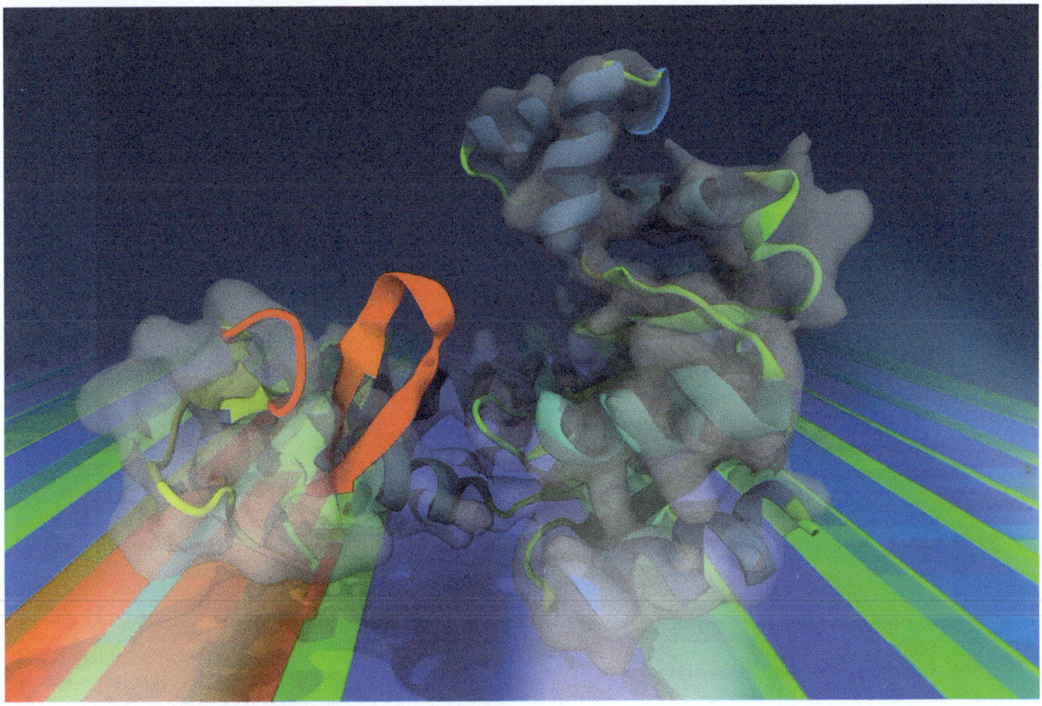

FARADAY DISCUSSIONS

Volume 169, 2014

ROYAL SOCIETY OF CHEMISTRY

The Faraday Division of the Royal Society of Chemistry, previously the Faraday Society, founded in 1903 to promote the study of sciences lying between Chemistry, Physics and Biology.

Faraday Discussions (Print ISSN 1359-6640, Electronic ISSN 1364-5498) is published 8 times a year by the Royal Society of Chemistry, Thomas Graham House, Science Park, Milton Road, Cambridge, UK CB4 0WF.

Volume 169 ISBN-13: 978-1-78262-143-0

2014 annual subscription price: print+electronic £827, US $1542; electronic only £785, US $1465. Customers in Canada will be subject to a surcharge to cover GST. Customers in the EU subscribing to the electronic version only will be charged VAT. All orders, with cheques made payable to the Royal Society of Chemistry, should be sent to RSC Order Department, Royal Society of Chemistry, Thomas Graham House, Science Park, Milton Road, Cambridge, CB4 0WF, UK.
Tel +44 (0) 1223 432398; E-mail www.orders@rsc.org

If you take an institutional subscription to any RSC journal you are entitled to free, site-wide web access to that journal. You can arrange access via Internet Protocol (IP) address at www.rsc.org/ip. Customers should make payments by cheque in sterling payable on a UK clearing bank or in US dollars payable on a US clearing bank.

Printed in the UK

MIX
Paper from responsible sources
FSC www.fsc.org FSC® C013604

Faraday Discussions documents a long-established series of Faraday Discussion meetings which provide a unique international forum for the exchange of views and newly acquired results in developing areas of physical chemistry,
biophysical chemistry and chemical physics.

Molecular Simulations and Visualization

Faraday Discussions

www.rsc.org/faraday_d

A General Discussion on Molecular Simulations and Visualization was held in Nottingham, UK on the 7th, 8th and 9th May 2014.

RSC Publishing is a not-for-profit publisher and a division of the Royal Society of Chemistry. Any surplus made is used to support charitable activities aimed at advancing the chemical sciences. Full details are available from www.rsc.org

CONTENTS

ISSN 1359-6640; ISBN 978-1-78262-143-0

Cover
Image courtesy of John E. Stone, Ryan McGreevy, Barry Isralewitz, and Klaus Schulten, University of Illinois, USA.

INTRODUCTORY LECTURE

PAPERS AND DISCUSSIONS

SPONSORS

Janssen
PHARMACEUTICAL COMPANIES OF Johnson & Johnson

NVIDIA.

SUPPORTING ORGANISATIONS

Labex Dynamo

MGMS molecular graphics & modelling society

CCPBioSim

British Biophysical Society

IOP Institute of Physics
Molecular Physics Group

Faraday Discussions

PAPER

Molecular simulations and visualization: introduction and overview

Jonathan D. Hirst,[a] David R. Glowacki[bcde] and Marc Baaden*[f]

Received 25th September 2014, Accepted 25th September 2014
DOI: 10.1039/c4fd90024c

Here we provide an introduction and overview of current progress in the field of molecular simulation and visualization, touching on the following topics: (1) virtual and augmented reality for immersive molecular simulations; (2) advanced visualization and visual analytic techniques; (3) new developments in high performance computing; and (4) applications and model building.

Introduction

It is increasingly the case that biology, chemistry and materials science make extensive use of computational methods. The applications are diverse, spanning a range of timescales and lengthscales, from the cellular level (*e.g.* in systems biology) all the way down to detailed atomistic simulations of molecular assemblies, materials or small molecules. The 2013 Nobel Prize in Chemistry awarded to Martin Karplus, Michael Levitt and Arieh Warshel for the development of multiscale models for complex chemical systems is an indication of the achievements of such simulations. Molecular simulations and visualization offer fertile territory where research in human–computer interaction (HCI) and virtual reality may interact with and provide substantial benefit to computational molecular sciences. One of the principal challenges of this research area is that it is inherently cross-disciplinary and therefore requires deep exchanges beyond the boundaries of each discipline. In what follows, we outline progress in this emerging field, and offer a glimpse of potential directions. The discussion that follows is broken down into four interconnected topics. The first topic focuses on the use of virtual and augmented reality in the context of immersive molecular simulations. Progress here requires (a) advanced visualization and visual analytic

[a]*School of Chemistry, University of Nottingham, Nottingham, NG7 2RD, UK*
[b]*School of Chemistry, University of Bristol, Bristol, BS8 1TS, UK*
[c]*Department of Computer Science, University of Bristol, BS8 1UB, UK*
[d]*PULSE Institute and Dept of Chemistry, Stanford University, Stanford, CA 94305, USA*
[e]*SLAC National Accelerator Laboratory, Menlo Park, California 94025, USA*
[f]*Laboratoire de Biochimie Théorique, CNRS, UPR9080, Université Paris Diderot, Sorbonne Paris Cité, 13 rue Pierre et Marie Curie, 75005, Paris, France. E-mail: baaden@smplinux.de*

techniques, and (b) the ability to harness new developments in high performance computing (*e.g.*, general-purpose-GPUs (GP-GPUs), clouds and more), which respectively form the second and third topics in our discussion. The final topic discusses a relatively young area where progress relies on fusing the previous three topics – namely, the development of applications and serious games: from docking to model building.

Virtual and augmented reality and immersive molecular simulations

Setting up, running, and analyzing a molecular simulation is generally a lengthy process that requires considerable user patience and expertise. In principle, interactive simulations provide a much lower latency approach for manipulating and exploring molecular structures.[1] Such methods provide a virtual or augmented reality framework for immersing the user within the simulation, with the aim that molecular interaction becomes as intuitive as possible.[2] One of the earliest and perhaps most obvious methods for facilitating a more immersive molecular interaction experience involves the sense of touch. Indeed, the use of physical models has a long history within both chemistry and biochemistry, perhaps most famously captured by Watson and Crick's physical model of DNA. With the advent of computing and robotics, touch is a sensory channel that continues to receive a great deal of attention, mostly through the use of haptic interaction strategies.[3–5] Touch is also exploited in systems that utilize tangible physical models[6] that may be augmented virtually[7] and which have shown promise in enhancing student learning.[8] Beyond touch, a variety of sensory channels related to visual and audio feedback may be used to enhance the immersive effect, and preliminary applications of such integrated methods have occurred in the context of docking problems.[9] The design and utilization of efficient and effective strategies for interaction with a molecular simulation relies on careful consideration of methods developed within the field of human–computer interaction.

In general, an interactive molecular simulation framework requires several different ingredients: (1) a mechanism for representing the molecular system – usually involving a screen-based visual display and/or a physical object; (2) a mechanism for interaction with the simulation – typically through dedicated peripherals like haptic devices, cameras, a mouse, *etc.*; (3) animation of the simulation using a physics-based simulation engine; and (4) a pipeline which allows low latency coupling of these elements – often utilizing some sort of network protocol. In contrast to many fields of scientific computation, latency is an important consideration for interactive systems, and important bottlenecks need to be ameliorated – *i.e.*, the simulations need to be fast enough to run at an interactive speed, and the visualization tools need to be sufficiently elaborate and flexible so that the user can tackle both very complex and heterogeneous data. Interactive simulations are presently supported in many software packages, including NAMD,[10] LAMMPS,[11] HOOMD,[12] ProtoMol[13] and OPEP/HiRE-RNA.[14] As far as the simulation engines are concerned, one of the principal concerns is that of performance: for a reasonable interactive experience, animation of the molecular system needs to occur with refresh rates of at least 24 Hz.

Parallelisation therefore plays an important role in interactive simulation, because it can give substantial increases in the performance of molecular dynamics simulation engines. Stream architectures like GP-GPUs (as described below) provide one way to exploit parallelisation and achieve performance increases.[12,15] Compared to standard single-core environments, data transfer tends to present a significant bottleneck in massively parallel environments, resulting in a number of challenges – *e.g.*, extracting atomic positions without degrading performance in a stream parallelized software package is not trivial. More generally, MD simulations can produce a large quantity of simulation data, and transfer of this data may generate bottlenecks for coupling with visualization and interaction modules. Interactive all-atom simulations have recently been reported for systems up to 2 million atoms.[16] For the exploration of large systems, coarse graining offers another well-balanced alternative to speed up the physical engine driving an immersive simulation experiment.[17]

There are a wide variety of potential applications for interactive molecular simulation frameworks, including structural modelling, conformational searching, and interpretation of the mechanisms that drive function in complex biological models. Drug design is another active area in which new interactive simulation strategies are under investigation (including haptics interaction, virtual reality, and 3D printing[18]), in part motivated by the very high cost of bringing new medicines to market. Interactive simulations have been used to facilitate forms of nano-manipulation and even to prototype and design nanorobots.[19,20] Interactive simulation frameworks already offer considerable potential for providing microscopic insight into experiments; however, an even closer linking with experiment is likely to emerge in the near future. For example, a multitude of biophysical techniques (SAXS, cryo-EM, FRET, mass spectrometry, *etc.*) nowadays routinely generate a huge amount of biological data, opening up the possibility of combining all the data to build increasingly accurate models.[21–23] It will soon be possible to extend immersive approaches to explore not only simulation data, but experimental information as well (*e.g.* from NMR spectroscopy[24]) and to subsequently build models from such data under direct human supervision.[25] Another application of interactive molecular simulation involves reconstituting molecular assemblies from cryo-EM data.[26] These sorts of applications open a whole new range of collaborative opportunities and questions.[27–29]

Advanced visualization and visual analytics

One of the cornerstones of modern molecular simulation concerns the visual representation of the structure of a molecule and its properties. Visual representation is particularly important in guiding the manner in which scientists think about atomic and molecular structure, which is partly a result of the fact that our human perception requires some form of augmentation in order to 'see' this world.[30] Attempts to conceptualize and visualize molecules reach far back in the history of chemistry. In terms of three-dimensional molecular structures, a notable milestone along this path goes back to the early 1940s when Roger Hayward depicted the arrangement of atomic assemblies in collaboration with Linus Pauling in both a scientifically accurate and aesthetically pleasing way.[31] Since this era (and in particular recently), technical progress has considerably improved our ability to visualize the molecular world. Nowadays, molecular graphics are ubiquitous and every scientist can display the structure of a biomolecule on his/her personal computer[32,33] or tablet device.[34] To ensure that the enormous quantity of information contained in molecular simulation data (interactive or not) furnishes maximum insight into microscopic phenomena, the investigation of new visualization and visual analysis methods is an area of active research. One of the primary focuses of this emerging area concerns the development of new ways to understand and rapidly process the radically expanding deluge of data which molecular simulations are capable of generating. Visualization assists in grasping the complexity of these data and identifying emerging properties.

It is now possible to interactively visualise very large molecular assemblies, and new developments (including the use of GPU programming) are driving performance gains and opening up new possibilities for visualization.[35] Nowadays, millions of atoms and their bonds can be depicted interactively,[36–38] with considerable speed-ups in secondary structure representation.[39] Calculating molecular surfaces on-the-fly is more demanding,[40–43] but realistic rendering including the effects of lighting and ambient occlusion[44] reaches real-time refresh rates.[45] Impressive progress has been achieved with interactive raytracing of molecular systems on the GPU (http://www.molecular-visualization.com/#!home/mainPage). Using ray-casted instancing, even whole-cell simulation data may be visualized smoothly[46] on stereoscopic displays,[47] allowing the reconstruction of 3D cellular complexes built from proteins and DNA molecules.[48]

The widespread availability of molecular simulation packages means that molecular visualization must have grown accustomed to depicting time-dependent dynamics, calling for on-the-fly visualization of atomic and molecular interactions (*e.g.*, hydrogen bonds),[37] molecular properties such as helix bending,[49] or the dynamics of molecular paths and cavities.[50–52] Visual analytics (http://www.visual-analytics.eu) have great potential to aid in understanding the increasing number of simulation datasets, and have been applied in a few cases.[53,54] Simplifying large quantities of complex data like those produced by MD simulations may be achieved by appropriate abstractions. In this context, techniques from scientific and medical illustration are helpful and have found their way into the visualization of chemical structures.[55–57,112] A stimulating recent example is the continuous abstraction of a molecular illustration[58] to yield a continuum of molecular depictions. Another challenge that arises in particular for the visualization of molecular simulations concerns the depiction of molecular flexibility.[59] In fact, chemical reactions themselves are difficult to render for many visualization tools. More generally, the visualization of dynamic molecular interaction networks is a very active field of research,[60–62] but is beyond the scope of this short introduction.

The ubiquity of molecular images and associated visualization tools is in part a consequence of the fact that it has been benefited from other high-growth economic areas. For example, tools that are traditionally dedicated to areas such as the video game industry or the film industry may be used for molecular visualization[63–65] or animation.[66–68] The availability of such tools has enabled the use of molecular visualization in collaborative structural biology, for example using TV-based[69] or web-based[70] solutions. A similar cross-fertilization is observed for GPUs, originally used in the consumer graphics market and nowadays omnipresent in high performance computing and scientific visualization.

Computing power revolution and new algorithms: GP-GPUs, clouds and more

The field of molecular simulation and visualization intrinsically depends on high-performance computing (HPC) to ensure the underlying calculations can be carried out in real time and on a broad range of hardware including commodity computers. In this context, general-purpose-GPUs,[71] cloud computing,[72] and many-core architectures[73] are finding their way into the molecular simulation community. Multi-core architectures are evolving quickly, with massive-parallelism and massive-threading available on machines like the 1.3 million thread Blue Waters supercomputer. Bespoke architecture development, like that available on the Anton machine, is similarly allowing researchers to push the boundaries of simulation[74] and new techniques based on cloud-based methods and ultrafast high-performance networking are just around the corner. These and likely future developments in HPC are making massively parallel computations viable, and are stimulating innovation across hardware, software, and hardware/software integration, much of which is aimed at tackling the main challenges of molecular simulation: the size of systems which can be simulated, the time-scales which it is possible to simulate, the ability to sample large regions of molecular phase space, and the rigor of the underlying physics within the models. Lane *et al.*[75] have touched on many of these aspects in the context of protein folding.

The advent of programmable GPUs[76] using high-level languages like C, in conjunction with the NVIDIA CUDA (Compute Unified Development Architecture) tools, OpenCL and other frameworks has been instrumental in porting software and developing new algorithms. The rise of GPUs offers another example of how advances in high-growth consumer markets (namely video gaming) has been exploited for the purposes of scientific simulation. As a result of the power of GPUs, and the fact that they are relatively inexpensive, much academic and commercial molecular dynamics (MD) software (*e.g.*, GROMACS, NAMD and AMBER)[77,113] has been GPU-accelerated. The enhancement in speed can vary, in part due to the specific algorithms used, and also as a result of the particular GPU hardware architecture on which the code is run, both of which lead to different scalability and execution time. Adding many-body terms to potentials used in classical simulations is a case where the computational cost has been mitigated through exploitation of new hardware, by the development of a shared-memory force-decomposition algorithm.[78] Calculations using ReaxFF, which is a reactive force field, have been accelerated using GPUs.[79] Aspects of reliability and reproducibility have been studied in the context of error-correcting code.[80]

For quantum chemistry, software adoption of GPUs has been slower than for MD simulations, but building on initial work[81–83] many electronic structure packages now have GPU-enabled codes, and there is significant interest in utilizing fast quantum chemical methods, for example, to investigate reaction mechanisms.[84] Recent work has investigated GPU acceleration in a range of contexts, for example: (1) double precision matrix multiply operations within legacy quantum chemistry codes;[85] (2) ONETEP, a linearly-scaling plane wave density functional theory (DFT) code;[86] (3) BigDFT, a hybrid DFT code based on Daubechies wavelets;[87] (4) VASP, GPU-accelerated electronic structure calculations;[88] (5) real-space DFT implementations within the Octopus code;[89] and (6) semiempirical methods.[90] Very recently, Sisto *et al.* have outlined fragment-based quantum chemical methods which rely on both distributed and shared memory GPU parallelism to carry out very large excited state time-dependent DFT (TDDFT) calculations using the TeraChem software framework.[91]

Cloud computing[92–95] is a relatively recent approach to molecular simulation that builds on distributed computing approaches like FightAIDS@home, SETI@home, and Folding@home. Distinct from other high-performance and distributed paradigms, it provides large-scale compute infrastructure on demand. In many respects, cloud-based approaches are still in their infancy, but are attracting growing attention. For applications of molecular simulation and modelling, cloud computing can offer large-scale data and compute capability for a short 'burst' phase. Cloud computing provides another example wherein molecular simulation benefits from exploiting approaches which have applications in other sectors: for example, cloud-based computing has appeal to small and medium biotech start-ups where continuous in-house HPC facilities would be under-utilised. Embarrassingly parallel tasks, like the generation of combinatorial databases, virtual screening of millions of compounds, and the analysis of the huge genome datasets, are well suited to existing cloud provisions. A workflow system called AutoDockCloud[96] enables distributed screening on a cloud platform using the molecular docking program AutoDock. For applications with greater demands for inter-processor communications, scalability is a key issue. A plugin[97] for the popular VMD software[98] (a front-end for NAMD[10]) allows one to (1) create a cloud-compute cluster on Amazon EC2; (2) submit a parallel NAMD job; (3) transfer the results back for subsequent post-processing; and (4) shutdown and terminate the compute cluster on Amazon EC2. These and other case studies of molecular modelling using cloud computing have been reviewed by Ebejer *et al.*[72]

Crowd-sourcing and serious games: from docking to protein folding

Molecular simulation, like many areas of computational science, involves a tradeoff between user control and automation. Users usually have a deeper understanding and context for the problem at hand, but limited speed and memory. Computational systems, on the other hand, excel in memory and speed, but are limited when it comes to understanding and context. Even with the tremendous advances in computation discussed in the previous section, it is likely the case that there will always be a limit to the size and accuracy of models that can be built for a particular system, and therefore some level of human understanding will always be required. It is therefore of fundamental interest to consider radically new approaches to molecular modelling – *i.e.*, utilizing paradigms that do not rely exclusively on ever-faster computational frameworks.

Very recently, there has been a great deal of interest directed at investigating whether human intuition and problem solving skills can be effectively mobilized (usually *via* the internet) as a new resource for solving research questions.[99] The interest in these solutions is such that participants may be stimulated by the prospect of being remunerated.[100] Success in this area requires that the research approach or proposition is cast in a way that is sufficiently engaging, entertaining, or educational. Along these lines, the Defense Advanced Research Projects Agency (DARPA) recently developed a challenge to see how quickly it is possible to involve a large number of people to fulfil a particular task.[101] Such 'crowd-sourced' research approaches[102] have received increasing attention. For example, one particularly successful example is the Galaxy Zoo[103] project, which transforms a potentially mundane, but difficult computer vision task (classifying images of galaxies) into an attractive challenge. When it comes to solving scientific research problems, collective and intrinsic motivation can marshal large communities of volunteers. This requires a high-level of visibility, which social media and modern communication technologies can successfully facilitate. Once a volunteer community is established, strategies and structures must be in place to maintain the ongoing engagement of the community. In many cases, crowd-sourced scientific computing paradigms raise interesting questions related to data analysis and data integrity.

Crowd-sourced approaches to research can generate useful insight owing to user intuition as a solution to cope with complex data and unveil emerging properties: a striking example is the game Foldit.[104,105] This project presents protein folding as a sort of three-dimensional jigsaw puzzle, where players are invited to shake and wiggle the three-dimensional structure of proteins in order to find the most stable conformations. Since May 2008, when the first beta version of this game was released, the project has gathered a large community. In some cases, Foldit players have been able to find optimal structures that automated search strategies failed to sample. Players do not necessarily require significant knowledge of biology to play the game and to find stable protein configurations. It is more a matter of spatial representation in three dimensions, as well as collaboration between players. The first 'levels' of the Foldit game are designed to train the players in order to accomplish increasingly complicated tasks. Interactions among players have led to remarkable results from a biological point of view[106–108] and also led players to collaboratively develop new algorithms to solve a particular problem.[109]

Not only do interactive and video game interfaces offer the potential for crowd-sourced research studies, they also offer an engaging medium for scientific education, helping students of all ages learn scientific principles and knowledge. As a consequence, educational games are flourishing. For example, the Spectral Game[110] seeks to teach quite advanced concepts in spectroscopy, specifically proton nuclear magnetic resonance (NMR). In addition to meeting specific targets, educational games and interactive molecular simulation platforms (like the distributed computing projects discussed above[111]) have a more general effect – *i.e.*, they engage the public and thereby increase public awareness and understanding of scientific problems. New channels for engaging the public with scientific ideas are also emerging in less traditional venues – *i.e.*, on the frontiers of aesthetic imagination and scientific visualization. As art moves increasingly toward digital mediums, artists have become fascinated with the glimpse into the invisible atomic world provided by molecular simulations and visualizations, to the extent that it has inspired new forms of artistic expression and aesthetic content.[30]

Conclusion

Molecular simulation and visualization represent a vibrant melting pot of many scientific disciplines that both benefits from and drives significant progress across a range of fields. New hardware architectures, new software algorithms, and new technological developments inspire this evolution and herald an exciting era of increasingly sophisticated and perhaps unconventional molecular simulations. The potential for these new simulation frameworks is extremely exciting: they will allow us to obtain unprecedented new research insights, develop new ways for interacting with and imagining the microscopic world, drive progress in HCI and computer science, and ultimately have profound effects beyond the scientific realm within the broader culture.

Acknowledgements

The authors thank Drs Matthieu Chavent, Antoine Taly and Alex Tek for stimulating discussions and exchanges that have nourished parts of this manuscript. M.B. acknowledges support by the "Initiative d'Excellence" program from the French State (Grant "DYNAMO", ANR-11-LABX-0011). J.D.H. thanks EPSRC for support under grant EP/I006559/1. D.R.G. acknowledges support as a Royal Society Research Fellow.

References

1 A. Tek, B. Laurent, M. Piuzzi, Z. Lu, M. Baaden, O. Delalande, M. Chavent, N. Ferey, C. Martin, L. Piccinali, B. Katz, P. Bourdot, and L. Autin, Advances in Human–Protein Interaction – Interactive And Immersive Molecular Simulations, in *Protein Interactions – Computational and Experimental Tools/ Book 2*, ed. W. Cai and H. Hong, Intech, Croatia, 2012.

2 J. E. Stone, A. Kohlmeyer, K. L. Vandivort and K. Schulten, Immersive molecular visualization and interactive modeling with commodity hardware, *Lect. Notes Comput. Sci.*, 2010, **6454**, 382–393.

3 A. M. Wollacott and K. M. Merz Jr, Haptic applications for molecular structure manipulation, *J. Mol. Graphics Modell.*, 2007, **25**, 801–805.

4 N. Zonta, I. J. Grimstead, N. J. Avis and A. Brancale, Accessible haptic technology for drug design applications, *J. Mol. Model.*, 2009, **15**, 193–196.

5 A. Ricci, A. Anthopoulos, A. Massarotti, I. Grimstead and A. Brancale, Haptic-driven applications to molecular modeling: state-of-the-art and perspectives, *Future Med. Chem.*, 2012, **4**, 1219–1228.

6 M. Francl, Tangible assets, *Nat. Chem.*, 2013, **5**, 147–148.

7 A. Gillet, M. Sanner, D. Stoffler and A. Olson, Tangible Interfaces for Structural Molecular Biology, *Structure*, 2005, **13**, 483–491.

8 G. E. Höst, C. Larsson, A. Olson and L. A. Tibell, Student learning about biomolecular self-assembly using two different external representations, *CBE Life Sci. Educ.*, 2013, **12**, 471–482.

9 N. Ferey, J. Nelson, C. Martin, L. Picinali, G. Bouyer, A. Tek, P. Bourdot, J. M. Burkhardt, B. F. G. Katz, M. Ammi, C. Etchebest and L. Autin, Multisensory VR interaction for protein-docking in the CoRSAIRe project, *Virtual Reality*, 2009, **13**, 273–293.

10 J. C. Phillips, R. Braun, W. Wang, J. Gumbart, E. Tajkhorshid, E. Villa, C. Chipot, R. D. Skeel, L. Kalé and K. Schulten, Scalable molecular dynamics with NAMD, *J. Comput. Chem.*, 2005, **26**, 1781–1802.

11 S. Plimpton, Fast Parallel Algorithms for Short-Range Molecular Dynamics, *J. Comput. Phys.*, 1995, **117**, 1–19.

12 J. A. Anderson, C. D. Lorenz and A. Travesset, General purpose molecular dynamics simulations fully implemented on graphics processing units, *J. Comput. Phys.*, 2008, **227**, 5342–5359.

13 T. Matthey, and J. A. Izaguirre, ProtoMol: A Molecular Dynamics Framework with Incremental Parallelization, in *Proc Tenth SIAM Conf Parallel Processing for Scientific Computing (PP01), Proceedings in Applied Mathematics*, 2001.

14 F. Sterpone, S. Melchionna, P. Tuffery, S. Pasquali, N. Mousseau, T. Cragnolini, Y. Chebaro, J. F. Saint-Pierre, M. Kalimeri, A. Barducci, Y. Laurin, A. Tek, M. Baaden, P. Hoang Nguyen and P. Derreumaux, The OPEP protein model: from single molecules, amyloid formation, crowding and hydrodynamics to DNA/RNA systems, *Chem. Soc. Rev.*, 2014, **43**, 4871–4893.

15 J. E. Stone, D. J. Hardy, I. S. Ufimtsev and K. Schulten, GPU-accelerated molecular modeling coming of age, *J. Mol. Graphics Modell.*, 2010, **29**, 116–125.

16 M. Dreher, M. Piuzzi, A. Turki, M. Chavent, M. Baaden, N. Ferey, S. Limet, B. Raffin and S. Robert, Interactive Molecular Dynamics: Scaling up to Large Systems, *Procedia Comput Sci.*, 2013, **18**, 20–29.

17 O. Delalande, N. Ferey, G. Grasseau and M. Baaden, Complex Molecular Assemblies at hand via Interactive Simulations, *J. Comput. Chem.*, 2009, **30**, 2375–2387.

18 G. A. Dalkas, D. Vlachakis, D. Tsagkrasoulis, A. Kastania and S. Kossida, State-of-the-art technology in modern computer-aided drug design, *Briefings Bioinf.*, 2013, **14**, 745–752.

19 M. Hamdi, A. Ferreira, G. Sharma and C. Mavroidis, Prototyping bio-nanorobots using molecular dynamics simulation and virtual reality, *Microelectron. J.*, 2008, **39**, 190–201.

20 C. T. Chen, S. Y. Chen, C. H. Liao and S. C. Zeng, An interactive nanomanipulation visualization based on molecular dynamics simulation and virtual reality, *Trans. Can. Soc. Mech. Eng.*, 2013, **37**, 991–1000.

21 T. Schwede, Protein Modeling: What Happened to the "Protein Structure Gap"?, *Structure*, 2013, **21**, 1531–1540.

22 E. Karaca and A. M. J. J. Bonvin, Advances in integrative modeling of biomolecular complexes, *Methods*, 2013, **59**, 372–381.

23 D. Schneidman-Duhovny, R. Pellarin and A. Sali, Uncertainty in Integrative Structural Modeling, *Curr. Opin. Struct. Biol.*, 2014, **28**, 96–104.

24 J. N. Block, D. J. Zielinski, V. B. Chen, I. W. Davis, E. C. Vinson, R. Brady, J. S. Richardson and D. C. Richardson, KinImmerse: Macromolecular VR for NMR ensembles, *Source Code Biol. Med.*, 2009, **4**, 3.

25 G. Da Costa, A. Bondon, O. Delalande, L. Mouret and J. P. Monti, Elucidation by NMR solution of neurotensin in small unilamellar vesicle environment: molecular surveys for neurotensin receptor recognition, *J. Biomol. Struct. Dyn.*, 2013, **31**, 809–817.

26 J. Heyd and S. Birmanns, Immersive structural biology: a new approach to hybrid modeling of macromolecular assemblies, *Virtual Reality*, 2009, **13**, 245–255.
27 J. Simard, M. Ammi and M. Auvray, Collaborative strategies for the search of 3D targets in molecular environments, *IEEE Trans. Syst. Man Cybern. C Appl. Rev.*, 2012, **42**, 1555–1565.
28 J. Simard and M. Ammi, Haptic communication tools for collaborative deformation of molecules, in *Proc. of EuroHaptics*, Tampere, Finland, 2012, pp. 517–527.
29 P. Romano, R. Giugno and A. Pulvirenti, Tools and collaborative environments for bioinformatics research, *Bioinformatics*, 2011, **12**, 549–561.
30 ZKM Center for Art and Media Karlsruhe, *Molecular Aesthetics*, ed. Peter Weibel and Ljiljana Fruk, The MIT Press, Cambridge, MA, U.S.A./London, England, Germany, 2013, p. 400.
31 L. C. Pauling and R. Hayward, *Architecture of Molecules*, W.H.Freeman & Co Ltd, 1970.
32 S. I. O'Donoghue, D. S. Goodsell, A. S. Frangakis, F. Jossinet, R. A. Laskowski, M. Nilges, H. R. Saibil, A. Schafferhans, R. C. Wade, E. Westhof and A. J. Olson, Visualization of macromolecular structures, *Nat. Methods*, 2010, **7**, S42–S55.
33 C. Mura, C. M. McCrimmon, J. Vertrees and M. R. Sawaya, An Introduction to Biomolecular Graphics, *PLoS Comput. Biol.*, 2010, **6**, e1000918.
34 E. Krieger and G. Vriend, YASARA View—molecular graphics for all devices—from smartphones to workstations, *Bioinformatics*, 2014, DOI: 10.1093/bioinformatics/btu426.
35 M. Chavent, B. Levy, M. Krone, K. Bidmon, J. P. Nomine, T. Ertl and M. Baaden, GPU-powered tools boost molecular visualization, *Briefings Bioinf.*, 2011, **12**, 689–701.
36 S. Grottel, G. Reina, C. Dachsbacher and T. Ertl, Coherent Culling and Shading for Large Molecular Dynamics Visualization, *Comput. Graph. Forum*, 2010, **29**, 953–962.
37 M. Chavent, A. Vanel, A. Tek, B. Levy, S. Robert, B. Raffin and M. Baaden, GPU-accelerated atom and dynamic bond visualization using hyperballs: A unified algorithm for balls, sticks, and hyperboloids, *J. Comput. Chem.*, 2011, **32**, 2924–2935.
38 N. Lindow, D. Baum and H. C. Hege, Interactive Rendering of Materials and Biological Structures on Atomic and Nanoscopic Scale, *Comput. Graph. Forum*, 2012, **31**, 1325–1334.
39 M. Wahle and S. Birmanns, GPU-Accelerated Visualization of Protein Dynamics in Ribbon Mode, in *Proceedings of SPIE, Visualization And Data Analysis 2011*, ed. P. C. Wong, J. Park, M. C. Hao, C. Chen, K. Borner, D. L. Kao and J. C. Roberts, 2011, vol. 7868.
40 M. Chavent, B. Levy and B. Maigret, MetaMol: High-quality visualization of molecular skin surface, *J. Mol. Graphics Modell.*, 2008, **27**, 209–216.
41 M. Krone, K. Bidmon and T. Ertl, Interactive Visualization of Molecular Surface Dynamics, *IEEE Trans. Visual. Comput. Graph.*, 2009, **15**, 1391–1398.
42 M. Krone, J. E. Stone, T. Ertl and K. Schulten, Fast Visualization of Gaussian Density Surfaces for Molecular Dynamics and Particle System Trajectories, in *EuroVis 2012 Short Papers*, 2012.

43 J. Parulek and I. Viola, Implicit Representation of Molecular Surfaces, in *Proceedings of the IEEE Pacific Visualization Symposium (PacificVis 2012)*, 2012, pp. 217–224.

44 M. Tarini, P. Cignoni and C. Montani, Ambient occlusion and edge cueing to enhance real time molecular visualization, *IEEE Trans. Visual. Comput. Graph.*, 2006, **12**, 1237–1244.

45 S. Grottel, M. Krone, K. Scharnowski and T. Ertl, Object-Space Ambient Occlusion for Molecular Dynamics, *IEEE Pacific Visualization Symposium 2012*, ed. H. Hauser, S. Kobourov and H. Qu, 2012, pp. 209–216.

46 M. Falk, M. Krone and T. Ertl, Atomistic Visualization of Mesoscopic Whole-Cell Simulations Using Ray-Casted Instancing, *Comput. Graph. Forum*, 2013, **32**, 195–206.

47 B. Sommer, C. Bender, T. Hoppe, C. Gamroth and L. Jelonek, Stereoscopic cell visualization: from mesoscopic to molecular scale, *J. Electron. Imag.*, 2014, **23**, 011007.

48 S. Hornus, B. Levy, D. Lariviere and E. Fourmentin, Easy DNA Modeling and More with GraphiteLifeExplorer, *PLoS One*, 2013, **8**, e53609.

49 A. C. E. Dahl, M. Chavent and M. S. P. Sansom, Bendix: intuitive helix geometry analysis and abstraction, *Bioinformatics*, 2012, **28**, 2193–2194.

50 N. Lindow, D. Baum, A. N. Bondar and H. C. Hege, Exploring cavity dynamics in biomolecular systems, *BMC Bioinf.*, 2013, **14**(Suppl 19):S5.

51 J. Parulek, C. Turkay, N. Reuter and I. Viola, Visual cavity analysis in molecular simulations, *BMC Bioinf.*, 2013, **14**, S4.

52 M. Krone, G. Reina, C. Schulz, T. Kulschewski, J. Pleiss and T. Ertl, Interactive Extraction and Tracking of Biomolecular Surface Features, *Comput. Graph. Forum*, 2013, **32**, 331–340.

53 D. Bromley and V. Daggett, Analyzing disease-associated protein structures with visual analytics, *AMIA Joint Summits Translat Sci Proc*, 2013, vol. 2013, p. 33.

54 S. Thakur and M. A. Pasquinelli, Adapting Visual-Analytical Tools for the Exploration of Structural and Dynamical Features of Polymer Conformations, *Macromol. Theory Simul.*, 2011, **20**, 286–298.

55 J. Weber, ProteinShader: illustrative rendering of macromolecules, *BMC Struct. Biol.*, 2009, **9**, 19.

56 S. Bruckner and M. E. Gröller, Style transfer functions for illustrative volume rendering, *Comput. Graph. Forum*, 2007, 715–724.

57 G. Cipriano, G. Wesenberg, T. Grim, G. N. Phillips and M. Gleicher, GRAPE: GRaphical Abstracted Protein Explorer, *Nucleic Acids Res.*, 2010, **38**, W595–W601.

58 M. van der Zwan, W. Lueks, H. Bekker and T. Isenberg, Illustrative Molecular Visualization with Continuous Abstraction, *Comput. Graph. Forum*, 2011, **30**, 683–690.

59 A. Bryden, G. N. Phillips and M. Gleicher, Automated Illustration of Molecular Flexibility, *IEEE Trans. Visual. Comput. Graph.*, 2012, **18**, 132–145.

60 D. C. Y. Fung, S. S. Li, A. Goel, S. H. Hong and M. R. Wilkins, Visualization of the interactome: What are we looking at?, *Proteomics*, 2012, **12**, 1669–1686.

61 T. Praneenararat, T. Takagi and W. Iwasaki, Integration of interactive, multi-scale network navigation approach with Cytoscape for functional genomics in the big data era, *BMC Genomics*, 2012, **13**, S24.

62 G. Agapito, P. H. Guzzi and M. Cannataro, Visualization of protein interaction networks: problems and solutions, *BMC Bioinf.*, 2013, **14**, S1.

63 G. T. Johnson, L. Autin, D. S. Goodsell, M. F. Sanner and A. J. Olson, ePMV Embeds Molecular Modeling into Professional Animation Software Environments, *Structure*, 2011, **19**, 293–303.

64 R. M. Andrei, M. Callieri, M. F. Zini, T. Loni, G. Maraziti, M. C. Pan and M. Zoppe, Intuitive representation of surface properties of biomolecules using BioBlender, *BMC Bioinf.*, 2012, **13**, S16.

65 Z. Lv, A. Tek, F. Da Silva, C. Empereur-mot, M. Chavent and M. Baaden, Game on, Science – how video game technology may help biologists tackle visualization challenges, *PLoS One*, 2013, **8**, e57990.

66 G. McGill, Molecular movies… Coming to a lecture near you, *Cell*, 2008, **133**, 1127–1132.

67 S. Bromberg, W. Chiu and T. E. Ferrin, Workshop on Molecular Animation, *Structure*, 2010, **18**, 1261–1265.

68 J. H. Iwasa, Animating the model figure, *Trends Cell Biol.*, 2010, **20**, 699–704.

69 R. Yennamalli, R. Arangarasan, A. Bryden, M. Gleicher and G. N. Phillips, Using a commodity high-definition television for collaborative structural biology, *J. Appl. Crystallogr.*, 2014, **47**, 1153–1157.

70 C. Marion, J. Pouderoux, J. Jomier, S. Jourdain, M. Hanwell and U. Ayachit, A Hybrid Visualization System for Molecular Models, *WEB3D 2013: 18th International Conference On 3d Web Technology*, ed. S. N. Spencer, 2013, pp. 117–120.

71 J. A. Baker and J. D. Hirst, Molecular dynamics simulations using graphics processing units, *Mol. Inf.*, 2011, **30**, 498–504.

72 J. P. Ebejer, S. Fulle, G. M. Morris and P. W. Finn, The emerging role of cloud computing in molecular modelling, *J. Mol. Graphics Modell.*, 2013, **44**, 177–187.

73 R. M. Farber, Topical perspective on massive threading and parallelism, *J. Mol. Graphics Modell.*, 2011, **30**, 82–89.

74 D. E. Shaw, P. Maragakis, K. Lindorff-Larsen, S. Piana, R. O. Dror, M. P. Eastwood, J. A. Bank, J. M. Jumper, J. K. Salmon, Y. Shan and W. Wriggers, Atomic-level characterization of the structural dynamics of proteins, *Science*, 2010, **330**, 341–346.

75 T. J. Lane, D. Shukla, K. A. Beauchamp and V. S. Pande, To milliseconds and beyond: challenges in the simulation of protein folding, *Curr. Opin. Struct. Biol.*, 2013, **23**, 58–65.

76 J. E. Stone, D. J. Hardy, I. S. Ufimtsev and K. Schulten, GPU-accelerated molecular modeling coming of age, *J. Mol. Graphics Modell.*, 2010, **29**, 116–125.

77 R. Salomon-Ferrer, A. W. Gotz, D. Poole, S. Le Grand and R. C. Walker, Routine Microsecond Molecular Dynamics Simulations with AMBER on GPUs. 2. Explicit Solvent Particle Mesh Ewald, *J. Chem. Theory Comput.*, 2013, **9**, 3878–3888.

78 W. M. Brown and M. Yamada, Implementing molecular dynamics on hybrid high performance computers-Three-body potentials, *Comput. Phys. Commun.*, 2013, **184**, 2785–2793.

79 M. Zheng, X. X. Li and L. Guo, Algorithms of GPU-enabled reactive force field (ReaxFF) molecular dynamics, *J. Mol. Graphics Modell.*, 2013, **41**, 1–11.

80 R. M. Betz, N. A. DeBardeleben and R. C. Walker, An investigation of the effects of hard and soft errors on graphics processing unit-accelerated molecular dynamics simulations, *Concurrency Comput. Pract. Ex.*, 2014, **26**, 2134–2140.
81 I. Ufimtsev and T. Martínez, Graphical Processing Units for Quantum Chemistry, *Comput. Sci. Eng.*, 2008, **10**, 26–34.
82 P. Brown, C. Woods, S. McIntosh-Smith and F. R. Manby, Massively Multicore Parallelization of Kohn–Sham Theory, *J. Chem. Theory Comput.*, 2008, **4**, 1620–1626.
83 K. Yasuda, Accelerating Density Functional Calculations with Graphics Processing Unit, *J. Chem. Theory Comput.*, 2008, **4**, 1230–1236.
84 M. P. Haag and M. Reiher, Real-time quantum chemistry, *Int. J. Quantum Chem.*, 2013, **113**, 8–20.
85 S. S. Leang, A. P. Rendell and M. S. Gordon, Quantum Chemical Calculations Using Accelerators: Migrating Matrix Operations to the NVIDIA Kepler GPU and the Intel Xeon Phi, *J. Chem. Theory Comput.*, 2014, **10**, 908–912.
86 K. Wilkinson and C. K. Skylaris, Porting ONETEP to Graphical Processing Unit-Based Coprocessors. 1. FFT Box Operations, *J. Comput. Chem.*, 2013, **34**, 2446–2459.
87 L. Genovese, M. Ospici, T. Deutsch, J. F. Mehaut, A. Neelov and S. Goedecker, Density functional theory calculation on many-cores hybrid central processing unit-graphic processing unit architectures, *J. Chem. Phys.*, 2009, **131**, 034103.
88 M. Hacene, A. Anciaux-Sedrakian, X. Rozanska, D. Klahr, T. Guignon and P. Fleurat-Lessard, Accelerating VASP electronic structure calculations using graphic processing units, *J. Comput. Chem.*, 2012, **33**, 2581–2589.
89 X. Andrade and A. Aspuru-Guzik, Real-Space Density Functional Theory on Graphical Processing Units: Computational Approach and Comparison to Gaussian Basis Set Methods, *J. Chem. Theory Comput.*, 2013, **9**, 4360–4373.
90 X. Wu, A. Koslowski and W. Thiel, Semiempirical Quantum Chemical Calculations Accelerated on a Hybrid Multicore CPU–GPU Computing Platform, *J. Chem. Theory Comput.*, 2012, **8**, 2272–2281.
91 A. Sisto, D. R. Glowacki and T. J. Martinez, Ab Initio Nonadiabatic Dynamics of Multichromophore Complexes: A Scalable Graphical-Processing-Unit-Accelerated Exciton Framework, *Acc. Chem. Res.*, 2014, **47**, 2857–2866.
92 R. Buyya, C. S. Yeo, S. Venugopal, J. Broberg and I. Brandic, Cloud computing and emerging IT platforms: Vision, hype, and reality for delivering computing as the 5th utility, *Future Generat. Comput. Syst.*, 2009, **25**, 599–616.
93 S. Zhou, R. Liao and J. Guan, When cloud computing meets bioinformatics: a review, *J. Bioinf. Comput. Biol.*, 2013, **11**, 1330002.
94 O. Korb, P. W. Finn and G. Jones, The cloud and other new computational methods to improve molecular modelling, *Expert Opin. Drug Discovery*, 2014, **22**, 1–11.
95 P. M. Kasson, Computational biology in the cloud: methods and new insights from computing at scale, *Pac. Symp. Biocomput.*, 2013, 451.
96 S. R. Ellingson and J. Baudry, High-throughput virtual molecular docking with AutoDockCloud, *Concurrency Comput. Pract. Ex.*, 2014, **26**, 907–916.
97 A. K. L. Wong and A. M. Goscinski, A VMD Plugin for NAMD Simulations on Amazon EC2, *Procedia Computer Science*, 2012, **9**, 136–145.

98 W. Humphrey, A. Dalke and K. Schulten, VMD: visual molecular dynamics, *J. Mol. Graphics*, 1996, **14**, 33–38.
99 B. M. Good and A. I. Su, Crowdsourcing for bioinformatics, *Bioinformatics*, 2013, **29**, 1925–1933.
100 K. R. Lakhani, K. J. Boudreau, P.-R. Loh, L. Backstrom, C. Baldwin, E. Lonstein, M. Lydon, A. MacCormack, R. A. Arnaout and E. C. Guinan, Prize-based contests can provide solutions to computational biology problems, *Nat. Biotechnol.*, 2013, **31**, 108–111.
101 G. Pickard, W. Pan, I. Rahwan, M. Cebrian, R. Crane, A. Madan and A. Pentland, Time-Critical Social Mobilization, *Science*, 2011, **334**, 509–512.
102 J. Surowiecki, *The Wisdom of Crowds: Why the Many Are Smarter than the Few and How Collective Wisdom Shapes Business, Economies, Societies and Nations*, Doubleday Books, New York, 2004, p. 336.
103 C. Lintott, K. Schawinski, S. Bamford, A. Slosar, K. Land, D. Thomas, E. Edmondson, K. Masters, R. C. Nichol, M. J. Raddick, A. Szalay, D. Andreescu, P. Murray and J. Vandenberg, Galaxy Zoo 1: data release of morphological classifications for nearly 900 000 galaxies, *Mon. Not. R. Astron. Soc.*, 2011, **410**, 166–178.
104 S. Cooper, F. Khatib, A. Treuille, J. Barbero, J. Lee, M. Beenen, A. Leaver-Fay, D. Baker, Z. Popovic and F. Players, Predicting protein structures with a multiplayer online game, *Nature*, 2010, **466**, 756–760.
105 B. M. Good and A. I. Su, Games with a scientific purpose, *Genome Biol.*, 2011, **12**, 135.
106 C. B. Eiben, J. B. Siegel, J. B. Bale, S. Cooper, F. Khatib, B. W. Shen, F. Players, B. L. Stoddard, Z. Popovic and D. Baker, Increased Diels-Alderase activity through backbone remodeling guided by foldit players, *Nat. Biotechnol.*, 2012, **30**, 190–192.
107 S. Cooper, F. Khatib, A. Treuille, J. Barbero, J. Lee, M. Beenen, A. Leaver-Fay, D. Baker, Z. Popovic and F. Players, Predicting protein structures with a multiplayer online game, *Nature*, 2010, **466**, 756–760.
108 F. Khatib, F. DiMaio, Foldit Contenders Group, *et al.*, Crystal structure of a monomeric retroviral protease solved by protein folding game players, *Nat. Struct. Mol. Biol.*, 2011, **18**, 1175–1177.
109 F. Khatib, S. Cooper, M. D. Tyka, *et al.*, Algorithm discovery by protein folding game players, *Proc. Natl. Acad. Sci. U. S. A.*, 2011, **108**, 18949–18953.
110 J. C. Bradley, R. J. Lancashire, A. S. I. D. Lang and A. J. Williams, The Spectral Game: leveraging Open Data and crowdsourcing for education, *J. Cheminf.*, 2009, **1**, 9.
111 J. Bohannon, Distributed computing. Grassroots supercomputing, *Science*, 2005, **308**, 810.
112 G. T. Johnson and S. Hertig, A guide to the visual analysis and communication of biomolecular structural data, *Nat. Rev. Mol. Cell Biol.*, 2014, **15**, 690–698.
113 P. Larsson, B. Hess and E. Lindahl, Algorithm improvements for molecular dynamics simulations, *Wiley Interdiscip. Rev.: Comput. Mol. Sci.*, 2011, **1**, 93–108.

Faraday Discussions

PAPER

3D molecular models of whole HIV-1 virions generated with cellPACK

Graham T. Johnson,*[abc] David S. Goodsell,[c] Ludovic Autin,[c] Stefano Forli,[c] Michel F. Sanner[c] and Arthur J. Olson[c]

Received 19th February 2014, Accepted 19th March 2014
DOI: 10.1039/c4fd00017j

As knowledge of individual biological processes grows, it becomes increasingly useful to frame new findings within their larger biological contexts in order to generate new systems-scale hypotheses. This report highlights two major iterations of a whole virus model of *HIV-1*, generated with the *cellPACK* software. cellPACK integrates structural and systems biology data with packing algorithms to assemble comprehensive 3D models of cell-scale structures in molecular detail. This report describes the biological data, modeling parameters and cellPACK methods used to specify and construct editable models for HIV-1. Anticipating that cellPACK interfaces under development will enable researchers from diverse backgrounds to critique and improve the biological models, we discuss how cellPACK can be used as a framework to unify different types of data across all scales of biology.

Introduction

Biological processes in their mesoscale context

Multiscale biological models that span from atoms to cells can greatly improve our understanding of the mechanisms of health and disease. However, methods do not yet exist to observe, visualize, or model the *mesoscale*, the intermediate scale of 10^{-7}–10^{-8} m bridging molecular and cellular biology, in full molecular detail. To address this need, we have developed a software framework called *cellPACK*, which integrates multiple types of data across scales into comprehensive 3D spatial models. This report describes a cellPACK pipeline used to create increasingly rigorous molecular models of HIV-1 (Human Immunodeficiency Virus) from structural components and systems biology data.

Structural modeling of HIV-1 poses a timely challenge. HIV is a membrane-enveloped lentivirus that approaches a cell in its complexity. The genome of HIV is packaged inside a protein capsid, which requires a sophisticated collection of

[a]*University of California, San Francisco, CA 94143, USA. E-mail: graham.johnson@ucsf.edu*
[b]*California Institute for Quantitative Biosciences (QB3), QB3 MC: 2522, 1700 4th Street, Byers Hall, Suite 214, San Francisco, CA 94158-2330*
[c]*The Scripps Research Institute, 10550 North Torrey Pines Road, La Jolla, CA 92037, USA*

viral and cellular molecules to deliver it to new hosting cells, to initiate replication of the genome, and to repackage the necessary viral components for successful infection.[1] NIGMS has recently created a series of Specialized Centers, bringing together dozens of research labs to characterize the structures and mechanisms across the entire lifecycle, and to identify rational targets for drug and vaccine design (http://www.nigms.nih.gov/Research/SpecificAreas/AIDSStructuralBiology/Pages/HIVspecializedcenters.aspx).

The goal of cellPACK is to create rigorous structural models of HIV and its interactions with host cells, which frame our current knowledge on the scale of the entire virion, and in its cellular context in a consistent format. These models will impel hypothesis-driven research by integrating diverse data types, identifying gaps in knowledge, and exploring the ranges of models that are consistent with current knowledge. The level of complexity of HIV is ideally suited to push the current capabilities of cellPACK and to uncover the development needs of the open-source cellPACK project as it evolves to model larger systems.

Methods to bridge from atoms to cells

Experimental techniques for integrative structural biology are now linking molecular structure and supramolecular structure, for instance, to explore the structure of HIV glycoprotein multimers[2] and HIV capsid.[3,4] However, methods to observe, visualize or model the larger end of the mesoscale in atomic-resolution detail do not yet exist. To make the next step, from atoms to significant portions of cells, many sources of data can be used to synthesize a view of this level. Constraint optimization modelers like IMP,[5] procedural and analytical modeling approaches like Molecular Silverware[6,7] and relaxation approaches like Brownian Dynamics modeling[8,9] are currently used to build large-scale models. However, they are designed to function locally and do not model ultrastructural features like organelle membranes or fibrous molecules like actin. Qualitative manual protocols that use a more artistic approach have typically been required to produce mesoscale models. For example, when a large multidisciplinary collaboration determined the recipe for an average synaptic vesicle, the assembly of a structural model used to visualize the results of this effort took weeks of work using a combination of manual positioning with physics solvers to minimize collision.[10,75]

For the past twenty years, we have taken a semi-quantitative approach for the creation of illustrations of the cellular mesoscale, which integrate data across the nano and microscales. *Recipes* consisting of structures, concentrations and locations of molecules, macromolecules, and organelles taken from published reports are arranged into mesoscale compositions consistent with ultrastructural data from light and electron microscopy.[11,12] The illustrations are created manually with a 2.5D metaphor, typically presenting viruses or cells in cross-section, with simplified representations for the individual molecules. Fig. 1A shows a recent model of HIV-1 created with this method for educational outreach at the Protein Data Bank.[13]

The data used to specify this illustrated model served as the basis for the HIV-1 cellPACK recipe and 3D model. cellPACK computationally automates the modeling steps of the semi-quantitative artistic approach to extend the results from 2D paintings into 3D models that can be explored, animated, simulated,

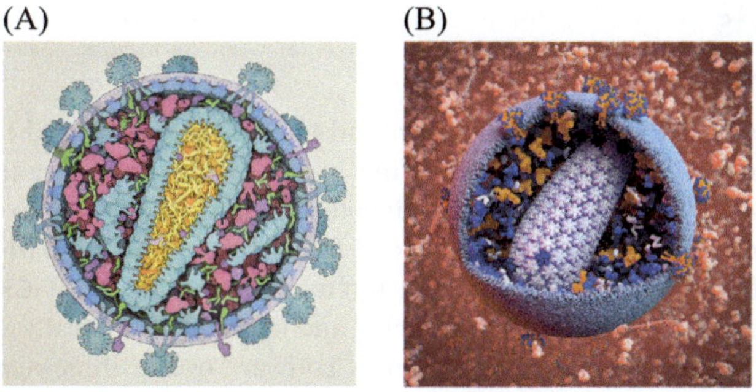

Fig. 1 Mesoscale Models of HIV-1. (A) A hand-drawn illustration of HIV-1 integrates data from structural biology, biophysics and microscopy. Data sources from this illustration were later used to create the digitized starting point recipe *HIV-1_0.1.0*. (B) A 3D model of HIV-1 enables exploration, analysis, animation and simulation, as well as easy variation, modification and updating. This model is based on the cellPACK recipe *HIV-1_0.1.4* nested in the larger environment recipe of *HIV_in_Blood_Plasma_0.1.2*, with a custom lipid bilayer added by the artist. The image was created using the cellPACK viewer plugin by digital artist Andrew McWhae for the *autoPACK Visualization Challenge 2013*, an outreach project which challenged artists to *"convey humanity's complex relationships with this virus, be they emotional, political, or intellectual… to excite general audiences with visuals that will help us spread interest in the search for a cure."* (http://autopack.cgsociety.org)

analyzed, and easily edited and updated (Fig 1B shows a rendering of a 3D model created from the fourth iteration of the digitized recipe called *HIV-1_0.1.4* (recipes use a 3 digit semantic versioning system to track them in a database where *HIV-1_0.1.4* stands for a recipe named "HIV-1" version 0.1.4).

cellPACK can model the mesoscale at the pace of HIV research

HIV structural biology is currently an area of intense research, with new breakthroughs reported each month. A major goal of cellPACK is to provide a nimble interface for integrating new structural results into the body of available data and evaluating their consequences on the integrative models. For instance, Fig. 4–6 describe two major iterative updates to the HIV-1 recipe, where *HIV-1_0.1.6* in Fig. 6 includes structural updates to the envelope glycoprotein, capsid, and other HIV proteins. In addition, cellPACK is designed to allow exploration of ambiguity or imprecision in the various components of the recipe, creating a range of models consistent with the current state of knowledge. Fig. 3, for example, shows 3D structures of HIV that result from varying the parameters affecting surface protein distributions and quantities.

Modeling an organism involves two conceptual steps: gathering of data to create a recipe for the model, and use of this recipe to build a virtual model. The goal of this report is to describe the creation of a recipe and model for a mature HIV-1 particle, which will become part of a larger *HIV-in-blood-plasma* recipe, including details on all of the viral and host macromolecules. This document presents the references used to support the structure, location and interactions of molecules depicted in the rendered models. Several recent review articles provided an overview to the topic and sources for additional information.[1,14–18]

Methods

HIV-1 cellPACK recipes

Data sources used to create the illustration in Fig. 1A were reused to create the cellPACK recipe *HIV-1_0.1.0* as a driving biological project for cellPACK development. Digitization of the recipe involved building 3D representations of molecular components and other needed files, and associating parameters that drive localization, concentration, and interaction into the cellPACK recipe file format. The cellPACK recipe was corrected and updated periodically through four iterations, while developing the cellPACK software over a three-year period, to produce the recipe *HIV-1_0.1.4* visualized in Fig 1B. A major update to the recipe, created with the semi-quantitative approach, was published in 2012;[19] this recipe was used to update *HIV-1_0.1.4* to create *HIV-1_0.1.5*, which is described in detail in the Results and in Fig. 4 and 5. Three major structural updates to the ingredients have since been published in 2013 and 2014, which were used, as noted in the Results, to update the recipe to *HIV-1_0.1.6* (detailed in Fig. 6).

cellPACK

A full description of the cellPACK framework has been submitted for publication, meanwhile, details are available on the project website and online documentation at www.cellpack.org. In summary, cellPACK unites existing and novel packing algorithms to generate, visualize and analyze comprehensive 3D models of complex biological environments that integrate data from multiple experimental sources. cellPACK currently generates probabilistic 3D models of large sections of cells that can contain dozens to millions of molecules. It packs molecules and organelles into positions that satisfy constraints consistent with the input data across the full range of scales. cellPACK additionally provides tools to store, visualize, analyze and interact with the results to make mesoscale models and mesoscale modeling accessible to a variety of audiences.

As shown in Fig. 2, cellPACK framework defines a desired volume and stochastically packs it with objects, called *ingredients*, according to a *recipe*. Ingredients of arbitrary shape are placed into the allowable 3D space with zero or minimal overlap and with random distribution. Constraints and agent behaviors may be associated globally or with selected ingredients to add specific modes of interaction. A grid-based approach is used to generate models quickly while tracking and confirming global parameters. *cellPACK's* database consists of mesoscale recipes (*e.g.*, cytoplasm, blood plasma, synaptic vesicles, HIV, *etc.*) and examples of models for each recipe.

To produce the models and visualizations shown in the figures, HIV-1 recipe ingredients (typically molecules) are stochastically packed into *containers* (polygonal meshes defining the *envelope* and *capsid* ultrastructures as described in the Results), up to densities provided for each ingredient, typically specific to a particular recipe. Each ingredient comes with associated properties, including a structural representation (such as a molecular surface[20] or volume occupancy shell generated in this case with the molecular modeling software plugin *ePMV*[21]), and behaviors such as a particular collision–detection method or a list of binding partners that will modify the ingredient's global and local packing. For example, in *HIV-1_0.1.5* and *v0.1.6*, the MA proteins are restricted to pack on the inside

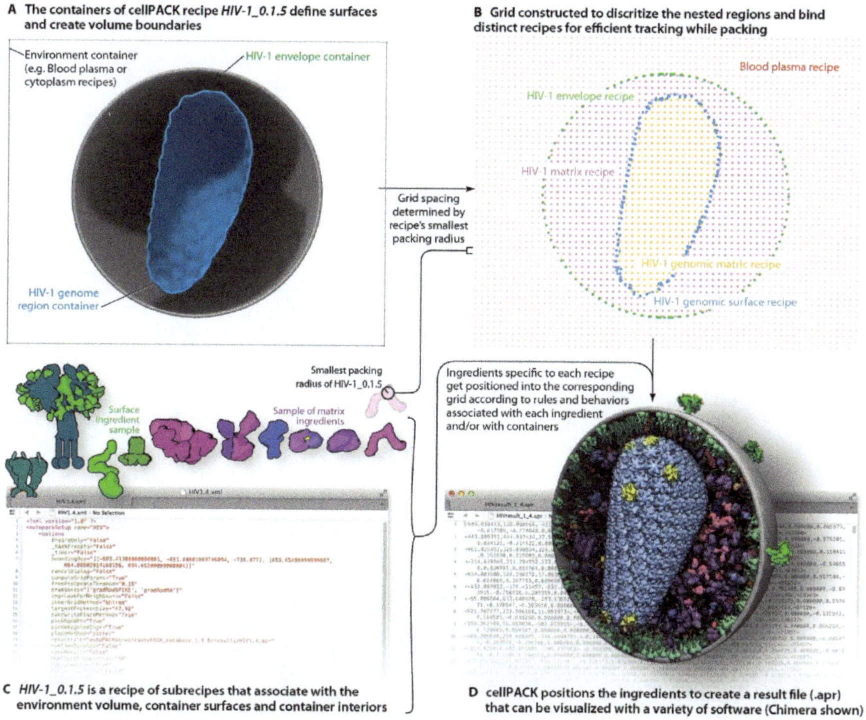

Fig. 2 cellPACK uses a grid-based method to track ingredients as they are efficiently packed into recipe-bounding containers. (A) Polyhedral mesh containers create boundaries to restrict the packing of HIV-1 envelope surface ingredients to the surface of the envelope mesh and HIV-1 matrix ingredients to the interior of the envelope and exterior of the capsid (HIV-1 genome region outer container). (B) An omniscient grid parameterizes the space to track ingredients as they pack and to weight regions for preferred packing based on global localization or distance measurements to ingredient interaction partners. The spacing of the grid is based on the radius of gyration of the smallest ingredient to pack. (C) The full recipe of ingredients is packed to (D) generate an .apr file with IDs and matrices for each ingredient packed. The .apr file is used by a variety of visualization (*e.g.* molecular viewers such as *Chimera* shown in D) or animation software packages (*e.g. Cinema 4D, Maya, Blender, 3D Studio Max, etc.*) to rebuild the scene in an interactive viewport where objects can be turned on/off, replaced with more detailed data such as the original PDB file for an ingredient, or analyzed with a collection of cellPACK analysis tools or default analysis tools native to each viewer.

surface and have a higher probability of packing towards one end of the envelope bilayer as a first-draft global packing constraint that can recapitulate the emergent properties of molecular interactions that have been observed as ENV clustering in fluorescence microscopy.[27] The polarization of ENV can be achieved by several methods currently under testing that include simplistic global constraints that force MA or ENV to pack preferentially towards some target location as demonstrated in Fig. 1B, as well as molecular scale interactions that vary association probabilities between MA–MA, MA–ENV, and ENV–ENV. Variations in the ingredient or recipe parameters can produce different results (Fig. 3). In the resultant model for a given recipe, each ingredient retains a connection to various other forms of data to enable deeper analysis, preparation for systems integration

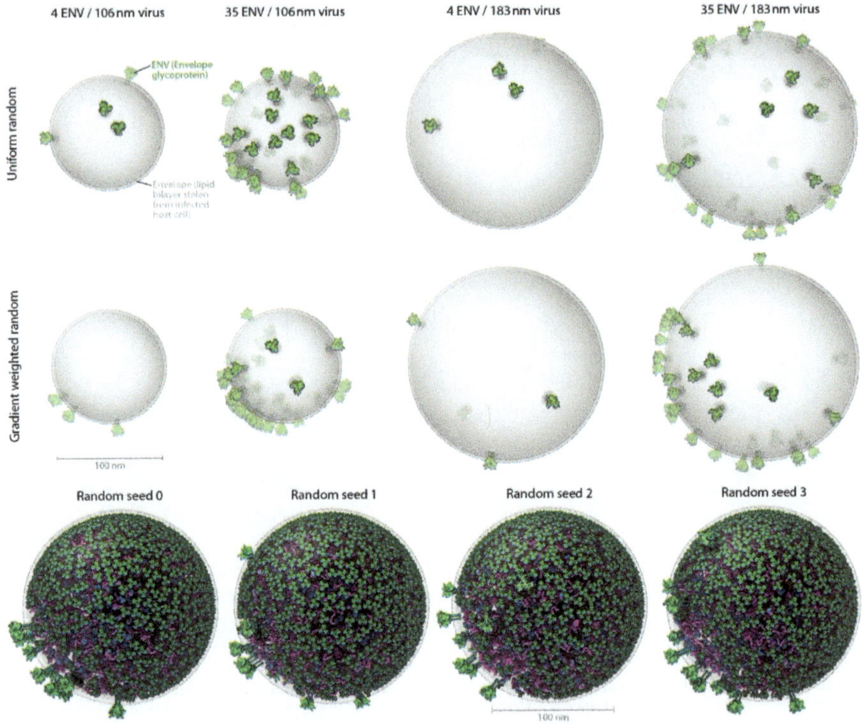

Fig. 3 Users can produce representative cellPACK models of HIV-1 by adjusting ingredient or recipe parameters. (A) Using *HIV-1_0.1.6* as a starting point, parameters for the ENV ingredient, the MA ingredient, and the envelope container were varied to experimental ranges to create models of the observed structural variability of the virion. The left two columns pack into a 106 nm envelope while the right two columns pack into a 183 nm envelope based on size ranges from electron tomography observations by Briggs *et al.*[25] (compare with the 130 and 145 nm diameters in Fig. 4). The first and third columns pack only 4 ENV per envelope container surface, while the second and fourth columns pack 35 ENV based on ranges from Zhu, *et al.*[26] The top row packs a uniform random distribution of MA, while the bottom row packs with a surface gradient weighted with a preference to deposit MA towards the upper right of the viewport. The dense gradient biased packing of MA excludes much of the upper-right surface of the envelope, forcing ENV to pack in the available space (primarily to the lower left). This simplistic polarization approach is one of several methods of packing and interaction parameter range settings currently being tested that can emulate the polarized clustering of ENV observed by Chojnacki *et al.*[27] (B) Four stochastic variations of *HIV-1_0.1.6* result from varying one global parameter– the *seed* used to generate the random number table.

or large-scale simulations, or for modifications of the ingredient representations. The representations use coarse molecular surfaces by default, but can be replaced with an interactive molecular modeling session to modify the representation style and to access the atomic details from the structural files, for example, from Protein Data Bank (PDB) files or Electron Microscopy Data Bank (EMDB) files.

Unique cellPACK recipes, created for each version of *HIV-1*, or *HIV_in_Blood_Plasma*, served as input to the cellPACK software. Packing the recipes with any version of cellPACK (stable release 1.0 and newer) produces the resulting models presented in this paper as *autoPACK Result (.apr)* files. An .apr file contains a list of ingredient IDs with positions and orientations as described in the online documentation (http://www.autopack.org/documentation/-apr-autopack-result-file-anatomy). As specified in the legends, visualization software including the molecular viewers PMV[22] and Chimera[23] and a collection of *cellPACK-uPy*[24] *plugin*-enabled 3D animation packages were used to load and visualize the model defined in the *HIV-1_0.1.5* and *v0.1.6* .apr files. The visualization software all use the same method of downloading the default geometric representation for each unique recipe ingredient followed by generating efficient copies of that representation on the computer's graphics card to enable real-time viewing and fast rendering to create static images, animated sequences, or online viewers.

Community consensus on theoretical mesoscale models

cellPACK provides scripting and GUI interfaces to access viewing, analysis, and modeling modules. A database, scheduled to go online in the Summer of 2014, will soon enable invited users to critique and modify recipes and ingredients while maintaining version histories directly through the project website. Having been through numerous internal edits, upon publication, the *HIV-1_0.1.6* recipe will become the first full-release (version 1.0) of HIV in the semantic versioning system (laterally named as *HIV-1_1.0.0*) and the first official model in the database to be opened for community feedback. Through web browser interfaces and leveraging social-networking technologies, the project website will ask the scientific community to analyze, critique and edit the models in an effort to produce confidence-voted community consensus models and to iteratively improve the recipes.

Results

cellPACK models track progress in HIV structural research

The modular and hierarchical architecture of cellPACK is designed to streamline the incorporation of the most current experimental results into representative models. The modules of cellPACK correspond to different experimental techniques with widely different data types. Ultrastructural data from electron microscopy is captured in the containers. Biophysical data on concentrations and localization are captured in the recipes. Structural data is captured in the models for each ingredient. This modularization allows new results to be efficiently added to an existing recipe.

We have included several examples of how this modularity can provide tools for exploring the structural biology of a biological system. Fig. 4 shows the models created with our two most recent recipes for HIV-1, *HIV-1_0.1.5* and *HIV-1_0.1.6*. Version *0.1.6* incorporates newly published results for the envelope glycoprotein (ENV), VIF, and capsid that were not available when the previous recipe was created. This figure visualizes the full HIV particles, with sections removed to reveal their internal components. Fig. 5 and 6 chart individual depictions of the molecular ingredient representations where Fig. 5 details the specifics of the *v0.1.5* recipe and Fig. 6 highlights the changes between the two versions.

Fig. 3A shows how cellPACK can be used to explore experimental uncertainties or ranges of parameters in experimental data. HIV is a pleomorphic organism, with a range of sizes and numbers of component molecules. cellPACK is able to create models that represent extrema in these ranges or that represent virions

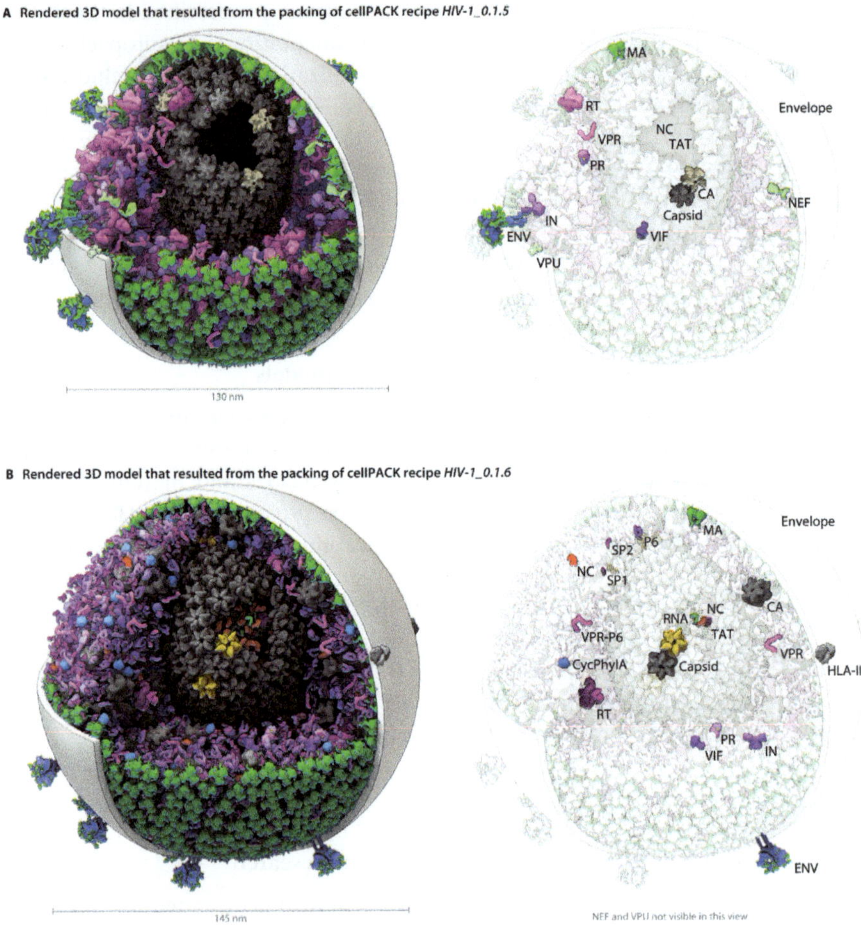

Fig. 4 A visual comparison of *HIV-1_0.1.5* and *HIV-1_0.1.6* reveals consistent, but stochastic packing where ingredients primarily change in their intramolecular structure and quantity rather than in location or localization. (A) The *HIV-1_0.1.5* recipe provides a substantial update to the *HIV-1_0.1.4* recipe shown in Fig. 1B with numerous ingredient corrections and additions. The model, visualized in a uPy-enabled 3D animation software package shows six different recipes (each a different color coding) unified into a single model. (B) Three major structural updates as well as more subtle quantification and envelope dimension changes were added to update *HIV-1_0.1.5* to *HIV-1_0.1.6*.

with average properties. Finally, Fig. 3B shows that cellPACK is able to create multiple representative models based on a given recipe, stochastically exploring the range of possible structures that are consistent with the given description. Packing times vary widely across recipes depending on the methods specified, the size and detail of the ingredients, the packing volume size, the total ingredient density, the complexity of the interactions, *etc.* Each of the four models in Fig. 3B pack the default *v0.1.6* recipe using a robust collision algorithm (the industry standard Bullet Physics Engine, bulletphysics.org) that allows zero ingredient surface overlaps. The models in Fig. 3B, which vary only in the random seed, each took < 267 seconds to pack on a standard consumer 2012 laptop computer.

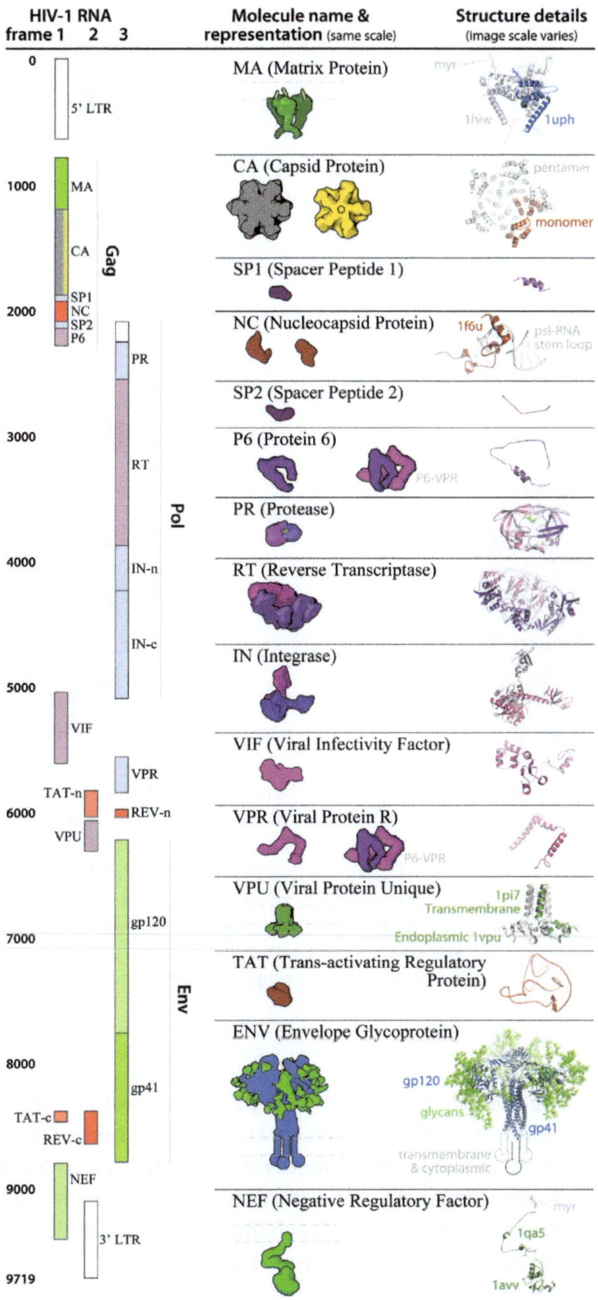

Fig. 5 *HIV-1_0.1.5* ingredient representations with structural summaries. The HIV genome landmark map on the left side (redrawn from www.hiv.lanl.gov/content/sequence/HIV/MAP/landmark.html) relates to the default coarse molecular surface representations and structural details of the HIV gene products in the center column. Many of the ingredients are updated in the newer iteration of this recipe, *HIV-1_0.1.6* as detailed in Fig. 6.

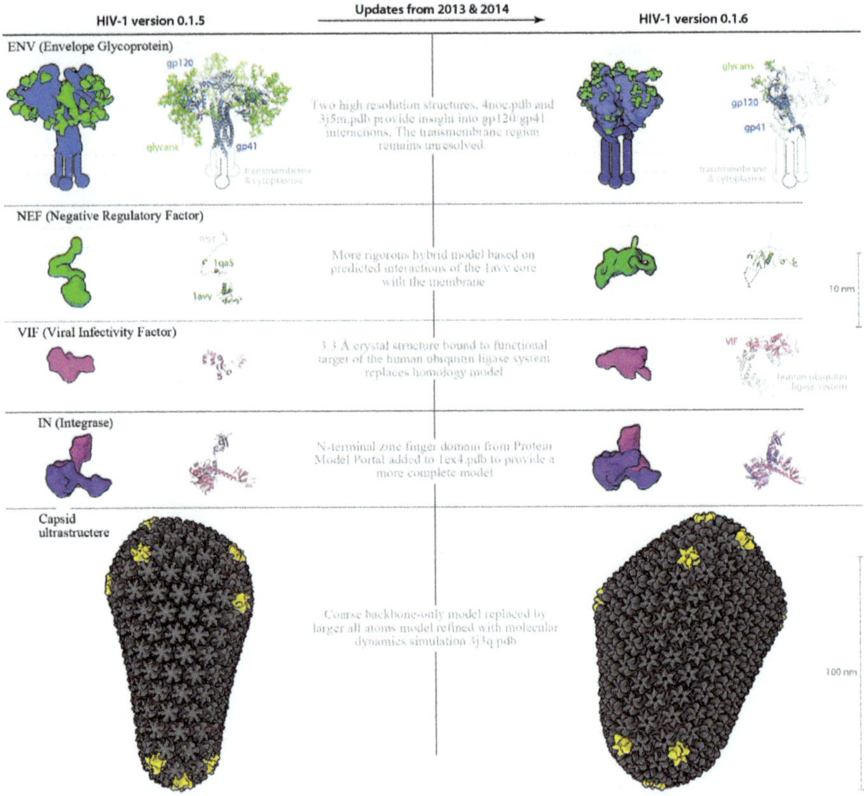

Fig. 6 Structure and parameter changes between *HIV-1_0.1.5* and *HIV-1_0.1.6*. Two ingredients ENV and VIF, and the ultrastructural capsid have been updated with new structures published in the past 9 months. The new structures provide important targets for vaccine and drug design, and because there may be steric hindrances imposed by their interactions or localizations, its important to visualize the ingredients in the context of the whole virus. Improved homology models for NEF and IN provide more complete representations for these two ingredients. The left column shows "older" structures and/or parameters from *HIV-1_0.1.5* that were changed to iterate the recipe to the "upgraded" *HIV-1_0.1.6*.

The remaining portion of the results will present the sources that were used for creation of the *HIV-1_0.1.5* and *v0.1.6* recipes, revealing some of the challenges for specifying this type of recipe and identifying several gaps in knowledge where approximations were required. This portion simultaneously includes a brief description of the known functions of each ingredient to provide a comprehensive resource that will be folded into the developing database.

Recipe biology, structure, and parameter details

Ultrastructure: Envelope container and matrix

HIV envelopes are lipid bilayers, captured from the host cell during release, that encapsulate the virus, display the viral ENV proteins, and shield most of the viral machinery from the host immune system. Based on electron micrographs, HIV

Table 1 Published viral envelope diameters

133 ± 17 nm ($n = 118$) (authentic immature virions)[28]
145 ± 25 nm ($n = 255$) (mature virions)[29]
139 ± 16 nm ($n = 197$) (immature virions)[29]
135.0 ± 17.4, 116.9 ± 8.5 ($n = 10$–15, mature virions, two cell lines)[30]
125 ± 14 nm (106–183) (mature virions)[25]
119 ± 207 nm (95% of mature virions)[31]
110 ± 8 nm (mature virions)[26]
126 ± 10 nm ($n = 30$ tomogram)[32]
130 ± 34 nm ($n = 150$ cryoEM)[32]
128 ± 32 nm ($n = 84$ STEM)[32]
100–150 nm (review article)[14]

particle envelopes are primarily spherical, but pleomorphic with occasional subtle lumps in the flexible lipid bilayer surface, and the envelope diameters range from 100 nm to 207 nm (Table 1). Early cellPACK recipes for HIV used distorted spheres (see Fig. 1B) to capture the observed variations in shape, but to simplify analysis (such as the angular distribution variance of ENV) and to describe more average HIV particles, the recipes for *HIV-1_0.1.5* and *v0.1.6* use perfect spheres as containers for the envelope packing surfaces. *HIV-1_0.1.5* envelope has a diameter of 130 nm, measuring to the center of the lipid bilayer. *HIV-1_0.1.6* has a diameter of 145 nm. The bilayer, modeled after MD simulated patches, ranges from 2–5 nm in thickness across the hydrophobic region (http://nbcr.ucsd.edu/data/sw/hosted/lipidwrapper). Most of the figures show the bilayer as a 3 nm thick spherical shell for visual clarity.

The envelope bilayer in both recipes is populated by both viral and host molecules. The region inside the envelope but excluded by the capsid is termed the *matrix*. In both of the models, the matrix is populated with a collection of soluble viral and host molecules.

Ultrastructure: Capsid container and genomic region

A distinctive cone-shaped capsid protects the HIV genome (2 strands of RNA visible in Fig. 4B) and delivers the genome to new host cells upon infection. The entire capsid is built from repeating units of a single capsid protein. The large N-terminal domain of a capsid protein binds to other copies of capsid proteins to form stable hexameric rings[33] and occasional pentameric rings.[34] The C-terminal domain links rings to neighboring rings to form the conic HIV capsid superstructure. HIV capsids have been studied by electron microscopy (EMDB entry 5639) and tomography[25] and have been reconstructed as atomic models using hybrid methods in two major publications. *HIV-1_0.1.5* uses a smaller coarse, backbone-only model provided by Pornillos and Yeager[34] that stands ~115 nm tall with 1056 monomers (166 hexamers and 12 pentamers, coordinates provided by the authors). *HIV-1_0.1.6* uses the larger of two all-atoms models that were refined with molecular dynamics, which stands ~118 nm tall, but is wider and contains 1356 capsid monomers (216 hexamers and 12 pentamers) comprised of 2 440 800 atoms[3] (PDB entry 3j3q).

The interior of the capsid holds the genome and associated proteins. In our models, we have termed this space the *genomic region*. Each capsid monomer has a unique conformation. To provide a coarser level of detail useful for interactive visualization, the *HIV-1_0.1.6* model uses a single representative monomer (chain A of 3j3q) aligned to each of the 1356 positions. The full detail PDB model with the unique conformers can be overlaid since the coarse model uses the same coordinate system.

Recipes: Molecules and interactions restricted to container boundaries

As new HIV particles bud from the infected host cell, the majority of the viral proteins are found in *Gag* (p55) and *GagPol* (p160) polyproteins (a note on nomenclature: the "p" and "gap" names are based on experimental molecular weights, and have largely been replaced by the more descriptive molecular names). Upon maturation, HIV protease (PR) cleaves each Gag and GagPol into 6 and 9 functional proteins respectively. Reports on Gag and GagPol quantification provide the most reliable abundances for proteins derived from Gag or GagPol in both mature and immature virus models. A wide range of values for the number and distribution of Gag and GagPol have been reported. Most current review articles have settled on a value of 2500 Gag polyproteins and a ratio of 20 Gag : 1 GagPol, with some primary sources listed in Table 2. *HIV-1_0.1.5* assumes 1710 Gag with 90 GagPol for a total of 1800 polyproteins. *HIV-1_0.1.6* uses 2375 Gag and 125 GagPol for a total of 2500 polyproteins.

Recipes from N to C: Molecules and interactions from Gag proteins

The Gag region of the HIV genome includes the proteins MA, CA, SP1, NC, SP2, and P6. *HIV-1_0.1.5* packs 1800 copies of each Gag region protein. *HIV-1_0.1.6* packs 2500 copies of each Gag region protein.

MA [envelope inner bilayer surface] – p17. In immature HIV-1 virions, the *matrix protein* (MA) localizes Gag and GagPol polyproteins to the host cell membrane, binding to the lipid bilayer through a covalently attached myristoyl chain. Mature MA forms trimers and remains associated with the membrane, but its major role in assembly appears to be the membrane attachment rather than the

Table 2 Gag and GagPol quantification and ratio reports Gag : GagPol ratios

20 Gag : 1 GagPol (quantified on gels[35])
89% Gag : 11% GagPol (shorter construct, quantified on gels)[36]
2000 Gag : 200 GagPol[28]
Number of Gag
2000 Gag[37]
1008, 1512 (calculated from experimental Gag : env ratio and counting of ENV on tomograms)[30]
4900 Gag (8 nm lattice at 60 nm radius in 145 nm particle, no gaps)[31]
2400 ± 700 (70% coverage, tomography)[32]
2400 (61% coverage, STEM)[32]
2500 Gag (review article)[14]
Distribution of Gag
40% coverage (28%–60%) (estimated visually from tomograms)[38]
60% ± 18% coverage ($n = 64$ cryoTEM, sucrose purified virion)[39]
82% ± 16% coverage ($n = 64$ cryoTEM, OptiPrep purified virion)[39]
2/3 coverage, continuous, with one gap (reviewed)[15]

ability to oligomerize.[14] *HIV-1_0.1.5* and *v0.1.6* use a hybrid model for MA assembled using two PDB files. An NMR monomer with N-terminal myristoyl (PDB entry 1uph) and more complete sequence was superimposed onto the truncated crystallographic structure of the MA trimer (PDB entry 1hiw). The covalently attached myristoyl group, which is bound in a surface pocket in the soluble NMR structure, was rotated about the myr–amino acid bond to allow insertion into the membrane. MA also interacts with the membrane *via* interaction with the cellular lipid PI(4,5)P2,[40] which is not included in the current models. MA is distributed on the inner leaflet of the envelope bilayer with a weighted envelope gradient that polarizes the distribution stochastically towards one side of the envelope in both *HIV-1_0.1.5* and *v0.1.6*. A packing module in development packs MA trimers into hexameric[41,42] and random clusters to recapitulate more recent observations[43] and to enable denser MA packing. *HIV-1_0.1.5* packs 600 MA trimers (from 1800 precursor monomers). *HIV-1_0.1.6* packs 833 MA trimers.

CA [capsid, matrix, genomic region] – p24. *Capsid protein* (CA) forms the cone-shaped capsid that protects and delivers the HIV genome. *HIV-1_0.1.5* uses structural models pulled from crystallographic structures of the hexamer and pentamer, with backbone-only structural files provided by Mark Yeager (from Pornillos *et al.*[34]). *HIV-1_0.1.6* uses subunits A, B, C, D, E, and F pulled from PDB entry 1vu4 (partial structure of the entire capsid 3j3q described under the Ultrastructures section above). *HIV-1_0.1.5* has a conic capsid with 1056 monomers leaving $1800 - 1056 = 744$ CA monomers in solution in the matrix, and *HIV-1_0.1.6* has a conic capsid with 1356 monomers leaving $2500 - 1356 = 1144$ CA monomers in solution in the matrix. The cellPACK recipes select from a library of hexamers, and various clusters of associated hexamers (partial capsids) that are packed into the matrix or genomic region until the respective limits are reached. A packing module under development, but not used here, can pack weighted assemblies of partial capsids on the fly directly from the hexameric subunits to enable the partial capsids to interact with other structures in the matrix and genomic region.

SP1 [matrix] – p2. *Spacer Peptide (SP1*, also known as P2) acts as a spacer between the CA and NC subunits in Gag and GagPol precursors. Although its role is poorly understood, SP1 is essential for viral assembly and infectivity.[44,45] *HIV-1_0.1.5* and *v0.1.6* model SP1 as an unstructured polypeptide 14 amino acids long. Structural studies are incomplete but suggest that SP1 may equilibrate[46] between a random coil and an α-helix.[47] In immature viruses, SP1 forms a 6-helix bundle in the Gag lattice that includes the last 12 CA residues.[38]

NC [genomic region bound to RNA] – p7. *Nucleocapsid protein* (NC) interacts specifically with packaging signals (see below) and nonspecifically with viral and host RNA. 5–7 RNA nucleotides per Gag is sufficient for assembly of virus-like particles of Gag[48] and a 5 nucleotide minimum binding site was identified by SPR on small oligonucleotides.[49] NC for both *HIV-1_0.1.5* and *v0.1.6* is assembled from PDB entries 1a1t (from a complex with a psi loop) and 1f6u (from a complex with an SD stem loop). A total of 1700 NC pack into the genomic region of *HIV-1_0.1.5*, with 100 in the matrix and a total of 2300 NC pack into *HIV-0.1.6* with 200 in the matrix.

SP2 [matrix] – p1. *Spacer Peptide (SP2*, also known as P1) separates NC from P6 and seems to be essential for viral infectivity. Deleting SP2 leads to defective capsid construction that may be related to kinetic effects in the C-terminal region of Gag.[50] *HIV-1_0.1.5* and *v0.1.6* use the same representation of SP2 modeled as an unstructured 17 amino acid polypeptide.

P6 [matrix] – p6. *P6* docks several viral and host-cell proteins and may associate with the envelope membrane.[51] Many studies suggest that P6 remains largely unstructured in the mature virion, and interacts with VPR.[37] *HIV-1_0.1.5* and *v0.1.6* split the number of P6 into two representations with different densities: one homology model (from the P6 FASTA sequence templated to PDB entry 2c55 *via* Swiss-Model[52]) with a partially helical and mostly unstructured chain of 51 amino acids (1500 copies), and a second representation with this homology model proximal to VPR (200 copies).

Recipes: Molecules and interactions from the pol region of GagPol

Roughly 5–11% of the Gag polyproteins are extended by a translational frameshift to create the GagPol polyproteins[36] which contain three additional functional subunits (PR, RT, and IN) that are processed by HIV-1 protease to create the mature enzymes. *HIV-1_0.1.5* packs 90 copies of each pol region protein. *HIV-1_0.1.6* packs 125 copies of each pol region protein.

PR [matrix]. *Protease* (PR) is a dimer of identical subunits, and is involved in the processing of Gag and GagPol during the process of maturation. *HIV-1_0.1.5* and *v0.1.6* use a crystallographic structure of the dimer from PDB entry 7hvp.

RT [matrix and genomic region] – p51. *Reverse transcriptase* creates a DNA copy of the HIV-1 RNA genome, which is then inserted into the host genome and used to create progeny virions. HIV-1 reverse transcriptase is a heterodimer of the RT protein and a truncated version of the protein. *HIV-1_0.1.5* and *v0.1.6* use a crystallographic structure of the heterodimer from the PDB entry 3hvt.pdb. Both recipes locate 25 of the 125 RT in the genomic region.

IN [matrix and genomic region] – p31. *Integrase* enzyme (IN) splices the DNA copy of the viral genome into host cell chromosomes. Full-length integrase, which functions as a homotetramer, has yet to be crystallized. *HIV-1_0.1.5* uses a partial model of IN based on the core structure in PDB entry 1ex4, which is missing the 450-amino-acid zinc finger domain from the N terminus. *HIV-1_0.1.6* uses a composite model with the N-terminal zinc-finger region from Protein Model Portal entry a510eadb0df840e958233f484781dde7_7 superimposed onto the C-terminal and core structure of PDB entry 1ex4. Both recipes locate 25 of the 125 IN in the genomic region.

Recipes: Molecules and interactions from the envelope region

Envelope glycoprotein ENV [envelope bilayer surface] – gp160. *Envelope glycoprotein* (ENV) forms trimers that protrude outward from the envelope surface. ENV contains the cellular receptor binding sites and membrane fusion machinery that power entry of the viral contents into the cytoplasm of a new host cell upon infection.[53] After binding to CD4 receptors on the surface of a cell, ENV snaps into a different conformation that pulls the viral envelope membrane close to the plasma membrane of the cell until the two bilayers fuse.

Three identical subunits trimerize to form each ENV spike and each trimer has ~81 glycosylation sites that cover ENV with a shielding layer of carbohydrates that comprises up to 50% of the mass of each spike. *HIV-1_0.1.5* uses an older chimeric model templated from 2x7r, 1g9m, 3dnn and glycosylated with glycprot server (http://www.glycosciences.de/modeling/glyprot/php/main.php). *HIV-1_0.1.6* uses a mix of two structures of the ectodomain, a crystal structure (PDB entry 4nco), and a single-particle electron microscope structure (PDB structure 3j5m). The new structures contain part of the gp41 stalk in the prefusion state, which provides insights into trimer assembly and the quaternary interaction between gp120 and gp41. The structures also contain bound broadly neutralizing antibodies as a focus for vaccine design. Some of the representations in the model include these antibody (Fab) fragments. The transmembrane region has not been resolved experimentally and is represented by geometric shapes of approximate size and shape based on molecular mass and positioned in close proximity to one another, based on electron tomography.[54]

The number and structure of ENV projecting from the virion surface has been difficult to quantify (see Table 3 for reported variations). Current reviews state that there are about 10 trimers on the surface.[15] *HIV-1_0.1.5* packs 21 ENV while *HIV-1_0.1.6* reduces the number to 12 for the 145nm virus. Fig. 3 shows models that result from varying this number across the published extremes (from 4 to 35).

The arrangement of ENV on the surface is a topic of current research. Fluorescence studies found a single cluster in mature virions, but not in immature virions.[27] *HIV-1_0.1.5* and *v0.1.6* pack MA along a preferential gradient that polarizes the distribution of ENV stochastically toward one side by stochastically excluding much of the surface otherwise available to ENV.

Interactions with molecules inside the virion also remain controversial. One study found that cellular protein TIP47 bridges MA and ENV and is required for packaging,[57] but a later paper found that MA and TIP47 interact, but is not needed for ENV incorporation.[58] Based on a current report, the "mechanism of incorporation still remains unclear": they present genetic evidence for ENV and MA interaction, but also state "it would seem plausible that passive incorporation could be sufficient to account for the ~10 copies",[59] therefore the recipes and resulting models do not yet include any specific interactions.

Recipes: Molecules and interactions of HIV-1 accessory proteins

VPR [matrix]. *Viral Protein R* (VPR) is an accessory protein with multiple functions critical for effective infection of CD4 cells and macrophages. VPR transports the HIV-1 pre-integration complex into the nucleus, arrests the cell cycle, and initiates transcription of HIV-1 LTR.[60] The 96 amino acid VPR representation for *HIV-1_0.1.5* and *v0.1.6* is based on PDB entry 1esx. VPR is incorporated into virions through interaction with Gag at a concentration of 1 VPR : 7 Gag[61] equaling 257 VPR in *HIV-1_0.1.5* and 357 VPR in *HIV-1_0.1.6*. 200 of the VPR in the matrix, remain bound to P6.[37]

Table 3 Published quantities of ENV/virus

70–80 "projections"[55]
7–14 trimers (measured Gag : ENV 60 : 1, assume 1200–2500 Gag)[56]
8, 10 (range of 3–12, 5–17, counted on tomogram from two cell lines)[30]
14 spikes (range 4–35, $n = 40$, in several clusters, counted on tomogram)[26]

Table 4 NEF quantification in HIV

<70 full length + cleaved per virion (quantified on gels[64])
1 NEF : 10–20 pol, 1 NEF : 333–200 Gag, 5–10 NEF per viron (assuming 2500 Gag)[35]

NEF [envelope bilayer inner surface, matrix]. *Negative Regulatory Factor* (NEF) modifies host cell structures and functions to favor replication and pathogenicity. *HIV-1_0.1.5* uses a hybrid model that positions the myristoyl group from the N-terminus of PDB entry 1qa5 in the membrane and positions the C-terminal core of PDB entry 1avv near the distal end. *HIV-1_0.1.6* uses a homology model interacting more closely with the bilayer,[62] generated with Modeller[63] based on the crystallographic structure of full-length NEF in PDB entry 3tb8, which includes only backbone atoms. The myristoyl group was taken from PDB entry 1qa5 and manually aligned into the bilayer using PyMol. Several studies have addressed the amount of NEF incorporated into virions (see Table 4):

NEF appears to be bound to the membrane in the immature virion, but the "majority"[64] or 50–70%[35] is cleaved to soluble form in mature virion, as quantified on gels. *HIV-1_0.1.5* packs 35 envelope bound NEF and 30 soluble NEF. *HIV-1_0.1.6* reduces this to 5 envelope bound NEF and 5 soluble.

VPU [envelope bilayer]. *Viral Protein Unique* (VPU) promotes CD4 degradation and release of the immature virus from the host cell plasma membrane. VPU forms a pentameric channel through the membrane. *HIV-1_0.1.5* and v*0.1.6* use a chimeric model assembled from the transmembrane region of PDB entry 1pi7 and the cytoplasmic region in PDB entry 1vpu. VPU is found primarily in the endoplasmic reticulum of the infected cell where it performs its job and little to none may escape with the immature virion, so both *HIV-1_0.1.5* and *v0.1.6* include only 20 copies (4 pentamers).

TAT/TAR [genomic region]. *Trans-activating Regulatory Protein* (TAT) increases viral replication efficiency by mediating the phosphorylation of cellular factors that increase the transcription of all HIV genes. TAT binds to the trans-activating response element (TAR) on the 5′ end of HIV transcripts and alters the transcription machinery to increase production of RNA. The TAT structure is taken from model 1 of the 86 amino acid NMR structure in PDB entry 1jfw. Both recipes pack TAT at a similar density, with 45 TAT in *HIV-1_0.1.5* and 54 TAT in the larger capsid of *HIV-1_0.1.6*.

VIF [matrix]. *Viral Infectivity Factor* (VIF) neutralizes host defenses by targeting the host enzyme APOBEC for ubiquitination and subsequent proteasomal degradation. *HIV-1_0.1.5* fixes an incorrect structure from *HIV-1_0.1.4*, which had been surfaced from the full model of PDB entry 3dcg, which includes only a small fraction of VIF and large sections of a human elongation factor protein. *HIV-1_0.1.5* uses a homology model from ModBase model ID 369ab672da66dee-cac38810f962b4e53, templated from PDB entry 1f0xA that covers 1–174 out of 192 amino acids in VIF. *HIV-1_0.1.6* improves further on the VIF model with a larger fragment of VIF (aa 1-176) from a crystallographic structure in PDB entry 4n9f,[65] where it was crystallized bound to the large human ubiquitin ligase system (its functional target). The literature reports 7–100 VIF per virion[37] and both *HIV-1_0.1.5* and *v0.1.6* pack at a density of 100 VIF/145nm virion.

Viral RNA [genome encased in the capsid]. The genome of HIV has 9749 nucleotides[66] detailed on the left side of Fig. 5. The genome has several regulatory regions in addition to the coding regions, including the Psi region that directs selective packaging and dimerization and the major packaging signal with SL1, which contains the dimerization initiation site (DIS), and SL3.[67] A prediction of secondary structure of entire genome has also been reported.[68] HIV-1.6 includes a very coarse structure modeled by Goddard

Table 5 A few specific interactions have been reported

Cyclophilin A 1 : 10 Gag,[70] shown in the illustration based on the PDB entry 1ak4 (cyclophilin A and N-terminal domain of CA)
Hsp70 1 : 20 Gag,[69] bound to Gag
INI1 1 : 2 integrase, bound to integrase[69]
Ubiquitin 1 : 10 Gag[69]
Monoubiquinated proteins 1 : 50 Gag[69]
Lys tRNA synthetase, bound to Gag
TIP47 bound to MA[57,58]
ABCE1 bound to NC[71]
Staufen1 (ds RNA binding protein) bound to NC[72]
Annexin2 bound Gag (in macrophages)[73]

[http://tinyurl.com/ChimeraRNA] incorporating only the predicted secondary structure. This RNA model is included for size and to give an idea of the complexity of packing, but does not specifically interact with the capsid and is not a plausible packing usable for research purposes. Modeling of RNA is an active area of development for cellPACK.

Cellular RNA. Cellular RNA is incorporated into the virion through interaction with NC.[16] For instance, 7SL (RNA from signal recognition particle) is found in a ratio of 7 : 1 with the HIV genome. *HIV-1_0.1.5* and *v0.1.6* do not include cellular RNA.

Cellular proteins. Many cellular proteins are incorporated into the virion (see Table 5), including ICAM1, HLA-II, and tetraspanins in the membrane, and soluble proteins such as thioltransferase, Hsp70, actin and actin-binding proteins, ubiquitin and lys tRNA synthetase.[69]

The *HIV-1_0.1.6* recipe includes these examples, but is not yet comprehensive. We plan a detailed literature search and modeling effort in the next iteration to include the host proteins into the recipe and model. This is an area of intensive research, as the community is uncovering the roles of host proteins in the viral lifecycle.

Conclusions

cellPACK is designed to form a bridge between the approaches of reductionist biology and systems biology. As computational and experimental techniques probe into the mesoscale from smaller and larger scale levels, frameworks like cellPACK, that will unify and contextualize the diverse data, are needed to serve as platforms for consistent communication to facilitate interdisciplinary research. cellPACK models can serve as hypothesis-generating tools, as starting points for mesoscale simulation techniques like Brownian Dynamics,[9] and have already been applied extensively for education and outreach.

Modeling a virus is a relatively simple task compared to modeling an entire eukaryotic cell in molecular detail. A complex virus like HIV serves as an ideal short-term goal, which now provides a much needed framework for the HIV research community to assemble their data and to better understand lentivirus mechanics. HIV simultaneously serves as a stepping-stone to modeling organisms of increasing complexity. The up-to-date release of *HIV-1_0.1.6* demonstrates that cellPACK can keep up with the fast pace of HIV research.

Hierarchical cellPACK models

The HIV-1 models presented here seek to characterize a single representative virion with full molecular detail. The hierarchical architecture of cellPACK allows this model to be used directly as an ingredient inside a larger environment recipe. Fig. 1B for example shows a rendering of *HIV_in_Blood_Plasma_0.1.2*, which incorporates *HIV-1_0.1.4* as a complete ingredient packed amongst other plasma ingredients. Recipes can nest inside of other recipes in this manner, where a whole HIV particle is deposited as a single "large" ingredient according to its target density, or nested recipes can mix their ingredients to pack the HIV molecules at the same time as the blood plasma molecules, which allows depositing ingredients to weight and affect one another as they interact during packing. This specification of models with hierarchical recipes, rather than hierarchical models, offers great potential: In the process of building the model, nested recipes allow the ingredients in the different entities to interact and affect the final form of the model. For instance, antibodies in the blood plasma model can form interactions with the HIV model during creation of a larger model, which can restrict or induce other specific interactions based on densities and kinetics.

Future plans

We have initiated several projects to experiment with the HIV-cellPACK system, for example, to explore the vast ranges of interacting parameter space across all of the ingredients. An integrated fluorescence microscopy simulation tool, similar to the approach described by Gardner *et al.*,[76] was recently integrated into cellPACK to enable cellPACK models to be compared more directly to microscope images and to begin to test competing hypotheses for distribution of the Env glycoprotein in immature HIV-1 virions. Upcoming experiments will incorporate probabilities for specific interactions between known binding partners, and will fully explore the structures and interactions of host proteins as the research community identifies them. To further improve the HIV-1 mature virion model and to explore other aspects of the lifecycle, we have initiated a project to model the immature virion. Will exploring the vast parameter space of a cellPACK recipe in an automated fashion with results clustered to higher-order structural scoring functions enable cellPACK to reliably recapitulate the less well-described interaction details from better-understood neighboring molecular building blocks alone? These new approaches have begun to address this question.

Having refined the cellPACK software with the driving biological project of modeling HIV-1, it becomes a relatively short step to model a slightly larger and more complex organism such as *Mycoplasma genitalium.* A systems biology project called WholeCellViz has provided a computational recipe of the hypothetical quantities, localizations, and interactions of every gene product of *M. genitalium*, dynamically over a full life cycle.[74] The WholeCellViz parameters are being transposed into cellPACK recipes that are completed by generating structural representations for each gene product. When complete, the recipe will be packed to generate the first 3D structural model of a whole cell and the first structural snapshots of a whole cell dynamically changing over an entire cell cycle. On a broader scale, because it collects Systems Biology and Structural Biology results into a unified framework, cellPACK can support big data projects like the 3D Virtual Cell project (http://3dvcell.org), where teams of biologists will merge efforts to construct virtual cells that can be probed experimentally to predict outcomes ranging from molecular mechanisms to cell-wide effects.

Acknowledgements

We would like to thank Megan Riel-Mehan for writing suggestions; Andrew McWhae for his award-winning image of HIV_in_Blood_Plasma_0.1.2; Thomas Goddard for efficient/coarse models of 3j3q and HIV RNA; Barbie K. Ganser-Pornillos, Owen Pornillos, Mark Yeager, Juan R. Perilla and Klaus Schulten for additional unpublished versions of their respective HIV capsid models. This work was supported in part by a predoctoral fellowship from NSF (NSF 07576), a grant from the NIH (P41 RR08605), a QB3 Fellowship grant from the California Institute for Quantitative Biosciences, qb3@UCSF, and a UCSF School of Pharmacy, 2013 Mary Anne Koda-Kimble Seed Award for Innovation. This is manuscript 27000 from the Scripps Research Institute.

References

1 B. K. Ganser-Pornillos, M. Yeager and W. I. Sundquist, *Curr. Opin. Struct. Biol.*, 2008, **18**, 203–217.

2 D. Lyumkis, J. P. Julien, N. de Val, A. Cupo, C. S. Potter, P. J. Klasse, D. R. Burton, R. W. Sanders, J. P. Moore, B. Carragher, I. A. Wilson and A. B. Ward, *Science*, 2013, **342**(6165), 1484.

3 G. Zhao, J. R. Perilla, E. L. Yufenyuy, X. Meng, B. Chen, J. Ning, J. Ahn, A. M. Gronenborn, K. Schulten, C. Aiken and P. Zhang, *Nature*, 2013, **497**, 643–646.

4 J. M. Grime and G. A. Voth, *Biophys. J.*, 2012, **103**, 1774–1783.

5 D. Russel, K. Lasker, B. Webb, J. Velazquez-Muriel, E. Tjioe, D. Schneidman-Duhovny, B. Peterson and A. Sali, *PLoS Biol.*, 2012, **10**, e1001244.

6 M. Blanco, *J. Comput. Chem.*, 1991, **12**, 237–247.

7 T. Byholm, M. Toivakka and J. Westerholm, *Powder Technol.*, 2009, **196**, 139–146.

8 T. AndoJ. Skolnick, *Proceedings of the International Conference of the Quantum Bio-Informatics IV.*, ed. L. Accardi, W. Freundenberg and M. Ohya, World Scientific Publishing Co., 2011, 28, pp. 413–426.

9 S. R. McGuffee and A. H. Elcock, *PLoS Comput. Biol.*, 2010, **6**, e1000694.

10 S. Takamori, M. Holt, K. Stenius, E. A. Lemke, M. Gronborg, D. Riedel, H. Urlaub, S. Schenck, B. Brugger, P. Ringler, S. A. Muller, B. Rammner, F. Grater, J. S. Hub, B. L. De Groot, G. Mieskes, Y. Moriyama, J. Klingauf, H. Grubmuller, J. Heuser, F. Wieland and R. Jahn, *Cell*, 2006, **127**, 831–846.

11 D. S. Goodsell, *Biochem. Mol. Biol. Educ.*, 2009, **37**, 325–332.

12 D. S. Goodsell, *Biochem. Mol. Biol. Educ.*, 2011, **39**, 91–101.

13 D. S. Goodsell, http://www.pdb.org/pdb/education_discussion/educational_resources/hiv-animation.html, RCSB Protein Data Bank, 2011.

14 B. K. Ganser-Pornillos, M. Yeager and O. Pornillos, *Adv. Exp. Med. Biol.*, 2012, **726**, 441–465.

15 J. A. Briggs and H. G. Krausslich, *J. Mol. Biol.*, 2011, **410**, 491–500.

16 W. I. Sundquist and H. G. Krausslich, *Cold Spring Harbor Perspect. Med.*, 2012, **2**, a006924.

17 A. A. Waheed and E. O. Freed, *AIDS Res. Hum. Retroviruses*, 2012, **28**, 54–75.
18 M. Balasubramaniam and E. O. Freed, *Physiology (Bethesda)*, 2011, **26**, 236–251.
19 D. S. Goodsell, *Biochem Mol Biol Educ*, 2012, **40**, 291–296.
20 M. F. Sanner, A. J. Olson and J. C. Spehner, *Biopolymers*, 1996, **38**, 305–320.
21 G. T. Johnson, L. Autin, D. S. Goodsell, M. F. Sanner and A. J. Olson, *Structure*, 2011, **19**, 293–303.
22 M. F. Sanner, *J. Mol. Graph. Model.*, 1999, **17**, 57–61.
23 E. F. Pettersen, T. D. Goddard, C. C. Huang, G. S. Couch, D. M. Greenblatt, E. C. Meng and T. E. Ferrin, *J. Comput. Chem.*, 2004, **25**, 1605–1612.
24 L. Autin, G. Johnson, J. Hake, A. Olson and M. Sanner, *IEEE Comput. Graphics Appl.*, 2012, **32**, 50–61.
25 J. A. Briggs, K. Grunewald, B. Glass, F. Forster, H. G. Krausslich and S. D. Fuller, *Structure*, 2006, **14**, 15–20.
26 P. Zhu, J. Liu, J. Bess, Jr., E. Chertova, J. D. Lifson, H. Grise, G. A. Ofek, K. A. Taylor and K. H. Roux, *Nature*, 2006, **441**, 847–852.
27 J. Chojnacki, T. Staudt, B. Glass, P. Bingen, J. Engelhardt, M. Anders, J. Schneider, B. Muller, S. W. Hell and H. G. Krausslich, *Science*, 2012, **338**, 524–528.
28 T. Wilk, I. Gross, B. E. Gowen, T. Rutten, F. de Haas, R. Welker, H. G. Krausslich, P. Boulanger and S. D. Fuller, *J. Virol.*, 2001, **75**, 759–771.
29 J. A. Briggs, T. Wilk, R. Welker, H. G. Krausslich and S. D. Fuller, *EMBO J.*, 2003, **22**, 1707–1715.
30 P. Zhu, E. Chertova, J. Bess, Jr., J. D. Lifson, L. O. Arthur, J. Liu, K. A. Taylor and K. H. Roux, *Proc. Natl. Acad. Sci. U. S. A.*, 2003, **100**, 15812–15817.
31 J. A. Briggs, M. N. Simon, I. Gross, H. G. Krausslich, S. D. Fuller, V. M. Vogt and M. C. Johnson, *Nat. Struct. Mol. Biol.*, 2004, **11**, 672–675.
32 L. A. Carlson, J. A. Briggs, B. Glass, J. D. Riches, M. N. Simon, M. C. Johnson, B. Muller, K. Grunewald and H. G. Krausslich, *Cell Host Microbe*, 2008, **4**, 592–599.
33 O. Pornillos, B. K. Ganser-Pornillos, B. N. Kelly, Y. Hua, F. G. Whitby, C. D. Stout, W. I. Sundquist, C. P. Hill and M. Yeager, *Cell*, 2009, **137**, 1282–1292.
34 O. Pornillos, B. K. Ganser-Pornillos and M. Yeager, *Nature*, 2011, **469**, 424–427.
35 R. Welker, H. Kottler, H. R. Kalbitzer and H. G. Krausslich, *Virology*, 1996, **219**, 228–236.
36 T. Jacks, M. D. Power, F. R. Masiarz, P. A. Luciw, P. J. Barr and H. E. Varmus, *Nature*, 1988, **331**, 280–283.
37 A. D. Frankel and J. A. Young, *Annu. Rev. Biochem.*, 1998, **67**, 1–25.
38 E. R. Wright, J. B. Schooler, H. J. Ding, C. Kieffer, C. Fillmore, W. I. Sundquist and G. J. Jensen, *EMBO J.*, 2007, **26**, 2218–2226.
39 N. Kol, M. Tsvitov, L. Hevroni, S. G. Wolf, H. B. Pang, M. S. Kay and I. Rousso, *J. Virol. Methods*, 2010, **169**, 244–247.
40 J. S. Saad, J. Miller, J. Tai, A. Kim, R. H. Ghanam and M. F. Summers, *Proc. Natl. Acad. Sci. U. S. A.*, 2006, **103**, 11364–11369.
41 A. Alfadhli, R. L. Barklis and E. Barklis, *Virology*, 2009, **387**, 466–472.
42 A. Alfadhli, D. Huseby, E. Kapit, D. Colman and E. Barklis, *J. Virol.*, 2007, **81**, 1472–1478.

43 J. A. Briggs, M. C. Johnson, M. N. Simon, S. D. Fuller and V. M. Vogt, *J. Mol. Biol.*, 2006, **355**, 157–168.

44 S. C. Pettit, M. D. Moody, R. S. Wehbie, A. H. Kaplan, P. V. Nantermet, C. A. Klein and R. Swanstrom, *J. Virol.*, 1994, **68**, 8017–8027.

45 H. G. Krausslich, M. Facke, A. M. Heuser, J. Konvalinka and H. Zentgraf, *J. Virol.*, 1995, **69**, 3407–3419.

46 J. L. Newman, E. W. Butcher, D. T. Patel, Y. Mikhaylenko and M. F. Summers, *Protein Sci.*, 2004, **13**, 2101–2107.

47 M. A. Accola, S. Hoglund and H. G. Gottlinger, *J. Virol.*, 1998, **72**, 2072–2078.

48 I. P. O'Carroll, F. Soheilian, A. Kamata, K. Nagashima and A. Rein, *Virus Res.*, 2013, **171**, 341–345.

49 R. J. Fisher, A. Rein, M. Fivash, M. A. Urbaneja, J. R. Casas-Finet, M. Medaglia and L. E. Henderson, *J. Virol.*, 1998, **72**, 1902–1909.

50 A. de Marco, A. M. Heuser, B. Glass, H. G. Krausslich, B. Muller and J. A. Briggs, *J. Virol.*, 2012, **86**, 13708–13716.

51 S. M. Solbak, T. R. Reksten, F. Hahn, V. Wray, P. Henklein, P. Henklein, O. Halskau, U. Schubert and T. Fossen, *Biochim. Biophys. Acta, Biomembr.*, 2013, **1828**, 816–823.

52 T. Schwede, J. Kopp, N. Guex and M. C. Peitsch, *Nucleic Acids Res.*, 2003, **31**, 3381–3385.

53 D. Lyumkis, J. P. Julien, N. de Val, A. Cupo, C. S. Potter, P. J. Klasse, D. R. Burton, R. W. Sanders, J. P. Moore, B. Carragher, I. A. Wilson and A. B. Ward, *Science*, 2013, **342**, 1484–1490.

54 J. Liu, A. Bartesaghi, M. J. Borgnia, G. Sapiro and S. Subramaniam, *Nature*, 2008, **455**, 109–113.

55 H. R. Gelderblom, E. H. Hausmann, M. Ozel, G. Pauli and M. A. Koch, *Virology*, 1987, **156**, 171–176.

56 E. Chertova, J. W. Bess, Jr., B. J. Crise, I. R. Sowder, T. M. Schaden, J. M. Hilburn, J. A. Hoxie, R. E. Benveniste, J. D. Lifson, L. E. Henderson and L. O. Arthur, *J. Virol.*, 2002, **76**, 5315–5325.

57 S. Lopez-Verges, G. Camus, G. Blot, R. Beauvoir, R. Benarous and C. Berlioz-Torrent, *Proc. Natl. Acad. Sci. U. S. A.*, 2006, **103**, 14947–14952.

58 M. A. Checkley, B. G. Luttge, P. Y. Mercredi, S. K. Kyere, J. Donlan, T. Murakami, M. F. Summers, S. Cocklin and E. O. Freed, *J. Virol.*, 2013, **87**, 3561–3570.

59 M. A. Checkley, B. G. Luttge and E. O. Freed, *J. Mol. Biol.*, 2011, **410**, 582–608.

60 M. Kogan and J. Rappaport, *Retrovirology*, 2011, **8**, 25.

61 B. Muller, U. Tessmer, U. Schubert and H. G. Krausslich, *J. Virol.*, 2000, **74**, 9727–9731.

62 X. Ren, S. Y. Park, J. S. Bonifacino and J. H. Hurley, *eLife*, 2014, **3**, e01754.

63 N. Eswar, B. Webb, M. A. Marti-Renom, M. S. Madhusudhan, D. Eramian, M. Y. Shen, U. Pieper and A. Sali, *Current protocols in bioinformatics / editoral board, Andreas D. Baxevanis… [et al.]*, 2006, Chapter 5, Unit 5 6.

64 M. W. Pandori, N. J. Fitch, H. M. Craig, D. D. Richman, C. A. Spina and J. C. Guatelli, *J. Virol.*, 1996, **70**, 4283–4290.

65 Y. Guo, L. Dong, X. Qiu, Y. Wang, B. Zhang, H. Liu, Y. Yu, Y. Zang, M. Yang and Z. Huang, *Nature*, 2014, **505**, 229–233.

66 L. Ratner, W. Haseltine, R. Patarca, K. J. Livak, B. Starcich, S. F. Josephs, E. R. Doran, J. A. Rafalski, E. A. Whitehorn, K. Baumeister, *et al.*, *Nature*, 1985, **313**, 277–284.
67 N. Jouvenet, S. Laine, L. Pessel-Vivares and M. Mougel, *RNA Biol.*, 2011, **8**, 572–580.
68 J. M. Watts, K. K. Dang, R. J. Gorelick, C. W. Leonard, J. W. Bess, Jr., R. Swanstrom, C. L. Burch and K. M. Weeks, *Nature*, 2009, **460**, 711–716.
69 D. E. Ott, *Rev. Med. Virol.*, 2008, **18**, 159–175.
70 E. K. Franke, H. E. Yuan and J. Luban, *Nature*, 1994, **372**, 359–362.
71 J. E. Dooher, B. L. Schneider, J. C. Reed and J. R. Lingappa, *Traffic*, 2007, **8**, 195–211.
72 L. Chatel-Chaix, K. Boulay, A. J. Mouland and L. Desgroseillers, *Retrovirology*, 2008, **5**, 41.
73 E. V. Ryzhova, R. M. Vos, A. V. Albright, A. V. Harrist, T. Harvey and F. Gonzalez-Scarano, *J. Virol.*, 2006, **80**, 2694–2704.
74 J. R. Karr, J. C. Sanghvi, D. N. Macklin, M. V. Gutschow, J. M. Jacobs, B. Bolival, Jr., N. Assad-Garcia, J. I. Glass and M. W. Covert, *Cell*, 2012, **150**, 389–401.
75 B. G. Wilhelm, S. Mandad, S. Truckenbrodt, K. Krohnert, C. Schafer, B. Rammner, S. J. Koo, G. A. Classen, M. Krauss, V. Haucke, H. Urlaub and S. O. Rizzoli, *Science*, 2014, **344**, 1023–1028.
76 M. K. Gardner, D. J. Odde and K. Bloom, *Methods*, 2007, **41**, 232–237.

Faraday Discussions

PAPER

Innovative interactive flexible docking method for multi-scale reconstruction elucidates dystrophin molecular assembly†

A.-E. Molza,[a] N. Férey,[b] M. Czjzek,[c] E. Le Rumeur,[a] J.-F. Hubert,[a] A. Tek,[d] B. Laurent,[d] M. Baaden[d] and O. Delalande*[a]

Received 13th December 2013, Accepted 12th February 2014
DOI: 10.1039/c3fd00134b

At present, our molecular knowledge of dystrophin, the protein encoded by the *DMD* gene and mutated in myopathy patients, remains limited. To get around the absence of its atomic structure, we have developed an innovative interactive docking method based on the BioSpring software in combination with Small-angle X-ray Scattering (SAXS) data. BioSpring allows interactive handling of biological macromolecules thanks to an augmented Elastic Network Model (aENM) that combines the spring network with non-bonded terms between atoms or pseudo-atoms. This approach can be used for building molecular assemblies even on a desktop or a laptop computer thanks to code optimizations including parallel computing and GPU programming. By combining atomistic and coarse-grained models, the approach significantly simplifies the set-up of multi-scale scenarios. BioSpring is remarkably efficient for the preparation of numeric simulations or for the design of biomolecular models integrating qualitative experimental data restraints. The combination of this program and SAXS allowed us to propose the first high-resolution models of the filamentous central domain of dystrophin, covering repeats 11 to 17. Low-resolution interactive docking experiments driven by a potential grid enabled us to propose how dystrophin may associate with F-actin and nNOS. This information provides an insight into medically relevant discoveries to come.

[a]*Institut de Génétique et Développement de Rennes, CNRS, UMR6290, Univ. Rennes 1, Campus Santé, 2 av du Pr Léon Bernard, 35043 Rennes Cedex, France. E-mail: olivier.delalande@univ-rennes1.fr*

[b]*Laboratoire d'Informatique pour la Mécanique et les Sciences de l'Ingénieur, CNRS, UPR3251, Univ Paris XI, Bât 508, 512 et 502 bis, 91403 Orsay Cedex, France*

[c]*Laboratoire Végétaux Marins et Biomolécules, CNRS, UMR7139, Univ. Pierre et Marie Curie-Paris 6, Station Biologique, 29682 Roscoff, France*

[d]*Laboratoire de Biochimie Théorique, CNRS, UPR9080, Univ. Paris Diderot, Sorbonne Paris Cité, 13 rue Pierre et Marie Curie, 75005 Paris, France*

† Electronic supplementary information (ESI) available: Figure representing Attract docking starting points and table for the SAVES structural analysis. See DOI: 10.1039/c3fd00134b

Introduction

Dystrophin is an essential protein of skeletal muscle[1] encoded by *DMD*, the largest human gene. Given that the cell membrane ruptures when dystrophin is lacking,[2] as seen in the lethal genetic disease Duchenne Muscular Dystrophy (DMD), dystrophin function is clearly crucial for muscle survival.[3] By interacting with cytoskeleton proteins and membrane lipids, dystrophin enables shear stress resistance in muscle cells.[4,5] Nevertheless, in the absence of a more exhaustive structural characterization, dystrophin's precise role at the molecular level remains poorly understood. The protein's filamentous state and huge size preclude easy answers, but understanding dystrophin at a molecular level, including its structure, dynamics, and interactions, constitutes a key step in the optimization of gene therapies for treating DMD.

Dystrophin is made up of four domains: a N-terminal actin-binding domain; a central rod domain comprising 24 spectrin-like repeats (R1–24);[6] a Cys-rich domain involved in the membrane dystrophin-associated glycoprotein complex; and the C-terminal end. These domains are connected by four hinges. The region of repeats 11 to 17 (R11–17) of the central rod domain is involved in crucial interactions with filamentous actin (F-actin), nitric oxide synthase (nNOS), and membrane lipids (Fig. 1). It has been previously established that the R11–15 fragment is associated with F-actin[7,8] and that the R16–17 fragment interacts with neural nNOS.[9,10] Both regions interact with membrane lipids.[11,12] However, few structural data are available for the dystrophin protein, and none of these concern the R11–17 region. Interestingly, this region is involved in the hotspot of the in-frame mutations leading to Becker Muscular Dystrophy (BMD).[13] Nevertheless,

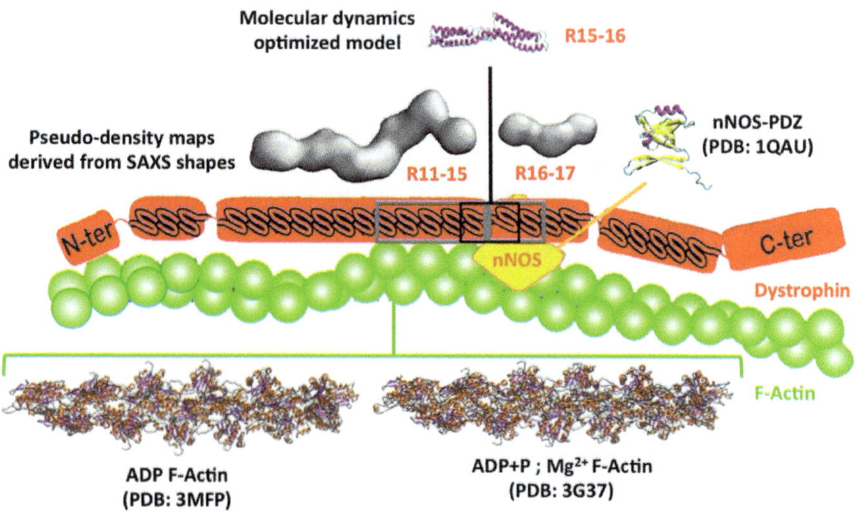

Fig. 1 The interaction between dystrophin, nNOS and F-actin. Structural information for the nNOS PDZ domain and F-actin filaments are respectively provided by X-ray diffraction and cryo-EM experiments. PDB codes are 1qau for nNOS and 3mfp or 3g37 for F-actin. Dystrophin structural data were obtained in solution from Small-angle X-ray Scattering (SAXS) or *in silico* from homology modelling followed by molecular dynamics optimization.

failure of nNOS recruitment at the membrane is observed in BMD patients, leading to a more severe phenotype.[14]

In the present study, we developed a multi-scale reconstruction approach, integrating various experimental and modelling methods. As a proof of concept, the approach was successfully applied, resulting in the structural characterization of the central R11–17 region of dystrophin and its interaction with major protein partners.

With that aim in mind, we first acquired Small-angle X-ray Scattering (SAXS) data from multi-repeat dystrophin fragments. To elucidate the atomic structure of these fragments, we developed theoretical approaches (BioSpring experiments) based on an interactive docking concept.[15] This approach uses an augmented Elastic Network Model (aENM) combining the spring network with non-bonded atom or pseudo-atom terms.[16] Interactive docking is a user-centred approach that uses advanced 3D interaction devices featuring force feedback (haptics) during a docking simulation to allow application of real-time constraints to the proteins. Details on interactive aspects, haptics, computation of interaction forces and user contributions are provided in the Materials and methods section and examined in the Discussion section. Complementary to traditional fully-automatic docking approaches, our interactive docking software allows the user to manipulate and drive several flexible proteins to a bound state using her or his own knowledge and expertise of the biomolecular system, but all the while remaining constrained by the underlying biophysical interactions.

We started by acquiring dystrophin structural data using the aENM coupled to the volumetric data provided by the SAXS experiments. Thereafter, the macromolecular complex was constructed using the aENM joined to the electrostatic potential grid computed from experimental structures. Interactive driving of the ligand to the receptor was enhanced using contact data obtained experimentally either from footprint-like mapping (trypsin digestion of lipid–protein complexes) or from *in vitro* and *in vivo* cellular experiments on mutant proteins. SAXS-based models provided either ligand (*versus* F-actin) or receptor (*versus* PDZ-nNOS) models for dystrophin docking with these two main partners. Beforehand, we performed a rigid low-resolution docking (Attract program) to extensively explore the most probable association modes for the starting structures later considered in the interactive docking step. BioSpring enables both the interactive optimization of the Attract poses following experimental contact mapping, and the evaluation of the effects of introducing flexibility in the molecular models upon their association.

Finally, we present here the first ternary complex composed of F-actin–dystrophin–nNOS proteins, revealing structural features that could be related to the dysfunction of the truncated dystrophin forms observed in BMD patients.

Experimental

Materials and methods

Molecular structures. Similar to our previous structural study,[17] homology models for the R11–15 and R16–17 dystrophin fragments were obtained thanks to the I-TASSER web server,[18] using an updated template library that includes the structures of the first repeat of both dystrophin and utrophin.[19] The molecular dynamics (MD) optimized model of the R15–16 fragment was taken from our previous work.[17] The actin filaments used to perform the docking experiments were 12-actinG polymers built from the cryo-EM structures available in the Protein Data Bank (PDB codes 3MFP and 3G37). The nNOS-PDZ crystallographic structure (PDB code 1QAU) was completed using homology modelling to fit Lai's experimental sequence.[10] This PDZ structure was submitted to normal modes analysis-driven deformations using the elNémo server[20,21] to produce several conformations. All structural data are shown in a schematic cellular context in Fig. 1.

Low-resolution rigid docking. Rigid low-resolution Attract docking calculations were performed for both R11–15–F-actin and PDZ-nNOS–R16–17. The most relevant associations between the structures, that could be further associated interactively, were predicted through exhaustive Attract searches. Newly-developed parameters for the Attract force field were used during the docking calculations.[22] Placing of the ligands, R11–15 or PDZ-nNOS for each experiment, was achieved thanks to density grids of 25 Å and 10 Å. A total of 157 starting points for the dystrophin R11–15 fragment were set all around the actin filament, creating a 140 Å-high cylinder with a 150 Å radius (see ESI†). The PDZ-nNOS had 381 starting points around R16–17. For each starting point, the standard 258 rotations were computed for the ligand. The final data was a result of 40 506 poses and 98 298 poses, respectively, for the R11–15–F-actin and PDZ-nNOS–R16–17 docking experiments. Ranking of the best poses was performed by sorting by complex energy and clustering over the ligand position with a root mean square deviation (RMSD) of 60 Å (F-actin docking) and 10 Å (PDZ-nNOS). The best theoretical complexes were thus selected for their low energy levels (Attract function) and because they were the most populated clusters. Molecular contacts were defined with a cut-off distance between the ligand and receptor of less than 5 Å.

Augmented elastic network. The coarse-graining of initial homology models and experimental PDB structures was performed in accordance to the rules defined by Zacharias.[23] The augmented Elastic Network Model (aENM) combines the usual ENM model (harmonic functions) and non-bonded terms (Attract coarse-grain force field) as per our previous work[16] and similar to that proposed by Periole *et al.*[24] or more recently by Globisch *et al.*[25] The aENM model supports the use of multiple cut-off distances to designate a final spring network (Fig. 2A). It thus allows the assignation of variable weight to secondary or ternary structures as needed. In this study, a cut-off distance of 9 Å was used for the backbone particles involved in secondary structure elements, and it was reduced to 5 Å for other particles that belong to loops or side chains. The two resulting spring networks were merged into a final aENM model after the particles of both previous aENM models were linked by a spring if they were closer than the cut-off distance of 5 Å. All-atom aENM was used for high-resolution reconstruction of the coarse-grain models. Double cut-off ENM had distances of 9 Å for backbone atoms and 3 Å for side chain and loop atoms, and 3 Å was used as the distance cut-off to merge the first two ENM models. Atomic parameters, including charge and radius, were extracted from the AMBER99 force field.[26]

Potential grid. Electrostatic potential maps were obtained from APBS calculations[27] achieved for the structure of the receptor molecules (actin or dystrophin R16–17 in each docking experiment) using the Amber force field and 50 mM sodium and chloride ions. APBS-provided electrostatic potential grid data in the OpenDX format are compatible with the BioSpring reading routine.

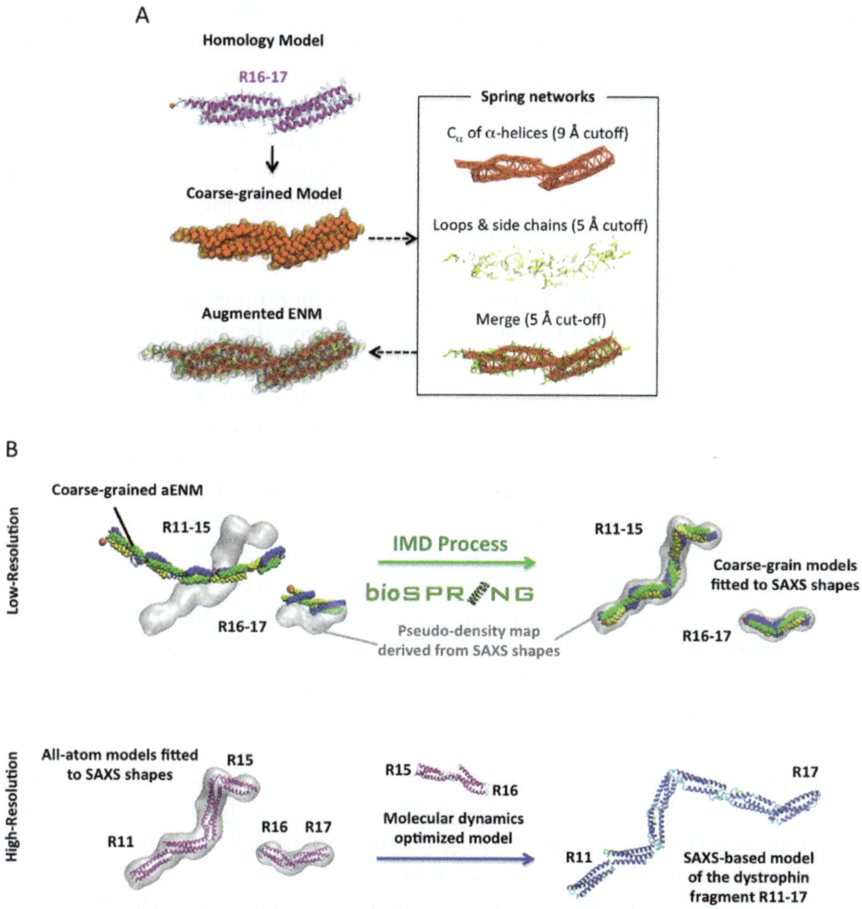

Fig. 2 (A) Protocol for building a coarse-grained augmented Elastic Network Model (aENM) of the R16–17 repeat fragment of dystrophin. From the all-atom homology model provided by the I-TASSER server, a low-resolution model was obtained using the Attract force field parameters developed by Zacharias *et al.*,[23] a multi-cut-off spring network can be designed to discriminate between the most flexible and the most rigid regions. Here, we chose to stabilize secondary (α-helices) and ternary (coiled-coil) structures. The completion of an aENM model is possible through combining the steric and charge parameters assigned to the pseudo-atoms with the springs defining the ENM. (B) (Top row) Schematic representation of BioSpring's interactive molecular dynamic processing of the R11–15 and R16–17 dystrophin fragments at low-resolution. For aENM, helices HA are yellow, helices HB are blue, and helices HC are green. The red spot represents the N-terminal end; coils are gray. SAXS shapes were obtained from the GASBOR program of the ATSAS suite then converted to pseudo-density maps for driving the aENM model into the volumetric data. (Bottom row) Fragment reconstruction of an all-atom model is performed by superimposition onto the coarse-grain model then finalized by an optimization into the pseudo-density maps. The full-length filament is obtained by bridging the two SAXS-based models with a molecular dynamics-derived model of the R15–16 fragment, unavailable experimentally.

Experimental volume maps. SAXS scattering curves were analyzed using the ATSAS software suite.[28] Twenty molecular shapes were generated using Gasbor's method which is based on Monte-Carlo draws of hard-sphere distributions (one bead per amino acid) paired with cross-validation *via* scattering curve simulation. The OLIGOMER program was used for weighing the 20 shapes (linear combination solution). OLIGOMER ranking showed that the best SAXS molecular shapes selected for both R11–15 and R16–17 fragments achieved a high rank (#1 for both) and high weight percentage (43% and 31% contribution to the scattering curve, respectively) as well as a low χ^2 value (1.3 and 2.5, respectively). The volumetric data for the Gasbor molecular shapes were converted into a pseudo-density grid using the Molecular Dynamics Flexible Fitting (MDFF) applet of the VMD program[29] at a resolution of 20 Å.

Interactive human-guided simulations. As in our previous studies,[16,30] the MDDriver library was used for data exchange between the simulation engine, in this case BioSpring calculations, and the visualization front-end, here the VMD viewer. The interactive simulations were computed on a 24-core desktop computer using a PHANTOM Omni haptic device (SensAble) for advanced 3D interaction. The 3D aspects of this device make it easier to pick an atom or a coarse-grained particle within the protein, which is intrinsically a spatially complex 3D object. Haptic features allow the user to feel the forces involved in the simulation through force feedback (see next paragraph for a more detailed discussion). The BioSpring software manages the aENM model and guides it through a scalar density field (read from data in the OpenDX format) by computing gradients to derive the force to be applied to a particle according to its mass (SAXS grid) or charge (APBS grid). As these simulations are user-guided, they need to be repeated several times in order to account for the variability of the interaction.

Haptic interactions. To provide haptic feedback to guide the user, forces must be computed and rendered at very high frequencies, usually between 300 Hz and 1000 Hz. This frame rate is clearly incompatible with typical molecular simulation frame rates, especially for complex structures, because each timestep is computed at a relatively low frequency, at most still an order of magnitude below the haptic requirements. To overcome this issue, an intermediate high frequency "simulation" is asynchronously used to provide haptic feedback. It consists of simulating springs virtually connecting the avatar of the haptic probe device in the simulation space and all the particles selected by the user. A sum of the spring forces is then rendered to the user through the haptic device, and each force on the particle is sent to the actual BioSpring simulation engine at each timestep as an external simulation constraint. The user-felt forces thus indirectly originate from the molecular simulation *via* the connecting springs. This solution provides usable and useful haptic feedback. As the perception system of the user naturally integrates force feedback over time, (s)he is able to feel the flexibility or rigidity of the area that is haptically manipulated. Likewise, using springs allows the user to finely tune the global stiffness of these springs to render the bonded and non-bonded interactions computed in the molecular simulation more accurately.

Interactive docking experiments. A strict protocol was set up and used for the six experiments carried out for each scenario (envelope, orientation) to prevent *a priori* bias that might occur for the interactive fit simulations of driving models into SAXS volumetric data (for instance, placing a specific coiled-coil repeat in a chosen spot of the SAXS envelope). Predefined backbone particle selections of each α-helix were used to translate the aENM model into the volumetric grid. The first step of the protocol consisted of the orientation of the aENM filament parallel to the molecular envelope. The entire aENM was then driven into the envelope. The aENM model was then stretched beyond both terminal ends of the molecular envelope, then was allowed to relax on is own, without interactive intervention, into the volumetric grid and over 20 000 calculation steps. In the case of interactive docking driven by an APBS potential grid, starting positions for the ligand were either set randomly or by using one of the best poses predicted from the preliminary Attract docking calculations. To complete efficient interactive simulations at an all-atom resolution on desktop computers, the OpenCL implementation newly developed by us for the BioSpring software was used.

Structure validation and representation. Validation of SAXS-based coarse-grained models was performed by volume overlap calculations in the Chimera program.[31] The final atomic models were evaluated *via* the SAVES metaserver (http://services.mbi.ucla.edu/SAVES/), which combines different programs for molecular structure quality control. All figures were realized with VMD or Chimera.

Results

Interactive fit of homology models into SAXS volumetric data: a quick and efficient method for producing reliable SAXS-based models

An interactive volume fitting procedure (Fig. 2B) enabled us to reconstruct the dystrophin central domain in less than a minute. For the short fragment, a SAXS-based model was produced without human-added driving force pushing the aENM into the volumetric data. In this particular case, haptic rendering alone provided essential information about the rapid convergence of the fit during the simulation. The good quality (see ESI†) of the models built following the interactive fitting procedure is based on two main features. Firstly, the robustness of the aENM allows us to maintain the secondary structure as well as the particle packing, thus avoiding the collapse of the structure so that reliable solutions could finally be obtained. Indeed, the basic fold of spectrin-like repeats in the triple coiled-coil[17,19] was maintained for each dystrophin repeat model. The flexibility of the model still enables individual α-helices to kink or coiled-coil repeats to rotate towards each other as needed. Secondly, online visualization allows a good evaluation of the volumetric fit between the aENM and the SAXS envelope. This enables the user to discard bad subdomain placement in the experimental volumetric data. It notably made it possible to discriminate between the different N-terminal to C-terminal or C-terminal to N-terminal orientations stemming from the absence of linear shape polarisation of the SAXS volumetric data. The final interactive reconstruction of a SAXS-based protein structure model was done by superimposing the all-atom homology model (aENM format) onto the final aENM position obtained by the preliminary coarse-grain interactive fitting, then by calculating the flexible volumetric fit with the all-atom aENM. The final volume overlaps obtained for the R11–15 dystrophin fragment were, respectively, 60%, 86%, and 92% for the static homology model, flexible coarse-grain model, and flexible all-atom model. Overlap percentages for the R16–17 fragment were 84% (homology), 87% (coarse-grain aENM), and 91% (all-atom aENM).

Given the fact that the R15–16 dystrophin fragment is unable to be produced *in vitro*, thus precluding SAXS data acquisition, we used the previously obtained MD models[17] to merge the two SAXS-based all-atom models and to propose the first model of the R11–17 fragment of dystrophin. The final SAXS-based atomic models revealed the dog-legged form of dystrophin's filamentous structure. It is worth noting that α-helical structures and coiled-coil arrangements can be maintained despite the sharp kinks between several consecutive repeats (for instance, between R14 and R15) that have been observed experimentally (SAXS molecular shapes).

Thanks to the interactive exploration and evaluation of the two possible relative orientations (N- to C-terminal end or C- to N-terminal end) within the SAXS envelopes, we found that the largest kinks are located at the R12–13, R14–15 and R16–17 junctions (linker regions). These linkers are either characterized as straight α-helices or as helices bearing a kink in its main axis. Indeed, SAXS volumetric data correspond to a time-averaged molecular shape, and the kinks we observed should correspond to a mean position between two existing extreme conformations. In fact, the kinks could have the wrong amplitude but impose the correct relative direction between two consecutive repeats. These flexibility points were considered as "preferential adjustable regions" for further dystrophin–F-actin docking interactions.

Interactive fit of SAXS-based models into APBS potential grids: how to create a theoretical complex using guidelines produced by systematic low-resolution docking combined with footprint-like experimental data

We explored the association modes between dystrophin and F-actin by analyzing the two available isoforms of filamentous actin when associated with ADP (PDB code3MFP) or with ADP, PO_4^{2-} and Mg^{2+} (PDB code 3G37). We considered these two conformational states to be representative of changes that could occur *in vivo* in a stretched or relaxed muscle cell, and each was used as a static receptor conformation for the interactive docking experiments.

The automated preparatory low-resolution Attract docking experiments provided evidence for the most probable F-actin molecular surfaces that could interact with the R11–15 SAXS-based model. A parallel orientation of both filaments seemed favoured, and the dystrophin fragment was inserted in a F-actin hydrophobic groove and surrounded by small electronegative patches. This latter point is in agreement with the observation that the R11 to R15 repeats are among the most positively-charged dystrophin repeats.[17] It appears that the three best complexes obtained by Attract, independently from the F-actin structures, *i.e.* with or without Mg^{2+}, showed very similar contact patterns. The first type of association (clusters ranked #1 and #2) showed preferential contacts between dystrophin's R14 and R15 repeats and F-actin, while R11 to R13 showed only poor contacts with the actin filament. The second association mode (clusters ranked #3) highlights preferential contacts between R11 and R12 repeats and F-actin. We completed contact frequency mapping by recording the contacts between both proteins with an interparticle distance $\leq$ 5 Å in the first 20 molecular complex clusters obtained by the rigid docking search. We used this mapping to visualize the most probable contact surface on the F-actin. This first docking using Attract did not allow dystrophin flexibility and likely avoids a proper contact between the two filaments, which would explain the discrepancy between the two best solutions with two different regions of dystrophin.

Next, we therefore used a flexible and human-guided docking approach with BioSpring. The driving force in interactive docking with BioSpring was provided by an APBS electrostatic potential grid computed on the F-actin experimental isoforms. This docking strategy seems highly relevant because of the predicted importance of electrostatics in the interaction of dystrophin with F-actin.[12] As an APBS map is calculated for experimental static structures, we considered the F-actin to be a static molecular shape, thus avoiding the miscorrelation in the spatial distribution of volume and electrostatics that would occur if we had proposed a flexible model (aENM) for the receptor filament. Interactive BioSpring docking enabled a flexible adaptation of dystrophin to the static F-actin molecular surface following the identified interaction groove predicted by the Attract-provided static low-resolution search. As shown in Fig. 3, graphical rendering of the molecular surface potentials of F-actin during the interactive docking proved to be a powerful aid in the interpretation of the dynamic behaviour of the R11–15 dystrophin ligand driven into the potential grid of the F-actin receptor.

The interactive flexible simulation enabled us to obtain a macromolecular complex in agreement with experimental data. In fact, we had previously[33] performed a MS/MS analysis after trypsin digestion of R11–15 in interaction with liposomes: the peptide fragments accessible to trypsin proteolysis were predicted to be free of phospholipid interactions and therefore to be potentially involved in

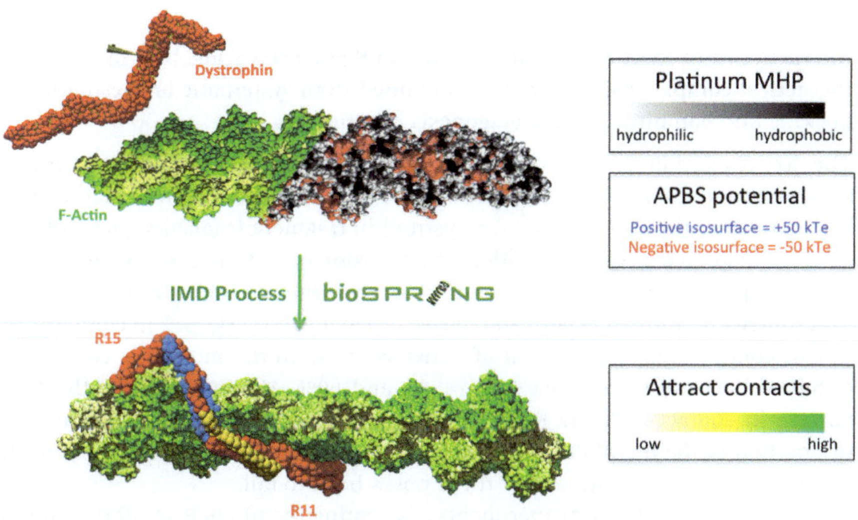

Fig. 3 (Top) Starting point for interactive flexible low-resolution docking. A dystrophin fragment (R11–15) is modelled by aENM shown here as red van der Waals spheres (coarse-grain particles) and yellow sticks (springs). The actin filament's molecular surface (PDB code 3g37) is shown with shading from yellow to green by increasing the probability of contact with dystrophin as indicated by the Attract docking results, or from white (hydrophilic) to black (hydrophobic) according to the molecular hydrophobicity potential (MHP) provided by the Platinum server.[32] The electrostatic potentials calculated using APBS software[27] are shown for ±50 kTe isosurfaces and are blue (+) or red (−). These visual representations can be used to guide the user in the interactive experiments. (Bottom) Best pose of the R11–15 fragment after optimization of the aENM model into the potential grid computed for the actin filament structure. For dystrophin fragment (red spheres), yellow and blue colors are indicating respectively the regions that are or that are not protected against trypsin digestion upon the interaction of the protein with small unilamellar vesicles.

another interaction, such as with F-actin. The conclusion then was that helices B and C of repeat R13 and helices A and B of repeat R14 could be involved in an interaction with the actin filament.

We selected the two best association modes proposed by Attract docking as our starting point for the interactive docking simulations. Using these, it appeared that matching the experimental contact restraints would impose a change in the R12 and R13 positioning towards the F-actin filament (Fig. 3). Indeed, in all the Attract poses, a contact between helix C of R12 is observed and should be avoided, as the region is expected to preferentially interact with sarcolemma phospholipids. The torsion of the R12–13 linker region allowed us to interactively match the previously-observed experimental contacts, and led to a decrease in the kink angle observed in the SAXS-based model. Looking at the macromolecular complex, it appears that dystrophin showed a modified topography when compared to the SAXS-based model obtained in solution, as the succession of repeats from R11 to R13 show a rather linear filamentous structure. On the contrary, the kink observed at the R14–15 linker region in the SAXS-based model of the R11–15 fragment was maintained in the dystrophin–F-actin complex. Interactive docking led to a continuous contact of both filaments on a dystrophin track defined from R11 helix B to R12 helix A, and from R13 helix B to R14 helix C. These results are in agreement with the experimental data collected earlier.

Interactive fit of SAXS-based models into APBS potential grids: how to create a theoretical complex using guidelines obtained from systematic low-resolution docking and combined with mutagenesis experiments

The low-resolution and rigid explorations of the association between nNOS structures and the R16–17 SAXS-based models showed three preferential relative orientations of the PDZ towards the dystrophin fragment (ranked #1, #2 and #6). All three displayed a good correlation with the contact surface identified indirectly on dystrophin by our recent theoretical study[34] and through the mutagenesis experiments performed *in vitro* and *in vivo* by Lai and coworkers.[10] In comparison to this contact data, the fraction of native contacts in the molecular complexes obtained here are respectively 53%, 43%, and 60%. However, even if the PDZ surface in contact with the R16–17 fragment seemed well-defined, the relative orientation of the nNOS PDZ subdomain towards the dystrophin fragment itself remained unfocused, with two of these poses head-to-tail.

The interactive docking experiments associating the nNOS PDZ subdomain to the R16–17 dystrophin fragment (Fig. 4) used the protocol previously detailed for dystrophin–F-actin docking, but with a major change. This time, in addition to dystrophin's flexibility, we also considered the relative flexibility of the receptor (R16–17), as both proteins are represented by an aENM. To avoid sliding or rotation of the receptor aENM model towards its APBS potential grid, the terminal-end C_α atoms of the dystrophin fragment were fixed (spatial restraint). The use of such a rigid-flexible heterogeneous aENM model enabled, upon interactive docking, a smooth modulation of the kink characterized experimentally in the SAXS shape at the R16–17 linker region. Long-range driving of the PDZ ligand to the target surface highlighted experimentally is not trivial given the poor contribution of electrostatics expected in the association of the two proteins.[34] In these conditions, the three best poses produced by rigid docking were used as

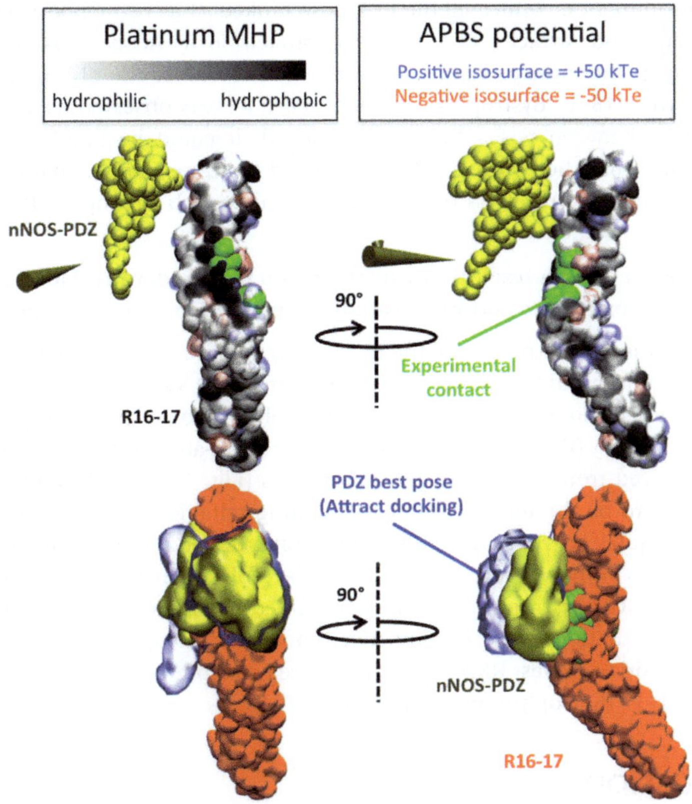

Fig. 4 (Top) Starting point for interactive flexible low-resolution docking. The dystrophin fragment (repeats R16–17) is shown with a molecular surface shading from white (hydrophilic) to black (hydrophobic) according to Platinum MHP.[32] APBS potentials[27] are ±50 kTe isosurfaces shown, respectively, in blue and red. The nNOS PDZ subdomain is simulated by an aENM in yellow van der Waals spheres. The green spheres highlight the contact regions defined by mutagenesis performed *in vitro* and *in vivo* by Lai *et al.*[10] (Bottom) The nNOS–dystrophin complex obtained after interactive BioSpring docking, showing the best correlation with the experimental contact mapping. The blue phantom envelope corresponds to one of the best poses, based on complex energy and weighted after clustering, obtained from systematic rigid low-resolution docking (Attract).

starting positions for the PDZ subdomain. Therefore, interactive experiments operate as a precise exploration of the alternative poses, allowing for rapid and accurate convergence on the most probable association mode, completed with online visual checks. The flexible adaptation of both molecular surfaces to each other was observed in each interactive simulation. This interactive flexible approach enabled a better match of the complex initially ranked #6 by the Attract function score. The fraction of native contacts in this final molecular complex is 80%, corresponding to an increase of 20% over the corresponding Attract pose. This result underlines the usefulness of flexible modelling in the docking approach.

Upon the interactive exploration of the three best alternative Attract poses, we tried to rank them by perturbation–relaxation trials. Online visualization of the docking simulation confirmed the PDZ pose ranked #1 as being poorly stabilized despite its good Attract score. The pose ranked #6 by Attract was selected by BioSpring as the most probable lowest energy position of the nNOS subdomain towards the R16–17 dystrophin fragment. Pose #2 was observed as an intermediate stage between poses #1 and #6. Indeed, it could be highly relevant to hypothesize that the two less stabilized poses are transient orientations of the PDZ, enabling it to reach its best association mode with dystrophin R16–17.

Interactive fit of high-resolution structures into pseudo-density data: a generic and efficient method to perform all-atom reconstruction of coarse-grain models

Final all-atom reconstruction of the full ternary complex was completed either by (i) a fragment-building strategy using a lowest-RMSD superimposition onto the coarse-grained models previously obtained, followed by a short energy minimization step, or by (ii) a volumetric data fit using an all-atom aENM positioned into a grid derived from the coarse-grained models (Fig. 5A). The second approach turned out to be the most efficient for yielding reliable models in the shortest time. This quick coarse-grain to all-atom reconstruction method using BioSpring enabled us to propose structurally-stable high-resolution models of dystrophin–F-actin and nNOS–dystrophin complexes obtained from low-resolution interactive docking simulations. The ternary macromolecular complex was obtained by superimposition (lowest RMSD on C_α) of these two atomic models onto the SAXS-based R11–17 dystrophin model (Fig. 5B).

Discussion

Exploring molecular plasticity: modelling biomolecules under the control of experimental data as if they were clay

Using the BioSpring software as a powerful model builder enabled us to confirm the robustness of the aENM approach to a user-imposed local deformation of the driving and shaping of biomolecules through interactive simulation. Molecular object inertia or heterogeneous response to the forces applied uniformly to a set of particles can be overcome by preselecting the interactive ensemble. An approach consisting of preferential interaction on a selection of protein backbone particles leads to less deformed structures and to a better haptic response, as the particles are more closely connected due to the higher particle density in their neighbourhood. The models' quality was further enhanced by a multi-cut-off spring-network design avoiding molecular distortion to a very high degree. We verified that the non-bonding terms (steric and electrostatic) set in the aENM model efficiently counterbalance the tendency of ENMs to remain very close to their initial conformation. For instance, the flexibility of the R11–15 model enabled us to induce conformational changes to dystrophin upon its complexation with F-actin.

We should emphasize that molecular packing and folding is not only rendered by the steric term associated to a particle's volume, but structurally encoded by springs in the aENM. Steric clashes, and more serious structural collapse, are thus easily avoided upon molecular docking or volumetric data fitting when compared to what can be observed with other fitting methods such as MDFF.[35] BioSpring interactive docking into volumetric data enables the

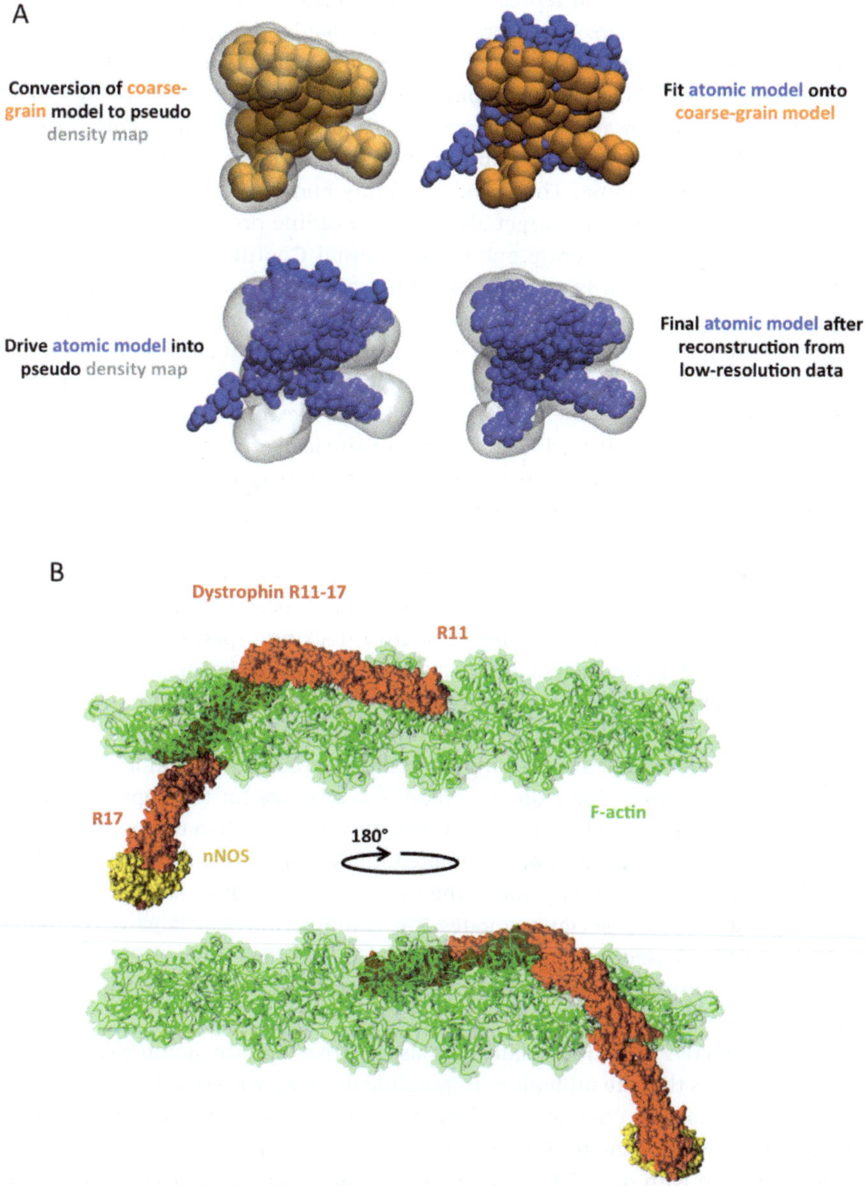

Fig. 5 (A) BioSpring protocol for the atomic reconstruction of the low-resolution model obtained from interactive flexible docking. (Top left) The final position of the coarse-grain aENM in the molecular complex converted into a pseudo density map. (Top right) The atomic aENM is superimposed (C_α RMSD) onto this same coarse-grain aENM final position. (Bottom left) The atomic aENM is driven into the pseudo density map. (Bottom right) The final atomic model is obtained after a few seconds of interactive BioSpring simulation. (B) Final ternary complex obtained by the association of dystrophin to f-actin and nNOS. Dystrophin is red, F-actin (PDB code 3g37) is green, and nNOS is yellow. The complex was developed by combining theoretical and experimental biostructural data with an interactive flexible approach.

calculation of either intramolecular or intermolecular interactions using non-bonded terms of the aENM representation, in contrast to a simple volume match guided by a density map. This representation considerably improves the quality of the flexible fit under experimental volumetric restraints, notably aiding in the judicious placing of loops in biomolecular models. We tested BioSpring experiments as an efficient and quick generic method for multi-scale reconstruction by generalizing the approach, combining flexible molecular models and a spatial distribution fitting frame. The speed reliability and convergence of fitting the models into the volumetric target also make an offline procedure possible. This is comparable to the recently published Internal Coordinates Flexible Fitting (ICFF) method,[36] which is several orders of magnitude faster than a Molecular Dynamics approach such as MDFF. However, detailed information about flexible regions is required before using the ICFF method to its full efficiency. Our more generic aENM-based approach allows limited flexibility in amplitude but does not necessarily impose the set of a predefined flexible ensemble, since the regions bearing an increased flexibility are intrinsically related to regions with a lower density of particles. Furthermore, BioSpring is quicker than MDFF of a corresponding atomistic model thanks to coarse-graining and code optimization (OpenCL).

Long-range driving of the ligand towards the receptor was essential to assist the scientist's gesture through haptic rendering. A possible limitation in electrostatic grid guidance during the interactive docking experiments could arise either from the biomolecular system size and complexity or from a lack of importance of electrostatics in the association of the two studied partners. Displaying experimental contacts on the ligand and/or the receptor provides essential cues for the interactive docking procedure, revealed to be more rapid and more accurate than without them. Exploration of space for the approach of the ligand towards the receptor thus could be reduced to less than twenty per cent of an exhaustive positioning by an automated method. By using an exhaustive low-resolution docking method prior to the interactive procedure, we prevented the oversight of putative association modes. The combination of exhaustive computer and knowledge-based human approaches led to us efficiently reaching the best solution in accordance with the interaction surfaces defined by mutation or footprint experiments.

The interactive procedure enabled us also to drive the association by proposing complex moves that are difficult or impossible to set up with traditional or offline docking approaches. For instance, during dystrophin–actin docking, the rotation of a part of the filament around its main axis (R12–13) was necessary to achieve a reliable match with data proposed by experimental contact mapping. Moreover, distancing a ligand subdomain from an unwanted position in the proximity of the receptor, as performed for the R15 repeat towards F-actin, is not easy to achieve in traditional docking programs, usually because they intrinsically force the matching of both the ligand and the receptor.

It is logical to note that the correlation between the experimental data increased if the flexibility was also extended from the ligand to the receptor contact surface, simply keeping its backbone fixed. The modelling of biomolecular matter can be easily set up by the design of a rigid/flexible heterogeneous construction from the BioSpring software commands, which distinguish between static and dynamic particle sets. This BioSpring option enabled us to considerably speed up the interactive calculation rate for very large systems, with the simulation being faster in proportion to a smaller size of the dynamic particle set.

Overall, this mixture of human interaction based on precalculated guidelines with experimental data and an underlying physical model has proven very efficient. It remains, however, challenging to characterize, describe and quantify the human contribution that may vary from one run to another. Therefore, several runs are systematically carried out until a convergence or a consensus is achieved. Haptics make the interaction natural, as would be the manipulation of a physical model, *e.g.* made of clay. In particular, guidance by the interaction forces, firstly electrostatics, is an invaluable cue. Human (and precalculated) knowledge allows the user to significantly reduce the search space and is critical to supervise the refinement of the models. This raises the question whether the solutions that were found are unique. The subjective assessment of the authors is that no other – significantly different – models compatible with the experimental constraints could be built with the underlying physical model.

Exploring dystrophin's interactome: modeling the protein complex formed in the region of a hotspot of a genetic pathology

In the final ternary complex, three major observations brought crucial understanding to the macromolecular system involving dystrophin. The first concerns the weak contacts of the R15 repeat with F-actin and of the R16 repeat with nNOS. These two repeats seem to be important for structuring the dystrophin filament, but also emphasize a structural frontier between the contact positions of each protein partner. This may be related to the atypical feature of the R15–16 linker region, a loop twelve residues long, which is probably at the origin of its *in vitro* instability. The second observation highlighted by our macromolecular model deals with the relative positioning of dystrophin and actin filaments with respect to each other. While the N-terminal end of the dystrophin R11–15 fragment is laid out parallel to the F-actin, mainly interacting with only one of the two actin filaments, the C-terminal end kinks at the linker regions of the final repeats, disengaging from the F-actin. This contact pattern extends along a distance of 150 Å and with a coiling of both dystrophin and actin filaments over an angle of 180°. This observation is compatible with both of the F-actin isoforms considered in our study, and indicates that dystrophin does not seem to bridge the two dimerizing filaments of F-actin. The third main observation concerns the relative positioning of F-actin towards the nNOS PDZ subdomain. In the final macromolecular complex, the dystrophin R11–17 contact surfaces interacting with F-actin and nNOS are oriented in a similar manner. This could emphasize the structuring of another interaction subdomain, that of the potential regions in contact with phospholipids. In the truncated BMD forms of dystrophin, this polarised 3D organisation could be affected, resulting in the loss of efficiency in dystrophin's essential role as a scaffold protein for the muscle cell membrane.

Conclusions

The interactive flexible method based on the BioSpring software presented here allows one to easily swap from a one-residue coarse-grain level (SAXS molecular shape in the Gasbor method) to the Attract coarse-grain level, finally coming up with a high-resolution molecular description. The approach is a highly reliable multi-resolution model builder for the construction of large molecular assemblies. In the future, the interactive simulations of the fitting of molecular models to SAXS data could be greatly improved by an instant display of the scattering curve calculated for a given state of the aENM.

We present here the first models of dystrophin fragments in solution, based on SAXS measurements. We also associate, using an interactive flexible docking approach, the dystrophin R11–15 SAXS model with the F-actin atomic structure, obtaining a final macromolecular complex that correlates well with experimental contact mapping. We also propose the first molecular model for the association of dystrophin R16–17 to the nNOS PDZ subdomain, which is well supported by *in vitro* and *in vivo* data. The ternary complex involving the BMD hotspot region of dystrophin (R11–17) with two of its main protein partners was modelled by combining theoretical and experimental information, leading to a highly reliable structural model.

Funding

The Association Française contre les Myopathies (AFM) and Rennes Métropole provided financial support, and Anne-Elisabeth Molza PhD fellow is funded by an AFM grant. Computer time was provided by GENCI (IDRIS and CINES, grant DYSIM 076816). M.B. thanks the French Agency for Research (Grant "ExaViz", ANR-11-MONU-003) and the "Initiative d'Excellence" program from the French State (Grant "DYNAMO", ANR-11-LABX-0011) for support.

Availability

The BioSpring software is an open source program available online at http://sourceforge.net/projects/biospring.

Acknowledgements

The authors are indebted to beamline staff for precious help on the different SAXS beamlines: Adam Round and Petra Pernod at ID14-EH3, ESRF, Grenoble, France and Javier Perez and Pierre Roblin at SWING, Soleil, St Aubain, France.

Notes and references

1 M. Koenig, A. P. Monaco and L. M. Kunkel, *Cell*, 1988, **53**, 219.
2 B. J. Petrof, J. B. Shrager, H. H. Stedman, A. M. Kelly and H. L. Sweeney, *Proc. Natl. Acad. Sci. U. S. A.*, 1993, **90**, 3710.
3 M. Koenig, E. P. Hoffman, C. J. Bertelson, A. P. Monaco, C. Feener and L. M. Kunkel, *Cell*, 1987, **50**, 509.
4 E. Le Rumeur, J. F. Hubert and S. J. Winder, *FEBS Lett.*, 2012, **586**, 2717.
5 E. Le Rumeur, S. J. Winder and J. F. Hubert, *Biochim. Biophys. Acta, Proteins Proteomics*, 2010, **1804**, 1713.
6 S. J. Winder, T. J. Gibson and J. Kendrick-Jones, *FEBS Lett.*, 1995, **369**, 27.
7 K. J. Amann, B. A. Renley and J. M. Ervasti, *J. Biol. Chem.*, 1998, **273**, 28419.
8 I. N. Rybakova, K. J. Amann and J. M. Ervasti, *J. Cell Biol.*, 1996, **135**, 661.

9 Y. Lai, G. D. Thomas, Y. Yue, H. T. Yang, D. Li, C. Long, L. Judge, B. Bostick, J. S. Chamberlain, R. L. Terjung and D. Duan, *J. Clin. Invest.*, 2009, **119**, 624.
10 Y. Lai, J. Zhao, Y. Yue and D. Duan, *Proc. Natl. Acad. Sci. U. S. A.*, 2012, **110**, 525.
11 S. Legardinier, C. Raguenes-Nicol, C. Tascon, C. Rocher, S. Hardy, J. F. Hubert and E. Le Rumeur, *J. Mol. Biol.*, 2009, **389**, 546.
12 J. Sarkis, V. Vié, S. J. Winder, A. Renault, E. Le Rumeur and J. F. Hubert, *FASEB J.*, 2013, **27**, 359.
13 S. Tuffery-Giraud, C. Béroud, F. Leturcq, R. Ben Yaou, D. Hamroun, L. Michel-Calemard, M. P. Moizard, R. Bernard, M. Cossée, P. Boisseau, M. Blayau, I. Creveaux, A. Guiochon-Mantel, B. de Martinville, C. Philippe, N. Monnier, E. Bieth, P. Khau Van Kien, F. O. Desmet, V. Humbertclaude, J. C. Kaplan, J. Chelly and M. Claustres, *Hum. Mutat.*, 2009, **30**, 934.
14 C. Gentil, F. Leturcq, R. Ben Yaou, J. C. Kaplan, P. Laforet, I. Penisson-Besnier, C. Espil-Taris, T. Voit, L. Garcia and F. Pietri-Rouxel, *Hum. Mol. Genet.*, 2012, **21**, 3449.
15 A. Tek, M. Chavent, M. Baaden, O. Delalande, P. Bourdot and N. Férey, Advances in Human-Protein Interaction - Interactive and Immersive Molecular Simulations in *Protein-Protein Interactions - Computational and Experimental Tools*, ed. Dr Weibo Cai, ISBN: 978-953-51-0397-4, InTech, 2012DOI: 10.5772/36568.
16 A. Saladin, C. Amourda, P. Poulain, N. Férey, M. Baaden, M. Zacharias, O. Delalande and C. Prévost, *Nucleic Acids Res.*, 2010, **38**, 6313.
17 B. Legrand, E. Giudice, A. Nicolas, O. Delalande and E. Le Rumeur, *PLoS One*, 2011, **6**, e23819.
18 A. Roy, A. Kucukural and Y. Zhang, *Nat. Protoc.*, 2010, **5**, 725.
19 M. Muthu, K. A. Richardson and A. J. Sutherland-Smith, *PLoS One*, 2012, **7**, e40066.
20 M. Delarue and Y. H. Sanejouand, *J. Mol. Biol.*, 2002, **320**, 1011.
21 F. Tama and Y. H. Sanejouand, *Protein Eng., Des. Sel.*, 2001, **14**, 1.
22 S. Schneider, A. Saladin, S. Fiorucci, C. Prévost and M. Zacharias, *Methods Mol. Biol.*, 2012, **819**, 221.
23 M. Zacharias, *Proteins: Struct., Funct., Bioinf.*, 2005, **60**, 252.
24 X. Periole, M. Cavalli, S. J. Marrink and M. Ceruso, *J. Chem. Theory Comput.*, 2009, **5**, 2531.
25 C. Globisch, V. Krishnamani, M. Deserno and C. Peter, *PLoS One*, 2013, **8**, e60582.
26 D. A. Case, T. E. Cheatham III, T. Darden, H. Gohlke, R. Luo, K. M. Merz Jr., A. Onufriev, C. Simmerling, B. Wang and R. Woods, *J. Comput. Chem.*, 2005, **26**, 1668.
27 N. A. Baker, D. Sept, S. Joseph, M. J. Holst and J. A. McCammon, *Proc. Natl. Acad. Sci. U. S. A.*, 2001, **98**, 10037.
28 M. V. Petoukhov, D. Franke, A. V. Shkumatov, G. Tira, A. G. Kikhney, M. Gajda, C. Gorba, H. D. T. Mertens, P. V. Konarev and D. I. Svergun, *J. Appl. Crystallogr.*, 2012, **45**, 342.
29 W. Humphrey, A. Dalke and K. Schulten, *J. Mol. Graphics*, 1996, **14**, 33.
30 O. Delalande, N. Férey, G. Grasseau and M. Baaden, *J. Comput. Chem.*, 2009, **30**, 2375.
31 E. F. Pettersen, T. D. Goddard, C. C. Huang, G. S. Couch, D. M. Greenblatt, E. C. Meng and T. E. Ferrin, *J. Comput. Chem.*, 2004, **25**, 1605.

32 T. V. Pyrkov, A. O. Chugunov, N. A. Krylov, D. E. Noldel and R. G. Efremov, *Bioinformatics*, 2009, **25**, 1201.
33 J. Sarkis, J. F. Hubert, B. Legrand, E. Robert, A. Chéron, J. Jardin, E. Hitti, E. Le Rumeur and V. Vié, *J. Biol. Chem.*, 2011, **286**, 30481.
34 E. Giudice, A. E. Molza, Y. Laurin, A. Nicolas, E. Le Rumeur and O. Delalande, *Biochemistry*, 2013, **52**, 7777.
35 K. Y. Chan, J. Gumbart, R. McGreevy, J. M. Watermeyer, B. T. Sewell and K. Schulten, *Structure*, 2011, **19**, 1211.
36 S. C. Flores, *Nucleic Acids Res*, 2014, **42**, e9, DOI: 10.1093/nar/gkt906.

Faraday Discussions

PAPER

A GPU-accelerated immersive audio-visual framework for interaction with molecular dynamics using consumer depth sensors†

David R. Glowacki,*[abcd] Michael O'Connor,[c] Gaetano Calabró,[e] James Price,[c] Philip Tew,[b] Thomas Mitchell,[bf] Joseph Hyde,[g] David P. Tew,[a] David J. Coughtrie[a] and Simon McIntosh-Smith[c]

Received 4th February 2014, Accepted 19th March 2014
DOI: 10.1039/c4fd00008k

With advances in computational power, the rapidly growing role of computational/simulation methodologies in the physical sciences, and the development of new human–computer interaction technologies, the field of interactive molecular dynamics seems destined to expand. In this paper, we describe and benchmark the software algorithms and hardware setup for carrying out interactive molecular dynamics utilizing an array of consumer depth sensors. The system works by interpreting the human form as an energy landscape, and superimposing this landscape on a molecular dynamics simulation to chaperone the motion of the simulated atoms, affecting both graphics and sonified simulation data. GPU acceleration has been key to achieving our target of 60 frames per second (FPS), giving an extremely fluid interactive experience. GPU acceleration has also allowed us to scale the system for use in immersive 360° spaces with an array of up to ten depth sensors, allowing several users to simultaneously chaperone the dynamics. The flexibility of our platform for carrying out molecular dynamics simulations has been considerably enhanced by wrappers that facilitate fast communication with a portable selection of GPU-accelerated molecular force evaluation routines. In this paper, we describe a 360° atmospheric molecular dynamics simulation we have run in a chemistry/physics education context. We also describe initial tests in which users have been able to chaperone the dynamics of 10-alanine peptide embedded in an explicit water solvent. Using this system, both expert and novice users have been able to accelerate peptide rare event dynamics by 3–4 orders of magnitude.

[a]*School of Chemistry, University of Bristol, Bristol, BS8 1TS, UK. E-mail: drglowacki@gmail.com*
[b]*Pervasive Media Studio, Watershed, 1 Canons Rd, Bristol BS1 5TX*
[c]*Department of Computer Science, University of Bristol, BS2 9LG, UK*
[d]*Department of Chemistry, Stanford University, Stanford, CA, 94305, USA*
[e]*EaStCHEM School of Chemistry, University of Edinburgh, Edinburgh EH9 3JJ*
[f]*Computer Science and Creative Technologies, University of the West of England, Bristol BS16 1QY*
[g]*School of Music and Performing Arts, Bath Spa University, Bath BA2 9BN*

† Electronic supplementary information (ESI) available. See DOI: 10.1039/c4fd00008k

1. Introduction

With advances in computational power and the improvement of software tools for exploiting modern parallel architectures, scientific models and the data sets they generate are rapidly increasing in size and dimensionality. In many cases, it is possible to design computer algorithms to analyze data sets, and thereby identify important features and trends. However, for a wide class of problems – *e.g.*, where the data sets involve extremely high-dimensional spaces and non-linear relationships, and identification of interesting phenomena requires some qualitative judgment – it is often the case that human subjects can identify important trends faster than computers. This makes visualization an increasingly important tool for researchers to quickly see trends and behavior that may be difficult to identify otherwise – *i.e.*, using standard mathematical or analytical algorithms to process large digitized data sets.[1]

The field of molecular simulation highlights many of these points. In particular, for the simulation of complex systems – *e.g.*, in materials science or biochemistry – the systems under investigation typically have thousands of degrees of freedom, and a single simulation run is easily capable of generating hundreds of gigabytes of data. For complex systems, molecular simulation is increasingly being used to conduct what is best described as 'computational experiments', where the system complexity is large enough that simulation results are not necessarily clear from the outset. Researchers often carry out such simulations in the hope of gleaning qualitative insight, usually linked to understanding the mechanism by which a particular molecular ensemble accomplishes its function. It can be a challenge to find appropriate algorithmic descriptors for these qualitative mechanistic insights, especially if they are unknown at the outset. However, by visualizing simulation results, and using what computational chemistry researchers frequently refer to as 'chemical intuition', it is often the case that humans are more efficient than computers at identifying qualitative features of a simulation and relating them to the predominant ideas within their subject area. Often, human insight gleaned from visualization of the system subsequently guides development of an appropriate algorithmic descriptor, or guides the setup of subsequent simulations (*e.g.*, to bias the simulations and increase the likelihood that they produce a desired outcome, or to terminate simulations which appear unlikely to yield interesting insight).

Visualization strategies have consequently become an indispensable tool in the arsenal of modern computational chemistry, offering a sort of virtual microscope that lets us see the behavior and dynamics of the atomic and molecular world – both accelerating research insight and facilitating efficient communication between researchers. Most of the time, visualization of molecular simulations takes place 'off-line' – *i.e.*, the researcher runs a simulation, and upon conclusion of the simulation, loads the results into a visualization program for viewing, generating snapshots and/or movies. With improved computational architectures, and more efficient software tools, recent years have seen the development of systems with 'on-the-fly' visualizations that are dynamically updated while the simulation is running. These systems allow humans to watch simulation progress generated from molecular dynamics (MD) simulations,[2] molecular docking,[3,4] hybrid structure prediction tools,[5] course-grained models,[6] and even quantum chemistry methods.[7] 'On-the-fly' visualization naturally led a number of groups to investigate interactive interfaces for molecular simulation.[8] Broadly speaking, there are three levels at which interactivity has been introduced within MD simulations:

(1) *Graphics Rendering*, giving the user control over a range of parameters controlling both the graphics rendering and viewing perspective. Beyond standard mouse and keyboard interfaces, these systems have utilized a wide range of interface options, including face tracking, stereoscopic displays, and virtual reality gloves.[9–13]

(2) *Simulation parameters*, giving the user control over any of a variety of general parameters that impact the molecular simulation's overall propagation algorithm (*e.g.*, temperature, pressure, time step, *etc.*).[14,15]

(3) *Molecular substituents*, where the user can pinpoint particular atoms or functional groups and manipulate them with an external force, thereby 'steering' the simulation program's internal propagation, similar to the sort of manipulations which are possible using atomic force microscopy (AFM) experiments.[16] Keyboard and mouse interfaces are utilized in such systems,[5,17] but the most popular interface has been haptic devices,[3,4,7,18–28] which offer up to six degrees of freedom (compared to two for a mouse). As such, they are well suited to facilitating user interaction with 3D molecular simulations. Additionally, the force-feedback that they provide allows users to 'feel' the force interactions of a given molecular system.

Initial research efforts into interactive molecular simulations have nearly all been aimed at a single user, to expand the utility of molecular simulation methodologies in both research and educational contexts. For example, exciting early progress using haptic interactive dynamics provided insight into the mechanisms of binding specificity in the enzyme glycerol kinase and transport specificity in the aquaporin membrane channel protein GlpF.[29] In this study, the haptic system allowed the researchers to carry out rapid exploration of vast regions of configuration space that would not have been accessible in a conventional simulation. Because haptic devices have not yet become widespread within the consumer market, their use has mostly been confined to specialist institutions devoted to interactive technology and molecular research.

With the design of distributed computing infrastructures to tackle scientific research questions over the last twenty years,[30–32] keyboard and mouse interfaces have been widely exploited to allow crowds to participate in a range of research tasks.[33,34] For certain tasks, human creativity and judgment can outperform automated classification and search algorithms. Recent years have seen exciting mergers between interactive molecular simulation and ideas within crowd-sourced human–computer interaction. For example, using a 'gamified' interface called Foldit, crowds of non-specialists are able to manipulate Rosetta, a protein structure prediction tool with a hybrid approach that utilizes stochastic and deterministic algorithms along with a combination of template assembly, template-based modeling, and all-atom refinement. Recent studies have shown that the strategies Foldit players use to solve complex non-linear optimization problems are distinct from automated algorithms, and sometimes superior.[5] In some cases, networked crowds can produce useful new strategies for molecular optimization tasks.[35] These exciting studies, at the interface of human–computer interaction, distributed computing, and molecular science, raise the prospect that distributed infrastructures along with new interface technologies can utilize the power of the internet along with crowd intelligence to solve scientific problems, simultaneously engaging the public with fundamental research questions.

In this paper, we outline an integrated hardware setup and algorithmic framework for carrying out interactive MD using depth sensors, which is scalable to an arbitrary number of users and has been adapted to large-scale immersive spaces. The fundamental idea guiding this framework is to utilize new hardware (an array of consumer-priced infrared depth sensors[36,37] that utilize structured light[38] to carry out real-time 3D imaging) in conjunction with new software that interprets the human form as an energy landscape. Together, the hardware and software provide an interactive interface for embedding users in a molecular simulation, which responds to the real-time motion of their fields. User interaction with the system results in feedback, which has both a visual and audio component: projections or screen displays allow users to see their energy fields embedded within the MD simulation. Simultaneously, we utilize a set of structural and spectral analysis algorithms for detecting transient fluctuations within the ensemble dynamics, for the purposes of sonification.[39] At present, all published systems for interactive molecular dynamics have exclusively relied on technologies which focus on a small number (usually one or two) of single interaction points – *i.e.*, the human–computer interaction is usually focused on very specific objects or properties within a simulation, *e.g.*, grabbing, moving, and releasing an atom using a haptic system, or mouse and keyboard events. The system outlined herein is somewhat of a departure from these systems insofar as it focuses on interactions which: (1) are far more nonlocal, allowing the user to interact with large subsets of atoms simultaneously, and (2) do not require tangible intervening objects for the user to interact with the simulation. In this sense, the system described herein builds on ideas first introduced by Myron Krueger, which attempted to go beyond 'a seated man poking at a machine with his fingers or waving a wand over a data tablet',[40] so that the focus of intention is on the action rather than the technology.

To date, we have referred to this system as 'danceroom Spectroscopy' (dS for short). This name might seem unconventional, but it actually has well-defined origins that arise from two different observations. First, chemists and biochemists often utilize dance and choreographic analogies when describing dynamical phenomena in the research literature, a conclusion which can be easily verified by inspecting the titles returned from a simple search for recent articles which contain 'chemistry' and 'dance' in the title or abstract. A few recent examples are ref. 41–49, which refer variously to 'molecular dancefloors',[43] 'single molecule dances',[41] 'polymer dances',[47] 'enzyme choreography',[42] 'radical dances',[48] *etc.* Second, one of the methods utilized by our system to generate audio feedback for the user(s) involves a spectral decomposition technique[50] – namely, fast Fourier transform of the ensemble averaged velocity–velocity autocorrelation function, which is discussed in further detail below. dS initially began as a digital art installation, and subsequently found application as the basis for an interactive dance performance, where dancers' motion generates both graphics and sound. In these early stages, our primary emphasis was aesthetic,[40,51] and the project received attention in public forums and media outlets across artistic and cultural sectors.

Effective implementation of the dS system relies on a suite of image processing and computer vision algorithms, a heterogeneous programming strategy built from a range of OpenCL GPU-accelerated computer algorithms, as well as algorithms from mixed classical-quantum molecular dynamics and vibrational spectroscopy. In this paper, we rigorously describe, for the first time, the dS framework in sufficient detail for it to be reproducible – including algorithms, technical details, and benchmarking. We also describe two new applications: (1) interactive simulation of Earth's atmosphere, and (2) interactive simulation of the 10-ALA peptide in an explicit water box with preliminary user studies that suggest acceleration of computational sampling by 3–4 orders of magnitude. The success of this system in engaging widely varied audiences in non-traditional contexts (*e.g.*, art, technology, and science education) raises a number of exciting possibilities – perhaps that its aesthetic appeal may be exploited to drive user participation in crowd-sourced molecular dynamics studies: *e.g.*, to accelerate rare event dynamics or to nudge complex molecular systems into rarely visited regions of phase space, providing information that may then be used to map the kinetic microstates of such systems, guided by the sort of 'chemical intuition' which is difficult to program into blind search algorithms.

2. Software and algorithms

2.1 Depth images and energy landscapes

3D capture systems typically return distance-to-target, or depth, z, as a function of pixel position within a two dimensional matrix indexed by pixels that span the x and y direction, as shown in Fig. 1, which was constructed by plotting the 640 × 480 pixel depth image matrix obtained from a Microsoft Kinect sensor. The plot in Fig. 1 shows a human form, where the intensity of the colors is linked to the magnitude of the local gradient vector on the image. The manner in which it is plotted suggests analogy with the concept of an energy landscape, which has become a fundamental idea guiding how chemists and physicists think about both kinetics and dynamics in a range of chemical systems, from small molecules to complex materials and biochemical systems.[52,53] An energy landscape is effectively a topological map of a system's potential energy, V, at a range of

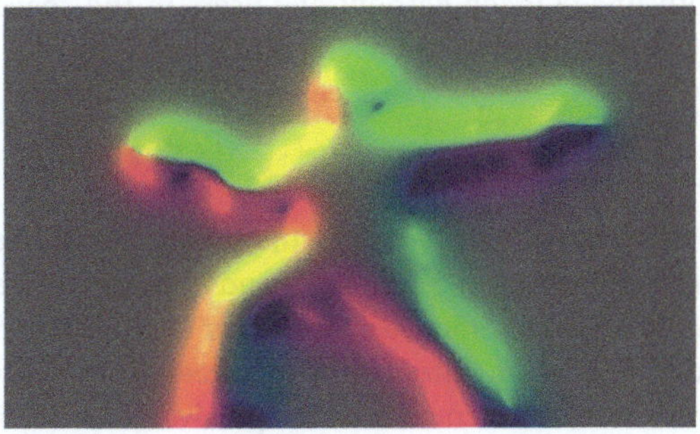

Fig. 1 Force topology map of the human form. The intensity of each color is related to the magnitude of the local force vector on the depth image. Color choice has been selected to effectively illustrate depth. Grey corresponds to a gradient of zero.

different configurations. Within any localized region of the energy landscape, the gradient of the energy, $\mathrm{d}V/\mathrm{d}\boldsymbol{q}$, relates the topology of the energy landscape to the classical forces felt by a particular molecular configuration. dS interprets people's movements as perturbations on a virtual energy landscape.

2.2 Interactive molecular dynamics with depth sensors

In its present form, dS carries out an MD simulation involving N atoms, each of which may move in a virtual coordinate system defined by Cartesian x, y, and z directions. Hamilton's equations of motion, commonly used to discuss the dynamics of molecular systems in both classical and quantum frameworks, provide a useful vantage point for describing how the system works. They are as follows:

$$\mathrm{d}\mathbf{p}/\mathrm{d}t = -\mathrm{d}H/\mathrm{d}\mathbf{q},\ \mathrm{d}\mathbf{q}/\mathrm{d}t = \mathrm{d}H/\mathrm{d}\mathbf{p} \tag{1}$$

where $\mathbf{p}$ and $\mathbf{q}$ are the momentum and coordinate vectors of each atom in the ensemble, and H is the so-called Hamiltonian function describing the total system energy – *i.e.*:

$$H = \sum_{i=1}^{N} \frac{m_i v_i^2}{2} + V \tag{2}$$

where i is an index that runs over a collection of N total atoms, m is the mass of an atom, and v is its velocity. The first term in eqn (2) describes the total kinetic energy of the system while the second, V, describes the total potential energy. Within dS, there are two different contributors to V:

$$V = V_{\mathrm{int}} + V_{\mathrm{ext}} \tag{3}$$

Like many MD programs, the most computationally expensive aspect of dS is associated with calculating the internal potential energy, V_{int}. As discussed further below, we have recently implemented a wrapper which allows dS to call the GPU-accelerated OpenMM library whenever a force evaluation is required, allowing a wide range of force interactions, including bonds, angles, torsions, non-bonded Lennard-Jones interactions and electrostatic interactions.[54]

The external potential energy, V_{ext}, in eqn (3) is calculated as a sum over the difference between a raw depth matrix at time t, $V_{\mathrm{ext}}(x_i,y_i,t)$, and an average background depth image taken without any users in the space, $\langle V_{\mathrm{ext}}(x_i,y_i,0)\rangle$, as follows (angled brackets indicate an average):

$$V_{\mathrm{ext}} = C_i \sum_{i=1}^{N} [V_{\mathrm{ext}}(x_i, y_i, t) - \langle V_{\mathrm{ext}}(x_i, y_i, 0)\rangle] \tag{4}$$

where the term in square brackets represents the potential energy that an atom 'feels' as a consequence of people's motion, and C_i is a variable scaling constant applied to a specific atom. Interactive control over C_i allows the user to determine how strongly any given atom interacts with forces from the users' fields, and whether a person's field is 'attractive' or 'repulsive'. Eqn (4) is responsible for coupling human motion to the atomic dynamics, allowing humans to sculpt the potential energy landscape felt by the atomic ensemble, and thereby chaperone the system dynamics.

In Hamiltonian mechanics, the energy function, H, remains constant for any closed dynamical system, in line with the conservation of energy required by the first law of thermodynamics.[55] However, the eqn (2) Hamiltonian is not subject to this constraint because of the V_{ext} term, which effectively makes the system open rather than closed. Fluctuations in the depth data arise as a consequence of noise in the depth images, and also from people's motion within the space mapped by the depth sensors. Both of these effects result in fluctuations of the total system energy, introducing significant instabilities into the Velocity Verlet[55] scheme used to propagate the time-dependent system dynamics in eqn (1). Such instabilities are a consequence of the fact that depth images, unlike standard molecular force fields, can give large forces which are not smoothly varying in time and space. Standard dynamics propagation schemes (utilizing reasonably sized time steps), which usually rely on being able to express the force as a low-order Taylor series expansion, are not always well-equipped to deal with the ill-behaved forces that arise from an interactive simulation. This can lead to explosions in the dynamical propagation as a consequence of rapid numerical error accumulation. To address this, and avoid the numerical explosions associated with such instabilities, we have implemented a modified Berendsen thermostat (described in detail in the ESI†), in which the atomic velocities are scaled by some factor λ to ensure that the instantaneous system temperature T_t approaches some desired temperature T_0 with a first order rate

$$\frac{\mathrm{d}T_t}{\mathrm{d}t} = \frac{1}{\tau}\cdot(T_0 - T_t) \tag{5}$$

Eqn (5) depends on a user-specified rate coefficient ($1/\tau$) and how far the system is from T_0. We found the standard Berendsen scheme to be unreliable for ensuring the stability of dS when exposed to users. Stability was considerably improved by looping over the atomic velocities to ensure that none of the atoms within the simulation have a velocity more than two standard deviations larger than the average atomic velocity (prior to determining the value of T_t required for calculating the atomic velocity scale factor λ). This procedure then gives a good compromise between computational efficiency, interactive fluidity, and system stability. It eliminates numerical instabilities that can arise when user motion suddenly 'injects' energy into the system Hamiltonian.

2.3 Smooth interactivity: frozen Gaussian dynamics

The vector of forces acting on a set of atoms $\mathbf{F}(t)$, can be written in terms of the system's potential energy – *i.e.*:

$$\mathbf{F}(t) = -\mathrm{d}H/\mathrm{d}\mathbf{q} = -\mathrm{d}V/\mathrm{d}\mathbf{q} \tag{6}$$

Substituting eqn (3) into eqn (6) gives

$$\mathbf{F}(t) = -\frac{\mathrm{d}V_{\mathrm{int}}}{\mathrm{d}\mathbf{q}} - \frac{\mathrm{d}V_{\mathrm{ext}}}{\mathrm{d}\mathbf{q}} = \mathbf{F}_{\mathrm{int}} + \mathbf{F}_{\mathrm{ext}} \tag{7}$$

where $\mathbf{F}_{int}$ and $\mathbf{F}_{ext}$ are the force vectors arising from the internal energy and the external field, respectively.

For the purposes of translating the depth map shown in Fig. 1 to external forces that act on those simulated atoms, there are two important differences between depth maps and the sorts of energy landscapes typically utilized in molecular simulation – both of which present complications. First, molecular energy landscapes represent space as a continuum; whereas depth matrices are discretized into pixels with a finite spatial extent. Second, whereas molecular energy landscapes are generally well-behaved, continuously differentiable functions, this is not necessarily the case for depth matrices. For example, Fig. 1 illustrates the abrupt change in the z-coordinate that distinguishes the human from the background. In our setup, we have found that discontinuity in depth images generally arises for a number of reasons: (1) there are abrupt changes in the 'distance-to-target' for different components of a particular scene; (2) depth capture for a particular scene is incomplete owing to particular objects within the scene casting an infrared shadow; and (3) as a result of random noise in the depth image, which may have any of several origins, including variations in the infrared light source, variations in detector response, and variations in the optical environment.

Initially, our intention was to propagate the system dynamics using eqn (1) with a purely classical approach, *i.e.*, the atoms represented as point particles, and purely local forces acting on any given atom, from both $\mathbf{F}_{\text{int}}$ and $\mathbf{F}_{\text{ext}}$. However, this approach resulted in choppy motion, unsatisfactory interactivity, and numerically unstable dynamics simulations. The cause of these problems arose for the reasons outlined above, and their subsequent effects on $\mathbf{F}_{\text{ext}}$. Achieving more fluid dynamics with improved interactivity and stability required that we introduce some sort of non-locality into our dynamics propagation strategy, so that $\mathbf{F}_{\text{ext}}$ depended on a local average within the space of the pixels. To incorporate this averaging in an efficient manner, we implemented an algorithm inspired by the so-called 'frozen Gaussian' approach to semiclassical dynamics,[56] which forms the basis for a number of approaches that approximately model the quantum dynamics of molecular systems. dS has two distinct coordinate spaces: (1) the Cartesian simulation space spanned by q_x, q_y, and q_z, denoted by the vector $\mathbf{q}$, where atom i has a position $[q_{\text{ix}}, q_{\text{iy}}, q_{\text{iz}}]$; and (2) the depth image pixel coordinate space spanned by r_x and r_y, in which each atom i is characterized by its centre $[r^c_{\text{ix}}, r^c_{\text{iy}}]$. The external forces acting on atom i as a result of the r_x and r_y directions of pixel space are obtained by integrating over a Gaussian function $G_{\text{i}}(r_x,r_y)$ as follows:

$$\left\langle \frac{\mathrm{d}V_{\text{ext}}}{\mathrm{d}r_{\text{i}\alpha}} \right\rangle = \int_{-\infty}^{+\infty}\int_{-\infty}^{+\infty} \frac{\mathrm{d}V_{\text{ext}}}{\mathrm{d}r_{\text{i}\alpha}} G_{\text{i}}(r_x,r_y)\mathrm{d}r_x\mathrm{d}r_y \tag{8}$$

where the angled brackets indicate an average force, $\alpha \in x,y$, and

$$G_{\text{i}}(r_x,r_y) = K\exp\left[-\frac{1}{2\sigma^2}\left(\left(r_x - r^c_{\text{ix}}\right)^2 + \left(r_y - r^c_{\text{iy}}\right)^2\right)\right] \tag{9}$$

where σ is the Gaussian width parameter, and the normalization constant $K = 1/(2\pi\sigma^2)$. Using integration by parts, eqn (8) may be rewritten as

$$\left\langle \frac{\mathrm{d}V_{\text{ext}}}{\mathrm{d}r_{\text{i}\alpha}} \right\rangle = \frac{1}{\sigma^2}\int_{-\infty}^{+\infty}\int_{-\infty}^{+\infty} \left(r_{\text{i}\alpha} - r^c_{\text{i}\alpha}\right) G_{\text{i}}(r_x,r_y)V_{\text{ext}}\mathrm{d}r_x\mathrm{d}r_y \tag{10}$$

Compared to eqn (8), which requires evaluating the gradient of dV_{ext} with respect to $dr_{i\alpha}$, eqn (10) only requires evaluation of V_{ext}, for which fast numerical interpolation routines are available using software libraries associated with OpenCL image types. Accurately solving integrals like those in eqn (10) may be accomplished using numerical methods like Gauss–Hermite quadrature. However, an accurate and efficient Gauss–Hermite quadrature implementation depends on being able to represent V_{ext} with relatively low order polynomial functions, and this is not always the case with the images returned from user interaction with a depth sensor. In practice, eqn (10) is evaluated numerically over a tiled set of squares with dimensions $\Delta x \cdot \Delta y$, with equally spaced centre points that lie within $\pm 3\sigma$ of $r^c_{i\alpha}$, *i.e.*:

$$\left\langle \frac{\mathrm{d}V_{\mathrm{ext}}}{\mathrm{d}r_{\mathrm{i}\alpha}} \right\rangle \approx \frac{1}{\sigma^2} \sum_{r^c_{\mathrm{ix}}-3\sigma}^{r^c_{\mathrm{ix}}+3\sigma} \sum_{r^c_{\mathrm{iy}}-3\sigma}^{r^c_{\mathrm{ix}}+3\sigma} \left(r_{\mathrm{i}\alpha} - r^c_{\mathrm{i}\alpha}\right) G_{\mathrm{i}}\left(r_x, r_y\right) V_{\mathrm{ext}} \Delta_x \Delta_y \tag{11}$$

Two key parameters in eqn (11) are the Gaussian width parameter σ and the number of tiles used for integration (both of which then determine $\Delta x \cdot \Delta y$). We typically set σ to be equal 10, giving a grid of 30×30 points, which generally gives smooth interactivity and satisfactory user control over the simulation. A grid of this size will be nearly three orders of magnitude more expensive than a purely local approach, where the force acting on a particular atom is determined by the numerical gradients calculated from only those pixels adjacent to that which contains $[r^c_{ix}, r^c_{iy}]$. However, by saving the image as a byte array, and handling it as an OpenCL image texture, we have been able to speed up this calculation enough that it is not a significant bottleneck, as discussed in further detail below.

2.4 System design and implementation

A schematic of the dS setup is shown in Fig. 2. The multi-sensor array is shown in the figure, with depth capture from only three sensors, and graphics output to

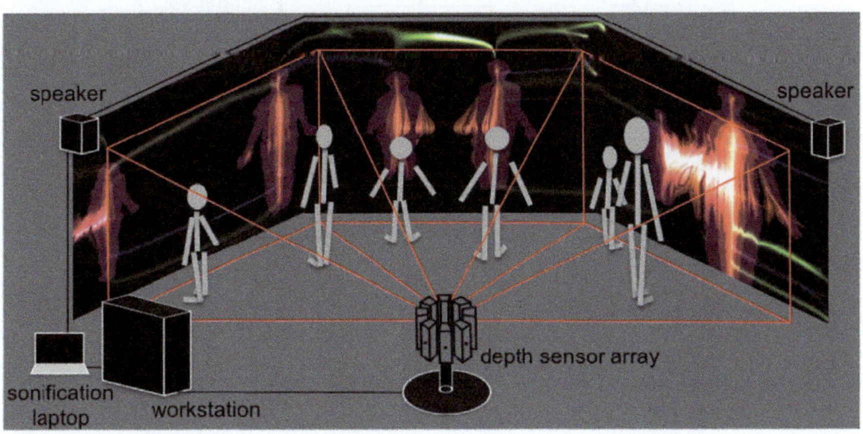

Fig. 2 Schematic of the dS setup. An array of eight depth sensors is shown; however for the sake of simplicity, we have only shown image capture from three sensors, and rendering to three screens.

three displays. A flow chart breaking down data flow and system execution is shown in Fig. 3. Below, we detail the different aspects of the system, beginning with the optical mount that enables depth capture.

2.4.1 Optical mount. Because we originally designed dS to be compatible with immersive and 360° environments, it has the capability to run with simultaneous depth matrix capture from up to ten sensors. We typically run with eight sensors, positioned within the optical mount shown in Fig. 4. Both Microsoft Kinect and Asus Xtion Pro sensors have a 43° × 57° (horizontal × vertical) field of view, so that eight vertically oriented sensors give ∼354° of coverage. The mount shown in Fig. 4 is portable, lightweight, sturdy, and quick to set up. It is also useful for setups that utilize conventional displays, where we typically run with 1–3 sensors. The mount consists of eight housings arranged around a central axis, each of which fits snugly around a depth sensor's outer casing. Vertical orientation of the sensors minimizes interference between the infrared sources on each camera, and also minimizes edge overlap effects, both of which simplify the image processing required to merge multiple depth matrices into a composite V_{ext}, shown in Fig. 3.

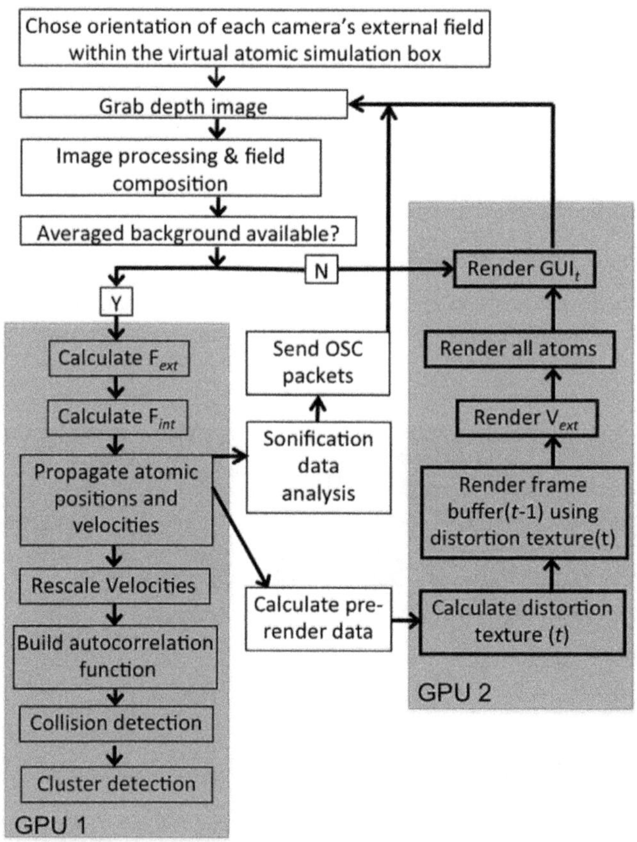

Fig. 3 Flow chart showing system operation. The grey blocks indicate those portions of the system that we accelerated on GPUs; the remaining functionality is executed on the CPU.

Fig. 4 Customized optical mount which houses the vertically oriented array of depth sensors used for 3D capture.

2.4.2 Workstation. dS has been run and tested on a range of workstations. The highest performance workstation on which it runs (required to run in the 360° immersive projection environments described below) is a customized 64-bit workstation with: (1) an Intel i7 3.2 GHz hexacore hyperthreaded CPU; (2) an Asus IV Rampage motherboard which has six separate USB hubs, and is fitted with a HighPoint RocketU 4 port PCI express USB card, letting us simultaneously run up to ten depth sensors; (3) an NVIDIA GTX Titan graphical processing unit (GPU) with 2688 floating point units, reserved for accelerated physics computations as detailed in Fig. 3 (GPU 1 in Fig. 3), and (4) an Asus HD 7970 Direct CU II GPU with 2048 floating point units and six graphics outputs which allows rendering across multiple displays (GPU 2 in Fig. 3).

The workstation runs code which is written in C# (~50 000 lines) built on Windows 7 in Visual Studio 2010. The code interface to the depth sensors utilized the OpenNI C# wrappers. Graphics rendering was carried out using DirectX 11. Code ported to the GPUs for accelerated compute operations utilized the OpenCL programming language. We devoted considerable effort to making the code general, flexible, and user-friendly so that important parts of the system can be modified directly by non-specialists. In addition to giving the user control over all of the graphics parameters described below, the multi-tab GUI allows real-time control over several other aspects of the system, including: depth matrix capture, image processing, background calibration; relative orientation, position, and blending of each camera's depth matrix within the composite field that makes up V_{ext}; and parameters important to the sonification algorithms described below.

2.4.3 Data capture and physics propagation. After connecting the depth sensors *via* USB cables to our workstation, our software allows us to interactively determine the orientation of different depth images to form the composite external field felt by the atoms. This includes the edge blending between different depth images. Once the orientation and edge blending is set, we obtain a background image, required in order to calculate V_{ext} in eqn (4). If the software is unable to load a saved background, or if the user requests a fresh background, the system carries out a number of depth matrix grabs to calculate and save an averaged background.

Once the background is available, the system begins solving the aforementioned equations of motion for an ensemble of atoms whose initial geometry and force field connectivity are specified by the user. System stability is ensured by rescaling the atomic velocities as detailed in eqn (5) above. Propagating the coordinates and velocities of each atom in the ensemble requires evaluating both F_{int} and F_{ext}. F_{int} may be evaluated using one of two approaches: (1) our own GPU-accelerated algorithms which include non-bonded Lennard-Jones, bonding, and angle terms; or (2) the OpenMM GPU accelerated software library, called from our code using a set of fast C# wrappers. Evaluating F_{ext} requires grabbing depth matrices from each sensor, integration of these depth matrices into a composite V_{ext}, and application of the frozen Gaussian equations described in eqn (8)–(11). Both F_{int} and F_{ext} are evaluated at a rate of 60 Hz, while depth matrix grabs occur at either 30 or 60 Hz, which are the operational frequencies of the depth sensors.

Important physics parameters which the user can interactively modify include: (1) the scaling factor C_i which determines whether an atom's motion is affected by V_{ext}, whether V_{ext} is attractive or repulsive, and how strongly V_{ext} affects the atomic dynamics; (2) σ, which determines any given atom's Gaussian width over V_{ext}; (3) T_0, the desired temperature of the ensemble; and (4) τ, which determines how strongly the thermostat drives the ensemble to T_0 at each dynamics step. It is also possible for 'one-click' switching between different molecular simulations.

2.4.4 Graphics output. Graphics rendering with dS takes place exclusively on GPU 2 using DirectX 11. The aims of the graphics are two-fold: to quickly provide information that allows the users to utilize their energy landscape to manipulate the simulation, and to provide an engaging aesthetic experience capable of both initiating and sustaining user engagement. Thus, the graphics system has been designed with a certain degree of flexibility – *i.e.*, it offers users a range of different graphics options for seeing how their motion perturbs V_{ext}, and thereby affects the atomic dynamics. A schematic flow chart indicating the graphics rendering pipeline within dS is shown in Fig. 5 for a two-sensor setup.

Fig. 5 begins with two raw depth images obtained from the sensors. The background from each image is subsequently subtracted, and then processed with a hole-removing occlusion algorithm from OpenCV.[57] Using linear blending at the edges of each image, the processed depth images are then combined to form the composite external field in eqn (4). V_{ext} may then be directly rendered to the frame buffer by mapping it into any of a number of color palettes, which the dS operator may interactively select. This direct rendering results in rather literal silhouettes of users as they perturb V_{ext}.

Other graphics renderings of the users embedded in the simulation are obtained by applying a distortion texture to V_{ext} prior to its rendering in the frame buffer at time t. The distortion texture, shown in Fig. 5, is effectively a measure of the extent of local variations within V_{ext}: large gradients in V_{ext} give small values in the distortion texture, and small gradients give small values within the distortion texture. The distortion texture at time t is calculated by adding the 2D gradient of V_{ext} to the previous distortion texture from time $t - 1$. The final rendered frame buffer is then obtained by combining the time t distortion texture with the

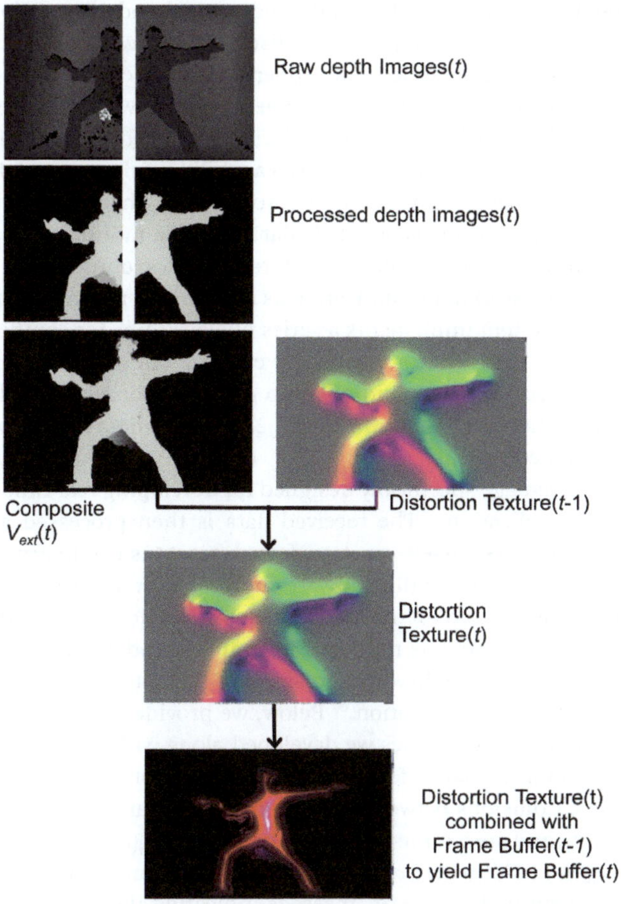

Fig. 5 Pipeline showing how graphics rendering is done in dS. The figure shows how images from two depth sensors are combined to yield V_{ext}, the distortion texture, and the final frame buffer. An example of the final frame buffer is shown, with only V_{ext} and no atoms.

time $t - 1$ frame buffer. The frame buffer itself is actually constructed from two buffers: a V_{ext} buffer and an atom rendering buffer. Feeding forward the distortion texture and frame buffers in the fashion described above is important for creating dynamic distortion effects which react to user motion, and has obvious analogues with a Velocity Verlet dynamics propagation strategy.

In the same way that dS allows users interactive control of important physics parameters, it also allows control over a range of graphics parameters related to rendering of both V_{ext} and the atoms within the final frame buffer, to tailor the dS graphics output as they like. In addition to being able to select the colors used in graphics rendering, the user may control how strongly the distortion texture is applied, how strongly the distortion texture and frame buffer from time $t - 1$ are fed forward to those at time t, and the intensity of V_{ext}. There are an enormous number of graphics parameter combinations, each of which produces distinctly different graphical states, one of which is shown in Fig. 5, and others which are shown throughout this article. Extending the atomic rendering options available in dS is an active area of development, as discussed below. At present, the atomic rendering in dS utilizes circles and spheres, whose colors may be selected according to a range of criteria. For example, Fig. 6 shows a dS image of the 10-ALA peptide embedded in an explicit solvent comprised of water molecules. A particularly useful aspect of dS's graphics capabilities is a tool which allows the user to interactively magnify the atomic resolution to whatever zoom level he or she desires, and thereby focus on particular parts of a molecular system.

2.4.5 Sonification. As a complement to graphics output, and also as an experiment in the use of data sonification as a means for reporting on simulation progress, the dS system implements a series of algorithms for sonification of the atomic dynamics. This is accomplished by carrying out a range of analyses and transmitting the results to an audio laptop *via* Open Sound Control (OSC) data structures over ethernet,[58] as shown in Fig. 2. The OSC data is then parsed using software developed within Cycling 74's Max/MSP environment,[59] a visual programming language specifically designed for developing real-time audiovisual processes and applications. The received data is then processed and sonified either within Max/MSP itself, or transferred *via* Musical Instrument Digital Interface (MIDI) to digital audio software such as Ableton Live to generate real-time audio feedback based on the atomic dynamics. A detailed description of the audio aspects of dS as well as the Max/MSP patches and Ableton interfaces that facilitate audio feedback is beyond the scope of this paper, but a more detailed account is currently in preparation.[60] Below, we provide a very brief outline of a few of the sonification processes we developed along with dS.

The simplest form of sonic feedback involves collision detection between non-bonded atoms. Collisions between non-bonded atoms are triggered by analyzing the distances r_{ij}, and are counted as having occurred if $r_{ij}(t-2) > r_{ij}(t-1)$, $r_{ij}(t-1) < r_{ij}(t)$, and $r_{ij}(t) < \sigma 2^{1/6}$, where σ is the Lennard-Jones van der Waals radius. Each collision is transmitted as an OSC message indicating the collision coordinates, velocity and atom type, which is used to trigger an arbitrary sound. Collision data is very high resolution, and fluctuates on fast timescales. From the perspective of audio composition, it is best suited to small numbers of atoms (*i.e.*, less than 250). Otherwise, it can grow cacophonous. It is possible for the dS operator to interactively specify a filter on the maximum number of collisions per time step which are transmitted; however, this can diminish user perceptions of interactivity.

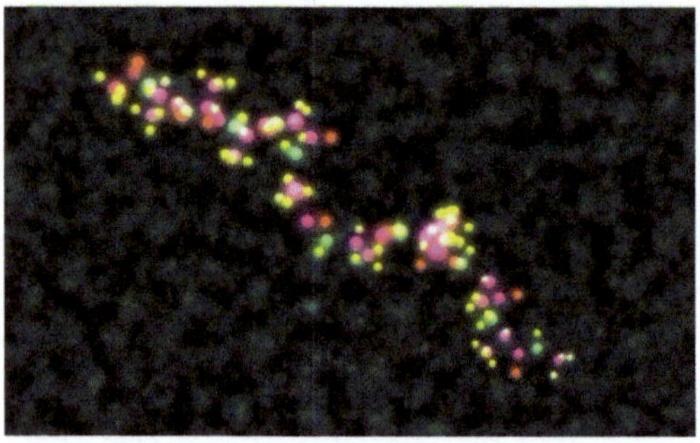

Fig. 6 10-Alanine peptide embedded in an explicit water solvent. Only atoms are rendered, and not bonds.

Second, we have developed a clustering algorithm which performs analysis of the ensemble coordinates and detects the formation of stable atomic clusters, tracking their average position, velocity, and size. Since the algorithm is sensitive to the particulars of the simulation (*e.g.*, number of atoms, interactions per unit time, T_t, and C_i), the dS software allows interactive modulation of parameters which impact algorithm performance.

Third, we measure the spectrum of the atomic ensemble, allowing us to track periodic atomic motion, arising from both the intrinsic atomic dynamics as well as periodic perturbations in the external field caused by human movement. The algorithm we use to accomplish this relies on maintaining a moving time history of the atomic velocity vector, **v**, to calculate the velocity autocorrelation function (VAC), defined as:

$$\mathrm{VAC}(t) = \mathbf{v}(t + t_0)\cdot\mathbf{v}(t_0) \tag{12}$$

where t is the elapsed time following some previous time point t_0. The VAC is a time series which measures how $\mathbf{v}(t_0 + t)$ projects onto $\mathbf{v}(t_0)$. At the FFT analysis stage shown in Fig. 3, we carry out Fast Fourier Transform (FFT) of the VAC to yield the real-time vibrational spectrum of the atomic dynamics, $F(\omega)$, *i.e.*:

$$F(\omega) = \frac{1}{\sqrt{2\pi}}\int_{-\infty}^{+\infty} \mathrm{d}t e^{i\omega t}\mathrm{VAC}(t) \tag{13}$$

where ω is the frequency in Hz. Characteristic vibrational frequencies within the atomic dynamics appear as peaks in $F(\omega)$. Following FFT, $F(\omega)$ is fit using a basis set of 50 cubic spline functions to facilitate an automatic peak identification algorithm that then sends frequency and amplitude data *via* OSC for subsequent sonification.

2.5 Performance and system latency

Our target refresh rate for frame rendering and dynamics propagation is 60 Hz (17 ms latency), with an allowed minimum of 30 Hz (33 ms latency). For the sake of fluid interactivity, it is more important for fluctuations in the refresh rate to be small, even if it drops to slightly less than 60 Hz. With serial CPU power alone, meeting the 60 Hz target limited us to interactive dynamics within fewer than 3000 atoms on a single display setup, using a single depth sensor. In immersive 360° environments, however, where there is significant additional computational overhead involved rendering on up to six graphics displays and capturing depth matrices from up to ten sensors, the serial CPU implementation suffered serious performance setbacks, and reduced by 1–2 orders of magnitude the number of atoms which we could typically simulate at the desired refresh rate. The GPU acceleration scheme outlined in Fig. 3 allowed us to considerably improve the system performance in 360°, and also led to corresponding performance enhancements in more conventional setups: *i.e.*, with a single display and fewer sensors. GPU acceleration was carried out by profiling our code and subsequent use of OpenCL to accelerate the most intensive computational tasks on the NVIDIA GTX Titan as shown in Fig. 3. The AMD HD7970 we reserved solely for GPU-accelerated DirectX graphics rendering over multiple video outputs.

To better characterize and understand the performance of the dS code, we carried out a range of profiling tests (shown in Fig. 7 and 8) using an MD simulation box (volume of 572 280 Å^3) with 917 O_2 molecules, 3654 N_2 molecules, 98 CO_2 molecules, and 194 H_2O molecules (ratios which approximately correspond to the molecular composition of the terrestrial troposphere). These tests utilized two depth sensors and two output displays. F_{int} was calculated using the parameters from the MM3 force field: Lennard-Jones for the non-bonded terms, bonding terms which included anharmonicity up to fourth order, and angle terms which included anharmonicity up to sixth order. Force constants for bonds and angles were chosen so as to approximately reproduce the vibrational frequencies of these molecules from vibrational spectroscopy experiments (published online in the NIST Chemistry WebBook).

We tested two different implementations of dS: the first utilized our own OpenCL accelerated routines for evaluating the F_{int} terms listed above, and the second utilized a wrapper around OpenMM, to test the performance of its own OpenCL routines for evaluating F_{int}. The components of F_{int} described above are not available by default in OpenMM, and were specified using customized OpenMM syntax that allows users to specify their own force terms. Each implementation was tested on two different architectures for parallelizing calculation of F_{int}: (1) utilizing the GTX Titan GPU and (2) utilizing our hexacore Intel i7 3.2 GHz CPU.

One of the most significant and well-known efficiency bottlenecks in molecular dynamics arises from calculating non-bonded interactions in the internal force vector F_{int}, which have a formal scaling that is quadratic with system size (*i.e.*, $N(N - 1)/2$, where N is the number of atoms in the simulation). Fig. 7a and 7b confirm that calculating F_{int} is indeed one of the largest computational costs in all the tests that we ran, with an expense that is comparable to the costs associated with image capture, graphics rendering, and calculation of F_{ext}. VAC and collision detection, both of which have a formal scaling that is quadratic with system size,

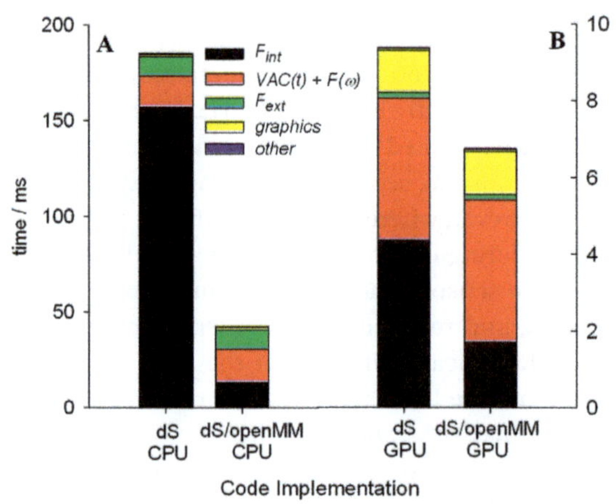

Fig. 7 Benchmark data showing computational time spent in different parts of the code during dS execution. Results were obtained from an atmospheric MD simulation with two sensors. We compared OpenCL accelerated code performance on the CPU and GPU architectures. On each architecture, we tested the performance of: (1) our own OpenCL accelerated routines for evaluating F_{int}; (2) the dS/OpenMM OpenCL code for evaluating F_{int}.

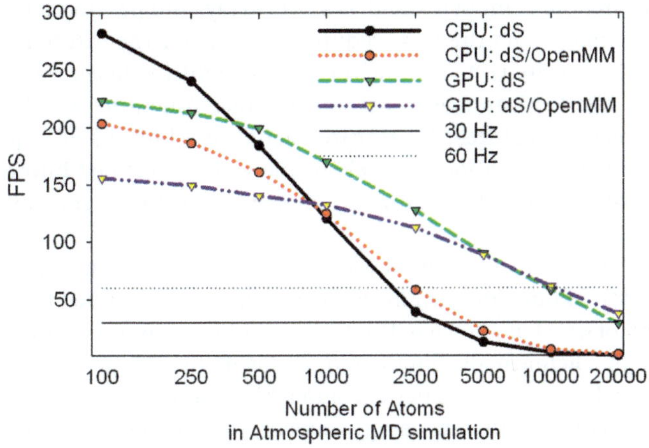

Fig. 8 Benchmark data showing frames per second obtained for the same tests described in Fig. 7. The 30 and 60 Hz limits are shown for reference.

can account for up to 57% of the execution time in the dS/OpenMM GPU configuration. Fig. 8 shows that, for small simulations (less than 1000 atoms on the CPU, and less than 5000 atoms on the GPU), the dS OpenCL routines are faster for calculating F_{int} than the corresponding OpenMM wrappers. For larger numbers of atoms, OpenMM has a factor of 2–3 better performance as a result of the internal protocol it uses for handling non-bonded interactions, which has improved scaling with system size. Comparison of Fig. 7a and 7b, along with Fig. 8, shows that parallelization of the non-bonded force evaluations on the GPU considerably improves the performance for simulations with larger numbers of atoms. Fig. 7a and 7b show that the GPU-parallelized dS code is nearly a factor of 35 faster than the corresponding CPU-parallelized code. This dramatic speed-up arises in part from the fact that we have mostly focused our efforts on optimizing the code for the GPU rather than the CPU, preferring recomputing rather than data storage (*e.g.*, in the distances required to calculate non-bonded force terms), and also optimizing for local memory access. OpenMM, on the other hand, has better performance portability, with the CPU implementation a factor of 8 slower than the GPU implementation. Fig. 7b demonstrates the power of the GPU-parallelized dS/OpenMM code: a single frame requires ~7 ms, well within our 17 ms target.

3. Preliminary applications

3.1 Atmospheric simulation in 360°

As discussed above, the efficiency of the dS/OpenMM interface allows us to run the atmospheric simulation described above, and leaves adequate time for additional computational tasks. In 360° immersive environments, these requirements are formidable, since they involve capture from multiple depth sensors, integration of these depth matrices into a single composite field, blending of the render data to accommodate non-linear surfaces, and subsequent output to six displays (five displays within the dome, and one control screen). Saving the composite field as a byte array, and passing it to the GPU as an OpenCL image cuts the time required to transfer the external field image from the CPU to the GPU by a factor of four, and furthermore, the OpenCL image type takes advantage of several member functions which exploit the GPU architecture features: (1) texture cache, in which pixel requests cause the GPU to cache neighboring pixels in the vicinity; (2) fast bilinear interpolation at arbitrary points within the image; and (3) efficient out of bounds handling.

The most intensive application which we have recently run using dS is a 360° interactive MD simulation with the atmospheric setup described in section 2.5, a photograph of which is shown in Fig. 9. This simulation was used to teach general principles of chemical dynamics and atmospheric chemistry to high-school students, including: atmospheric composition, atomic and molecular force interactions, collision theory, energy transfer, the relationship between temperature and molecular degrees of freedom (rotations, translations, and vibrations), and vibrational spectroscopy. Sonification of the atmospheric simulation using eqn (13) lets students 'hear' the differences in the vibrational periods of different molecules, providing qualitative insight into their infrared absorption, and corresponding radiative forcing efficiencies. The format in which these activities took place consisted of four 10-minute lectures, with 15 min in between lectures for the students to undertake guided and unguided interaction with the system.

Assessing outcomes of the atmospheric simulation shown in Fig. 9 is difficult, since there was no specific user 'task' to measure; rather, the goal was to utilize dS

Fig. 9 Interactive atmospheric molecular dynamics simulation in an immersive 21 m 360° projection dome, run as part of a recent chemistry and physics education event held in Bristol, UK. The array of depth sensors shown in Fig. 4 is located in the centre of the dome, but is hidden behind one of the students in the foreground. Approximately 20 students are holding hands in a circle around the dome. The result is a dynamic ring of molecules attracted to their fields on the surrounding projection screens. (Photo by Paul Blakemore).

in order to educate high-school students about molecular dynamics. In an attempt to assess the quality of the interactive experience, we obtained feedback from 67 of the student attendees, and obtained the following results: 76% of respondents said that they had fun; 81% indicated that the interaction helped them better understand the lecture content; and 86% indicated that the event would help them with their upcoming science exams. One of the most consistent criticisms was that more time should have been reserved for the students to interact with the dS system, with a more varied range of molecular simulations beyond the atmosphere.

3.2 Peptide chaperones

The largest simulation we have run using the dS/OpenMM code is for a 298 K simulation of the 10-alanine peptide explicitly embedded in 8296 TIP3P H_2O molecules in a box with volume 249 600 Å^3, giving a density of 1 g cm^{-3}. The peptide simulation was fully flexible, carried out using the algorithms described above, with a time step of 0.1 fs. The LEaP program in Amber 12[61] was used to design the 10-Alanine system and the Amber ff99SB were used to define the OpenMM force fields. Standard harmonic potentials were applied to the bond and angle interactions, while periodic functions were used to describe the dihedral potential energy. Electrostatic interactions were handled using a reaction field with a cutoff distance of 10 Å. Lennard-Jones interactions were truncated at the same distance. Fig. 10 shows two users embedded in the aforementioned simulation, cooperatively using their energy fields to chaperone the 10-ALA peptide. The solvent dynamics are unaffected by the users; only the atoms in the peptide

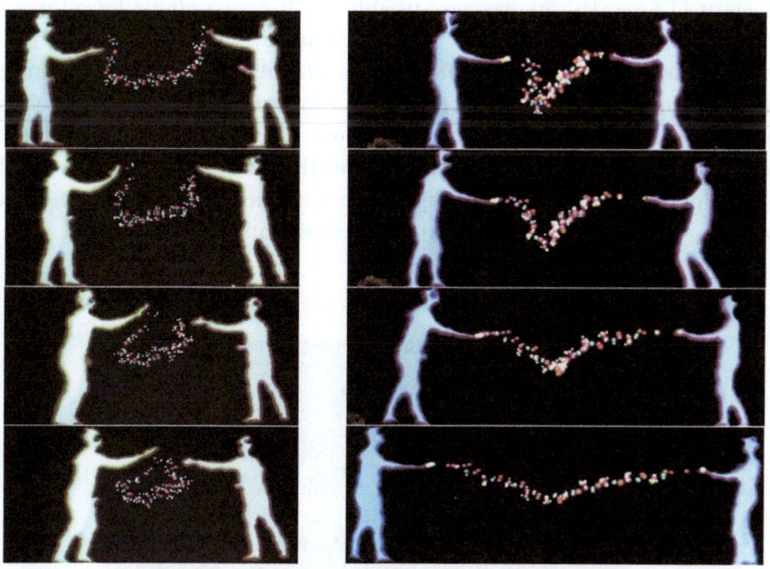

Fig. 10 Sequence showing two users chaperoning a 10 Alanine peptide. The left-hand panel shows the players as they chaperone the peptide's transition to the looped conformation. The right-hand panel shows the simulation progress as the players chaperone the peptide's transition to the stretched conformation.

feel the users' energy fields. In the example shown, the users alternately form the peptide into a loop, and subsequently stretch it out again. This is a prototypical rare-event process for which the kinetics and free energy have been investigated in detail during previous work,[62,63] some of which was carried out by one of us.[64–66] Previous workers have described systems which allow multiple users to manipulate molecular visualization viewing perspectives;[67] however, to the best of our knowledge, Fig. 10 presents the first platform which allows multiple users to actually manipulate the molecular dynamics.

The sequence shown in Fig. 10 took approximately five minutes in real-time.[68] In addition to the multi-user interaction framework shown in Fig. 10, manipulation of the peptide structure by a single user (as occurs with haptic devices) is also possible. For example, Fig. 11 shows how it is possible for a single user to use their hands to manipulate the 10-ALA peptide's energy landscape. To evaluate more systematically the efficiency with which users were able to manipulate the peptide, we carried out a limited number of tests on two different user groups: experts (those who were involved in writing the code and building the system), and novices (users with little experience of the dS interfaces, and in some cases, little experience of video games and molecular simulation). These two user groups were taken as representative limiting cases. The mean number of time steps ($\pm$ standard deviation) required for expert users to chaperone the peptide from a stretched to a loop configuration was 5961 $\pm$ 619 (596.1 $\pm$ 61.9 fs), compared to 11 093 $\pm$ 4923 time steps (1109.3 $\pm$ 492.3 fs) for novices. The mean number of steps required for going from a looped to a stretched configuration was 7918 $\pm$ 2158 time steps (791.8 $\pm$ 215.8 fs) for expert users and 7831 $\pm$ 2216 time steps (783.1 $\pm$ 221.6 fs) for novice users. To determine the time scale for spontaneous loop formation, we ran simulations under identical conditions with no user input. Over 2.65×10^6 time steps (2.65×10^5 fs), we were unable to observe a single loop formation event. Our failure to observe a single folding event over this timescale is compatible with previously published results obtained using a 10-alanine implicit solvent model where the average time to loop formation was determined to be on the order of $(0.943 \pm 0.160) \times 10^6$ fs.[65,69] Not including the additional computational tasks (*i.e.*, beyond the calculation of V_{int}), the timescale for loop formation in the chaperoned simulations is between 3–4 orders of magnitude faster than that for spontaneous loop formation.

Fig. 11 The same as Fig. 10, except that a single user's hands are shown manipulating the peptide.

4. Conclusions

In this paper, we have outlined a new immersive high-performance framework for carrying out interactive molecular dynamics using arrays of consumer-priced depth sensors scalable up to 360° immersive spaces. The fundamental idea driving the system lies in interpreting the human form as an energy landscape, and subsequent embedding of that energy landscape in a real-time molecular dynamics simulation. The system allows multiple users to simultaneously chaperone a molecular dynamics simulation, and relies on a suite of GPU-accelerated algorithms to accomplish the following tasks at 60 Hz: depth sensor image processing, construction of a composite external field, graphics rendering, and evaluation of the internal forces. The flexibility of our platform has been considerably enhanced by wrappers that facilitate fast communication with the GPU-accelerated routines available in OpenMM.

In addition to educational applications, initial tests utilizing this new system show that both expert and novice users have been able to use it to accelerate peptide rare event dynamics by 3–4 orders of magnitude. This massive acceleration is substantially larger than the additional (factor of 2) computational expense incurred through the additional computational overhead required by the system. This raises the exciting prospect that this system may be exploited to accelerate the sampling of states that are otherwise visited only rarely in simulations of complex molecular systems. For example, if we imagine several simultaneous instances of dS being run by a range of users (with a system in place for uploading data to a central analysis server), it may be possible to quickly identify important kinetic hubs and traps within a given biomolecule's conformational space.[70,71] During interactive journeys through molecular configuration space, users would have a higher statistical likelihood of visiting hubs and traps compared to other configurations.

A second exciting possibility for such a system is that it could be exploited to accelerate finding dynamical pathways through configuration space. This is a significant challenge for conventional simulation methodologies, especially when seeking pathways that are difficult to describe using standard order parameters and progress variables. A particularly tantalizing prospect is that users will be able to efficiently generate complex pathways that would be difficult to generate otherwise (*i.e.*, using blind search algorithms or acceleration schemes with simple progress parameters), allowing us to build kinetic network maps of complex molecular systems (*e.g.*, Markov state models,[72] master equation models,[73] or BXD models[66]), and obtain time constants and rate coefficients that could be directly compared with experiment. Quick qualitative discrimination between paths could likely be achieved through simply analyzing the difference between the cluster of highest and lowest energy points along a path, giving an effective path barrier. Quantitative path discrimination achieving such an outcome would require a reliable means for unbiasing the user-generated pathways, and could be achieved by subsequent analysis of the user-generated path using any of a range of methods, including transition path sampling,[74] the string method,[75] the nudged elastic band approach,[76] or BXD.[64] Coupling user-generated pathways to the above methods may be a subject worth investigating in its own right, given that generating an initial dynamical pathway is often a non-trivial challenge for these methods.

In both cases discussed above – conformational sampling, and path sampling – one could imagine a 'scoring' function which rewarded users for finding new low-energy states or new low-energy pathways. The idea behind such a gamified approach would be twofold: first, to incentivize users' improving their chemical intuition with increased playing, so that they develop an increasingly good feel for the conformational and dynamical preferences of complex systems; and second, to quickly sample a wider range of user intuition. Reasons for optimism that the dS system might provide a crowd-sourced platform for tackling research questions include the fact that: (1) the dS system utilizes commodity hardware; (2) when it has been deployed in an educational context, it has shown the ability to engage students well, and received positive responses, and (3) its aesthetic merit has been recognized by installations in major international art and cultural institutions.

To date, attempts at molecular dynamics have largely utilized interaction strategies focused on a very small number of local interaction sites – *i.e.*, keyboards, mice, and haptic devices. These technologies allow precise control over particular atomic components of a given system. dS, on the other hand, facilitates interactions which are significantly more nonlocal. For example, users can interact with and manipulate large subsets of atoms, and have the ability to control which of the atoms in a particular system will respond to their 'energy fields'. In this respect, it is worth investigating whether a system like dS is better suited to coaxing molecular systems to undergo large-scale conformational changes, or to follow complex reaction coordinates which require the concerted motion of large subsets of atoms.

In future work, we hope to address a range of pertinent issues required to extend system performance, and thereby increase its utility. These include: (1) using more than two GPUs to enhance overall system performance; (2) increasing the portability of the system so that it can run on a range of platforms (*e.g.*, Mac/Linux using OpenGL), and so that it is optimized over a range of different CPU/GPU architectures; (3) calculating z-direction forces from users' energy fields using a time history of V_{ext} in the $\pm z$ direction; (4) expanding the available rendering options to accommodate stereoscopic rendering, and to easily recognize molecular features like α-helices and β-sheets; (5) investigating the use of a new generation of depth sensors (*i.e.*, LEAP) which are specifically tailored to accurate hand tracking; (6) extension of the methodologies described in this paper to include coarse-grained approaches; (7) investigating additional means for how best to use audio feedback to provide useful insight into simulation progress; (8) testing additional high-performance strategies to further increase the speed of force evaluations on the GPU;[77] (9) detailed user studies to understand how to best exploit and improve nonlocal interaction strategies in the case of molecular simulation – *i.e.*, whether it is possible to define a set of human actions that correspond to molecular structural changes; and (10) investigating whether the form of nonlocal interaction employed by dS may be productively combined with more local interaction approaches, to exploit the best of both. Tackling these challenging problems will require an interdisciplinary approach with expert input from researchers across a range of fields including computational chemistry, computer science, human computer interaction, and human aesthetics.

Acknowledgements

DRG acknowledges funding as a Royal Society research fellow. DPT acknowledges funding as a Royal Society research fellow. DJC acknowledges a studentship from the EPSRC. Additional support has been provided through EPSRC grant EP/I017623/1, the University of Bristol, the Royal Society of Chemistry, NVIDIA, the Watershed Digital Media Centre, the University of the West of England, and Arts Council England. A full list of contributors to the dS project can be found at www.danceroom-spec/people

References

1 K.-L. Ma, *SIGGRAPH Comput. Graph.*, 2004, **38**, 4–7.
2 R. Kobler, T. Kockerbauer, U. Omasits, M. Neumann, W. Schreiner and J. Volkert, *Computer Aided Systems Theory' EUROCAST 2007*, 2007, pp. 443–447.
3 G. A. Dalkas, D. Vlachakis, D. Tsagkrasoulis, A. Kastania and S. Kossida, *Briefings Bioinf.*, 2013, **14**, 745.
4 X. Hou and O. Sourina, in *Transactions on computational science XII*, Springer, 2011, pp. 98–117.
5 S. Cooper, F. Khatib, A. Treuille, J. Barbero, J. Lee, M. Beenen, A. Leaver-Fay, D. Baker, Z. Popovic and F. Players, *Nature*, 2010, **466**, 756–760.
6 O. Delalande, N. Ferey, G. Grasseau and M. Baaden, *J. Comput. Chem.*, 2009, **30**, 2375–2387.
7 M. P. Haag and M. Reiher, *Int. J. Quantum Chem.*, 2013, **113**, 8–20.
8 T. Schlick, *Biophys. J.*, 2003, **85**, 1–4.
9 S. Park, J. Lee and J. I. Kim, *Computational Science and Its Applications'ÄìICCSA*, 2005, **2005**, 183–196.
10 W. Humphrey, A. Dalke and K. Schulten, *J. Mol. Graphics*, 1996, **14**, 33–38.
11 M. Krone, K. Bidmon and T. Ertl, *IEEE Trans. Visualization Comput. Graphics*, 2009, **15**, 1391–1398.
12 M. Patriarca, A. Kuronen, M. Robles and K. Kaski, *Comput. Phys. Commun.*, 2007, **176**, 38–47.
13 T. J. Callahan, E. Swanson and T. P. Lybrand, *J. Mol. Graphics*, 1996, **14**, 39–41.
14 P. Knoll and S. Mirzaei, *Rev. Sci. Instrum.*, 2003, **74**, 2483–2487.
15 D. Rapaport, *Phys. A*, 1997, **240**, 246–254.
16 S. Izrailev, S. Stepaniants, B. Isralewitz, D. Kosztin, H. Lu, F. Molnar, W. Wriggers and K. Schulten, in *Computational molecular dynamics: challenges, methods, ideas*, Springer, 1999, pp. 39–65.
17 M. C. Surles, J. S. Richardson, D. C. Richardson and F. P. Brooks, *Protein Sci.*, 1994, **3**, 198–210.
18 A. Bolopion, B. Cagneau, S. Redon and S. Regnier, Haptic molecular simulation based on force control, *IEEE Conference on Advanced Intelligent Mechatronics*, 2010, pp. 329–334.
19 J. Stone, A. Kohlmeyer, K. Vandivort and K. Schulten, *Advances in Visual Computing*, 2010, 382–393.
20 J. E. Stone, J. Gullingsrud and K. Schulten, *presented in part at the Proceedings of the 2001 symposium on Interactive 3D graphics*, 2001.
21 M. Dreher, M. Piuzzi, A. Turki, M. Chavent, M. Baaden, N. Ferey, S. Limet, B. Raffin and S. Robert, in *2013 International Conference on Computational*

Science, ed. V. Alexandrov, M. Lees, V. Krzhizhanovskaya, J. Dongarra and P. M. A. Sloot, 2013, vol. 18, pp. 20–29.

22 Y. G. Lee and K. W. Lyons, *Comput.-Aided Des.*, 2004, **36**, 75–90.

23 A. Ricci, A. Anthopoulos, A. Massarotti, I. Grimstead and A. Brancale, *Future Med. Chem.*, 2012, **4**, 1219–1228.

24 M. Stocks, S. Laycock and S. Hayward, *J. Comput.-Aided Mol. Des.*, 2011, **25**, 203–211.

25 A. M. Wollacott and K. M. Merz, *J. Mol. Graphics Modell.*, 2007, **25**, 801–805.

26 A. Ricci, A. Anthopoulos, A. Massarotti, I. Grimstead and A. Brancale, *Future Med. Chem.*, 2012, **4**, 1219–1228.

27 M. P. Haag, K. H. Marti and M. Reiher, *ChemPhysChem*, 2011, **12**, 3204–3213.

28 O. B. Bayazit, G. Song and N. M. Amato, Ligand binding with OBPRM and user input, *IEEE International Conference on Robotics and Automation*, 2001, pp. 954–959.

29 P. Grayson, E. Tajkhorshid and K. Schulten, *Biophys. J.*, 2003, **85**, 36–48.

30 E. J. Korpela, SETI@home, BOINC, and Volunteer Distributed Computing, *Annual Review of Earth and Planetary Sciences*, ed. R. Jeanloz, 2012, vol. 40, pp. 69–87.

31 R. Das, Q. Bin, S. Raman, R. Vernon, J. Thompson, P. Bradley, S. Khare, M. D. Tyka, D. Bhat, D. Chivian, D. E. Kim, W. H. Sheffler, L. Malmstrom, A. M. Wollacott, C. Wang, I. Andre and D. Baker, *Proteins: Struct., Funct., Bioinf.*, 2007, **69**, 118–128.

32 A. L. Beberg, D. L. Ensign, G. Jayachandran, S. Khaliq and V. S. Pande, *presented in part at the Proceedings of the 2009 IEEE International Symposium on Parallel&Distributed Processing*, 2009.

33 D. Clery, *Science*, 2011, **333**, 173–175.

34 M. Helmstaedter, K. L. Briggman, S. C. Turaga, V. Jain, H. S. Seung and W. Denk, *Nature*, 2013, **500**, 168–174.

35 F. Khatib, S. Cooper, M. D. Tyka, K. F. Xu, I. Makedon, Z. Popovic, D. Baker and P. Foldit, *Proc. Natl. Acad. Sci. U. S. A.*, 2011, **108**, 18949–18953.

36 J. Han, L. Shao, D. Xu and J. Shotton, *IEEE Transactions on Cybernetics*, 2013, **43**, 1318–1334.

37 Z. Zhang, *IEEE Multimedia*, 2012, **19**, 4–10.

38 J. Geng, *Adv. Opt. Photonics*, 2011, **3**, 128–160.

39 G. Kramer, B. Walker, T. Bonebright, P. Cook, J. H. Flowers, N. Miner and J. Neuhoff, *Sonification report: Status of the field and research agenda*, Faculty Publications, Department of Pyschology, Paper 444, 2010, http://digitalcommons.unl.edu/psychfacpub/444.

40 M. W. Krueger, presented in part at the Proceedings of the June 13–16 1977 National Computer Conference, Dallas, Texas, 1977.

41 M. Y. Ivanov, *Nature*, 2012, **483**, 161–163.

42 J. Villali and D. Kern, *Curr. Opin. Chem. Biol.*, 2010, **14**, 636.

43 S. Bradforth, *Science*, 2011, **331**, 1398–1399.

44 E. S. Shamay, K. E. Johnson and G. L. Richmond, *J. Phys. Chem. C*, 2011, **115**, 25304–25314.

45 A. Tafiviz, L. A. Mirny and A. M. van Oijen, *ChemPhysChem*, 2011, **12**, 1481–1489.

46 Q. Wu, T. Ochi, D. Matak-Vinkovic, C. V. Robinson, D. Y. Chirgadze and T. L. Blundell, *Biochem. Soc. Trans.*, 2011, **39**, 1387–1392.

47 L. Rothberg, *Nat. Chem.*, 2011, **3**, 425–426.

48 A. Chatterjee, A. B. Hazra, S. Abdelwahed, D. G. Hilmey and T. P. Begley, *Angew. Chem., Int. Ed.*, 2010, **49**, 8653–8656.

49 D. R. Livesay, *Curr. Opin. Pharmacol.*, 2010, **10**, 706–708.

50 T. Delatour, in *Molecular Aesthetics*, ed. P. Weibel and L. Fruk, MIT Press, Cambridge, MA, 2013, pp. 293–311.

51 D. R. Glowacki, P. Tew, J. Hyde, L. Kriefman, T. Mitchell, J. Price and S. McIntosh-Smith, in *Molecular Aesthetics*, ed. P. Weibel and L. Fruk, MIT Press, Cambridge, MA, 2013, pp. 248–257.

52 D. R. Glowacki, J. N. Harvey and A. J. Mulholland, *Nat. Chem.*, 2012, **4**, 169–176.

53 D. J. Wales, *Philos. Trans. R. Soc. London, Ser. A*, 2012, **370**, 2877–2899.

54 P. Eastman, M. S. Friedrichs, J. D. Chodera, R. J. Radmer, C. M. Bruns, J. P. Ku, K. A. Beauchamp, T. J. Lane, L. P. Wang, D. Shukla, T. Tye, M. Houston, T. Stich, C. Klein, M. R. Shirts and V. S. Pande, *J. Chem. Theory Comput.*, 2013, **9**, 461–469.

55 D. Frenkel and B. Smit, *Understanding molecular simulation: from algorithms to applications*, Academic press, 2001.

56 E. J. Heller, *J. Chem. Phys.*, 1981, **75**, 2923.

57 http://opencv.org/.

58 M. Wright, *Organised Sound*, 2005, **10**, 193–200.

59 http://cycling74.com/products/max/.

60 T. Mitchell, J. Hyde, P. Tew and D. R. Glowacki, in preparation.

61 R. Salomon-Ferrer, D. A. Case and R. C. Walker, *Wiley Interdiscip. Rev.: Comput. Mol. Sci.*, 2013, **3**, 198–210.

62 M. J. Feige and E. Paci, *J. Mol. Biol.*, 2008, **382**, 556–565.

63 D. K. West, P. D. Olmsted and E. Paci, *J. Chem. Phys.*, 2006, **125**, 204910.

64 D. R. Glowacki, E. Paci and D. V. Shalashilin, *J. Phys. Chem. B*, 2009, **113**, 16603–16611.

65 D. R. Glowacki, E. Paci and D. V. Shalashilin, *J. Chem. Theory Comput.*, 2011, **7**, 1244–1252.

66 D. V. Shalashilin, G. S. Beddard, E. Paci and D. R. Glowacki, *J. Chem. Phys.*, 2012, **137**, 165102.

67 C. Forlines and R. Lilien, *presented in part at the Proceedings of the working conference on Advanced visual interfaces*, Napoli, Italy, 2008.

68 A video of the user-accelerated simulation is available at http://www.vimeo.com/81531449.

69 D. R. Glowacki, E. Paci and D. V. Shalashilin, *J. Phys. Chem. B*, 2009, **113**, 16603–16611.

70 G. R. Bowman and V. S. Pande, *Proc. Natl. Acad. Sci. U. S. A.*, 2010, **107**, 10890–10895.

71 A. Dickson and C. L. Brooks, *J. Am. Chem. Soc.*, 2013, **135**, 4729–4734.

72 G. R. Bowman, K. A. Beauchamp, G. Boxer and V. S. Pande, *J. Chem. Phys.*, 2009, **131**, 124101.

73 F. Noe, *J. Chem. Phys.*, 2008, **128**, 244103.

74 J. Juraszek, J. Vreede and P. G. Bolhuis, *Chem. Phys.*, 2012, **396**, 30–44.

75 E. Vanden-Eijnden and M. Venturoli, *J. Chem. Phys.*, 2009, **130**, 17.

76 G. Henkelman, B. P. Uberuaga and H. Jonsson, *J. Chem. Phys.*, 2000, **113**, 9901–9904.

77 P. Gonnet, *J. Comput. Chem.*, 2012, **33**, 76–81.

Faraday Discussions

PAPER

Studying chemical reactivity in a virtual environment

Moritz P. Haag and Markus Reiher*

Received 25th February 2014, Accepted 28th February 2014
DOI: 10.1039/c4fd00021h

Chemical reactivity of a set of reactants is determined by its potential (electronic) energy (hyper)surface. The high dimensionality of this surface renders it difficult to efficiently explore reactivity in a large reactive system. Exhaustive sampling techniques and search algorithms are not straightforward to employ as it is not clear which explored path will eventually produce the minimum energy path of a reaction passing through a transition structure. Here, the chemist's intuition would be of invaluable help, but it cannot be easily exploited because (1) no intuitive and direct tool for the scientist to manipulate molecular structures is currently available and because (2) quantum chemical calculations are inherently expensive in terms of computational effort. In this work, we elaborate on how the chemist can be reintroduced into the exploratory process within a virtual environment that provides immediate feedback and intuitive tools to manipulate a reactive system. We work out in detail how this immersion should take place. We provide an analysis of modern semi-empirical methods which already today are candidates for the interactive study of chemical reactivity. Implications of manual structure manipulations for their physical meaning and chemical relevance are carefully analysed in order to provide sound theoretical foundations for the interpretation of the interactive reactivity exploration.

1 Introduction

A detailed understanding of the atomic rearrangements occurring during a chemical reaction is at the heart of chemistry. Computational investigations based on first principles of quantum mechanics can provide such a microscopic picture. They also complement the findings of experiments by making information on reaction steps accessible for which experimental data are not available.

Usually, the elucidation of reaction mechanisms starts with the set up of structures which are likely to be close to either the reactants, the products or the transition state structures. Subsequent optimisation of the electronic structure and of the atomic arrangement is computationally expensive, and it is by no means guaranteed that these optimisations produce the desired mechanism. A trial-and-error protocol has to be repeated until all potentially relevant reaction

Laboratorium für Physikalische Chemie, ETH Zürich, Vladimir-Prelog-Weg 2, CH-8093 Zürich, Switzerland. E-mail: markus.reiher@phys.chem.ethz.ch; Fax: +41 44 633 15 94; Tel: +41 44 633 43 08

steps have been identified. In most cases the build-up of the starting structures is carried out in conventional molecule editors using a computer mouse and a three dimensional ball-and-stick representation of the structure.

Despite the remarkable progress in the development of fast algorithms for electronic structure calculations and the continuously increasing computer power, it is still a challenge to predict reaction mechanisms of molecular systems consisting of a few hundred atoms. The reason for this is the inherently high complexity induced in the potential energy surface by the large number of degrees of freedom of such systems. Already for an elementary reaction step, a transition-state search requires starting structures from which local optimisations can be initiated and such optimisations may easily fail. Then, only little additional information is provided on how to improve on such a failed search.

In addition, technical hurdles need to be overcome. Firstly, the build-up of three dimensional chemical structures with two dimensional input devices is a slow and cumbersome procedure. Secondly, manipulating the system and receiving the resulting response from the calculations are sequential and delayed processes that often require different computer programs. This separation is not only due to the electronic structure optimisation being computationally expensive, but also due to the fact that there exists currently no general implementation to have the manipulation of the system and the instantaneous presentation of the quantum chemical reactivity response in the same program.

Both issues can efficiently be resolved by having all the processes united in a single virtual environment. Such a virtual environment allows for a deeper immersion of the scientist into the problem. In addition to the visual presentation the complex output of quantum chemical calculations can be perceived more intuitively by addressing further human senses such as the haptic sense. The main advantage then is the seamless combination of manipulation and output presentation. It allows the chemist to exploit prior knowledge about the system and his/her 'chemical intuition' for the exploration of reaction mechanisms.

Studying complex phenomena in a more intuitive way is not entirely new to chemists and biologists. For molecular dynamics,[1,2] protein docking[3] and structural biology[4] interactive approaches and even haptic interactions have been applied. The idea of an interactive build-up of molecular assemblies with continuous energy minimisation in the background for a quick assembly of reasonable structures has recently been implemented for parametrised analytical potentials and classical force fields in the molecule editors SAMSON[5] and AVOGADRO.[6]

In 2007, we initiated a research program to explore chemical reaction mechanisms interactively with quantum chemical approaches. We started with the haptic exploration of interpolated Born–Oppenheimer potential energy surfaces within the framework of what we call Haptic Quantum Chemistry.[7] During such a haptic exploration scientists are able to feel the quantum mechanical forces exerted on the atoms, when they are moved with an input device capable of providing force-feedback. We further extended this approach to larger systems and presented a protocol to refine potential energy surfaces guided by the haptic exploration.[8] To be able to treat structural relaxation in complex systems we also investigated the possibility to calculate the necessary forces directly.[9] Recently, we have investigated real-time reactivity exploration[10] in the framework of the SAMSON environment, for which a fast semi-empirical method for structure optimisations of hydrocarbons is already available.[11,12]

In this work, we explore the physical meaning of externally controlled configurational changes in molecular assemblies. We work out the requirements for treating molecular assemblies in virtual environments and how much of the underlying physical theory can be recovered in an interactive exploration of chemical reactivity. The goal is to provide an intuitive and at the same time chemically meaningful experience of a reactive molecular system.

The interactive study of reaction mechanisms in a virtual environment as described in this work is envisaged as being part of an even larger framework comprising additional automatic or semi-automatic tools supporting the investigation of complete reaction networks. We, therefore, start in section 2 with a description of this bigger picture to clarify the context of the methodology described in this work. The main ingredient for an interactive study of chemical reactivity is a real-time availability of first-principles forces, which will be detailed in section 3. We continue then by dissecting the main components of a reactivity exploration in a virtual environment in section 5. In section 6 the manipulation and the presentation of molecular systems are discussed in detail. This is then followed by an analysis of the physical laws governing a molecular system in a virtual environment in section 7. We then conclude the work and provide an outlook at the end of this paper.

2 Virtual environment for the investigation of reaction mechanisms

The interplay between the various components of the virtual framework introduced in this work for the investigation of reaction networks is summarised in

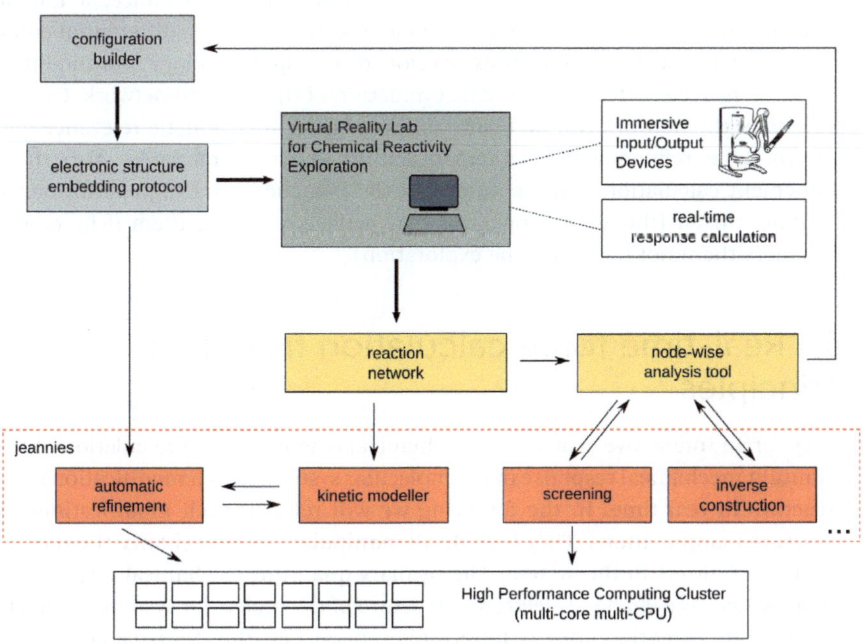

Fig. 1 Design of a virtual environment for the investigation of reaction mechanisms and networks.

Fig. 1. The figure highlights the essential components of the framework and serves as a blue print for a quantum chemical tool for the exploration of reaction networks.

The components of this virtual reactivity laboratory are as follows: The terminal in green colour is a symbol for the visual interface at which the relevant data can be displayed to the operator. The display may be a simple screen that could be placed right next to a bench in a wet lab or it can be as advanced as a virtual reality cave. The 'relevant data' will be the reaction network consisting of nodes and links, which is undergoing a rolling improvement that does not depend on whether the operator is in front of the terminal or not. Instead, service programs take care of as many tedious tasks and decisions as possible. We call these service programs 'jeannies'. They are supposed to run constantly in the background and independently take as many decisions as possible. The decisions taken are displayed to the user in order to give him or her the opportunity to steer these tasks when deemed necessary.

Information on the nodes and links of the reaction network are graphically displayed upon clicking on these entities. They will then open in a new frame or window on the terminal. The most important frame is the molecule editor which allows the investigation and manipulation of molecular structures to be taken from or eventually integrated into the reaction network. For the manipulation of molecular configurations, new jeannies are set up (denoted in grey and red in Fig. 1), which may fulfil highly system-dependent tasks — from supplementing heavy-atom only PDB structures with hydrogen atoms to the automated construction of reactive complexes, that might lead to new nodes in the reaction network, to inverse and rational design tools[13,14] for the construction of totally new reactants and structures.

Many other advanced jeannies can be envisaged. For instance, a kinetic modeller that is able to solve large systems of coupled kinetic differential equations — analogously to algorithms developed in systems biology and bio-informatics — is decisive for an automatic evaluation of the reaction network. Crucial regions in the network can be identified in terms of their kinetic relevance and can then be refined with accurate quantum chemical methods. Also these refinement calculations can be automatically launched on a high-performance compute-cluster (though the operator may actively suppress them if he or she considers them not useful for the exploration).

3 Real-time force calculation from first principles

A key for an interactive exploration of chemical reactivity is the calculation of the quantum mechanical response of the molecular system upon a modification of its structure in real time. In the following we will refer to such modifications as structure manipulations. They include all manipulations that change the relative position of atoms in the system. The primary quantum mechanical response to such changes is a changed electronic structure of the molecular system leading to a different potential energy (electronic energy in the Born–Oppenheimer approximation). From the electronic structure energies and forces can be obtained. The secondary response is the change in the arrangement of the atoms not manipulated, if structural relaxation is allowed. It is important to emphasise that only electronic-structure methods based on the first principles of quantum mechanics are suitable for reactivity studies because bonds are broken and formed, which cannot be reliably predicted by the pre-defined potentials of a (classical) force-field.

To obtain the quantum mechanical response, the electronic structure has to be calculated first. The forces can then be obtained analytically from the results of this optimisation procedure. The response of the atomic structure is calculated by minimising the forces on the unmanipulated atoms. To provide the result in real-time there is an upper time limit for all steps. However, in most cases the electronic structure optimisation is the limiting factor. Depending on external factors — such as the details of the optional structure optimisation procedure, the user-defined speed of structure manipulation or the required adiabaticity of the manipulation — the limit can vary considerably. However, an electronic structure (and force) calculation requiring on the order of one hundred milliseconds will be sufficiently fast.

In order to make the electronic structure optimisation sufficiently fast, usually additional approximations have to be introduced. A general rule is: The faster the optimisation is, the more severe are the approximations. Thus, in addition to the time limit it is necessary to specify a lower limit for the quality. To do so we recall the goals of an instantaneous exploration of chemical reactivity in a virtual environment, namely to explore the potential energy surface and identify the structures corresponding to the reactants, products, and transition states of elementary reactions, keeping in mind that they are only a starting point for further refinement using the standard methodology developed in quantum chemistry. Hence, the potential energy surface, *i.e.*, the electronic structure method employed to obtain it, should reproduce all main features of the exact surface. Thus, local minima and transition states should be detectable, while their absolute position and depths or heights are not required with high accuracy.[8] Clearly, not in every situation are all of the features needed. A first exploration to familiarise with the molecular system can be of low quality and only later, parts of the surface are examined more accurately. More information on this iterative character of a haptic exploration of potential energy surfaces to continuously refine them can be found in ref. 8.

3.1 Pre-calculated *ab initio* surfaces and interpolation

The interpolation of pre-calculated potential energies is a fast option to obtain the forces. We found[7] that the interpolating moving least squares (IMLS) method[15–18] is very well suited for haptic rendering. Other possible schemes are based on splines,[19] the modified Shepard method[20] or neural networks.[21] IMLS is a local method which performs the fitting at each required position giving configurations in the proximity higher weights. This feature enables the introduction of a cut-off procedure to reduce the number of data points needed for the interpolation, which is beneficial if the data set for the whole system is very large. Another advantage is that, because of the analytical expressions obtained at each point, the calculation of the forces is trivial as it involves only derivatives of a polynomial. However, an interpolation solely based on gradients is also possible. Furthermore, IMLS provides a good starting point for an automated refinement procedure of the surfaces.[22] To refine the surfaces of contiguous haptic explorations an efficient procedure is available that is guided by the findings of previous explorations.[8]

No matter what technique is chosen, interpolation allows one to achieve a very high accuracy of the potential energy surface and the forces because the raw data can be obtained with high accuracy *ab initio* methods. However, the application of interpolation is also limited. If the dimensionality of the hyper-surface increases, too many points have to be calculated and the interpolation becomes slow. In addition, if relaxation is allowed, the configuration space spanned by the manipulation is no longer unique but depends on the history of the exploration. Usually, addition or abstraction reactions involving only atoms or diatomic molecules in systems where relaxation does not play a major role are good candidates for the haptic exploration based on interpolation of pre-calculated surfaces.[8]

3.2 Classical reactive force fields

For cases which require a direct calculation of the forces due to large relaxation effects (introducing significant exploration history effects) during the reaction, reactive classical force fields are the fastest methods available. The bond-order potentials like the Tersoff potential,[23] the Brenner potential[24] or the Finnis–Sinclair potentials[25] are examples for such methods. Other general reactive force fields are the empirical valence bond potential by Warshel and co-workers[26] and the ReaxFF[27,28] force field. For a recent overview see ref. 29. All of these methods have in common that their simple formalism permits a very fast force calculation for a given atom arrangement. However, the reactive force fields, like all classical force fields, are highly parametrised and of limited transferability.

3.3 Standard semi-empirical methods

The next step towards a first-principles treatment of the electronic structure are semi-empirical methods. Since very many different variants have been developed in the past few decades, we focus on the ones suitable for the interactive study of reactive systems.

The simplest method of this family is the atom superposition and electron delocalization molecular orbital (ASED-MO) method,[30] which is based on the extended Hückel method[31] to treat the valence electrons of atoms. It applies atomic-repulsion corrections[32] in order to obtain correct structures, which is the main deficiency of the extended Hückel method. This method has already been shown to be feasible for the interactive study of structures containing only carbon and hydrogen.[11,12] It has the advantage of being non-iterative as the Fock operator does not depend on the molecular orbital coefficients owing to the severe approximation involved. Moreover, only a few simple integrals have to be evaluated.

The next more complex group of semi-empirical methods invokes the neglect of differential diatomic overlap (NDDO)[33,34] and the modified neglect of diatomic overlap (MNDO)[35,36] approximations. The PM6[37] and OMx[38] family of methods are recent members of this group. They are based on the Hartree–Fock theory but apply approximations such as neglecting electron–electron interaction integrals and replacing many of the remaining ones by empirically adjusted parameters.

Also density functional tight-binding (DFTB) methods like the most recent DFTB3[39–41] approach are suitable semi-empirical methods for a fast and quite reliable force calculation. Although DFTB methods involve fit parameters they are not truly semi-empirical in a strict sense since all these parameters are determined from full Kohn–Sham density functional theory (DFT) calculations. As the name suggests DFTB methods can be understood as a DFT analogue to NDDO/MNDO-based methods.[42]

All these semi-empirical methods are electronic structure methods (in contrast to reactive force field approaches). From the electronic structure (eventually, from the orbitals) quantum chemical properties can be calculated. Furthermore, all apply a finite basis set expansion of the orbitals. Hence, the optimised molecular orbital coefficients provide a reasonable estimate for more elaborate electronic structure methods to refine the results of the exploration. The connection to more accurate methods is therefore straightforward.

3.4 Minimal Hartree–Fock and density functional methods

The main idea behind this group of methods is to reduce the computational cost of standard quantum chemical calculations based on Hartree–Fock or density functional theory to a justifiable minimum. Since this group of methods does not introduce system-dependent parameters, they are generally applicable and, with increasing computational effort, gradually improvable towards the reference method (*i.e.*, towards the Hartree–Fock limit and the Kohn–Sham DFT result at the basis-set limit free of numerical inaccuracies). It is possible to apply algorithms developed in the field of linear-scaling electronic structure methods. However, to achieve very fast calculations mainly the size of the basis set is decreased, since this provides an easy to apply and very efficient way[9] to reduce the computational cost with a built-in natural strategy for systematic improvement. Moreover, effective core potentials (ECPs) reduce the number of explicitly treated electrons and thus also the number of basis functions.

We have explored this methodology taking DFT with a minimal basis set and ECPs and applying the usual integral screening and density-fitting techniques to accelerate the calculations. We have applied this minimal set-up first to calculate data points for surface interpolation[8] and then to show that reasonable energies and forces can be calculated for small systems even in real time.[9] Interestingly, such minimal approaches are experiencing a renaissance in quantum chemistry. The corrected small basis set Hartree–Fock method proposed by Sure and Grimme[43] for the fast calculation of mass spectroscopy data takes Hartree–Fock theory as a starting point with a minimal basis set called MINIX. To minimise the errors electronic core potentials and three different atom pair-wise correction terms with pre-fitted parameters have been applied.

The recent development in the field of quantum chemistry to utilise the computer power offered by graphical processing units or other specially tailored hardware, will make first-principles force calculations possible with increasing accuracy. Elaborated screening techniques to avoid unnecessary integral evaluations and sophisticated algorithms to evaluate the remaining integrals will also contribute to this development.[9]

4 Analysis of contemporary semi-empirical methods

In the following we discuss timings for state-of-the-art implementations and accuracy studies of semi-empirical and density functional tight-binding methods.

These results complement our previous work in which we demonstrated the general applicability of interpolation approaches[7,8] and minimal one-determinant Hartree–Fock and DFT methods.[9] With semi-empirical and density functional tight-binding methods the size of molecules that can be considered in real-time calculations is clearly much larger than the size of the reactive systems studied in ref. 9 (Br_2 approaching ethene and a chlorine anion approaching H_3CF).

4.1 Runtime study

In order to study contemporary semi-empirical methods and density-functional tight-binding methods for their potential to provide real-time energies and forces, we performed single-point energy calculations for different test systems for a variety of NDDO-based semi-empirical Hamiltonians (AM1,[44] RM1,[45] PM3,[46] PM6,[37] PM6-D3,[37,47] PM7[48]) and the third-order self-consistent-charge density-functional tight-binding method DFTB3.[39–41] Since the possibilities to obtain timings from the available implementations are limited and are not comparable among different programs, the timings were taken by measuring the overall execution time. To average out the influence of varying background tasks performed by the operating system 100 consecutive runs were carried out and averaged.

Our set of test molecules of different sizes and different elemental composition is shown in Fig. 2. The two alanine oligopeptides represent typical biological molecules. The system denoted as FeGP (iron-guanylylpyridinol) is a model for the active site of [Fe] hydrogenase, where the methylenetetrahydromethanopterin (HMPT) is a cofactor which we chose to model by structure e in Fig. 2. Both structures have been taken from ref. 52. The so-called compound A (*cf.* structure A in Fig. 3, of ref. 53) is a large organic molecule. The Schrock complex with an approaching N_2 molecule[54–60] is an example of a large transition metal complex employed in homogeneous nitrogen-fixation catalysis.

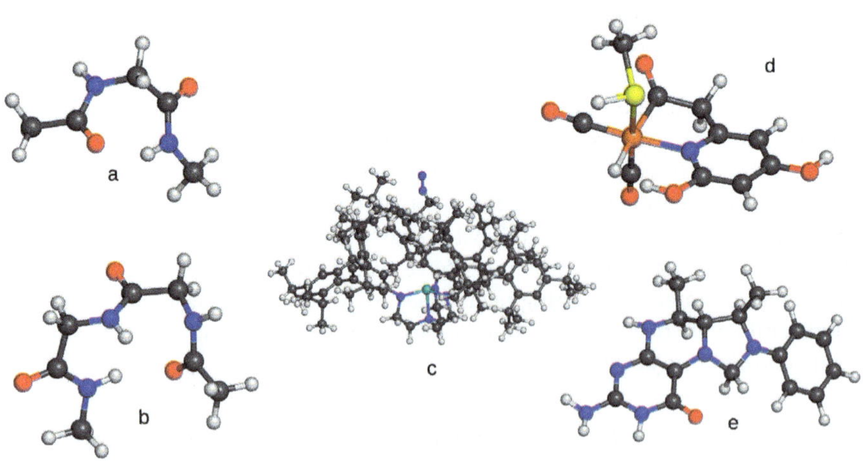

Fig. 2 Molecular systems used for the timings. (a) Alanine dimer, (b) Alanine trimer, (c) Schrock trisamidoamine Mo complex with N_2, (d) FeGP, (e) model for HMPT. Structures a, b, e have been obtained from a DFTBA[49,50] optimisation in Gaussian 09[51] Rev. d.01. Structure d has been taken from ref. 52, structure c from ref. 8.

All calculations started from the default molecular orbital (MO) coefficients guess provided by the respective program. The timings obtained are compiled in Table 1. Those molecules containing only C, N, O and H atoms required a similar number of iterations to reach self-consistency. However, a comparison of the required iterations for Ala_3 and the FeGP model system shows that the iteration number has only a small influence on the execution time. Although the times measured here contain the overhead of reading and writing to disk, printing information to output files and setting up the SCF iterations, they are all within the time range of what we would require for real-time calculations. Only the very large Schrock complex requires several seconds to optimise an electronic structure, which is mainly due to its size. The standard deviation tends to be larger for the large molecular systems and reflects the difficulty of determining a precise overall execution time.

The timings given here are only upper limits. In an implementation tailored to the needs of real-time quantum chemistry execution time can be reduced by restructuring the algorithms considering the following points.

• Separate all system specific initialisation and perform it at the beginning.

• Take MO coefficients from previous optimisations as a starting point for the next.

• Eliminate every output that is written to disc or to the console.

• Avoid any reading from or writing to disc.

• Perform as many steps as possible in-core.

• Choose convergence thresholds according to the resolution of the output device(s).

Table 1 Overall execution times in milliseconds averaged over 100 consecutive single point energy calculations on a Linux workstation with an Intel® Xeon® CPU @ 3.40 GHz. The internal standard guess MOs have been adopted. The number of cycles to reach self-consistency are given in parentheses. MOPAC2012[61] for AM1, RM1, PM3, PM6, PM6-D3 and PM7. DFTB+ [62] v.1.2.2 for DFTB3 with the 3OB[63] parameter set. Missing timings (denoted by '–') are due to a lack of appropriate parameters for certain elements of the periodic table. The standard deviation is given after each timing.

	AM1	RM1	PM3	PM6	PM6-D3	PM7	DFTB3
Ala_2 19 atoms C, H, N, O	248 ± 5 (11)	249 ± 9 (11)	249 ± 7 (11)	248 ± 9 (11)	251 ± 6 (11)	249 ± 10 (15)	344 ± 37 (11)
Ala_3 26 atoms C, H, N, O	249 ± 5 (14)	249 ± 6 (14)	250 ± 6 (14)	252 ± 5 (14)	252 ± 6 (14)	252 ± 5 (14)	342 ± 32 (13)
FeGP 29 atoms Fe, S, C, O, N, H	– (–)	– (–)	– (–)	258 ± 6 (236)	259 ± 6 (236)	259 ± 5 (61)	– (–)
HMPT 43 atoms C, H, N, O	255 ± 6 (14)	255 ± 4 (14)	255 ± 7 (17)	256 ± 4 (20)	257 ± 4 (20)	257 ± 5 (15)	354 ± 40 (12)
Compound A 84 atoms C, H, N, O	275 ± 21 (13)	276 ± 29 (13)	273 ± 6 (17)	289 ± 54 (25)	295 ± 63 (26)	279 ± 34 (14)	562 ± 32 (10)
Schrock-N_2 280 atoms C, H, N, Mo	– (–)	– (–)	– (–)	9197 ± 498 (142)	9414 ± 574 (142)	5719 ± 103 (95)	– (–)

Since all methods investigated compare similarly in terms of providing real-time energies, they may be selected according to their accuracy, their ability of being systematically improvable, and to the possibility of including them in a subsystem approach. Because of its close connection to DFT the DFTB approach is well suited for subsystem approaches using DFT.

4.2 Quality assessment

The ability of an electronic-structure method to yield qualitatively correct potential energy surfaces is of paramount importance for the study of chemical reactivity. As already stated, absolute barrier heights, depths, and accurate positions of local minima and first-order saddle points are not ultimately important as long as these features are present and can be detected.

To investigate the semi-empirical and density-functional tight-binding methods for their ability to yield such qualitatively correct surfaces, we chose a rotor-like molecule with three anthracene blades that undergoes a rearrangement with a challenging plateau-like barrier and that has recently been studied in our group with more accurate first-principles methods.[53] The energy profile from ref. 53 serves as a reference. The reference potential energy surface features two local minima (A and B) which are connected by two transition states (TS1 and TS2) with a shallow minimum in between (MinAB). For the corresponding structures see Fig. 3.

Since every semi-empirical Hamiltonian produces a different (approximate) potential energy surface the energies have been calculated after a structure optimisation employing the respective electronic structure method (*cf.* Table 2). The NDDO-based calculations were performed with MOPAC2012.[61,64] The DFTB3 structures were optimised in Gaussian09 d.01[51] with the DFTBA[49] method and the

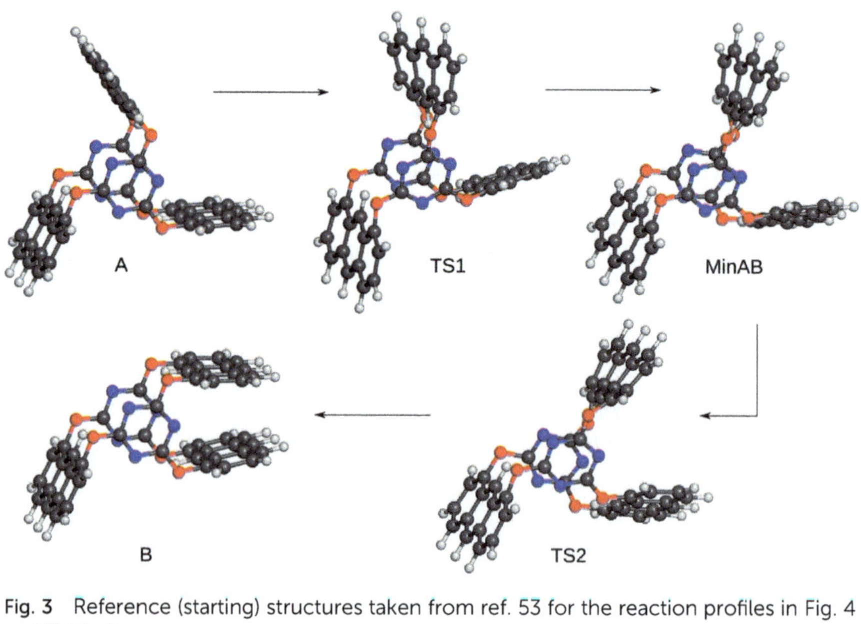

Fig. 3 Reference (starting) structures taken from ref. 53 for the reaction profiles in Fig. 4 and Table 2.

DFTB3[39–41] values were obtained from a single-point energy calculation on these structures with the DFTB+[62] (v.1.2.2) program. The results are collected in Table 2 and selected energy profiles are shown in Fig. 4.

Except for AM1 all semi-empirical methods show the local minima and saddle points of the reference surface (*cf.* Table 2). As expected, the heights or depths vary from method to method but the overall structure of the energy profile is conserved throughout. The transition state structures deviate more from the reference structures than the local minima, as is expected for semi-empirical methods.[48] The correct sign of the energy difference between product and reactant is only reproduced by PM6-D3[37,47] and PM7.[48] This is caused by a strong dispersion interaction effect, which is explicitly treated in a post-SCF manner in PM6-D3 and is directly incorporated in the PM7 Hamiltonian.[37,47,48] All NDDO-based methods have the first transition state energetically below the second one, as in the reference calculation.

PM6-D3 shows the best performance of all methods analysed. Although the heights of the TS1, MinAB and TS2 are significantly lower compared to the DFT reference the relative energies of the transition states and the middle minimum as well as the relative energy between product and reactant are reproduced. The best method for obtaining the correct structures of reactant and product (A and B) is PM7 with RMSD values below 0.1Å.

Interestingly, it was not possible to locate the stable minimum MinAB with the DFTB methods, but they reproduced the two transition states and the minima A and B quite well.

The overall performance of all methods studied here is astonishingly good, *i.e.* they perfectly meet the quality requirements stated in the beginning. The fact that the semi-empirical approaches without dispersion correction overestimate the

Table 2 Energy profile of the reaction A → TS1 → MinAB → TS2 → B relative to the energy of A in kcal mol^{-1}. Below each energy the root mean square deviation (RMSD) with respect to the reference structure is given in Å. The BP86, D3, def2-TZVP reference profile has been taken from ref. 53

		A	TS1	MinAB	TS2	B
BP86-D3/def2-TZVP[53]	E	0.0	20.1	19.6	20.5	−3.7
AM1	E	0.0	17.3	0.4	22.0	0.4
	RMSD	0.118	0.354	1.334	0.265	0.506
RM1	E	0.0	23.4	15.1	32.0	0.6
	RMSD	0.099	0.405	0.276	0.268	0.359
PM3	E	0.0	16.2	13.7	18.3	0.4
	RMSD	0.135	0.314	0.392	0.194	0.612
PM6	E	0.0	13.2	12.6	13.8	0.5
	RMSD	0.123	0.196	0.128	0.123	0.488
PM6-D3	E	0.0	13.0	12.3	13.3	−1.4
	RMSD	0.115	0.196	0.123	0.125	0.361
PM7	E	0.0	18.3	12.1	30.9	−8.3
	RMSD	0.085	0.405	0.237	0.273	0.090
DFTBA	E	0.0	21.8	0.0	18.6	0.3
	RMSD	0.122	0.731	1.199	0.194	0.353
DFTB3	E	0.0	19.2	0.0	17.9	0.2

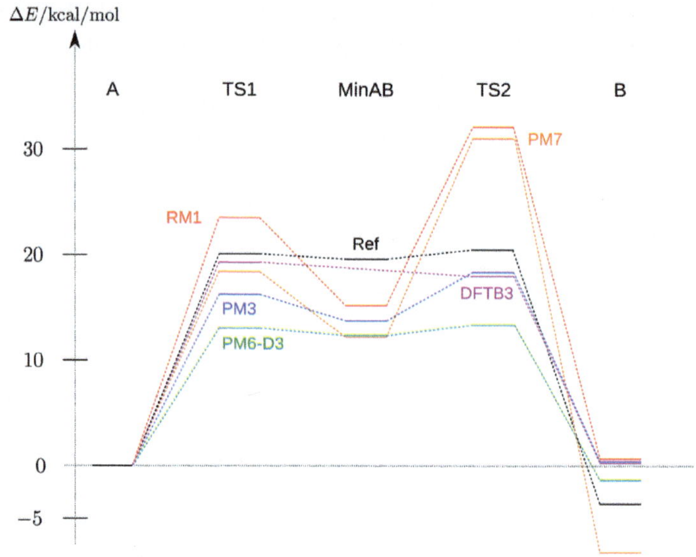

Fig. 4 Energy profile of the reaction A → TS1 → MinAB → TS2 → B relative to the energy of A of the corresponding method. Only a selection of the methods from Table 2 is shown. The reference BP86, DFT-D3, def2-TZVP profile has been taken from ref. 53.

stability of the MinAB structure is no drawback as the detection of a potential intermediate is feasible. Applied in an interactive chemical reactivity study these methods would show the desired features of the potential energy surface and would guide the user to the correct structures. The relatively small RMSD values with respect to the DFT reference (*cf.* Table 2) indicate that, if the user is able to correctly locate the minima, they will be perfectly suited as starting structures for a refinement employing a more accurate method.

5 Reactivity exploration in virtual environments

In the following we describe in detail how a virtual reactivity exploration experience should proceed. We work out what is needed from the chemist's perspective in order to successfully study chemical reactivity.

5.1 Preparatory steps

The procedure starts with the construction of the molecular structure of interest. This construction requires the definition of the type and position of atoms and the determination of the overall charge and spin multiplicity. To arrive at a stable configuration the energy can be continuously minimised in the background while the molecular structure is constructed.[5,6] However, applying a first-principles method can be problematic for placing additional atoms one by one. It may produce unwanted structural distortions upon continuous structure optimisation or may already show a non-converging electronic structure optimisation and artificially large gradients are the result. Classical force fields can be more suitable at this stage.

5.2 The exploration procedure

After the build-up of the structure the operator exploits *ab initio* force calculations to start the exploration. A possible first step is to grab one or more atoms and distort the structure by pulling at them. If the device with which the manipulation is performed features force–feedback functionality, the force acting on the manipulated atoms and hence the reactivity can be directly felt (haptic quantum chemistry[7]). Without haptic interaction the user has to rely on the visual information to infer the reactivity. In that case the user needs to be informed about how likely the applied distortion is by visual elements like force arrows on the atoms or a panel that indicates high energy configurations, which is neither a direct sensation nor is it practical for large molecular systems (the user would drown in a sea of force vectors).

In the case of intermolecular reactions, the starting point is a stable configuration with no intermolecular interactions, *e.g.*, at the dissociation limit. This means that many orientational changes are possible without a change in potential energy and therefore without forces. The reactive fragments are oriented such that a suitable initial position is obtained to start the reactive exploration. From there the scientist may record all configurations visited during the exploration. The recording can be stopped and restarted several times which results in a multitude of different paths. Usually both, the initial and the end configuration, are local minima that are stored separately from the paths. During the exploration the chemist is guided by the forces and by the structural changes occurring as a result of the structure manipulations. The exploration is therefore driven by a combination of the operator's prior knowledge, the information provided through the forces, and the structural changes upon structural manipulations.

Such an exploration can be performed to study the shape of the potential energy surface in a localised region or to directly find possible reaction paths between already known stable configurations (local minima).

5.3 Output for reaction networks

The structures recorded during the exploration together with the associated energies and gradients form the primary (raw) output data. They are what we recently defined as 'core quantities' in our definition of Real-time Quantum Chemistry.[9] Additional data such as dipole moments, the electron density and other quantum chemical properties can also be part of the output if calculated on the fly.

The structures, energies and gradients can be analysed and the distinct points with zero gradients can be extracted. They can be considered as nodes of a reaction network that emerges successively during an exploratory study. The paths between the structures with zero gradients (stationary points) connect the nodes as links. If they correspond to a minimum energy path, the associated transition state structure node can be marked as such. They are usually the main objective of the exploration. If, however, the stationary points are far apart in configuration space, no unique transition structure will be found for any connecting path, which can then be taken as a hint to search for additional local minimum structures in between (new nodes). This then produces a more fine-grained reaction network for which new connecting paths are defined. They are the starting point for further transition state searches—either by haptic exploration or by automatised protocols launched in the background.

5.4 Output for reaction paths

Obtaining all relevant nodes will be mandatory if a complete reaction network is needed. For the study of reaction mechanisms, however, the links connecting the nodes are the prime objective, since they contain kinetic information about how the atoms rearrange during a reaction. The analysis of these paths can be seen as an interactive analogue to the Reaction Path Hamiltonian methods[65–67] or the Unified Reaction Valley[67,68] methods. A very close connection exists to the Reaction Force approach[69] since the operator can follow these paths directly if a haptic device is employed. The output of the exploration can be used as a starting point for a more detailed analysis of the reaction dynamics or kinetics with standard approaches[70–72] eventually applying more accurate electronic structure methods.

5.5 Experimental realisation of manual force exertion

The virtual exploration of chemical reactivity is in fact connected to experimental realisations for the manual manipulations of molecular matter. To study chemical reactivity the resisting forces rendered to the user are an indirect descriptor of the topology of the potential energy surface. Such forces can be directly measured in special experiments. In these experimental setups mechanical forces can be applied to single molecules to change their structure and even to alter their reactivity. There exists a variety of experimental techniques for the realisation of this mechanochemistry. Examples are optical/magnetic tweezers, the biomembrane force probe technique, hydrodynamic techniques and atomic force microscopy. The manipulations applied by the user in the virtual environment can thus be directly linked to exerting real forces on parts of the molecular assembly. For a review of these experimental techniques see ref. 73. In fact, such experiments allow the characterisation of chemical bonds in a direct way[74,75] and the virtual environment can become a convenient means to steer such experiments.

The mechanical manipulation of molecular matter is called mechanochemistry. Theoretical work on mechanochemistry has been performed on ring opening reactions.[76,77] In this example it has been shown how additional mechanical forces alter the potential energy profiles of reactions that involve the rupture of chemical bonds. A review on the theoretical concepts and the computational tools has recently been provided by Ribas-Arino and Marx.[78]

6 Interaction with molecular systems in virtual environments

The core objective of an interactive and real-time treatment of molecular systems is the instantaneous calculation of energy and gradients, *i.e.* the response of a given atom arrangement.[9] Since a plethora of structures, energies and gradients are produced very quickly and in a cumulative manner, they can no longer be mapped out in huge data files as these can hardly be condensed to intuitive insights. Immersion into the virtual environment of the molecular system allows the scientist to perceive the vast amount of data more easily and intuitively.[79] Immersion refers to the process of creating a perception in a virtual reality.

6.1 Immersive perception

The visual presentation of the structure plays an important role for immersion. Hence, it should be done in the way elaborated in chemistry more than hundred years ago, namely in terms of balls as atoms and sticks as bonds. This is the default way to present the structures as applied in molecular visualisation tools like VMD,[80] PyMOL[81] or Chimera[82] to name only a few. Also conventional molecular structure builders like GaussView,[83] the graphical user interface of ADF,[84] Avogadro[6] and SAMSON[5] can serve as examples.

Suitable visual output devices for an efficient immersion are devices that are able to give a comprehensive and intuitive picture of the virtual world. The larger the visual devices are, the more they exclude the perception of reality around the user. Panoramic screens or head-mounted displays[85–87] are best suited for this purpose. Rendering the atoms, bonds and all volumetric data with ray tracing is very helpful in this regard. Also current three dimensional presentation techniques like polarised or active shutter systems are beneficial, since they allow for a much easier perception of the spatial configuration of the molecular system.

To intensify the immersion by addressing the tactile sense requires the rendering of the occurring forces by haptic devices. Haptic pointer devices, for instance, can render forces in three to six dimensions. They can be employed to probe the forces acting on different sites of the molecular assembly. The scientist picks an atom and feels the resulting force response of the system. Also two or more haptic pointer devices can be utilised simultaneously to understand the interplay of forces. Such devices can be purchased at modest prices, which makes them attractive for virtual environments employed in chemical laboratories.

A more elaborate way to render forces and to address the human tactile sense are external skeletons. Since they render the force directly for each finger, they provide an even more natural experience for the tactile sense. However, these devices are still very expensive and currently not useful for a large scale deployment in chemical laboratories.

Additional data like force vectors on atoms not probed by a haptic device, the total energy, partial charges, isosurfaces of electron densities or molecular orbitals need to be seamlessly integrated as well. The most intuitive form to do so is to visualise them in three dimensions together with the molecular structure. Since these quantities as well as the structures change very fast during the manual exploration efficient implementations using graphical processing units and multi-core central processing units are needed.[88,89]

Global scalar quantities like the electronic energy could be rendered by addressing even more senses. For example, the auditory sense could be utilised for this purpose. High-frequency sounds could signal a configuration with very high energy and thus regions in configuration space that are not accessible in a given thermal setting. However, a change of background colour achieves the same and is certainly less annoying. Accordingly, the tactile sense remains the most important one besides human vision.

6.2 Immersive interaction

Human–computer interaction is a key ingredient in immersive technologies as it allows the user to participate in the virtual environment. For the study of chemical reactivity, interaction is mandatory because it drives the reactivity exploration.

Input devices facilitate the human–computer interaction and therefore play an important role. In an immersive virtual environment they allow for a manipulation of the virtual objects. Therefore, the real-world position of the device is transferred into virtual-world coordinates of a virtual representation of the device. With that the virtual objects, in our case, atoms or atom groups, can be moved. A general requirement for devices to be suitable for immersion is that a physical movement of the operator is transferred directly into a movement in the virtual world. Transformations applied in between, *e.g.* to avoid range limitations of the physical device or to stabilise motions, need to be transparent for the operator, since the positions of virtual objects are controlled with them.

Since the forces resulting from such manipulations can be very sensitive with respect to position, a precise position control is most important. A lack of precision renders the control over the manipulation nontransparent which means that physical movements of the device are not consistent with the virtual movement. This also implies that the devices should not loose accuracy if the user quickly moves the devices. Fluctuations in the position due to instabilities caused by the user is also a reason for a less precise position. Especially freely movable input devices require attention to avoid unintentional small amplitude motions. The stabilisation procedure can take place already in the hardware, but also on the software side. Stabilisation can be efficiently achieved by scaling down the input movement.

The interaction with large molecular assemblies in a virtual environment is challenging because they are complex and dense (many objects in a small space). Moving in and interacting with such an environment should be done with devices that allow a manipulation directly in three dimensions. Input devices with less dimensions, like an ordinary computer mouse, allow only a two dimensional position control and are therefore less intuitive to handle. Such input devices acting in two dimensions require additional visual cues and sometimes also additional input with another device to be able to manipulate three dimensional objects.

Three dimensional input devices can be divided into two categories. To the first category belong devices in which a physical object like a ball or a pen is moved. The haptic pointer device, which we have been using in the framework of haptic quantum chemistry, features a pen connected *via* joints to a machine which calculates position and orientation and sends them to the computer. Such devices allow a relatively precise control over position in three dimensions (below 0.01mm according to manufacturer specifications[90,91]).

The second category contains devices which do not require to move a physical object. They directly capture the motion of the user's hand. Such motion sensing devices provide an even more natural form of three-dimensional input. Modern motion sensing devices can have an accuracy down to the sub-millimetre range.[92] A problem with such devices is the detection of two fingers very close to each other. Hence, grabbing atoms with such devices is difficult, since it involves two fingers coming close to each other. Pointing at the atoms and moving them by just one finger is easier to implement, but is also less intuitive. The accuracy of the finger tip positions can be improved by wearing passive or active markers that make the detection easier.

Currently haptic pointer devices are more suitable since input and output can be combined in one such device.

6.3 Rendering forces in reactive potentials

The rendering of the forces is a very important component of the virtual environment proposed in this work, since they convey significant information. One has to distinguish between the force rendered by the device, which is felt by the user, and the force calculated by the chosen electronic structure method. The former we may call 'haptic force' and the latter 'molecular force'.

In a straightforward implementation both forces would be exactly the same. However, the molecular forces are in the range of 1 nN, whereas the forces which can be felt by the user are on the order of 1 N. It is immediately clear that both forces cannot be the same but have to be transformed. This transformation needs to fulfil two requirements in order to be suitable for studying chemical reactivity. As a guide through the configuration space avoiding unphysical configurations the rendered forces need to be intuitive and smooth. Yet, the transformation between the molecular and the haptic force needs to be as transparent as possible, since from the force direction and magnitude the user infers the shape of the potential energy surface. The most transparent transformation is to multiply the molecular force by a constant factor to obtain the haptic force. This is how we transform the nanoscale molecular forces to a haptic force that is experienced by the user.

The shape of potential energy surfaces occurring in reactive chemical systems requires additional transformations. Shallow and deep minima, steep walls and flat hills may lie in close proximity on a reactive potential energy surface. This results in a wide range of forces that need to be rendered. Since the range of forces which can be rendered by the devices is limited and also the tactile sense has a limited resolution, the forces need to be scaled. For instance, in strongly repulsive regions huge forces quickly exceed the limits of the devices. Hence, the forces need to be scaled down. In almost flat regions, however, where shallow minima need to be detectable the force requires to be scaled up. A constant force scaling[93] as mentioned before is therefore not suited. A variable gain scheme[94] can account for both situations without the need to change the scaling manually.

The third reason why one needs to transform forces is more technical. To avoid instabilities and to compensate for slow simulation loops often frequency filters are applied. However, a too dominant filter can lead to an in-transparent force rendering. To obtain a stable control scheme usually the position is not directly controlled but the user applies an additional force that is incorporated into the simulation.[95,96] This, however, hides the true forces acting on the manipulated atom from the user, which makes it unsuitable for studying chemical reactivity, where the forces transport a substantial part of the desired information.

6.4 Moving objects *vs.* probing reactivity

For the manipulation of molecular assemblies one may distinguish between two different intentions. The first intention is to move parts of the assembly around to either see them from a different perspective or to arrange them in a way better suited to start the reactivity exploration. During such a rearrangement the internal structure of the molecules stays intact. The second intention is to probe the chemical reactivity, which involves the breaking and/or formation of chemical bonds. The virtual environment has to accommodate both scenarios.

A seamless transition from moving a molecule to probing reactivity is needed when atoms come closer and the forming or breaking of bonds becomes possible. The problem can be illustrated by considering the abstraction of an atom from a molecule *versus* just pulling at an atom to move the molecule when no further constraints are applied. Whether pulling at an atom results in an abstraction or just in a movement of the whole molecule depends on how fast the system responds in terms of the relaxation described at the beginning of this section. A manipulation within an infinitely fast relaxing system would correspond to a reversible adiabatic process. If the virtual environment is designed to mimic this scenario there would be no net change in energy and therefore also no reactivity. A non-adiabatic process is emulated when the relaxation rate is slower than the manipulation by the user. In this scenario moving a molecule by just pulling at an atom would not be possible without breaking a chemical bond.

This dilemma can be resolved in different ways depending on the limits imposed by the underlying force calculations: either by adjusting the relaxation rate or by adding additional constraints. The first is only possible if the relaxation (structure optimisation) is always faster than the manipulation. Then one can adjust the rate such that the users can switch between adiabatic and non-adiabatic manipulations by adjusting their manipulation speed. Pulling fast at an atom induces the abstraction, while pulling slowly just moves the molecule. Complementary to that would be to change the relaxation rate instead.

If an adjustment of the relaxation rate is not possible or unwanted, additional constraints can be introduced. Possible constraints are to fix atoms in space, preserve internal structures or pin the orientation and centre of mass of a molecule. The user can realise them employing additional input devices (*e.g.*, one hand holds a metal complex in position and the other probes the reactivity of a possible ligand) or by keeping the corresponding degrees of freedom fixed during the relaxation procedure.

In both solutions employing additional input devices is the preferred way since then the constraints can easily be altered or removed if necessary. By contrast, changing the parameters of the relaxation procedure, *e.g.*, changing the rate or the constraints, requires interruption of the exploration. Accordingly, this should only be done if the constraints are expected to remain constant during the exploration, *e.g.*, when certain parts of the molecular assembly are known to be non-reactive.

6.5 Restricting the manipulation to avoid high energy configurations

If in a molecular structure every atom is freely movable, then the manipulation can easily lead to configurations with very high energy. For instance, if bond lengths are decreased below the equilibrium distance of two fragments, very high electronic energies result. Since high energy configurations may not be accessible by the system under real conditions at finite temperature such manipulations have to be avoided or at least made hard to perform.

A natural way of restriction is provided by force-feedback devices. The rendered force guides the user and allows only reasonable manipulations, since a rapid increase in energy is associated with a steep gradient pushing the user in the opposite direction. By this an artificial blocking of certain manipulations can be avoided.

Another possibility is to perform a very fast relaxation of the remaining degrees of freedom. The energy is thereby removed from the system so that the system always maintains a configuration with a reasonable energy. For the physical implications of such an energy removal see next section.

7 Physical laws in virtual environments

Clearly, any useful virtual environment designed to immerse the chemist into the reactive molecular assembly has to obey certain physical laws. They are a trade-off between an intuitive presentation and the 'true' physics. For the former most of the requirements have already been outlined in the sections above. Here, we will show which consequences for the design of the virtual environment follow from the physical theory.

7.1 Potential energy and force field of the molecular systems

Although the atoms are visually represented as hard spheres they are described as classical point-like objects in the calculations. Also the interaction of the operator with the molecular system necessitates a classical treatment of the atoms, since at all times they have a completely determined position in space, *i.e.*, their movement corresponds to classical trajectories. Accordingly, the virtual environment implies a 'macroscopic' picture of the molecular entities which is classical by definition. Hence, quantum mechanical effects of the atoms, such as tunnelling, are not directly representable. One may account for them by discontinuous atom position dislocations when induced by a tunnelling probability measure in separate studies if the potential energy surface is sufficiently well known. However, this will only be necessary for potential proton transfer steps.

The dynamics of the electrons are not explicitly represented in the virtual environment, but they largely define the interactions between the atoms. This separated treatment of the electron and atom dynamics is based on the Born–Oppenheimer approximation.[97] In fact, the electrons give rise to the potential in which the atoms move. The electronic energy is then defined by the electronic Schrödinger equation

$$\hat{H}_{\mathrm{el}}\Psi_{\mathrm{el}} = E_{\mathrm{el}}\Psi_{\mathrm{el}} \tag{1}$$

and the (approximate) nuclear Schrödinger equation reads

$$[\hat{T}_{\mathrm{nuc}} + E_{\mathrm{el}}]\Psi_{\mathrm{nuc}} = E_{\mathrm{nuc}}\Psi_{\mathrm{nuc}}. \tag{2}$$

The electronic energy $E_{\mathrm{el}}(\{\boldsymbol{R}_I\})$ is a parametric function of the nuclear coordinates $\{\boldsymbol{R}_I\}$. A first-principles treatment of the electrons requires that $E_{\mathrm{el}}(\{\boldsymbol{R}_I\})$ has to be calculated by solving equation (1). This has to be done at discrete configurations of the atoms.

Despite its calculation as a discrete function, E_{el} can be considered as a continuous function of all atom coordinates. We may define a continuous variable $\boldsymbol{x}$ in $3M$ dimensional space as

$$\boldsymbol{x} = (\boldsymbol{R}_1, \boldsymbol{R}_2, \ldots, \boldsymbol{R}_M)^{\mathrm{T}} \text{ with } \boldsymbol{R}_I = (x_I, y_I, z_I)^{\mathrm{T}}. \tag{3}$$

Accordingly, $E_{el}(\boldsymbol{x})$ is a hyper-surface in a 3M-dimensional Cartesian configuration space spanned by the coordinates of all atoms. There is a conservative force field connected to this surface. It describes the forces acting on each atom A and is given by the negative gradient $\boldsymbol{g}_A$ of the potential energy $E_{el}(\boldsymbol{x})$

$$\boldsymbol{f}_A(\boldsymbol{x}) = -\boldsymbol{g}_A \text{ with } \boldsymbol{g}_A = \nabla_A E_{el}(\boldsymbol{x}). \tag{4}$$

The gradients (or forces) can be calculated directly from the electronic structure. Only this analytic differentiation is sufficiently fast for immersive techniques.

From this setting it follows naturally that the molecular system has to be described classically with respect to the atom movements and quantum mechanically with respect to the electronic structure and derived properties—a standard and well-investigated approximation for mechanistic studies in complex molecular systems. Interestingly, this partition into a classical and a quantum mechanical description reveals another way to determine which information has to be presented visually and which information needs to be rendered by addressing other senses. The part described by classical mechanics is presented visually and the highly non-intuitive quantum behaviour of the electrons is expressed and presented as forces rendered to the tactile sense.

A fully quantum mechanical treatment of the atom movement would also be possible, but the visualisation of such a situation cannot be easily realised in a virtual environment where molecules are represented by ball-and-stick arrangements. Instead, nuclear probability distributions need to be considered, which are related to the molecular structures in the Born–Oppenheimer approximation.[98,99] However, the tremendous computational demands for such a full quantum treatment are prohibitive, apart from the fact that the force on an atom is then no longer straightforwardly defined.

7.2 Coordinates of the configuration space

The operator manipulates the molecular structures by changing the Cartesian coordinates of one or a few atoms at a time. Hence, the Cartesian coordinate system employed above is the most natural way to define the configuration space. Alternatively, coordinates based on the internal degrees of freedom of the molecular system can be chosen. But for the discussions to follow, the coordinate system does not need to be specified.

We assume that the collection of coordinates can be separated into different sets (subsystems). The first separation follows naturally from the way the user interacts with the system. For every manipulation we can define a set of coordinates $\boldsymbol{s}$ which are strictly defined by the manipulation (system coordinates) and the set of all remaining (environment) coordinates $\boldsymbol{e}$,

$$(\boldsymbol{R}_1, \boldsymbol{R}_2, \ldots, \boldsymbol{R}_M) \rightarrow (\boldsymbol{s}, \boldsymbol{e}). \tag{5}$$

Since the set of environment coordinates is not constrained by the manipulation, different ways to treat them are possible. They may be completely frozen, they may be subdivided into subsystems with reduced degrees of freedom or they may be optimised. Completely frozen inactive coordinates are only reasonable if they are not very much affected by the manipulation (detected by forces defined as gradient components with respect to these coordinates). Although the environment is frozen the active (system) part movement is not just ballistic, since interactions can still occur. Partly frozen coordinates (frozen subsystems) are useful, when some groups of atoms are expected to be internally rigid. The most general treatment is to optimise the inactive coordinates according to a certain protocol, which is discussed in the following section.

7.3 Energy deposition, redistribution and dissipation

Most of the manipulations performed to study reactivity involve a change in electronic energy. If a low temperature is to be maintained any excess energy has to be removed from the system. The energy deposited by the structure manipulation of the operator may be restricted according to a single measure proportional to room temperature. Any excess energy can result in a relaxation of all unconstrained coordinates. That is to minimise the energy by varying the unconstrained coordinates,

$$E_{\min}(\boldsymbol{s}) = \min_{\boldsymbol{e}} \ E_{\text{el}}(\boldsymbol{s}, \boldsymbol{e}) \ . \tag{6}$$

Removing the excess energy from the free coordinates is an artificial dissipation of energy. If always the direction with minimum resistance is chosen, the path is equal to the so called minimum energy path between reactants and products.

The redistributed energy gives rise to a motion of the atoms along the unconstrained degrees of freedom. Apart from structure relaxation, this motion can also be treated by the usual methods of molecular dynamics (MD) and corresponds then to a steered BO-MD[100,101] scheme similar to steered classical MD.[102] Under certain conditions, the work performed by the operator is equal to the free energy change of an equilibrium process according to the Jarzynski equality.[103]

7.4 Reaction path and reaction force

The structures $\boldsymbol{x}$ recorded during a virtual exploration correspond to points in the configuration space for a pseudo-one-dimensional reaction coordinate $\xi(\boldsymbol{x})$, the arc length of the reaction path. With each point in configuration space along the path ξ a set of forces is associated

$$\boldsymbol{x}(\xi) \rightarrow f(\xi). \tag{7}$$

If ξ describes a minimum energy path (MEP), there are only forces along the constrained coordinates and the forces in the (orthogonal) environment coordinates are reduced or zero by structure minimisation. Therefore, the work W along the MEP can be obtained from

$$W = \int_{\text{MEP}} f^{(\text{s})}(\xi) \ \mathrm{d}\xi \ , \tag{8}$$

where $f^{(\text{s})}$ denotes the forces along the constrained (system) coordinates.

7.5 Virtual inertia and friction

The correct physical dynamics of an object *in vacuo* requires that a force exerted on it accelerates it and the object moves then with constant velocity when the force is switched off. In the case of molecules, any structure manipulation is performed by exerting forces which would lead to a continuous linear movement after the force vanished at the end of the manipulation. The operator would be constantly occupied with stopping unintentionally accelerated molecules. The energy minimisation discussed before can remove excess energy from the inactive coordinates and therefore also from the overall translation. In any case, a molecule which had been pulled at should not move after it has been released by the operator (provided that no other forces are present).

The technical implementation of the energy dissipation process, however, gives rise to a sensation of viscosity when a molecule is grabbed and pulled around in the virtual environment. The removal of energy by relaxing all unconstrained degrees of freedom requires a certain amount of time. This delays the relaxation and, hence, pulling at a molecule shows a resisting force. This resisting force leads to a sensation of friction as if the molecule would be dragged through a viscous liquid. It appears that this artifact is in fact useful, since the user naturally expects inertia when accelerating atoms of fragments, although the haptic exploration does not consider any mass-dependent reaction to the exerted force (action).

7.6 Adiabatic and non-adiabatic manipulations

Depending on the rate of the relaxation process relative to the speed of manipulation, the manipulations are either adiabatic or non-adiabatic. In the adiabatic case the relaxation is faster than the manipulation and hence the path described by the manipulation including relaxation is a minimum energy path. If the relaxation is slower than the manipulation, a non-adiabatic manipulation is performed and no minimum energy path results.

7.7 Implications of modified forces

As has already been discussed in section 6 an intuitive interaction with molecular systems in virtual environments often requires the manipulation of the calculated forces before they are rendered to the user. In a more formal way the forces felt by the operator $\boldsymbol{f}_{\mathrm{h}}$ are a modified version of the molecular force $\boldsymbol{f}_{\mathrm{m}}$ as

$$\boldsymbol{f}_{\mathrm{h}} = C\,(|\boldsymbol{f}_{\mathrm{m}}|)\boldsymbol{f}_{\mathrm{m}}, \tag{9}$$

where C is a force-dependent scaling factor and is in general a function of the absolute value of the molecular force. Since the forces are derived from the potential, also the potential is implicitly modified. Hence, the scientist explores not the true potential energy surface.

As we have already discussed in our work on the haptic exploration of potential energy surfaces[8] the requirement for the scaled surface is that it preserves the main features like the positions of minima and first-order saddle points. The steepness of walls or valleys is not that important if their relative depth or height is approximately preserved. Especially, schemes with a constant scaling factor[93] for the forces (C = const) need to meet this requirement. Clearly, the reaction network explored in such a way can be refined by accurate quantum chemical methods in the background.

In variable gaining schemes[94] the scaling factor is a function of the absolute value of the haptic force. To preserve the existence of all features allowing only a modification of their relative heights or depths, the function C is subject to some conditions. (1) $C = 0$ only at $|\mathbf{f}_\mathrm{m}| = 0$ and nowhere else and (2) C must be strictly positive. Only then are no artificial saddle points created. With such a function, for instance, a shallow minimum is detectable but its depth may be exaggerated.

It is important to give the operator the freedom to change the scaling during an exploration. Since during an exploration all structures, gradients and energies can be recorded the true potential energy surface can be reconstructed afterwards.

7.8 Influence of interactive manipulations

Every input device implicitly introduces a constraint by keeping certain atoms at fixed positions. Additional constraints may be imposed by the parameters of the optimisation procedures. Hence, in the course of an exploration many constraints can be simultaneously active. In fact every constraint that is introduced restricts the part of configuration space that can be explored.

However, since every constraint introduces additional forces, the force on a certain atom, can be significantly different with and without a constraint. If, for instance, some atom positions are fixed either by an additional input device or by just excluding them from the relaxation procedure, new constraints are introduced. These constraints in turn alter the behaviour of the system and can be regarded as additional forces. It is clear that the design of such constraints is a crucial ingredient for the manual exploration of reaction mechanisms.

7.9 Local reactivity *vs.* conformational entropy and large amplitude motions

We envisage a virtual exploration of chemical reactivity that involves the making and breaking of chemical bonds featuring a bond energy that is at least one order of magnitude larger than the thermal energy. Only under these conditions can a manual exploration of individual molecular structures be meaningful. A typical example is the formation of coordinative bonds to a transition metal ion. The thermodynamics and kinetics of such a process in solution is usually dominated by the changes in electronic energy, while temperature and entropic effects may be neglected in a first step. This is the reason why computational studies on transition-metal catalysed reactions can be successfully carried out by stationary quantum chemical methods. We may call such situations dominated by 'local reactivity'. Temperature and entropic effects are then considered within simplified models for the degrees of freedom (*e.g.*, by harmonic potentials). Usually, they lead to small modifications of the electronic energy profile. Significant departures of the free energy surface from the Born–Oppenheimer potential energy surface can be observed if the number of reactants changes in an elementary reaction step in a gas-phase process because of non-negligible contributions from the translational entropy change. This situation changes when many weak contacts among conformationally flexible reactants and surrounding molecules become important such that entropy changes upon chemical transformations can pile up. Then, resorting to statistical methods as provided by molecular dynamics or stochastic simulation schemes becomes mandatory. Other chemical processes that may require such techniques are the occurrence of structural rearrangements on long time scales (such as large amplitude motions that may induce massive overall structural changes in a macromolecular assembly). However, sampling approaches that may account for these processes can be called upon within the virtual reactivity environment described here. In fact, the hardware demands are currently just exorbitantly larger. As a consequence, the first implementation of a virtual reactivity lab may only deal with typical local-reactivity problems.

8 General discussion and conclusions

The remarkable theoretical and algorithmic achievements in quantum chemistry in the past decades made it possible to assign an electronic energy and molecular properties to a molecular structure in a reasonable time. Traditional research along these lines seeks to improve on accuracy as well as on increasing the size of molecules for which such calculations are still feasible. We have argued that a different focus on algorithmic developments will lead to a paradigm shift in this field. This produces a virtual environment for the intuitive and direct study of chemical reaction mechanisms. Its ingredients are (i) the possibility to calculate quantum-mechanical information about a reactive system of at most a few hundred atoms in real time[9] and (ii) the possibility to explore large amounts of data by physically experiencing their contents through the tactile human sense.[7] The latter can be realised directly[9] or through an intermediate interpolation layer.[7,8] It is important to understand that such a reactivity exploration tool requires new concepts for the step-by-step exploration of a reaction path.[10] For reactivity exploration, algorithmic tools need to be developed[10] that are significantly advanced compared to interactive structure optimisations, which aim to aid the set-up of configurations in molecule editors. For the latter, the final result, *i.e.*, a minimum structure on the Born–Oppenheimer surface counts, while considering physical principles for the way in which such a minimum structure is found is not decisive.

In this work, we undertook the next step and elaborated on a new route of development in quantum chemistry towards ubiquitous computing that deeply immerses a chemist into his/her molecular target. This virtual environment poses new challenges for the organisation and presentation of tons of quantum chemical raw data. Generating alpha-numerical input files for complicated electronic structure calculations and processing the alpha-numerical output is one of the main obstacles prohibiting ubiquitous interactive computing in chemistry. This traditional, but pedestrian way of reactivity exploration is likely to die out, when ubiquitous computing, big-data handling, and immersive technology becomes available within a single implementation. Before such a virtual reactivity lab can be made available, a thorough analysis of its components is mandatory.

A central part is certainly how the quantum mechanical response is calculated in such an environment. The three main requirements for a suitable electronic structure method are: (1) it needs to be fast enough for a real-time execution, (2) it has to be based on the first principles of quantum mechanics for bond breaking and making and (3) the quality has to be such that the main features of the potential energy surface are detectable. As a first step one has to resort to approximate electronic structure methods. Hence, we discussed and evaluated contemporary semi-empirical methods suitable to study chemical reactivity in real time. With respect to both, time and accuracy, the semi-empirical methods investigated here permit an interactive study of chemical reactivity within a virtual environment. Combining the results of this work with the results we obtained in our previous studies[7–9] we now have a range of methods at hand, which allow the calculation of real-time quantum mechanical responses. These methods range from interpolation techniques, semi-empirical methods up to minimal DFT and HF calculations.

Furthermore, we describe in detail here the design of a virtual environment tailored for the exploration of chemical reaction mechanisms and the subsequent study of reaction networks. We show how the quantities presented in the proposed virtual environment are connected to those manifest in a quantum mechanical description. The introduced approximations, which are necessary to allow for an interactive study of chemical reactivity have been analysed and the inherent limits have been worked out. We also discuss routes on how these limits can be partly alleviated for the incorporation of, *e.g.*, tunnelling or entropic effects. The scaling of the forces (and therefore also of the potential energy surface) in a virtual environment has been analysed, too. Carefully chosen, they do not impair the physics of the system but can even enhance the experience. In fact, an interactive exploration of the Born–Oppenheimer potential energy surface given by the electronic energy can be carried out with a minimal quantum chemical model that recovers the rough position of minima and first-order saddle points and thus the specific structure of the surface (*i.e.*, without introducing artifacts like spurious or missing stable intermediates).

The operator does not need to analyse any intermediate results. Instead, good candidates for stable structures are stored as nodes of a reaction network, which the operator sees emerging on the screen. Connecting lines of these nodes can be supplemented with a barrier height from an automated transition state search, which can be launched in the background by one of many service programs, which we have called 'jeannies' here. If the jeannie realises that there is no unique transition state, the connecting line will most likely not represent an elementary reaction step and this fact can be visualised. The jeannie may hand this information to another service program responsible for the optimisation of local minimum structures (within the structural interval defined by the two nodes in the reaction network). Since real-time quantum chemical data is a prerequisite for the whole virtual environment, the local-minimum and the transition-state searches do not need to be very efficient, but they need to be very stable. This poses a whole new set of constraints on such algorithms.

While the operator observes the expanding reaction network on the screen and can investigate structures on nodes as well as the connecting transition-state structure by a simple mouse click, he/she may choose to steer the construction of the emerging network. Certain branches of the network may thus be shut down and excluded from further exploration at will when deemed necessary. Accepted nodes may be refined by more accurate quantum chemical calculations, automatically launched in the background by another jeannie. While the accurate optimisation of a minimum structure at a node is certainly valuable for obtaining reliable structural information, the associated electronic energy is only of secondary importance. Instead, the energetical position of the node within the reaction network is of primary importance. In this systems-chemical view of complex chemical processes, a rolling kinetic modelling of all elementary reaction steps emerging in the network will determine, which nodes and links are most important. These are the ones for which more accurate quantum chemical calculations are then launched by the jeannie.

While we have already realised some of the components necessary for the chemical virtual reality lab, many ideas described here await their implementation and we will continue to finally provide such a tool to the chemistry and materials science communities. Currently, the chemical reactivity of hydrocarbons can already be explored in real time and reaction networks for these compounds can be mapped out.[10] If the fully-fledged tool is completed, it will represent a new type of predictive and creative tool for chemical research *in silico*. Imagine a catalytic reaction mediated by some catalyst. Modelling a catalytic cycle is currently a very laborious and time-consuming undertaking. Because of the huge effort required, side reactions, of which there are orders of magnitude more than there are reaction steps in the catalytic cycle, are usually omitted. The virtual reality lab would be able to deal with side reactions, too. The data can be naturally absorbed into the components of the reaction network introducing new configurations to be constructed and finally added to the existing nodes. Natural reaction partners of the intermediates in the catalytic cycle are all molecules occurring in the cycle (including the intermediates themselves) as well as standard molecules from the environment like solvent molecules (especially water) and dioxygen. The plethora of structures needed for such a screening of potentially degradative reactions can again be set up and screened automatically by jeannies. Clearly, another useful ingredient for this endeavour is a jeannie that aids designing structures for reactive systems with specific functionality.[13,14]

Hence, the virtual environment envisaged here will be much more than a simple tool to control and manage quantum chemical calculation (as by standard graphical user interfaces to quantum chemistry program packages). Instead, it will be a new tool for truly creative work in chemistry that can challenge the work of experimental chemists.

Acknowledgements

This work has been supported by ETH Zurich (grant number: ETH-08 11-2).

References

1 J. F. Prins, J. Hermans, G. Mann, L. S. Nyland and M. Simons, *Future Gener. Comput. Syst.*, 1999, **15**, 485–495.

2 A. Bolopion, B. Cagneau, S. Redon and S. Régnier, *Intelligent Robots and Systems*, 2009. IROS 2009. IEEE/RSJ International Conference on, 2009, pp. 237–242.

3 N. Férey, J. Nelson, C. Martin, L. Picinali, G. Bouyer, A. Tek, P. Bourdot, J. Burkhardt, B. Katz, M. Ammi, C. Etchebest and L. Autin, *Virtual Reality*, 2009, **13**, 273–293.

4 J. Heyd and S. Birmanns, *Virtual Reality*, 2009, **13**, 245–255.

5 M. Bosson, S. Grudinin, X. Bouju and S. Redon, *J. Comput. Phys.*, 2012, **231**, 2581–2598.

6 M. D. Hanwell, D. E. Curtis, D. C. Lonie, T. Vandermeersch, E. Zurek and G. R. Hutchison, *J. Cheminf.*, 2012, **4**, 1–17.

7 K. H. Marti and M. Reiher, *J. Comput. Chem.*, 2009, **30**, 2010–2020.

8 M. P. Haag, K. H. Marti and M. Reiher, *ChemPhysChem*, 2011, **12**, 3204–3213.
9 M. P. Haag and M. Reiher, *Int. J. Quantum Chem.*, 2013, **113**, 8–20.
10 M. P. Haag, A. C. Vaucher, M. Bosson, S. Redon and M. Reiher, *ChemPhysChem*, submitted [arXiv: 1405.4036].
11 M. Bosson, C. Richard, A. Plet, S. Grudinin and S. Redon, *J. Comput. Chem.*, 2012, **33**, 779–790.
12 M. Bosson, S. Grudinin and S. Redon, *J. Comput. Chem.*, 2013, **34**, 492–504.
13 T. Weymuth and M. Reiher, *Int. J. Quantum Chem.*, 2014, **114**, 838–850.
14 T. Weymuth and M. Reiher, *Int. J. Quantum Chem.*, 2014, **114**, 823–837.
15 R. Dawes, D. L. Thompson, Y. Guo, A. F. Wagner and M. Minkoff, *J. Chem. Phys.*, 2007, **126**, 184108.
16 Y. Guo, L. B. Harding, A. F. Wagner, M. Minkoff and D. L. Thompson, *J. Chem. Phys.*, 2007, **126**, 104105.
17 G. G. Maisuradze and D. L. Thompson, *J. Phys. Chem. A*, 2003, **107**, 7118–7124.
18 G. G. Maisuradze, D. L. Thompson, A. F. Wagner and M. Minkoff, *J. Chem. Phys.*, 2003, **119**, 10002–10014.
19 J. M. Bowman, J. S. Bittman and L. B. Harding, *J. Chem. Phys.*, 1986, **85**, 911–921.
20 M. A. Collins, *Theor. Chem. Acc.*, 2002, **108**, 313–324.
21 L. M. Raff, M. Malshe, M. Hagan, D. I. Doughan, M. G. Rockley and R. Komanduri, *J. Chem. Phys.*, 2005, **122**, 084104.
22 R. Dawes, D. L. Thompson, A. F. Wagner and M. Minkoff, *J. Chem. Phys.*, 2008, **128**, 084107.
23 J. Tersoff, *Phys. Rev. B*, 1988, **37**, 6991–7000.
24 D. W. Brenner, *Phys. Rev. B: Condens. Matter*, 1990, **42**, 9458–9471.
25 M. W. Finnis and J. E. Sinclair, *Philos. Mag. A*, 1984, **50**, 45–55.
26 A. Warshel and R. M. Weiss, *J. Am. Chem. Soc.*, 1980, **102**, 6218–6226.
27 A. C. T. van Duin, S. Dasgupta, F. Lorant and W. A. Goddard, *J. Phys. Chem. A*, 2001, **105**, 9396–9409.
28 K. D. Nielson, A. C. T. van Duin, J. Oxgaard, W.-Q. Deng and W. A. Goddard, *J. Phys. Chem. A*, 2005, **109**, 493–499.
29 S. Mishra and M. Meuwly, *Kinetics and Dynamics*, Springer Netherlands, 2010, vol. 12, pp. 137–155.
30 A. B. Anderson, *Int. J. Quantum Chem.*, 1994, **49**, 581–589.
31 R. Hoffmann, *J. Chem. Phys.*, 1963, **39**, 1397–1412.
32 A. B. Anderson, *J. Chem. Phys.*, 1975, **62**, 1187–1188.
33 J. A. Pople, D. P. Santry and G. A. Segal, *J. Chem. Phys.*, 1965, **43**, S129–S135.
34 J. A. Pople, D. L. Beveridge and P. A. Dobosh, *J. Chem. Phys.*, 1967, **47**, 2026–2033.
35 M. J. S. Dewar and W. Thiel, *J. Am. Chem. Soc.*, 1977, **99**, 4907–4917.
36 M. J. S. Dewar and W. Thiel, *J. Am. Chem. Soc.*, 1977, **99**, 4899–4907.
37 J. J. P. Stewart, *J. Mol. Model.*, 2007, **13**, 1173–1213.
38 W. Weber and W. Thiel, *Theor. Chem. Acc.*, 2000, **103**, 495–506.
39 M. Elstner, D. Porezag, G. Jungnickel, J. Elsner, M. Haugk, T. Frauenheim, S. Suhai and G. Seifert, *Phys. Rev. B: Condens. Matter*, 1998, **58**, 7260–7268.
40 Yang, H. Yu, D. York, Q. Cui and M. Elstner, *J. Phys. Chem. A*, 2007, **111**, 10861–10873.
41 M. Gaus, Q. Cui and M. Elstner, *J. Chem. Theory Comput.*, 2011, **7**, 931–948.
42 N. Otte, M. Scholten and W. Thiel, *J. Phys. Chem. A*, 2007, **111**, 5751–5755.

43 R. Sure and S. Grimme, *J. Comput. Chem.*, 2013, **34**, 1672–1685.
44 M. J. Dewar, E. G. Zoebisch, E. F. Healy and J. J. Stewart, *J. Am. Chem. Soc.*, 1985, **107**, 3902–3909.
45 G. B. Rocha, R. O. Freire, A. M. Simas and J. J. P. Stewart, *J. Comput. Chem.*, 2006, **27**, 1101–1111.
46 J. J. P. Stewart, *J. Comput. Chem.*, 1989, **10**, 209–220.
47 S. Grimme, J. Antony, S. Ehrlich and H. Krieg, *J. Chem. Phys.*, 2010, **132**, 154104.
48 J. J. P. Stewart, *J. Mol. Model.*, 2013, **19**, 1–32.
49 G. Zheng, H. A. Witek, P. Bobadova-Parvanova, S. Irle, D. G. Musaev, R. Prabhakar, K. Morokuma, M. Lundberg, M. Elstner, C. Köhler and T. Frauenheim, *J. Chem. Theory Comput.*, 2007, **3**, 1349–1367.
50 G. Zheng, M. Lundberg, J. Jakowski, T. Vreven, M. J. Frisch and K. Morokuma, *Int. J. Quantum Chem.*, 2009, **109**, 1841–1854.
51 M. J. Frisch, G. W. Trucks, H. B. Schlegel, G. E. Scuseria, M. A. Robb, J. R. Cheeseman, G. Scalmani, V. Barone, B. Mennucci, G. A. Petersson, H. Nakatsuji, M. Caricato, X. Li, H. P. Hratchian, A. F. Izmaylov, J. Bloino, G. Zheng, J. L. Sonnenberg, M. Hada, M. Ehara, K. Toyota, R. Fukuda, J. Hasegawa, M. Ishida, T. Nakajima, Y. Honda, O. Kitao, H. Nakai, T. Vreven, J. A. Montgomery, Jr., J. E. Peralta, F. Ogliaro, M. Bearpark, J. J. Heyd, E. Brothers, K. N. Kudin, V. N. Staroverov, R. Kobayashi, J. Normand, K. Raghavachari, A. Rendell, J. C. Burant, S. S. Iyengar, J. Tomasi, M. Cossi, N. Rega, J. M. Millam, M. Klene, J. E. Knox, J. B. Cross, V. Bakken, C. Adamo, J. Jaramillo, R. Gomperts, R. E. Stratmann, O. Yazyev, A. J. Austin, R. Cammi, C. Pomelli, J. W. Ochterski, R. L. Martin, K. Morokuma, V. G. Zakrzewski, G. A. Voth, P. Salvador, J. J. Dannenberg, S. Dapprich, A. D. Daniels, O. Farkas, J. B. Foresman, J. V. Ortiz, J. Cioslowski and D. J. Foxls, *Gaussian 09*, Gaussian Inc. Wallingford CT 2009.
52 A. R. Finkelmann, M. T. Stiebritz and M. Reiher, *J. Phys. Chem. B*, 2013, **117**, 4806–4817.
53 M. J. Kory, M. Bergeler, M. Reiher and A. D. Schlüter, *Chem.–Eur. J.*, 2014, DOI: 10.1002/chem.201400364.
54 D. V. Yandulov and R. R. Schrock, *Science*, 2003, **301**, 76–78.
55 R. Schrock, *Angew. Chem., Int. Ed.*, 2008, **47**, 5512–5522.
56 M. Reiher, B. Le Guennic and B. Kirchner, *Inorg. Chem.*, 2005, **44**, 9640–9642.
57 B. Le Guennic, B. Kirchner and M. Reiher, *Chem.–Eur. J.*, 2005, **11**, 7448–7460.
58 S. Schenk, B. Le Guennic, B. Kirchner and M. Reiher, *Inorg. Chem.*, 2008, **47**, 3634–3650.
59 S. Schenk and M. Reiher, *Inorg. Chem.*, 2009, **48**, 1638–1648.
60 S. Luber, J. Neugebauer and M. Reiher, *J. Chem. Phys.*, 2009, **130**, 064105.
61 J. J. P. Stewart, *MOPAC2012*, http://openmopac.net, Stewart Computational Chemistry.
62 B. Aradi, B. Hourahine and T. Frauenheim, *J. Phys. Chem. A*, 2007, **111**, 5678–5684.
63 M. Gaus, A. Goez and M. Elstner, *J. Chem. Theory Comput.*, 2013, **9**, 338–354.
64 J. D. C. Maia, G. A. Urquiza Carvalho, C. P. Mangueira, S. R. Santana, L. A. F. Cabral and G. B. Rocha, *J. Chem. Theory Comput.*, 2012, **8**, 3072–3081.
65 W. H. Miller, N. C. Handy and J. E. Adams, *J. Chem. Phys.*, 1980, **72**, 99–112.

66 B. A. Ruf and W. H. Miller, *J. Chem. Soc., Faraday Trans. 2*, 1988, **84**, 1523–1534.

67 E. Kraka, *Wiley Interdiscip. Rev.: Comput. Mol. Sci.*, 2011, **1**, 531–556.

68 Z. Konkoli, E. Kraka and D. Cremer, *J. Phys. Chem. A*, 1997, **101**, 1742–1757.

69 A. Toro-Labbé, S. Gutiérrez-Oliva, J. S. Murray and P. Politzer, *Mol. Phys.*, 2007, **105**, 2619–2625.

70 A. Laio and M. Parrinello, *Proc. Natl. Acad. Sci. U. S. A.*, 2002, **99**, 12562–12566.

71 C. Dellago, P. G. Bolhuis, F. S. Csajka and D. Chandler, *J. Chem. Phys.*, 1998, **108**, 1964–1977.

72 R. Alves, F. Antunes and A. Salvador, *Nat. Biotechnol.*, 2006, **24**, 667–672.

73 M. K. Beyer and H. Clausen-Schaumann, *Chem. Rev.*, 2005, **105**, 2921–2948.

74 M. Grandbois, M. Beyer, M. Rief, H. Clausen-Schaumann and H. E. Gaub, *Science*, 1999, **283**, 1727–1730.

75 S. W. Schmidt, A. Kersch, M. K. Beyer and H. Clausen-Schaumann, *Phys. Chem. Chem. Phys.*, 2011, **13**, 5994–5999.

76 M. T. Ong, J. Leiding, H. Tao, A. M. Virshup and T. J. Martinez, *J. Am. Chem. Soc.*, 2009, **131**, 6377–6379.

77 J. Ribas-Arino, M. Shiga and D. Marx, *Angew. Chem., Int. Ed.*, 2009, **48**, 4190–4193.

78 J. Ribas-Arino and D. Marx, *Chem. Rev.*, 2012, **112**, 5412–5487.

79 M. Schirski, C. Bischof and T. Kuhlen, *Proceedings of the 13th Eurographics conference on Virtual Environments*, Aire-la-Ville, Switzerland, Switzerland, 2007, pp. 69–76.

80 W. Humphrey, A. Dalke and K. Schulten, *J. Mol. Graphics*, 1996, **14**, 33–38.

81 Schrödinger, LLC, The PyMOL Molecular Graphics System, http://www.pymol.org.

82 E. F. Pettersen, T. D. Goddard, C. C. Huang, G. S. Couch, D. M. Greenblatt, E. C. Meng and T. E. Ferrin, *J. Comput. Chem.*, 2004, **25**, 1605–1612.

83 R. Dennington, T. Keith and J. Millam, *GaussView Version 5*, Semichem Inc., Shawnee Mission, KS, 2009.

84 G. te Velde, F. M. Bickelhaupt, E. J. Baerends, C. Fonseca Guerra, S. J. A. van Gisbergen, J. G. Snijders and T. Ziegler, *J. Comput. Chem.*, 2001, **22**, 931–967.

85 J. C. Chung, M. R. Harris, F. P. Brooks, H. Fuchs, M. T. Kelley, J. Hughes, M. Ouh-young, C. Cheung, R. L. Holloway and M. Pique, *Proc. SPIE–Int. Soc. Opt. Eng.*, 1989, **1083**, 42–52.

86 E. Patrick, D. Cosgrove, A. Slavkovic, J. A. Rode, T. Verratti and G. Chiselko, *Proceedings of the SIGCHI conference on Human Factors in Computing Systems*, New York, NY, USA, 2000, pp. 478–485.

87 T. Shibata, *Displays*, 2002, **23**, 57–64.

88 J. E. Stone, J. Saam, D. J. Hardy, K. L. Vandivort, W.-m. W. Hwu and K. Schulten, *Proceedings of 2nd Workshop on General Purpose Processing on Graphics Processing Units*, New York, NY, USA, 2009, pp. 9–18.

89 J. E. Stone, D. J. Hardy, I. S. Ufimtsev and K. Schulten, *J. Mol. Graphics Modell.*, 2010, **29**, 116–125.

90 SensAble Technologies, Inc., http://www.sensable.com.

91 Force Dimension, http://www.forcedimension.com.

92 F. Weichert, D. Bachmann, B. Rudak and D. Fisseler, *Sensors*, 2013, **13**, 6380–6393.

93 Y.-G. Lee and K. W. Lyons, *Computer-Aided Design*, 2004, **36**, 75–90.

94 A. Bolopion, B. Cagneau, S. Redon and S. Régnier, *World Haptics Conference (WHC)*, 2011 IEEE, 2011, pp. 469–474.
95 A. Bolopion, B. Cagneau, S. Redon and S. Régnier, *J. Mol. Graphics Modell.*, 2010, **29**, 280–289.
96 A. Bolopion, B. Cagneau, S. Redon and S. Régnier, *Advanced Intelligent Mechatronics (AIM)*, 2010 IEEE/ASME International Conference on, 2010, pp. 329–334.
97 M. Born and R. Oppenheimer, *Ann. Phys.*, 1927, **389**, 457–484.
98 E. Mátyus, J. Hutter, U. Müller-Herold and M. Reiher, *Phys. Rev. A: At., Mol., Opt. Phys.*, 2011, **83**, 052512.
99 E. Mátyus, J. Hutter, U. Müller-Herold and M. Reiher, *J. Chem. Phys.*, 2011, **135**, 204302.
100 D. Marx and J. Hutter, *Ab Initio Molecular Dynamics: Basic Theory and Advanced Methods*, Cambridge University Press, 2009.
101 B. Kirchner, P. J. Dio and J. Hutter, *Multiscale Molecular Methods in Applied Chemistry*, Springer Berlin Heidelberg, 2012, vol. 307, pp. 109–153.
102 J. E. Stone, J. Gullingsrud and K. Schulten, *Proceedings of the 2001 Symposium on Interactive 3D Graphics*, New York, NY, USA, 2001, pp. 191–194.
103 C. Jarzynski, *Phys. Rev. Lett.*, 1997, **78**, 2690–2693.

Faraday Discussions

PAPER

ExaViz: a flexible framework to analyse, steer and interact with molecular dynamics simulations

Matthieu Dreher,[a] Jessica Prevoteau-Jonquet,[c] Mikael Trellet,[d] Marc Piuzzi,[b] Marc Baaden,*[c] Bruno Raffin,*[a] Nicolas Ferey,[d] Sophie Robert[e] and Sébastien Limet[e]

Received 16th December 2013, Accepted 12th February 2014
DOI: 10.1039/c3fd00142c

The amount of data generated by molecular dynamics simulations of large molecular assemblies and the sheer size and complexity of the systems studied call for new ways to analyse, steer and interact with such calculations. Traditionally, the analysis is performed off-line once the huge amount of simulation results have been saved to disks, thereby stressing the supercomputer I/O systems, and making it increasingly difficult to handle post-processing and analysis from the scientist's office. The ExaViz framework is an alternative approach developed to couple the simulation with analysis tools to process the data as close as possible to their source of creation, saving a reduced, more manageable and pre-processed data set to disk. ExaViz supports a large variety of analysis and steering scenarios. Our framework can be used for live sessions (simulations short enough to be fully followed by the user) as well as batch sessions (long-time batch executions). During interactive sessions, at runtime, the user can display plots from analysis, visualise the molecular system and steer the simulation with a haptic device. We also emphasise how a CAVE-like immersive environment could be used to leverage such simulations, offering a large display surface to view and intuitively navigate the molecular system.

1 Introduction

Molecular dynamics (MD) simulations are nowadays routinely used to study the dynamics of biological complexes in structural biology. The complexity of the biological objects is increasing, involving multiple macromolecules, membranes, the cytoplasm and many other biologically relevant molecular components. The

[a]INRIA, LIG, France. E-mail: Marc.Baaden@ibpc.fr
[b]INRIA, LJK, France
[c]Laboratoire de Biochimie Théorique, CNRS, UPR9080, Univ. Paris Diderot, Sorbonne Paris Cité, 13 rue Pierre et Marie Curie, 75005, Paris, France
[d]Université d'Orsay, LIMSI, France
[e]Univ. Orléans, ENSI de Bourges, LIFO, EA 4022, F-45067, Orléans, France

corresponding system sizes can be tackled through simulation libraries with continuously improving parallelisation strategies and scalability. NAMD[1] and Gromacs[2] are two common packages able to use up to thousands of cores on large clusters. The data produced are massive and complex. For instance, the complete chemical structure of the capsid of the HIV-1 virus has recently been resolved.[3] The molecular model comprises a total of 64 million atoms. To simulate this model, scientists used the Blue Waters supercomputer, the simulation producing about 10 TB per run of 100 nanoseconds of simulated time. The traditional approach to analyse these results by transferring the trajectory files on the scientist's desktop computer is not adopted anymore. In this paper, we explore several alternative approaches to cope with this data deluge. We show that by relying on a modular and flexible software infrastructure, the same application base can be used and adapted for different usages, building a continuum between these approaches.

A promising paradigm to reduce the amount of data to save to disk, while speeding up the data post-processing, is to immediately treat the data as they are produced by the simulation. This also enables the scientist to monitor and steer the simulation execution. This approach is often called *in situ* or live analytics. In this paper, we describe, evaluate and discuss two *in situ* scenarios.

The course of MD simulations can be altered using techniques such as steered molecular dynamics (SMD), where external forces are applied to an object (atom, residue, protein, *etc.*) to probe its mechanical function or to accelerate a process that may be too slow to model otherwise. Such a system was proposed by ref. 4. External forces are defined *a priori* and applied at runtime but without the possibility for the user to change them. Interactive molecular dynamics simulations (IMD) enable more reactive adjustments of the forces applied. The scientist directly applies forces, usually with the help of a force feedback device.[5] Even though IMD has the benefit of parallel executions to speed up the simulation,[6] strong forces need to be applied to enable the scientist to experience the expected behaviour in a reasonable time frame. We present here an early attempt to replay a previously captured IMD session applying lower forces in an SMD mode.

Computational biologists are used to plot graphs from the results of their analysis on trajectories, and to visualise the spatial relationships in molecular systems in 3D. Visualisation is clearly a key component of the analysis process. In this article, we explore several features used to move from traditional graph plotting and 3D navigation in molecular systems towards visual analytics, "the science of analytical reasoning facilitated by visual interactive interfaces".[7]

To date, none of the common software packages in the field of computational biology such as VMD or Pymol permits one to perform such fully integrated analyses, where the results are intrinsically linked to the trajectory and can be acted upon. However, this is a workflow that scientists follow on a regular basis. They assemble time-consuming and cumbersome workflows, working on several trajectories, using several software libraries. We show that our framework complemented with features such as real-time graph plotting and direct trajectory frame selection in graphs, can simplify this process.

Immersive 3D visualisation environments are more and more accessible through 3D TVs, the emerging new generation of head-mounted displays, as well as large displays like cave automatic virtual environments (CAVES) that are being used more frequently in research environments. During the navigation of abstract scientific data, where no implicit or explicit orientations exist, several issues have to be taken into account when considering the user experience and efficiency. Simple navigation tasks can trigger cybersickness,[8] significantly reducing the user comfort and his efficiency to perform a specific task in an immersive environment. Here we try to fill the gap between what a scientist aims for when exploring a molecular complex and the ways he has to reach his targets. We propose several navigation algorithms aimed at lightening the navigation process, as well as to provide useful insights about the observed molecular complex. This can be achieved by basing our navigation algorithms on geometrical features that are present in large molecular assemblies.

2 Molecular dynamics use case scenarios

In this manuscript, we focus on two use case scenarios inspired by ongoing research projects related to the study of membrane proteins in a lipid bilayer environment. Two bacterial systems are investigated: 1) in the context of active iron transport and 2) related to ion permeation through channels.

2.1 The bacterial iron transporter FepA

Iron is a scarce yet essential nutrient for bacteria such as *Escherischia Coli*. Iron transporter membrane proteins attract and transport iron captured by small iron carrier molecules called siderophores. The ferric enterobactin receptor FepA, for instance, secretes the enterobactin siderophore to capture iron. Intriguingly, these transporters are 22-stranded beta-barrels with their N-terminus end folded within the barrel. Therefore, no obvious passages from extra- to intracellular regions are present. The goal in the numerical experiments is to find suitable paths for an iron–siderophore complex to go through the FepA channel. For this purpose and to mimic active transport, we guide the siderophore through the barrel lumen using steering forces. How such forces are monitored and adjusted will be described later on in this article. The FepA channel with the lipid bilayer and the iron–enterobactin complex is made of 69 953 atoms. It has been set up using the ffgmx force field,[9] SPC water model[10] with an ion concentration equivalent to 0.1M NaCl and a DMPC lipid bilayer. The enterobactin parameters and details on its parametrization are available for download at the URL http://www.baaden.ibpc.fr/pub/other/ebact.zip. We used a rectangular box shape with regular periodic boundary conditions. Long range electrostatics are handled with the particle mesh Ewald (PME) technique. Simulation temperature and pressure are controlled by the Berendsen algorithm at 310 K and 1 bar with 1.0 ps coupling constants.[11] A 2 fs integration time step was used. Globally, the simulation protocol was similar to that used in previous studies of related iron transporters such as FhuA.[12,13]

2.2 The pentameric ligand-gated ion channel GLIC

Signal neurotransmission in the brain is an essential biological process whose molecular basis remains to be fully understood. In particular, the family of pentameric ligand-gated ion channels plays a central role and their computational study has been enabled by recent crystal structures of prokaryotic members of the family. GLIC from the *Gloeobacter violaceus* bacterium is the best characterized system to date. It is an ion channel with a central pore formed by its five identical subunits. The channel lets ions flow through when it is in an open state. Its passage to a closed state is called gating and inhibits ion flow. The molecular mechanism of gating remains elusive. We recently observed a link between the orientation of the pore-lining M2 helices of the channel and the number of water molecules in the channel lumen. Here we characterize these phenomena as use cases for the ExaViz framework. The simulation setup of the full 1430 00 atom system uses the AmberGS forcefield[14] with a 5 fs time step. We used a rectangular box shape and regular periodic boundary conditions. Electrostatic interactions were computed using fast smooth PME. Constant temperature and pressure are maintained using the Berendsen algorithm. This setup follows a protocol used in previous studies.[6]

3 ExaViz framework overview

We introduce a unified and flexible framework to interactively explore and analyse large scientific data sets from molecular dynamics simulations. We use this framework to evaluate several strategies in terms of feasibility, efficiency and overall interest for dealing with large scale molecular dynamics simulations of biological systems. The ExaViz framework relies on a component-based software architecture that enables flexibility, reusability and efficiency. Fig. 1 illustrates the ExaViz framework, centered on both the trajectory of the atoms and the user interactions. ExaViz reads trajectories from files or *in situ* while the simulation is running. The 3D visualisation allows us to display the molecular system on various display devices, from desktop computers to Virtual Reality (VR) environments, including camera path computations to assist the user in exploring cluttered systems. The modularity of the framework makes it straightforward to support and integrate existing analysis libraries, to process multiple trajectories or time steps in parallel and to plot the results while the simulation and analyses are computed. The user can interact with the analysis through selections in the analysis graphs or with the simulation by applying forces on atoms through a haptic device. The same base application can be used to explore and refine the comprehension of an MD simulation through interactive *in situ* and postmortem analysis.

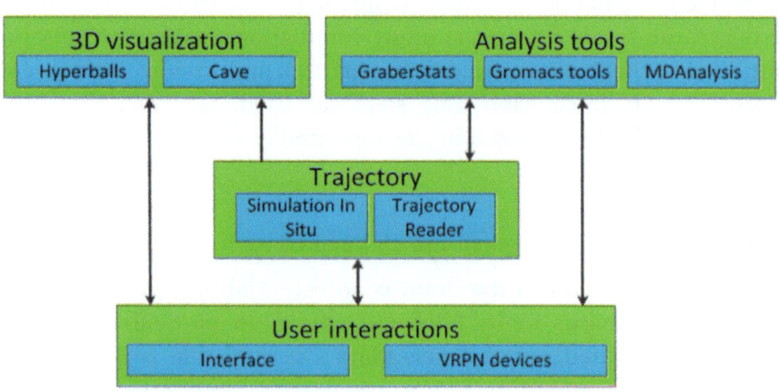

Fig. 1 ExaViz architecture overview.

3.1 A dataflow-oriented framework using the FlowVR middleware

ExaViz relies on the FlowVR middleware that offers a flexible yet efficient environment for designing, deploying and executing distributed analytics workflows on parallel computers. We briefly present this framework. Refer to ref. 15 for more details.

A FlowVR application is described as a dataflow graph where the edges are First In First Out (FIFO) communication channels, and vertices, called *modules*, are codes processing data received on input channels and producing results sent on output channels.

A module is a process or thread running an infinite loop. A module has input and output ports it relies on to send and receive messages in the FlowVR world. A module has no further view of the application. This componentisation enables the reuse of the same modules in different graphs without having to recompile them. A communication channel is a link between an input port and an output port.

The user assembles his application through a Python script. This consists in listing the modules involved and how they connect their ports. Note that a change in the Python script does not require any recompilation, allowing fast prototyping of complex applications.

The application is instantiated for a given target machine by assigning modules to compute nodes and optionally by specifying the cores where to pin them. The execution of the Python script generates various XML files that list the command lines to start the various modules and configure routing tables. To run the application, the user simply calls the *flowvr* command with the application name. The runtime takes care of the application deployment, module coordination and data exchanges, with various levels of optimisation. Because we tried to make FlowVR as unintrusive as possible, the user keeps a strong control and good understanding of the application behaviour.

Modules never communicate directly with each other, the FlowVR logic being implemented outside modules to enforce componentisation and to limit the intrusion into the module executables. A special process called the FlowVR daemon runs on each node hosting a module. Daemons take care of message exchanges between modules.

3.2 *In situ* and in transit live analysis modules

The goal of performing live analysis is to reallocate part of the simulation resources to perform the analysis while the simulation is running. Instead of saving raw data to disks for further post-processing as is classically done, the live analysis paradigm proposes performing data processing as closely as possible to where and when the data are produced.[16] The goal is first to reduce the amount of data to be transferred and stored, the weak link of current and emerging supercomputers. But it also appears that, by properly interleaving the simulation and analysis steps,[17,18] this approach enables us to better use the available resources of the machine. By reallocating part of the simulation resources to run live analysis, the simulation performance may be affected, but with an overall significant time saving when considering both the simulation and post-processing steps. Furthermore, this approach enables monitoring of the simulation state, and, if necessary, the option to take early measures to stop the simulation or change some parameters.[19] This feedback can be used to produce a periodic light report on the current state of the simulation.

We distinguish different mappings on resources for the processes of the analysis workflow: *in situ* running asynchronously on the same nodes as the simulation but often on dedicated *helper cores*, with *in transit* on *staging nodes* dedicated to analysis.

A typical scenario consists of extracting the data from each of the simulation processes, merging them locally on each node and performing some filtering and analysis computations *in situ* on a node-local helper core (Fig. 2). These computations are constrained to work on data from the local simulation processes unless they explicitly exchange data between nodes. For more global computations, the results are usually gathered on dedicated staging nodes (Fig. 3). Typically a ratio of 1/64 is applied between the staging nodes and the simulation nodes. Once the data are on the staging nodes, any kind of treatment can be performed without perturbing the simulation.

Our framework enables us to map modules on helper cores or staging nodes simply by specifying a host name and a core in the Python script. The FlowVR daemon takes care of transparently moving the data to where they are expected.

3.3 Coupling to the Gromacs molecular dynamics simulation engine

Our scenarios currently rely on MD simulations executed with Gromacs. Gromacs runs in parallel with MPI. It also supports other parallelisation paradigms (OpenMP, GPUs) not used in this paper. Gromacs[20] follows a master/slave organisation. The first MPI process (the master) loads the simulation, configures it and propagates all necessary information to the slaves. During the simulation, only the master can access the global state of the simulation. Each process is responsible only for the atoms contained inside its domain (called the *home atoms*). To write the trajectory and other backup files, a global state of the molecular system is gathered on the master node. This operation involves costly communications and synchronisations.

We instrumented each MPI process of Gromacs with FlowVR. Every x iteration of the simulation, each Gromacs process copies the home atom positions into a FlowVR message and puts it on an output port. When data are extracted with FlowVR, the native Gromacs communications to gather data on the master before being written to files are suppressed. We also set an input port per process to retrieve external forces to be imposed to a selection of atoms. This port is activated only for the scenarios requiring steering.

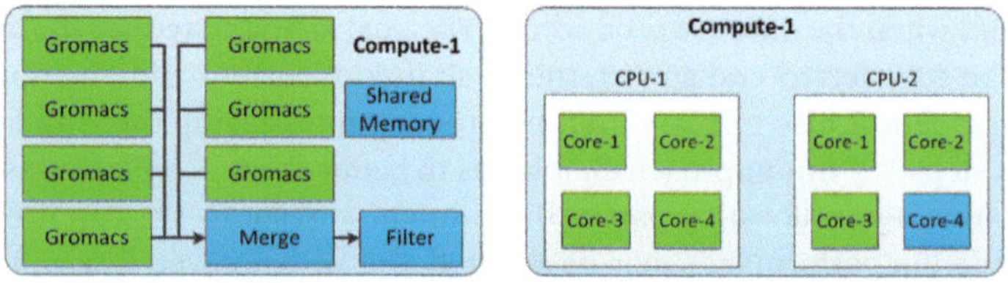

Fig. 2 A node running simulation and *in situ* processes. Each MPI (Message Passing Interface) process of the simulation (here Gromacs) is pinned to one core and the external analytics processes (filter and merge) and the FlowVR daemon share the dedicated helper core.

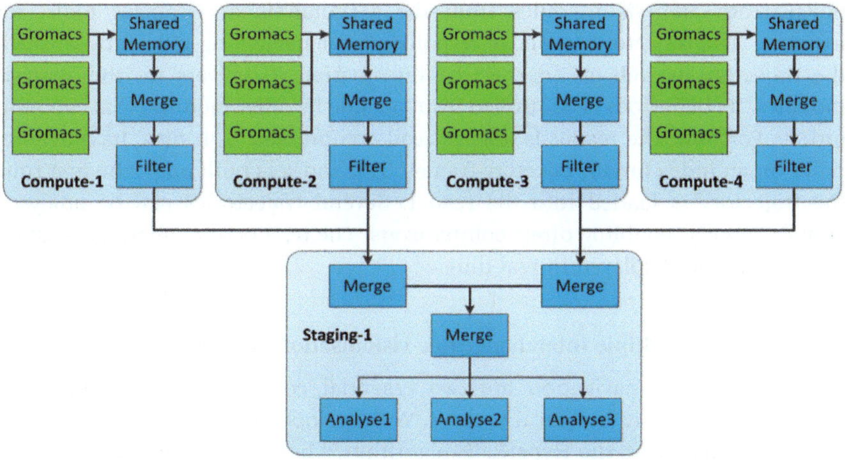

Fig. 3 Example of a simple yet classical live analysis deployment scenario using 4 compute nodes and 1 staging node. Data are first filtered at the node level (*in situ*), then they are sent to a staging node for further treatments (in transit). The output can be saved to disk or sent to a visualisation node to display the result.

3.4 Trajectory analysis with MDAnalysis and Gromacs tools

Software packages, such as VMD,[21] MDAnalysis[22] or the built-in Gromacs tools,[20] enable the analysis of molecular systems and simulation trajectories. MDAnalysis is a Python library providing various classical algorithms to analyse structural and dynamic properties as well as membrane parameters. It provides a powerful selection language inspired by the CHARMM syntax. The Gromacs tools are small executables mainly designed to perform stream-style post-treatments on a trajectory. It is possible, for instance, to filter a trajectory, recenter it on a particular residue or reconstruct a split system divided by the boundary conditions. We turned the MDAnalysis and the Gromacs tools into FlowVR modules to be able to use these tools in an ExaViz application. Both libraries read trajectories from files and write results in output files. We extended their respective file formats with a new format able to read and write data from/to a FlowVR input/output port. To do so, we mapped the FlowVR module API to the traditional file I/O scheme. For live analysis, we map these modules on staging nodes where all data are gathered, as these tools are expected to work on trajectories of the full molecular system.

3.5 Visual and postmortem analysis

Live processing is well suited for analysis that can be performed on a stream of time steps. For more general computations, it makes more sense to save the results to disk and post-process the data in a classical way. Postmortem processing is, of course, particularly useful to execute an analysis that the scientist did not anticipate.

The ExaViz framework can actually be used for such computations. A trajectory reader module has been developed to read trajectories from file and output the needed time steps as FlowVR messages, these messages being processed by downstream analysis modules from the MDAnalysis or Gromacs tools, for instance. The trajectory reader is responsible for navigating efficiently through the trajectory as well as filtering the information it delivers for the analysis modules. Its behaviour is controlled from outside by other modules that can send orders. For now, the reader is able to read compressed XDR files. Its efficiency relies on an index for all trajectory frames. This index is built once, the first time the trajectory is loaded into the reader. Several trajectories can be analysed simultaneously, enabling direct comparisons. The results from analysis modules are plotted and displayed in real-time.

3.6 Access to multiple interchangeable visualisation environments

Visualisation and navigation are two essential components that cannot be dissociated from any analytical process. We developed our own 3D visualisation component based on the HyperBalls algorithm,[23] but other 3D viewers could be supported, taking advantage of ExaViz modularity. Rapid prototyping of original visualisation algorithms can be performed within our UnityMol framework.[24] We primarily target desktop environments. But we also developed a 3D viewer for immersive environments, from stereoscopic devices to CAVE-like systems. Navigating into complex molecular systems is cumbersome. We propose camera positioning and path planning algorithms taking into account the geometric features of molecular complexes (see section 4.1).

4 Immersive and interactive simulations

In order to perform the scenarios presented in the previous section, especially during IMD experiments, providing assisted navigation into a complex 3D molecular system is a crucial point before steering a target. In this part, we present our contribution about navigation paradigms for molecular content exploration that we experimented with in a desktop as well as in a CAVE-like virtual environment.[25] Next, we especially point out the advanced techniques we applied to provide an immersive collaborative context for simulation analysis, and the methodological and technical aspects underlying the steering of a simulation in progress.

4.1 Content-aware/task-aware navigation

Many large biological 3D structures are constructed around specific symmetry layouts.[26] Symmetry is not only a spatial feature of molecular complexes, but it also plays a significant role for the stability and the function of such complexes. Symmetries found among large molecular complexes can be used as guidelines to extract pertinent navigation paths. A few examples of symmetric complexes found in the cell are illustrated in Fig. 4.

From these geometric features, we can anticipate certain specific navigation paths that would drive the user during the exploration of a molecular structure. The symmetry information we use all along our navigation paradigms are provided as a mandatory step right before any initiation of the visualisation process. It consists of a center or symmetry axis and a list of angles/translations linking each symmetry-related monomer of the molecular system studied.

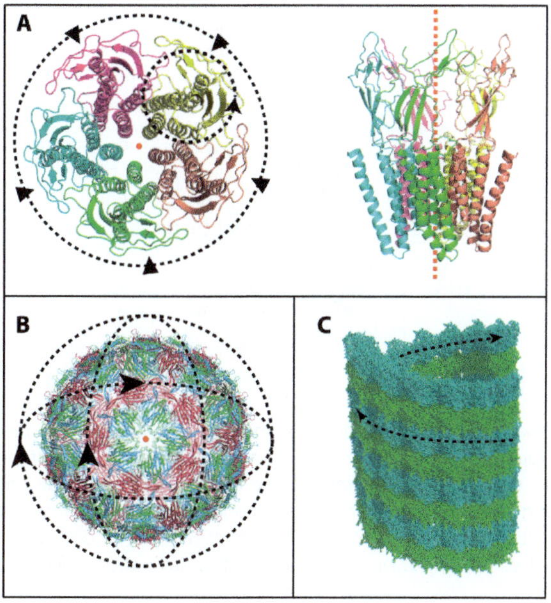

Fig. 4 Several types of symmetries found in molecular complexes. (A) GLIC, transmembrane protein composed of 5 components with a single central symmetry axis. On the left, top view of the protein, on the right, side view. (B) Virus capsid with two types of components presenting a symmetry center overlaying the gravity center of the virus. (C) Tubulins assembling into the shape of a microtubule with a screw-like symmetry.

Symmetry information can be manually extracted or calculated by off-line analysis tools such as SymD.[27]

4.1.1 Camera path for molecular exploration. To maintain a uniform and coherent point of view for the user, the symmetric nature of the complex is used. More particularly, a symmetry axis (or center for specific cases) is extracted from the 3D structure of each complex in a manual or automated way. The symmetry axis is considered as the link between the "floor" and the "roof" of the virtual scene and the protein of interest is observed and explored following the metaphor of a tower.[28]

Circular paths around the structure are set up based on the symmetry axis and the bounding box of the protein. The *up* vector of the camera stays parallel to the symmetry axis all along the navigation and always faces the symmetry axis/center. Camera orientations and movements are then directly set with respect to the relative position of the camera and the symmetry axis/center. Additional exploration paths along the symmetry axis may exist when a pore or a tunnel are present in the molecular complex. The GLIC pore example is illustrated in Fig. 5.

4.1.2 Semi-automatic point of view extraction. We developed an algorithm to compute the least occluded position to place the point of view and observe a given region of interest. Starting with the center of an area identified by the user, a sphere is created around this point with a radius matching a configurable distance between the camera and the target. From each element of the sphere surface, a ray is cast toward the target. If a ray crosses a cell of the grid with an atom in it, the grid element corresponding to a surface starting point is considered as occluded. At the end, each element of the surface is associated to a 2D

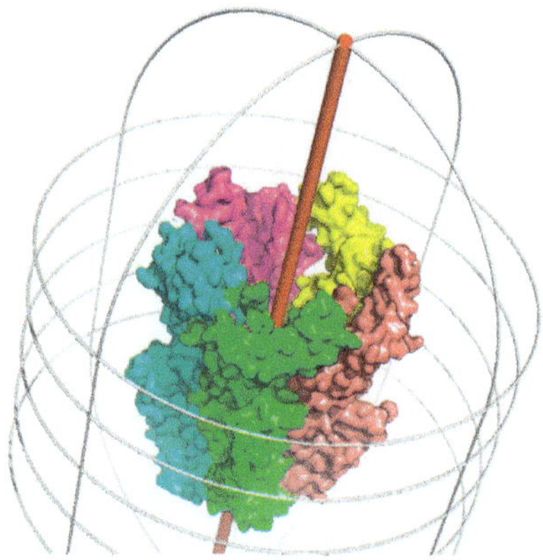

Fig. 5 View of navigation paths when a symmetry axis (in red) has been detected. The molecular complex is coloured by chain and several examples of possible paths are rendered in light grey. When constrained, the camera follows these paths for which the radius is changing according to the camera distance to the symmetry axis. At any moment, the camera is facing either the center of the symmetry axis or one of its end points when the camera reaches the upper or lower parts of the navigation area (semi-circular paths).

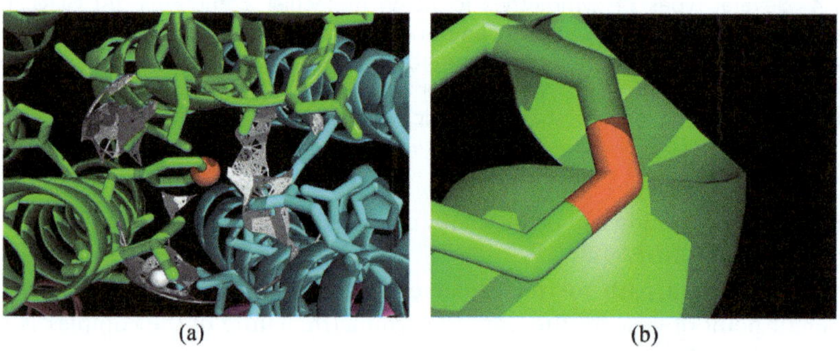

Fig. 6 View of the rendering process used to extract the best point of view when targeting a precise atom (in red) in a molecular system. (a) In white, all the points that belong to a sphere of a radius matching the distance between the target and the future camera position. The center of the largest cluster (white sphere) is output as the best point of view to visualise the target. (b) Camera view when its position is brought to the white sphere location shown in (a).

table and we extract the area with less occluded neighbours (Fig. 6(a)). The centers of these areas, once re-transformed into 3D coordinates, yield the observation points with the least occlusions to observe the target (Fig. 6(b)). The user is then able to choose among the optimal points of view by modifying the position and the orientation of the camera.

4.1.3 Navigation jumps between symmetry-related equivalent positions. A frequent exploration task is to visualise a specific region of a molecular system, repeated along symmetry-related positions, for example, the five monomers of the GLIC receptor. The movements to reach these regions involve quick and precise motions to be performed in order to simultaneously keep some spatial landmarks and reach the appropriate point of view in the middle of thousands of particles.

The possibility to jump from one monomer site to another in a smooth and quite instantaneous way is particularly interesting for the comparison of binding sites or dynamic structures involved in complex stability. Targeting a specific region of a monomer includes four steps implemented in our navigation paradigm: face the region of interest using guided navigation features (Fig. 4), selection of the interesting area before computing the best point of view (as presented in the previous section), save the point of view, and jump to the next monomer thanks to the available symmetry information.

4.2 Immersive and collaborative exploration

Most of the navigation features evoked in the previous section were motivated by the immersive context offered by virtual environments. These are especially adapted for interactive simulations and for *postmortem* simulation analysis for several reasons. The stereoscopic 3D rendering allows us to explore complex biological structures that are intrinsically 3D objects. Moreover head-tracked stereoscopy allows the user to turn around the 3D object and explore it from different points of view just by walking naturally in the immersive system, as in a real-life context. When navigation is not required, the head-tracked stereoscopy features allow the user to focus on his molecular structure without being disturbed by interaction devices required to navigate within a virtual scene. We highlight the fact that these navigation features were designed to be independent from the nature of the control device. In this way, the mouse and keyboard used as input devices to control navigation in a desktop context are substituted by virtual reality devices such as a 3D mouse, joystick and advanced head-tracking techniques to control navigation features in an immersive context.

It is important for a scientist to easily share an experimental result or observation. Immersive environments address this question by implementing multi-collaborative techniques that allow sharing of a working space between two users with their own point of view. To do so the active and passive filtering capability of a CAVE-like system has been used and developed in order to provide the collaborative capacity, while gathering the best navigation conditions to a group of users. Like the advanced systems proposed by Froehlich,[29] our immersive system is able to provide two head-tracked stereoscopic points of view, with each user having his/her own molecular representation of the same structure. This setup provides the first basic features for collaborative work where the real hands of the users are co-localised with the 3D pointed virtual object, and then the pointing place is coherent for the two users' points of view.

We can also extend our immersive environment with specific haptic devices for 3D interactive picking of atoms. A test case is shown in section 5.1.1 focusing on an interactive simulation. This provides proprioceptive feedback of the force that steers the simulation which in turn can drive user interaction with the simulation. We integrated an advanced haptic device that allows the user to access the whole workspace of the immersive environment, automatically following the user during the haptic picking of an atom (see Fig. 7(b)).

4.3 Human steering

A convenient task to perform interactively is to steer a system, possibly in an immersive context (see Fig. 7(b)). This allows the user to immediately adapt and

Fig. 7 (a) Two users in our collaborative CAVE-like system, each with their own head-tracked 3D point of view, allowing pointing to discuss about a shared 3D object. (b) Our advanced haptic device allows users to access the whole workspace of the immersive system.

react to the behaviour of the molecular system and draw benefit from human intuition and instantaneous decision-making. Hence even very complex manipulations can be carried out. However, the concept of IMD[5,30] is limited by the simulation speed and human time. Typically, with a simulation running at 500 Hz and an integration step of 2 fs, we can only simulate about 1 ns in a 15 min interactive session. This considerably limits the range of possible applications since most biologically relevant phenomena require at least several hundreds of nanoseconds or even much longer timescales to occur. To mitigate this problem, significant forces are sometimes applied to speed up the biological process with the risk of affecting the correctness of the simulation. This important issue will be discussed in more detail in subsequent paragraphs about *in situ* steering.

From a technical point of view, traditionally, an IMD system sends forces to one process of the simulation (called the master node) and retrieves the atom positions for visualisation. This scheme implies that the simulation broadcasts the user forces to all the other processes and gathers the atom positions at each step, which are two costly operations. Experiments have shown that this scheme does not scale up to large systems and/or high levels of parallelism. In previous work,[6] we used the coupling method described in section 3.3. The key feature of this method is that the broadcast of the forces and the gathering of the atom positions are done asynchronously outside of the simulation. This ensures a minimal impact on the simulation performance by avoiding costly blocking operations during the simulation execution and a severe bottleneck on the master node. With this system, we were able to steer molecular systems with up to 1.7 million atoms at about 25 Hz using 384 cores for the simulation with a cost of 17% on the simulation performance (compared to Gromacs at equal core count without I/O).

In contrast to human IMD steering, the SMD approach[4] uses a similar concept but the forces are described by a script or algorithm. This allows the running of long simulations in batch mode on a cluster. However, scripting complex paths can be very tedious compared to a direct interactive steering.

5 *In situ* steering, reporting and analysis

We present two live analysis scenarios. The first one draws a continuum between IMD and SMD approaches. We propose to automatically steer a molecular system based on a user-defined path captured through a preparatory IMD session. The second one is a more classical scenario. We show that it is possible to filter and save to disk a trajectory ready to use for postmortem analysis, with a significant performance gain compared to Gromacs' native I/O system. This step addresses the frequent tasks of either reducing the trajectory size or increasing the saving frequency for specific trajectories of the subsystems of interest.

5.1 Live SMD steering

5.1.1 Moving from IMD to SMD. Our method relies on three steps. First, the user performs an interactive session with our IMD system, most likely applying a level of force that is too high to be realistic. However, this session allows the user to already eliminate some possible paths. Our previous work[6] demonstrated that our IMD system is able to run large molecular simulations interactively, allowing a wide variety of scenarios. Next, based on the trajectory extracted from this interactive session, the user defines a series of key points on the path called targets. Once these targets are identified, the user can relaunch his simulation in SMD mode. In this mode, the steered complex is pushed successively toward the targets.

To automatically push the steered complex, we add to our IMD system a component called TargetSystem. This component has three main functionalities: tracking the steered complex, tracking the position of the targets and emitting new forces. At each output step, the TargetSystem receives the atom positions from the simulation by the same communication channel as the visualization in the IMD system. The current position of the steered complex and the targets are extracted from the atom positions using the selection capabilities of MDAnalysis. The direction of the emitted force is then defined by the difference in positions between the current active target and the center of mass of the steered complex.

During the automated sessions, several measures are displayed to provide a quick overview on the state of the steering process. Information such as the current level of applied forces, distance to the next target or the estimated time remaining until completion allow the user to see easily the behavior of the steering process and to decide quickly if the system can continue on its own. The guiding system also forces Gromacs to produce a checkpoint file when the steered particles reach a target, making it easily possible to resume an automated session at a certain point. In a batch mode session, this information is gathered inside a simple report file accessible by the scientist at any time of the simulation.

Note that this application can run both in batch mode or during an interactive session. In an interactive session, the system behaves like a traditional IMD system with visualisation and the use of a haptic device in addition to the TargetSystem component. The user has the possibility to activate the automatic steering or not, depending on his needs. In the case where the automatic steering is activated, the user still has the possibility to apply forces by himself, as the forces emitted by the haptic device are prioritised over the automated forces. This feature is very useful in case the steered complex gets stuck on its path. Manual guidance allows the user to complement the automated system and reposition the complex in the proper path, or find an alternative possible path. To run in batch mode, we simply disconnect the visualisation and haptic device from the application. The rest of the application can run without any intervention from the user.

5.1.2 Experiments on the FepA channel. We ran our system on the Grid'5000 testbed as a computational cluster plus a visualisation node. The hardware used for the visualisation is a Xeon E5530 quad core CPU at 2.40 GHz with a GeForce 680 GTX graphics card. The computational cluster used is composed of 72 nodes, each node running 2 Xeon E5520 quad core CPUs at 2.27 GHz, 24 GB RAM. Communications between the computational cluster nodes go through a QDR Infiniband fabric (40 Gbits/s) while we only have a DDR (20 Gbits/s) Infiniband interconnection between the computational cluster and the visualisation node.

As a use case, we extend our previous experiments on the FepA channel. In ref. 6, we found two possible paths with our IMD system, but those required a level of force considered too high to be realistic (2000 pN per atom). We tried interactively to follow a similar path with a lower amount of force, however this was difficult without a visual guide indicating the previously found path and because of the slow motion of the complex.

To run the simulation we used 16 nodes from the computational cluster with 7 cores per node for the simulation and 1 dedicated core. The dedicated core runs the FlowVR daemon, a merger to aggregate the data from the 7 local Gromacs processes, an atom filter to avoid sending unwanted atoms to the visualisation such as the water molecules, and an iteration filter to send only one frame every x frames. The combination of both filters at the node level ensures minimum network traffic minimising our impact on the simulation performance. The other components (visualisation, haptics, targeting system) run on the visualisation node, which is our staging node. With this setup, we obtain a simulation running at 230 Hz with data sent every 10 frames to the staging node.

In our first experiments, we launched our system with the predetermined path and the same amount of force as in our IMD session (2000 pN per atom). We have been able to reproduce this experiment multiple times automatically. Surprisingly the automatic system is faster than the human-guided one to perform the same experiment (3 min for the user against 2 min for the automated system). This is most likely due to the fact that the human constantly deviates from the path, whereas the automatic system stays on target. As a second step, we launch the same setup with 1500 pN per atom, a level of force for which we could not complete the process in a reasonable time interactively. Our system was able to follow the same path as the interactive one. However, the complete experiment took about 10 min to complete. During the experiment, we saw that the complex was stuck at certain steps. After a period of time, the side chains reoriented in a way that allowed the complex to continue on its path. As in the previous experiments, we were able to perform the same run multiple times without human intervention. We tried the same experiment with 1000 pN per atom using the same setup as before. In this case, we encountered some problems related to the deformation of the channel over time. After 15 min, the channel starts to translate and slightly modify its shape. As our path definition is for now absolute in space, the path becomes irrelevant after these transformations. However, we were able to push the complex inside the channel in the first period of the simulation where the channel is stable compared to its initial state. These observations tend to indicate that the complex could go through the entire path given enough time.

5.1.3 Discussion. With long simulations arise two major problems in our guiding system: conformational changes and jumps of the molecule due to the periodic boundary conditions. These issues pose mainly technical challenges. A more fundamental issue relates to the correctness and repeatability of such experiments.

To solve the conformational problem, we are currently working on a way to describe the target positions relative to some reference atom positions in the system. In this way, the target position would follow the same transformation as the reference atoms. However, there is no obvious way to describe a relative position which would be robust enough in all cases. For instance, if we take as a reference several residues of the backbone of the channel, our target will follow the global movement of the channel but will miss the local changes inside the channel. On the other hand, if we take local points only to describe the target, the target position might become unstable if one of the reference points moves significantly.

A mandatory condition to set up targets properly is to preserve the integrity of a reference structure. However, the periodic boundary conditions of the simulation can split this structure into one or several image boxes any time during the execution. This raises two major problems for the targeting system: first the computation of the target position must take into account the various image domains of the simulation, second the emitted force direction must also take into account this periodicity. To solve this problem outside of Gromacs, extra information is required: the bounding box shape of the simulation to detect and undo the jump into image space, and topological information of the system to preserve the integrity of a reference structure. The *trj_conv* tool from Gromacs is commonly used to post-process a trajectory based on this information. It possesses various useful functionalities to rearrange a trajectory-like centering, correcting the jumps and splits due to periodic boundary conditions, filtering atoms, *etc.* A promising solution to our problem is to integrate this tool in our framework and to plug it with the correct options between the simulation and the targeting system. This way, the targeting system can simply ignore the periodic boundary conditions since they are corrected by the *trj_conv* module. Note that this is possible because *trj_conv* only translates atoms but does not apply any rotation. Therefore the direction of the emitted forces can be directly sent to the simulation. Otherwise it would be necessary to undo the rotation of the emitted force before sending it to the simulation. This solution is currently tested on the FepA channel experiment. A fundamental question arises from the level of force applied to the simulation system. Although hundreds to a few thousands of pN have previously been used for similar experiments on ligand dissociation,[31] peptide membrane penetration,[32] glucose transport[33] or substate entry into an active site,[34] the applied forces remain almost an order of magnitude higher than the forces encountered in natural biological systems. Therefore, the correctness of the results should be checked very carefully, for instance, by carrying out numerical experiments with various levels of force and by repeating an experiment several times. Repetition is particularly important for human-guided experiments, and we typically carry out 3 to 10 repeats to account for the inherent variability of the human intervention.

5.2 Post-treatment and analysis with Gromacs and MDAnalysis tools

Handling of trajectory files is a major problem for long simulations. A trade-off must be found between the trajectory file size, the impact of I/O operations on the simulation performance and the timescale of the biological phenomena. With this application, we propose a possible solution for these three points. The goal is to enable the scientist to output a complete trajectory at a common interval and at the same time much smaller filtered trajectories but with a higher output frequency.

5.2.1 Application setup. In this experiment, we use the GLIC ion channel as an application example. One of the points of interest of this simulation is the presence or not of water inside the channel pore and the shape of the pore-lining M2 helices according to the presence of water. Based on this analysis, we would like to produce a lighter trajectory with only the M2 helices, the 1000 closest water molecules from the channel center and the ions in the system. The trajectory should be centered on a residue inside the channel to avoid shifting of the trajectory. Overall, each frame is reduced from 143 000 atoms to about 3000 atoms. To filter this trajectory, we use the Gromacs tools *trj_order* and *trj_conv* integrated within our framework. Simultaneously, a complete trajectory is produced at a normal output frequency since not all analyses could run with the light trajectory. We expect the *trj_order* tool to run slower than the simulation output rate since heavy operations such as atom sorting are required to select the 1000 closest water molecules. To tackle this problem, we create multiple instances of the *trj_order* module and add a rooting component. This rooting component dispatches incoming frames on the *trj_order* modules in a round-robin fashion. A similar component is placed after the *trj_order* modules to reorder the frames and send them to the *trj_conv* module.

The experiments ran on Froggy, a 138 compute nodes cluster from the Ciment infrastructure. Each compute node is equipped with 2 eight-core processors Sandy Bridge-EP E5-2670 at 2.6 GHz, 64GB of memory. Nodes are interconnected through a FDR Infiniband network. FlowVR 2.1 and Gromacs 4.6 are compiled with Intel MPI 4.1.0. For all experiments, Gromacs runs a simulation with the GLIC model composed of about 143 000 atoms and with PME nodes activated. We use up to 32 nodes for the simulation as compute nodes and one extra node as a staging node. On each compute node, we set 15 Gromacs MPI processes and one dedicated core. The dedicated core runs the FlowVR daemon and a merger. The data are then aggregated toward the staging node. On the staging node, we run a *trj_conv* instance with no specific option to output a complete trajectory in XTC format. Another *tjr_conv* instance and a *trj_order* filter are set up to produce the filtered trajectory. Every 10 simulation steps, atom positions are extracted from the simulation and sent to the staging node. The pipeline generating the filtered trajectory receives each of the extracted frames and writes them. The *trj_conv* module for the complete frame outputs a frame only every 100 received frames, which corresponds to a usual output frequency of every 1000 frames in Gromacs.

5.2.2 Output frequency evaluation. Table. 1 shows the Gromacs running frequency for various configurations. *Gromacs-0-16cores* (resp. *Gromacs-0-15cores*) indicates the performance of Gromacs running on 16 cores (resp. 15 cores) out of 16 available and without performing any I/O. These values set the upper

Table 1 Gromacs performance with native writing methods compared to our framework (higher is better). The GLIC model (143 000 atoms) is used with a fixed configuration of 16 nodes (256 cores max). *Gromacs-X-Ycores* write every *X* step with the Gromacs native writing method using *Y* cores per node. *Exaviz-full-X-filtre-Y-Zcores* writes a complete frame every *X* frames and a filtered frame every *Y* frames with our framework writing method using *Z* cores per node. Relative costs are given compared with the same core count

Setup	Performance (Hz)	Relative cost
Gromacs-0-16cores	272	—
Gromacs-0-15cores	280	—
Gromacs-1000-16cores	271	0.36%
Gromacs-100-16cores	261	4.04%
Gromacs-10-16cores	187	31.35%
Exaviz-full-10-15cores	273	2.5%
Exaviz-full-1000-filtre-10-15cores	273	2.5%

performance bounds we should try to stay close to when activating data saving. When Gromacs writes the atom positions to disk every 10 steps, the performance drops significantly (31% drop on *Gromacs-10-16cores*).

The line *Exaviz-full-10-15cores* reproduces the Gromacs file writing pattern with FlowVR. Data are gathered on the staging node and written in an XTC file with the Gromacs *trj_order* tool modified as described in section 3.4. The curve *Exaviz-full-1000-filtre-10-15cores* reproduces Gromacs file writing every 1000 steps and the writing of the filtered trajectory every 10 frames. Five *trj_order* tools were created in parallel to absorb the production rate. Both *Exaviz* tests have the same cost on the simulation, which is expected since the major cost is the data transfer that is the same in both cases. Overall, these results show that our writing method is very close to *Gromacs-0-15cores* (2.5% drop) despite the very high output frequency from the simulation (1 frame every 10 frames).

5.2.3 Scalability evaluation. In these experiments, we evaluate the scalability of our framework to write a trajectory at high frequency. Due to its relatively small size, the GLIC system is not suitable to scale well on a large number of cores. Instead we use a molecular system specifically designed to scale up to several thousands of cores. We run a Martini simulation[35] with a patch of 54 000 lipids representing about 2 100 000 coarse-grained particles. In the previous section, we demonstrated that our framework enabled us to produce extra trajectories at no extra cost. In this experiment, we produce one complete trajectory at high frequency.

Fig. 8 shows the Gromacs running frequency on our benchmark model. The curve notations follow the same conventions as previously. In this case, the curve *Gromacs-100-16cores* scales poorly beyond 512 cores due to the very important amount of global communication to write the trajectory. The curve *Exaviz-full-100-15cores* demonstrates that our framework is able to scale up to 1664 cores even with this size of molecular model and output frequency. We note a slight cost compare to *Gromacs-0-16cores* which is mainly due to the reserved dedicated core on each node and the extra network traffic to gather the data on the staging node.

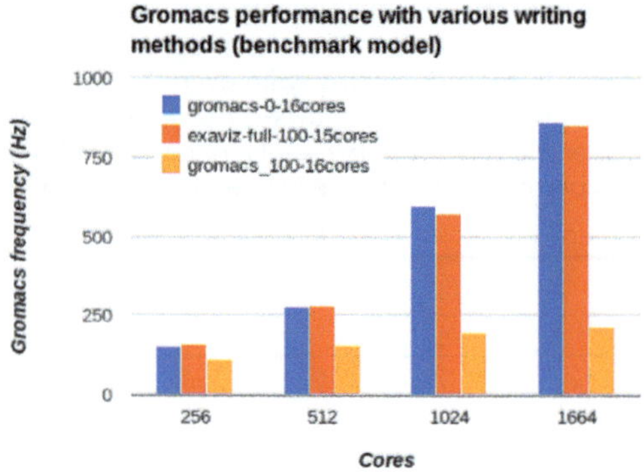

Fig. 8 Gromacs performance with various writing methods. The benchmark model (2M1 atoms) is used. *Gromacs-X-Ycores* write every *X* steps with the Gromacs native writing method using *Y* cores per node. *Exaviz-full-X-Ycores* write a complete frame every *X* frames with our framework method using *Y* cores per node.

5.2.4 Discussion. These experiments demonstrate two important features of our framework to help resolve the necessary tradeoff for trajectory writing. First, it is possible to write multiple trajectories with filtering at the same cost of writing one complete trajectory. This alleviates the trajectory size problem. Second, we are able to output trajectories at a high rate, even for the complete trajectory, at the cost of one dedicated core per node, considerably improving the Gromacs writing performance even on a high number of cores.

6 Visual and postmortem analysis

In addition to the 3D visualisation of the molecular system, the scientist can add several other windows to complement his visual analytics dashboard. The results of the analysis performed on staging nodes can be streamed to the visualisation node for on-line plotting. These features are available for live as well as postmortem analysis. For postmortem analysis, extra features enable navigation through or selection of frames on one or several trajectories.

6.1 Trajectory player

The trajectory player module is a graphical interface that allows the user to navigate through the trajectory. The user interface resembles a DVD player where it is possible to go forward and backward, to accelerate, slow down or pause the reading. It is possible to loop on a selected part of the trajectory. This module mainly controls the behaviour of the trajectory reader described in Section 3.5. The trajectory reader is connected to the analysis and visualisation modules and provides the same functionalities as in an *in situ* and interactive context for atom filtering. The quality of the postmortem analysis relies on the ability of the trajectory reader to quickly access the different trajectory frames. For now, we performed experiments on trajectories ranging from hundreds of Megabytes to several Gigabytes and we obtained a random access to the frames ranging from 0.01 s for small systems (about 10 000 atoms) to 0.2 s for several million atom systems. However, the current trajectory reader must be improved to handle very large trajectories of several Gigabytes. This can be achieved by using techniques such as those described by Stone *et al.*[36] within our framework.

6.2 Real-time plots

It is common to use families of function graphs in the visual analysis process.[37] Usually, when the scientist requests a trajectory analysis, he only gets the result at the end of the trajectory processing. If this may be acceptable for postmortem analyses, it does not enable monitoring of the simulation during its execution. ExaViz enables plotting of the streams of data received from analysis modules in real-time. The scientist gets the results as soon as they are available, enabling him to make putative decisions for further analyses or reparametrisation of the trajectory.

To implement this feature in the FlowVR layer, we connect a module for each analysis script and a module for displaying graphs called *GraberStats*. *GraberStats* is based on the Qwt library. In addition to a framework for 2D plots, Qwt provides scales, sliders, dials, compasses, thermometers, wheels and knobs to control or display values, arrays, or ranges of type double. It is possible to display multiple plots on the same graph to superimpose the results and to compare them quickly. Note that this module is available in any context except batch mode. A future version will allow writing of the graphs into images to be compatible with batch mode constraints.

An example was performed for the analysis of the GLIC ion channel. This structure has both open and closed states; the transition mechanism between the different states is unknown. Understanding this transition is crucial as potentiating molecules such as anaesthetics and alcohols are believed to affect transition barriers or the relative free energy of states. To investigate these dynamics, we have performed extensive simulations of GLIC. Pore hydration seems to be related to the pore organisation and mainly the pore lining helix M2. To characterise the conformation of this helix, we have implemented two MDAnalysis scripts.

First, we have computed the tilt angles, and decomposed them into two angles: the radial (θ) and lateral (π) tilt angles of M2 around its center of mass with respect to the overall channel pore axis (Fig. 9). These angles are useful to describe the state of the pore. Then, we counted the number of water molecules in the pore zone between the two hydrophobic residues I9′ and I16′ (Fig. 9). This region was the most affected by the pore dewetting. These two scripts have been applied on a trajectory during the reading by the trajectory player. The results were displayed on three graphs shown in Fig. 10. In the former there is a curve for each M2 helix indicating the lateral angle around its center of mass. In the second it is the same for the radial angle. Finally on the last are the numbers of water molecules within the pore. To speed up the analysis process, we use the same strategy as in Section 5.2.1 and deploy multiple instances of the analysis modules if enough CPU resources are available. Note that these analyses are performed in a post-mortem context for this example however they can be computed *in situ* as well.

6.3 "Smart" selections linking analysis data to trajectory frames

To make the analysis process less cumbersome, it is common to investigate subgroups of the full simulation dataset.[38] We integrated the possibility to select a

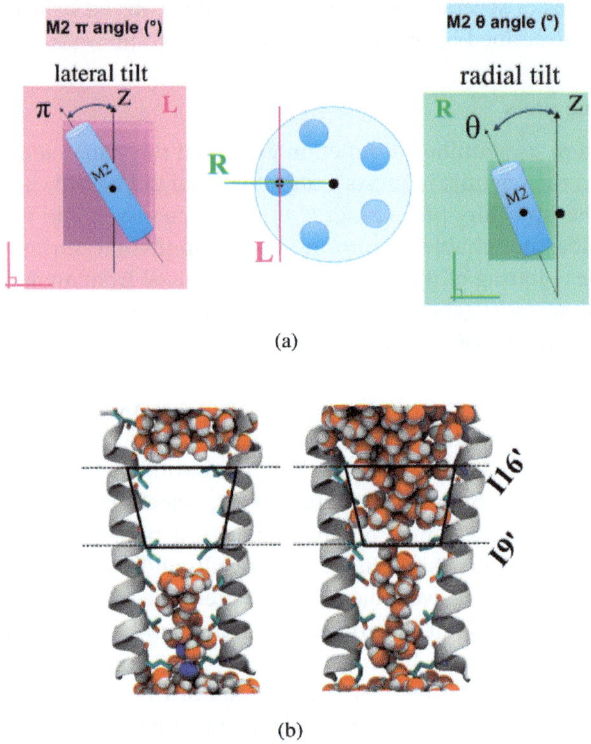

Fig. 9 Analysis examples for the GLIC ion channel case, which is an ion channel with a central pore formed by its five identical subunits as shown in Fig. 4(a).

pool of frames in a graph, and to trigger the computation of a new plot for this selection. Such selections can be repeated to refine the analysis. Selections can be done by choosing a range in the frame numbers (x-axis) and/or in the analysis values (y-axis) directly in a graph window (Fig. 10). Once the selection is made, zoom operates in all the graphs so that results are available only for the selection. The resulting frame numbers are sent back to the *TrajReader* to reduce the trajectory to the selected frames. This sub trajectory can be replayed without recomputing the analysis since the results do not change from the initial trajectory. To run a new analysis on this sub-trajectory, the scientist has to save the selected frame number and relaunch the application with a new analysis pipeline and the selected frames loaded.

In the GLIC example where we plotted the number of water molecules in the pore, the scientist could, for instance, select all the frames that do not have any water molecules in the pore. The resulting trajectory will therefore enable the visualisation of the protein in a closed and dewetted state, allowing for focusing on any interesting features.

6.4 Concurrent analyses on multiple trajectories

As shown in Fig. 10, the *GraberStats* module is able to display multiple curves in the same window. This allows an easy comparison between results. Based on this

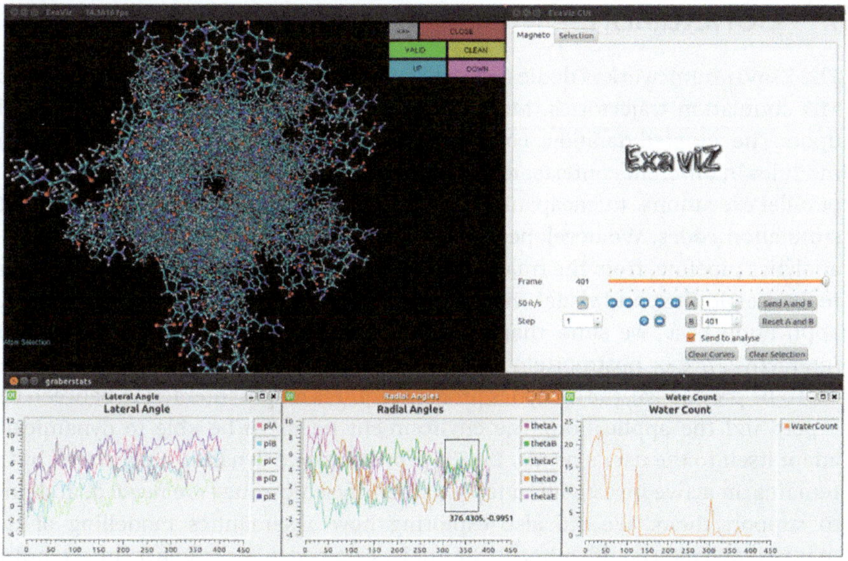

Fig. 10 ExaViz visual interface with the *TrajReader* and *GraberStats* modules, and two analysis scripts for the GLIC ion channel. The user selects several frames based on the frame numbers (*x*-axis) and the analysis values (*y*-axis).

feature, we are able to provide a multi-trajectory support. To do so, we simply duplicate the analysis pipeline segment from the *TrajReader* to the analysis modules and connect each output port of the analysis modules to the *GraberStats* instances. Then the scientist simply has to attach both data sources to the same window to obtain a comparative view of multiple trajectories.

Fig. 11 shows the application graph that computes the number of water molecules inside the GLIC channel from *N* different trajectories. The full analysis pipeline is duplicated as many times as trajectories are available for analysis. To support the selection of frames from the *GraberStats* window, the developer has to reconnect each selection output associated to a data source to its rightful origin, *i.e.* to the correct *TrajReader*. Note that this application scheme would be exactly the same for the *in situ* analysis of multiple simulations, except that the *TrajReader* would be replaced by multiple simulations.

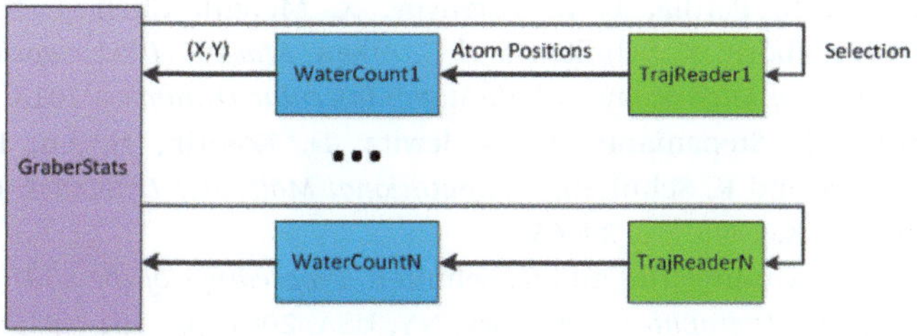

Fig. 11 Example of a multi-trajectory application. The number of atoms in the GLIC channel is counted for *N* simulations and sent to the *GraberStats* module. The analysis pipeline is simply duplicated to handle multiple trajectories.

7 Conclusion

The ExaViz framework is dedicated to the extraction, analysis and visualisation of MD simulation trajectories. Modularity is the key concept that ExaViz is built upon. The adopted dataflow, component-oriented approach enables the reuse of modules in different contexts, to distribute components on a large scale to enable parallel executions, to encapsulate existing libraries or to couple to existing MD simulation codes. We developed MD specific modules for different levels of the analytics pipeline, from the trajectory reader or *in situ* data extraction, to analysis, real-time plotting, and visual navigation in 3D molecular systems. From the same application base, we show that the scientist can set up different scenarios for interactive, live or postmortem trajectory analysis and steering.

To be productive, the analytics process requires a tight interaction between the expert and the application. The environment needs to be able to dynamically adapt itself to the user's needs. ExaViz provides some dynamic behaviours, but it remains an active investigation topic to identify which ones are needed, and how to support them. We are also exploring how a semantics modelling of the manipulated data could enhance the analytics process, for instance, by extending the concept of smart selection. We plan to publicly release the ExaViz code in the near future and hope it will foster the development of new scenarios and features.

Acknowledgements

This work was supported by the French Agency for Research (Grant "ExaViz", ANR-11-MONU-003) and the "Initiative d'Excellence" program from the French State (Grant "DYNAMO", ANR-11-LABX-0011). Experiments were performed on the Grid'5000 experimental testbed (https://www.grid5000.fr), and the Froggy platform of the CIMENT infrastructure (https://ciment.ujf-grenoble.fr//). We thank Dr Samuel Murail and Benoist Laurent for stimulating discussions about the analysis of the GLIC system.

References

1 J. C. Phillips, R. Braun, W. Wang, J. Gumbart, E. Tajkhorshid, E. Villa, C. Chipot, R. D. Skeel, L. Kal and K. Schulten, *J. Comput. Chem.*, 2005, **26**, 1781–1802.

2 B. Hess, C. Kutzner, D. van der Spoel and E. Lindahl, *J. Chem. Theory Comput.*, 2008, **4**, 435–447.

3 G. Zhao, J. R. Perilla, E. L. Yufenyuy, X. MengB. Chen, J. NingJ. Ahn, A. M. Gronenborn, K. Schulten and C. Aiken, *Mature HIV-1 Capsid Structure by Cryo-electron Microscopy and All-Atom Molecular Dynamics*, 2013.

4 S. Izrailev, S. Stepaniants, B. Isralewitz, D. Kosztin, H. Lu, F. Molnar, W. Wriggers and K. Schulten, *Computational Molecular Dynamics: Challenges, Methods, Ideas*, 1998, pp. 39–65.

5 J. E. Stone, J. Gullingsrud and K. Schulten, *Proceedings of the 2001 symposium on Interactive 3D graphics*, New York, NY, USA, 2001, pp. 191–194.

6 M. Dreher, M. Piuzzi, A. Turki, M. Chavent, M. Baaden, N. Férey, S. Limet, B. Raffin and S. Robert, *International Conference on Computational Science, ICCS 2013*, 2013.

7 J. Thomas, *Illuminating the Path: The Research and Development Agenda for Visual Analytics*, IEEE Computer Society Press, 2005.
8 J. J. LaViola, Jr., *ACM SIGCHI Bulletin*, 2000, **32**, 47–56.
9 E. Lindahl, B. Hess and D. van der Spoel, *J. Mol. Model.*, 2001, **7**, 306–317.
10 H. J. C. Berendsen, J. P. M. Postma, W. F. van Gunsteren and J. Hermans, *Intermolecular Forces*, 1981, pp. 331–342.
11 H. J. C. Berendsen, J. P. M. Postma, W. F. van Gunsteren, A. DiNola and J. R. Haak, *J. Chem. Phys.*, 1984, **81**, 3684–3690.
12 J. D. Faraldo-Gomez, G. R. Smith and M. S. P. Sansom, *Biophys. J.*, 2003, **85**, 1406–20.
13 J. D. Faraldo-Gomez, L. R. Forrest, M. Baaden, P. J. Bond, C. Domene, G. Patargias, J. Cuthbertson and M. S. Sansom, *Proteins: Struct., Funct., Bioinf.*, 2004, **57**, 783–791.
14 A. E. Garcia and K. Y. Sanbonmatsu, *Proc. Natl. Acad. Sci. U. S. A.*, 2002, **99**, 2782–2787.
15 J. Allard, J.-D. Lesage and B. Raffin, *Presence: Teleoperators and Virtual Environments*, 2010, **19**, 142–162.
16 H. Yu, C. Wang, R. Grout, J. Chen and K.-L. Ma, *Computer Graphics and Applications, IEEE*, 2010, **30**, 45–57.
17 F. Zheng, H. Zou, G. Eisnhauer, K. Schwan, M. Wolf, J. Dayal, T. A. Nguyen, J. Cao, H. Abbasi, S. Klasky, N. Podhorszki and H. Yu, *IPDPS'13*, 2013.
18 M. Dorier, G. Antoniu, F. Cappello, M. Snir and L. Orf, *CLUSTER - IEEE International Conference on Cluster Computing*, 2012.
19 W. Gu, G. Eisenhauer, K. Schwan and J. Vetter, in *Ninth International Conference on Parallel and Distributed Computing and Systems (PDCS'97)*, 1998, pp. 699–736.
20 S. Pronk, S. Pall, R. Schulz, P. Larsson, P. Bjelkmar, R. Apostolov, M. R. Shirts, J. C. Smith, P. M. Kasson, D. van der Spoel, B. Hess and E. Lindahl, *Bioinformatics*, 2013, **29**, 845–854.
21 W. Humphrey, A. Dalke and K. Schulten, *J. Mol. Graphics*, 1996, **14**, 33–38.
22 N. Michaud-Agrawal, E. J. Denning, T. B. Woolf and O. Beckstein, *J. Comput. Chem.*, 2011, **32**, 2319–2327.
23 M. Chavent, A. Vanel, A. Tek, B. Levy, S. Robert, B. Raffin and M. Baaden, *J. Comput. Chem.*, 2011, **32**, 2924–2935.
24 Z. Lv, A. Tek, F. Da Silva, C. Empereur-mot, M. Chavent and M. Baaden, *PLoS One*, 2013, **8**, e57990.
25 C. Cruz-Neira, D. J. Sandin, T. A. DeFanti, R. V. Kenyon and J. C. Hart, *Commun. ACM*, 1992, **35**, 64–72.
26 D. S. Goodsell and A. J. Olson, *Annu. Rev. Biophys. Biomol. Struct.*, 2000, **29**, 105–153.
27 C. Kim, J. Basner and B. Lee, *BMC Bioinformatics*, 2010, **11**, 1–16.
28 A. Hanson, E. Wernert and S. Hughes, *Scientific Visualization: Dagstuhl '97 Proceedings, IEEE Computer*, 1997, pp. 95–104.
29 A. Kulik, A. Kunert, S. Beck, R. Reichel, R. Blach, A. Zink and B. Froehlich, *ACM Trans. Graphics*, 2011, **30**(6), 188.
30 J. E. Stone, A. Kohlmeyer, K. L. Vandivort and K. Schulten, *Proceedings ISVC'10*, Berlin, Heidelberg, 2010, pp. 382–393.
31 D. Long, Y. Mu and D. Yang, *PLoS One*, 2009, **4**, e6081.
32 S. Yesylevskyy, S. J. Marrink and A. E. Mark, *Biophys. J.*, 2009, **97**, 40–49.

33 A. Sheena, S. S. Mohan, N. P. Haridas and G. Anilkumar, *PLoS One*, 2011, **6**, e25747.
34 S. K. Ludemann, V. Lounnas and R. C. Wade, *J. Mol. Biol.*, 2000, **303**, 813–830.
35 S. J. Marrink, H. J. Risselada, S. Yefimov, D. P. Tieleman and A. H. de Vries, *J. Phys. Chem. B*, 2007, **111**, 7812–7824.
36 J. E. Stone, K. L. Vandivort and K. Schulten, *ISVC (2)*, 2011, pp. 1–12.
37 Z. Konyha, K. Matkovic, D. Gracanin, M. Jelovic and H. Hauser, *IEEE Trans. Visualization Comput. Graphics*, 2006, **12**, 1373–1385.
38 C. Turkay, A. Lundervold, A. Lundervold and H. Hauser, *IEEE Trans. Visualization Comput. Graphics*, 2012, **18**, 2621–2630.

ROYAL SOCIETY OF CHEMISTRY

DISCUSSIONS

Virtual and augmented reality immersive molecular simulations: general discussion

DOI: 10.1039/c4fd90019g

Professor Hirst opened the discussion of the paper by Dr Johnson: The plan to make cellPACK available to the public seems like an excellent opportunity; what thoughts have you had with respect to assuring the quality of information provided by the community?

Dr Johnson answered: As cellPACK grows to model increasingly complex organisms, it will be important for large numbers of biologists to collaborate on building models. However, cellPACK's current user workflows are time-intensive to learn. These access barriers will deter most untrained biologists from using the software. One current goal is to make our useful, but difficult-to-use cellPACK modeling system more easily accessible to researchers through a simple web server interface and to incorporate mechanisms for collaboration and community curation within that interface.

Implementation: Although a hierarchical system, each recipe in the cellPACK database can be represented as a row in a spreadsheet. The recipe will include annotations from depositors and users that define the confidence of each value, such as: *cellPACK core-team approval rating*, *number of views*, *community confidence value average*, *community confidence value weighted by registered user trust-level*, *community critiques* (*list of comments*), *references*, *etc.* Additional parameters can be added at any time as the project evolves. Users will be able to view these parameters through the interface in spreadsheet form and they will be able to filter available models that match selected parameters from the community input parameter sets, for example, a novice user or outreach educator may stick to project-approved recipes, while an expert in the field of HIV may prefer to see all HIV models with an average community confidence vote $\geq$30%. Media wiki standards http://www.mediawiki.org will be used to track the participation and trust-weighting of registered users.

Dr Zheng remarked: The densities of proteins and other large molecules in a virus and in a cell are dramatically different. How do you estimate their densities in different environments?

Dr Johnson replied: Forming viruses or other regions of cells that collect molecules into assemblies can pack recruited molecules into loose or regular formations with localized densities that vary greatly from the overall density of the surrounding environment. In general, cellPACK incorporates measured densities from published sources where available. In the case of mature HIV-1, we base the density of Gag-derived viral proteins on ranges of the Gag and GagPol polyprotein precursors from immature HIV-1 sources, as cited in Table 2 in the paper. Other ingredient densities are based on ranges provided by the sources cited for each ingredient and we target a cytoplasmic density of the host-derived cytoplasm to approximately 30% protein by volume. The mature HIV-1_0.1.5 and 0.1.6 models assume that complete cleavage of these Gag precursor polyproteins into their mature counterparts has occurred to result in mixing the Gag-defined densities with the cytoplasmic derived densities into the confined space of the matrix. This final density can vary from recipe to recipe or, when defined by published research, it can become a target density, *i.e.*, a global packing constraint itself.

Dr Rozmanov asked: cellPACK seems to be a very powerful tool for model generation. Is it also suitable for the generation of initial systems for further molecular simulations? Or, is it mainly designed for visualization? How are the structural lipid components and solvent treated in models generated by cellPACK?

Dr Johnson replied: cellPACK models can be used as starting points for simulation. cellPACK result files are lists of ingredient IDs with matrices describing the position and orientation of each placed ingredient. The ingredient IDs point to the ingredient files, which contain data such as the original Protein Data Bank (PDB) file from which the ingredient input models derive. Thus, atoms from the PDB file or other simulation input data, such as coarse-grain models derived from the PDB file, can be positioned into the cellPACK model according to any provided result file. To demonstrate cellPACK's potential to setup simulations, BD_BOX, an open-source Brownian dynamics engine,[1] was recently used to perform a 9 ms (45 000 000 steps at 5 ps per step) relaxation simulation of a 105 nm cubic section of a nine-component Blood Plasma recipe result comprised of 351 molecules using BD_BOX's default hydrodynamic bead settings and the rigid body method. cellPACK integrates many Python modules and works across different scales. For example, lipids can be extracted as rigid bodies from MD simulations and packed explicitly, or third party software such as LipidWrapper (http://nbcr.ucsd.edu/data/sw/hosted/lipidwrapper) can be used to pack lipids around the proteins previously placed by cellPACK. cellPACK works at the resolution of the mesoscale where it is most common to describe solvent implicitly during the preparation of a post-packing dynamics simulation, but we have experimented with packing solvent explicitly as well.

1 M. Długosz, P. Zieliński and J. Trylska, *J. Comput. Chem.*, 2011, **32**, 2734–2744.

Dr Zoppè remarked: Thank you for the nice work. I think that cellPACK will be valuable for the development of models that allow us to actually study viral biology. As you surely know, no two viruses are the same; when we say that a virus is 110 nm in diameter, or that a virion holds 14 spikes, we implicitly talk about mean value. Also, the precise distribution of all proteins in the virion can change while maintaining the general attributes. So my question is: if you run your program with the same input data, do you always get identical results or is there some degree of elasticity in the packing model?

Dr Johnson responded: This is one of the major strengths of the method – cellPACK is specifically designed to allow the construction of models consistent with the current range of knowledge, or known ranges of values.

cellPACK allows the user to vary the randomness or parameter selection ranges with control. By default, a particular recipe such as HIV-1_0.1.6 will produce identical result files when run on any computing system. As described in Fig. 3 in the paper, changing a single parameter called the *random seed*, will produce a completely different but repeatable stochastic packing. This value can be changed by the user to ensure repeatability, or the seed can be replaced with an ever-changing variable such as their computer's *system time* to ensure that no two packings will be identical. Similarly, a recipe can include a specific number or a specific density of ingredients or it can use a fuzzy number or fuzzy density. In selecting a parameter such as the number of HIV-1 ENV (spike) proteins, the user can choose for example: exactly 7 ENV, 7 ± 5 with an even probability of picking the number of ENV from the range between 2 and 12, or 7 ± 5 with a Gaussian probability towards 7 between 2 and 12. Every numeric parameter in cellPACK can similarly have precise or fuzzy probabilities.

Dr Zoppè asked: When we infect a cell with a virus in the lab, we have to add about 100 times the number of viral particles to obtain the expected infections. This means that 99% of virions are actually incapable of establishing an active infection. There are many possible reasons (from insufficient binding efficiency to mutations in the integrated DNA). Yet all our physical measurements are made on the bulk of virions: how do we know that the virus that we model is a good one? If we could provide the recipe with the values randomized, according to known variability, we might be able to reproduce the variation observed in nature, and possibly get some insights into the determinants of infection efficiency.

Dr Johnson replied: cellPACK is capable of producing models that randomly sample known ranges of variability as described in the previous answer. The difficulty then becomes focused on: how can we analyze the results to predict which models might represent active forms of the virus? We have recently developed new integrated microscopy analysis tools for cellPACK. We used the cellPACK software to generate coarse models of immature HIV to investigate several hypotheses of molecular interactions that give rise to distributions of HIV ENV surface proteins observed by super-resolution microscopy.[1] We used the integrated microscopy analysis tools to simulate fluorescence images for direct comparison to the observed fluorescent images[1] following the approaches of Gardner *et al.*[2] Our preliminarily results demonstrate a promising new method that we and other researchers can use to investigate, define, and refine the molecular structure and molecular interaction parameter ranges that give rise to the complex emergent properties of whole viruses. We are eager to refine the packing modules that control molecular interaction in the cellPACK packing process and to refine the analysis tools to explore these putative immature HIV interactions more fully. The system will incorporate new analysis tools that will be used to directly compare the modeled results against statistics obtained from microscopy to facilitate the testing of hypotheses on molecular interactions that may give rise to virus-scale properties. Thus, as microscopy assays for virus particle activity emerge, cellPACK will be prepared to generate collections of putative molecular models. Competing cellPACK models that result from exploring ranges of ingredient interaction parameters, can be tested for consistency with experimental measurements (currently limited to STED[1] microscopy), biophysical principles and other computational approaches.

1 J. Chojnacki, T. Staudt, B. Glass, P. Bingen, J. Engelhardt, M. Anders, J. Schneider, B. Müller, S. W. Hell and H. G. Kräusslich, *Science*, 2012, **338**, 524–528.
2 M. K. Gardner, D. J. Odde and K. Bloom, *Methods*, 2007, **41**, 232–237.

Dr Zoppè asked: Finally, my understanding is that with this program we get a model of a virus that is static; do you think that cellPACK can be adapted to become a dynamic modeler? Can we build a virus (or a family) and observe its maturation and its activity in the infection cycle?

Dr Johnson responded: Generating dynamic models, *e.g.*, by producing multiple models of an organism or virus over time and using those as statistical constraints to guide organism-scale dynamics is a major long-term goal of the cellPACK project. Stay tuned.

Dr Hall remarked: In your manuscript, you state that you can include systems biology data in model building. Typical models in systems biology operate on different levels of abstraction relative to molecular models – could you elaborate on how you include this information, with specific examples from the HIV models?

Dr Johnson answered: System scale interaction affinities are programmed into the molecular ingredients as probabilities for packing interactions. In HIV, for example, the MA ingredients have a high probability to pack next to existing MAs, which can lead to an excluded volume that pushes ENV to interact in a more limited space. This is just one of many HIV surface molecular interaction parameter hypotheses that we are beginning to test with a new integrative fluorescence microscopy analysis tool that enables direct comparison of cellPACK's molecular models to lower-resolution statistical distributions.

Dr O'Donoghue said: The interactive visual models made by cellPACK are created by assembling a wide variety of experimental data. For the experimentalists who derived these data, the cellPACK models provide new and more integrated views of their own data that can potentially lead to novel insights into underlying biomolecular processes. Do you know of specific examples where novel insight has been derived from cellPACK models? If so, can you share them? Also, are there cases where experimentalists been inspired by the cellPACK initiative to gather new data specifically to improve future cellPACK models?

Dr Johnson replied: Our first publications describing the cellPACK system to researchers are in review. Our recent experiments to generate and compare models with different molecular interaction hypotheses and to compare them with integrated microscopy analysis tools now provide a pipeline for hypothesis generation and mathematical model testing. We have just begun working with super-resolution microscopy labs to begin applying this protocol. We are also developing an online version of cellPACK that will make it easier for researchers to participate in cellPACK related projects and to critique models as a community.

Dr Procter commented: The work described by Johnson and colleagues effectively democratises the process of initial parameterisation of biological system generation, by allowing whole research communities to critique and evaluate simulations in the light of new experimental data. This is a step-change in the way that researchers employ modelling and visualisation, since Johnson and colleagues have provided a standard means for generating interactive visualisations, rather than relying on each researcher's own expertise to employ complex 3D modelling and animation systems to consistently integrate available knowledge for communication and further analysis.

My first question is: In the application of the system to HIV, what were the most effective steps made to inform and engage the community and drive the process of critiquing and recipe refinement?

Dr Johnson answered: The models described in this paper were generated as the first-draft models for HIV by the paper authors. The growth of cellPACK to model larger and more complex structures will necessitate a shift to big data approaches. An easy to use online interface will complement cellPACK's standardization of mesoscale data, to allow whole research communities to visualize, analyze, critique and edit or add to the modeling recipes as new data becomes available. This democratization of data will require novel interfaces, mechanisms of curation and key indicators of trust and quality in the resulting models that can be filtered by end users. We are developing a system to incorporate community input and filtering through an online web-service. We will test this service with our new HIV biology collaborators and release this service with a follow-up paper to better answer this question.

Dr Procter asked: What do you consider to be the key indicators that a biomedical research community is sufficiently mature and able to effectively utilise this kind of shared model approach to provide an 'up to the minute' snapshot of available knowledge for use by all members of the community?

Dr Johnson responded: We believe that this is the right time to develop and share the kinds of models that cellPACK is designed to create. The research community is currently providing a wealth of disparate data from multiple experiments at multiple scales that needs to be integrated into holistic models of cellular environments. The recent Science paper on the synaptic bouton[1] clearly demonstrates this. Likewise, the advances in simulation of large molecular systems will need tools such as cellPACK to automatically generate the ensemble of starting models needed for examining and understanding complex behavior of such systems. Clearly the software to create such models is only the beginning of a community wide process of continual improvement of the models. It also requires building a user community through easy access and powerful user interfaces.

We expect to be able to answer this question more thoroughly after the new online cellPACK system has been released for community use and studied in practice. As a biological model, HIV provides a rare fit for the current capabilities of cellPACK. As other biological systems become similarly well-defined in structure, quantification, and function, they will become equally accessible for cellPACK, but we also expect to see new use cases from different communities that may require less rigorous inputs to address different types of questions that integrate these data types.

1 B. G. Wilhelm, S. Mandad, S. Truckenbrodt, K. Kröhnert, C. Schäfer, B. Rammner, S. J. Koo, G. A. Claßen, M. Krauss, V. Haucke, H. Urlaub and S. O. Rizzoli, *Science*, 2014, **344**, 1023–1028.

Dr Glowacki opened the discussion of the paper by Dr Delalande: In your paper, you discuss the complexities of the haptic algorithm required for BioSpring. Could you elaborate on how exactly the algorithm works, perhaps with some equations or a pseudo-code diagram. In particular: (1) how does the intermediate high frequency simulation layer work, and (2) do you sum the forces on the atoms, and let the user feel that? Do the haptic strategies that you use differ significantly from the algorithms used by other workers in the field (*e.g.*, haptic IMD as implemented in VMD). If so, how?

Dr Delalande responded: Details of the haptic algorithm used as well as corresponding equations and use cases are described in a dedicated open access book chapter on interactive and immersive molecular simulations.[1] Briefly, in order to reach the frequency required by the haptic loop (between 300 Hz and 1000 Hz), an intermediate high frequency simulation layer is generated by a spring connecting the atomic position update (at simulation frequency, on the atoms previously selected by the user) with the visualization and interaction module. In this way, our approach is directly inspired by haptic IMD as implemented in VMD. As the spring equation is calculated at almost no computational cost, it provides a universal decoupling of the haptic feedback frequency from the running simulation.

The forces, as explained in ref. 1, are not directly taken from the simulation but are the calculated spring reaction to the user's action. The user haptically feels the sum of all the forces involving each spring depending on the position of the probe and the atoms, and each force sent to the simulation layer. The user can tune the spring stiffness for the feedback, but also the spring stiffness used to compute the force sent to the simulation. The haptic algorithm is then quite similar to the approach used in VMD. However, as we use a home-made coarse-grained simulation code based on an augmented spring network, it allows us to reach the interactive time on complex biomolecular systems (about 30 to 60 Hz) and get better and more accurate haptic feedback, especially to feel flexibility and rigid areas on the protein. As these forces depend on how the simulation accommodates the user manipulation, the spring force indirectly reflects the balance of the intrinsic forces acting on the atoms.

Beyond the example provided in the present paper, using our coupled simulation and interaction tools, with VMD as visualization layer, we are able to feel rigid steric collision and electrostatic forcefield strength when manipulating an atom/ion into a structure, allowing us to identify new ion binding sites in the DNase I enzyme.[2] In this application, the strength of the force feedback is a direct measure of the strength of a given binding site, providing very intuitive feedback to the user.

A future extension may be to use haptic device capabilities more fully, for example as discussed in ref. 3 for a simple assembly task.

1 A. Tek, B. Laurent, M. Piuzzi, Z. Lu, M. Baaden, O. Delalande, M. Chavent, N. Férey, C. Martin, L. Piccinali, B. Katz, P. Bourdot, L. Autin, Advances in Human-Protein Interaction – Interactive And Immersive Molecular Simulations, in *Protein Interactions – Computational and Experimental Tools*, *Book 2*, ed. W. Cai and H. Hong, Intech, Croatia, 2012. (http://www.intechopen.com/download/pdf/34342)

2 O. Delalande, N. Férey, B. Laurent, M. Gueroult, B. Hartmann, M. Baaden, *Multi-resolution approach for interactively locating functionally linked ion binding sites by steering small molecules into electrostatic potential maps using a haptic device*, Pacific Symposium on Biocomputing 15, 205–215 (2010).

3 A. Tek, M. Baaden, N. Férey, P. Bourdot, *MANHaptic: a Haptic Adaptive Method for Precise Manipulation, Assembly and Navigation*, International Journal on Human Machine Interaction, 2013, **1**, 60–67.

Dr Glowacki commented: In the movie which you showed, where the haptic interface is used to fit the dystrophin fragment structures inside the SAXS cavities, the fitted parts appear to stop moving once they are inside the SAXS cavities. Is this actually the case? Can you elaborate on exactly why this occurs? What forces are implemented to guarantee that the fitted parts of the peptide remain in the cavity, and do not wiggle out during subsequent haptic manipulation? Do they stop moving entirely, or is it merely the case that their motion is restrained?

Dr Delalande replied: Our BioSpring program is indeed based on a spring network that approximates the bonded interactions (we can see it as an interactive Normal Mode Analysis approach), augmented by non-bonded interactions (steric and electrostatic) we call aENM. A BioSpring run is based on a molecular dynamics simulation loop, thus, particles are theoretically always in movement. A damping equation can be used to stabilize the aENM and to help quick and stable convergence to a steady state. Damping forces are described in ref. 1. This high level coarse-grained method allows us to manipulate a stable but flexible structure in real time, without Brownian movements typically observed using classical all-atom molecular simulations.

In the case of flexible fitting simulations, the particles of this aENM are then immersed into a potential density map derived from SAXS or CryoEM experiments. Being submitted to a gravitational field representing the molecular envelope, the particles are attracted and trapped into the most dense area, but are still interacting with all surrounding neighbour particles, according to the steric and electrostatic interactions computed using any common coarse-grained force field (Martini, Zacharias Attract model, Levitt, *etc.*). When a particle is driven into a vector field such as the one generated from SAXS data, it may thus appear at rest. In this case, the particle is trapped because of the convergence and balance of vectors of the surrounding cells, therefore the particle is not moving in regard to the potential or the pseudo-potential grid (vector field). It is still interacting with all surrounding neighbour particles, according to the defined cutoff for steric and electrostatic terms of the force field equation. The ensemble is further stabilized by the rapid spring network convergence and bias towards its initial state. Overall, the balance of forces is such that moderate user-applied forces are sufficient to extract the model from such a local minimum in order to explore other possible solutions. As we take into account steric and electrostatic interactions, bonded interactions approximated by springs, as well as a density field, we can compute an approximation of energies that allows us to compare and to evaluate the quality of the solution proposed by the user. This approach must be considered as an interactive modelling approach that allows immersion of a protein into a density map. In this context, the haptic device is especially useful to feel attractive areas and to reshape the protein, taking into account the plasticity of rigid and flexible areas embedded within the spring network.

1 A. Tek, B. Laurent, M. Piuzzi, Z. Lu, M. Baaden, O. Delalande, M. Chavent, N. Férey, C. Martin, L. Piccinali, B. Katz, P. Bourdot, L. Autin, Advances in Human-Protein Interaction – Interactive And Immersive Molecular Simulations, in *Protein Interactions – Computational and Experimental Tools, Book 2*, ed. W. Cai and H. Hong, Intech, Croatia, 2012. (http://www.intechopen.com/download/pdf/34342)

Dr Brancale asked: In your simulator, how do you calculate the forces for the haptic device? What feedback does the user experience?

Dr Delalande answered: The response to the earlier question of Dr David R. Glowacki above should be a clear and detailed answer to your question.

Mr Hall commented: Haptic devices can be expensive. This is particularly the case for haptic devices that have both rotational and translational degrees of freedom. Where do you envision haptics being incorporated in computational chemistry laboratories given that these devices can be expensive enough, even if they are only $2000 a piece, to put them out of the reach of most computational chemistry groups? What do you think will convince people that the expense is justified?

Dr Delalande responded: It will indeed be difficult to convince all theoretical chemists to buy a haptic device for satisfying occasional use of interactive molecular dynamics simulations, but the main question is not really the price. A first step will be to convince the community of the unique opportunities such devices offer; that it can be useful and efficient to address research problems in the theoretical biochemistry and chemistry fields. To do so, we designed our interactive modelling/simulation method including computer science skills, human–computer interaction skills, and theoretical and experimental chemistry/biochemistry skills. Working on the end user issues, we obtained promising results on numerous and various biomolecular systems, including predictive hypotheses that have been confirmed either by classical theoretical studies (Molecular Dynamics) or by experimental results.[1]

An international workshop dedicated to Interactive Simulations could be one solution to increase community awareness, given the fact that a test or a demo on a well-known familiar molecular system can sometimes replace pages of arguments. We are currently thinking of proposing such an event in the next couple of years.

Another aspect is to provide at least the possibility for computational chemists to easily test such a device. This requires both appropriate software and available hardware. Some of the devices are sufficiently robust and portable to give them away for a short while, such that another lab may test them. As these devices are not used continually, a joint acquisition might lower the initial cost for a single lab and enable several research groups to experiment with these approaches.

1 O. Delalande, N. Férey, B. Laurent, M. Gueroult, B. Hartmann, M. Baaden, Pacific Symposium on Biocomputing 15, 205–215 (2010); A. Saladin, C. Amourda, P. Poulain, N. Férey, M. Baaden, M. Zacharias, O. Delalande, C. Prévost, *Nucleic Acids Res.*, 2010, **38**, 6313–6323; O. Delalande, S. Sacquin-Mora, M. Baaden, *Biophys. J.*, 2011, **101**, 1440–1449; G. Da Costa, A. Bondon, O. Delalande, L. Mouret, J.-P. Monti, *J. Biol. Struct. Dyn.*, 2013, **31**, 809–817; F. Paquet, O. Delalande, S. Goffinont, F. Culard, K. Loth, U. Asseline, B. Castaing, C. Landon, *PLoS ONE*, 2014, **9**, e88809.

Dr Baaden commented: It was mentioned during the discussion that one limitation for a wider use of interactive approaches such as those described in this paper might be the cost of a haptic device. One should bear in mind however that the cost for experimental equipment lies orders of magnitude above the price of a haptic device. An institution that can afford a SAXS machine should therefore be able to spend the money for a haptic device without difficulty.

Mr Stone remarked: In response to the question regarding the utility of haptic devices and their associated costs, I wanted to briefly summarize some of my experiences. We initially developed the haptic feedback interface for interactive molecular dynamics simulations in VMD in 2001. At that time, there were few commercially available haptic input devices, and they initially cost over $10 000. Over the past decade, the prices of commercially available devices have dropped by roughly a factor of ten; the devices are physically much more robust, they include better system software, and they are supported by many more applications. While it is true that haptic devices are still somewhat expensive, my experience has been that, with few exceptions, they greatly outlast the computer workstations they are attached to. The haptic devices in our laboratory have seen four generations of desktop workstations come and go over roughly 12 years. In light of their longevity, they don't seem as expensive as they might at first glance. As the number of applications supporting haptic devices grows, the devices will become more broadly useful to scientists, and their cost should continue to drop over time. Another factor that I expect might help make haptic interfaces more popular now as opposed to ten years ago is the fact that GPUs are bringing computational capabilities to desktop computers that once were available only on clusters or parallel computers. The former requirement for clusters and parallel computers had an effect of limiting the potential user community for haptic interfaces and interactive simulation to those with significant computational resources and expertise. I expect that several of the factors I have mentioned will make haptic interfaces much more practical for scientists to use than ever before.

Dr Delalande responded: We totally agree with this comment by Dr Stone and share the same experience about haptic devices surviving for many more years than their initially dedicated pilot workstation. One should consider that buying a haptic device is a long term hardware investment (possibly over 10 years lifetime?)

Mr Stone remarked: I wanted to ask for additional details in regards to the statement in the paper that says that steric clashes and structural collapse are more easily avoided by aENM as opposed to other methods such as MDFF. I was wondering if there is a specific feature of aENM that confers these benefits, and if so why, or if this is perhaps an observation associated with a particular simulation scenario?

Dr Delalande replied: The fact that we observed collapsing events when using the MDFF method for the reconstruction of the dystrophin filament based on SAXS data may have two origins. First, the long filamentous nature of this particular protein, second the nature of the underlying SAXS volume data, that does not reach the high precision of recent CryoEM experimental data. Given that molecular shapes obtained from hard-spheres Monte Carlo simulations designed to fit the scattering curve are very coarse-grained (20–30 Å resolution), the smooth and rather homogeneous driving potentials we apply to the molecule to fit into the volume seem more adequate in the case of non-globular systems considered here. We should point out that our experience with SAXS-based model reconstruction in terms of molecular shape diversity is still very recent and requires more experiments to derive firm conclusions. The multi-scale stepwise refinement protocol we employ should also benefit an accurate high-resolution reconstruction: at coarse-grained level, sidechains are managed by harmonic function restraints that participate in diminishing putative collapsing events.

Professor Brooks remarked: Our experience has paralleled Mr Stone's. We were a beta test site for the first popular haptic device and many since. Our department has two 6-D devices and I think four 3-D ones. Margaret Minsky did her MIT dissertation in our lab, and designed a custom device and user studies on simulated gratings and sandpaper. She determined that the absolute minimum update rate for satisfactory motion is 500 updates per s, and that things feel noticeably better at 1000 updates per s.

Professor Olson commented: In fitting the SAXS envelope with the dystrophin fragment, you use a coarse grained potential that drives the model into the envelope. Dr Brooks has questioned the utility of the haptic device in facilitating this operation. Since the potential used can automatically guide the model into the density, I would postulate that the primary advantage of the haptic device in this case is its utility as an effective 3 dimensional positioning device, and that the haptic feedback is secondary.

Dr Delalande replied: Usually there are a lot of solutions to fit a protein into a density map (for CryoEM, but also especially for SAXS, because there is a large number of density maps that can be associated to a SAXS spectrum). Moreover, for very flexible or large biomolecular assemblies, using the user knowledge to find a solution, potentially comprising both to solve a docking and a folding problem, is essential in the process. Using interactive simulation is an approach that allows inclusion of human expertise during a simulation in progress. This approach supposes the use of well adapted interaction devices. The advanced 3D navigation aspect of the haptic device is indeed a key point for the efficiency of the interactive manipulation during a BioSpring simulation. However, haptic rendering seems essential for four main reasons: (i) despite 3D visual rendering, guiding a moving multi-particle object in a molecular scene remains a complex task. In our experience, translating a driving potential assisting the user into a haptic force metaphor is essential to quickly identify a reliable solution. More generally, the 3D mouse aspect of the haptic device to interact with a 3D simulation in progress is a key point for the efficiency of the interactive manipulation of the flexible 3D molecular complex; (ii) haptic feedback is also helpful to prevent molecular clashes in the case of molecular docking (*via* collision rendering), including the case of intramolecular interactions observed upon the deformation of the molecular object. More generally, haptic feedback is helpful to feel and better understand molecular forces involved in a simulation in progress; (iii) equilibrating user-provided forces with respect to simulation forces is sometimes rather delicate. Here, haptic rendering is used as a guide for the user to modulate (and moderate) his action (amplitude and rapidity) on a running simulation. This usage improves stability and prevents calculation crashes; and finally, to supplement the navigation point evocated at the first point (iv) a haptic device can be used to assist in picking a moving target within a complex molecular scene, using haptic force metaphors such as magnets placed on pickable atoms. This approach is essential to help the user to be more accurate and efficient during interaction and manipulation of the molecular target structure.

In our point of view, a haptic device is crucial in terms of enhancing the speed and the efficiency of interactive BioSpring simulations. We have not yet carried out a quantitative user study to document the level of enhancement that can be achieved with the present application. We refer to the study by Simard and Ammi who analyzed the improvement of efficiency for a molecular deformation task.[1]

1 J. Simard, M. Ammi, 'Haptic communication tools for collaborative deformation of molecules', in *Proc. of EuroHaptics*, Tampere, Finland, June 2012, pp. 517–527.

Professor Brooks commented: I agree completely with Dr Olson that the haptic device is indeed useful as a very natural 3-D pointer, and that that is its primary advantage. I doubt very much if the 3-D haptic feedback *per se* helps the guiding process measurably in a 6-D task. In my experience, the haptic display user perceives at best the strong forces. I should want to see a controlled experiment demonstrating the incremental usefulness of the haptic device, *e.g.* one group of users with forces on, and the control group with full use of the device and the guiding potential, but forces off.

Dr Chavent opened the discussion of the paper by Dr Glowacki: Is it possible to use a human body to create a particular shape? For example, mimic the cylindrical shape of a chaperonin protein? Similarly, is it possible to add particular properties to a part of the body? For example, hydrophobic properties on a hand or charged properties on a foot?

Dr Glowacki responded: The quick to answer this question is yes – in fact, the idea of using dS to 'chaperone' molecular simulations is an allusion to the role that so-called 'chaperonin' structures play in guiding protein dynamics. At the moment, it is possible for users to specify which atoms they want to be sensitive to their human Hamiltonian. I have not thought about adding the capability to selectively guide subsystems of proteins based on their hydrophilicity or hydrophobicity, but I see no reason why it would not be possible. One thing to note is this: because we are using a single camera, occlusion is an issue that inevitably arises. Hence, the complexity of shapes which users can form to guide simulation are limited to those which are not occluded from the perspective of the camera. It would be extremely interesting to adapt the dS system to the sorts of multi-camera virtual reality environments that Bruno Raffin has been designing in the course of his research, where occlusion is considerably reduced. In fact, one of the exciting aspects of dS, compared to localized haptic interaction strategies, is that it allows users to form shapes and thereby chaperone the dynamics of large subsets of atoms simultaneously. Recent work in biochemistry has shown that dynamics in complex systems is characterized by collective coordinates, which involve the coordinated motion of large groups of atoms. Compared to haptic approaches, dS seems to me well-suited for guiding these sorts of nonlinear collective atomic motions.

Mr Stone asked: The paper indicates that the time complexity for force calculations scales quadratically, but this would only be true if long-range electrostatics algorithms such as the Particle Mesh Ewald (PME) are not employed. Is the quadratic scaling behavior mentioned in the paper simply a reference to the behavior of the openMM library when simulating very small structures, as used for the work described in the paper?

Dr Glowacki answered: You are correct: quadratic scaling in the number of particles is the worst-case scenario, if one calculates all the non-bonded interactions without adopting clever screening strategies. The early tests that we conducted were preliminary, undertaken without any cutoffs simply in order to guarantee that our system was behaving correctly and conserving energy. The most important point is that the additional computational load introduced by the dS platform is small compared to the force calculations, which remain the primary computational bottleneck. At present, we are using OpenMM to implement more sophisticated approaches with significantly improved scaling properties, allowing us to simulate significantly larger systems at a reduced cost.

Mr Haag remarked: Your idea of incorporating the user as an additional potential in the simulation is very interesting. Does the accuracy of the body shape detection allow the user to control the position of single atoms or molecular fragments? Is the precision determined by the detection devices or are there limitations in the implementation?

Dr Glowacki replied: The dS platform allows users to specify which atoms, residues, or subsets thereof they guide with their 'forcefields'. In this way, it is possible for the user to manipulate specific atoms. Using these sorts of strategies, we carried out tests on both 'expert' and 'novice' users, instructing them to stretch and compress a small 10-ALA peptide (shown in the ESI to our paper), and found that both expert and novice users were able to complete these tasks in a reasonable amount of time. These tests indicate that users did not struggle too much with the body shape detection available in the depth sensors. Because the algorithm smooths over random noise in the depth images in a fashion that is intuitive and smooth, we have not found random noise in the depth images to be a major drawback. Improvements in the hardware – namely, the ability to capture higher resolution depth images, and increases in the number of frame captures per second – will certainly lead to a finer level of control, although it remains to be seen whether these lead to improvements in the user's ability to manipulate molecular systems.

In my experience, the precision of molecular control depends as much on how intuitive the interface is, as it does on the device *per se*. For example, with haptics, a user's entire body is collapsed into an abstract localized pointing device. For many users – myself included – this is not an intuitive interface, and difficult to use effectively. The ability of users to see their bodies as Hamiltonians in dS actually helps them to quickly orient themselves in the space, and allows for a rather intuitive interaction with the simulation – more akin to how we typically use our bodies day-to-day.

Mr Krone asked: Is your application intended to be used mainly in an educational or outreach scenario (like a serious game) or do you think it can also be useful for scientific analysis by experts? What are your experiences regarding the sensor accuracy; is it good enough to steer a scientific experiment?

Dr Glowacki responded: To date, our emphasis has been on designing an intuitive platform for both specialists and non-specialists to interact with rigorous molecular dynamics simulations using depth sensors. This has been partially motivated by the fact that depth sensors are cheap and widely available. It has also been motivated by our intention to develop a pervasive platform for gamified molecular dynamics simulations. We want to develop a system which satisfies two targets: (1) it is sufficiently intuitive and engaging to sustain the interest of non-specialists and drive them to further exploration, and (2) it is useful to experts.

We carried out preliminary user studies in our paper, instructing users to stretch and compress a 10-ALA peptide, which is a molecular system that we have previously benchmarked extensively.[1] Both expert and novice users were able to complete these tasks in a reasonable amount of time, indicating that (at least for one user task) our combined hardware/software framework was not a major drawback to achieving the specified molecular manipulation. Whether the users can achieve the same task faster using haptic devices remains to be seen, and we hope to undertake further tests to investigate this. In my experience, the precision of molecular control depends not only on the accuracy of the device, but also on how intuitive it is to use the device. For example, haptic devices collapse a user's entire body into an abstract localized pointing device. For many users – myself and John Stone included – this is not an intuitive interface, making it difficult to use effectively. The ability of users to see their bodies as Hamiltonians in dS actually helps users to quickly orient themselves in the virtual simulation space, and allows for a rather intuitive interaction with the simulation – more akin to how we typically use our bodies day-to-day. This is very much in line with ideas introduced by human–computer interaction (HCI) pioneers like Myron Krueger, attempting to go beyond 'a seated man poking at a machine with his fingers or waving a wand over a data tablet', so that the focus of intention is on intuitive action rather than the precision of the technology.[2] It seems to me that the ideal interface for molecular simulation may involve some balance between precision and intuition.

1 D. R. Glowacki, E. Paci and D. V. Shalashilin, *J. Phys. Chem. B*, 2009, **113**, 16603–16611; D. R. Glowacki, E. Paci and D. V. Shalashilin, *J. Chem. Theory Comput.*, 2011, **7**, 1244–1252; D. V. Shalashilin, G. S. Beddard, E. Paci and D. R. Glowacki, *J. Chem. Phys.*, 2012, **137**, 165102.
2 M. W. Krueger, 'Responsive Environments', presented at the Proceedings of the June 13–16, 1977, National Computer Conference, Dallas, Texas, 1977.

Mr Krone commented: The representation you used for the molecular data is different from standard models in molecular graphics; it seems to be more artistic and abstract. Do you think this will influence the users, especially the ones that are used to the more common representations?

Dr Glowacki answered: The question of how best to render molecules is a fascinating topic because it requires thinking about the frontiers where scientific imagination meets artistic representation. At its heart, molecular rendering touches on aesthetic questions – *i.e.*, what is the best aesthetics for representing molecules and thinking about their behavior? This is an absolutely fundamental issue for workers in our field, because the atomic and molecular domain in which we work, discuss, and carry out research is invisible to our eyes. In science, aesthetics is arguably most important when the objects of study are invisible. We cannot place a molecule on a table and talk about it; rather we rely on aesthetic motifs, representations, and models to both communicate with other scientists, and teach students the behaviors and properties of the atomic/molecular world. The more we rely on a particular aesthetic, the more that aesthetic guides how we imagine the mechanics of the invisible. This is why aesthetics is extremely important: in fact it guides our thinking process. Projects like that presented by Graham Johnson do far more than simply provide pretty pictures of cellular structures; in fact, they are manufacturing an aesthetic vocabulary for imagining and communicating the mechanics of the cellular world.

For small to medium-sized molecules, our aesthetics has been dominated by the 'ball-and-stick' model of molecular structure. For biomolecules, ribbons seem to be the most popular way of imagining molecular structure. However, these are two particular aesthetics amidst a mass of potential aesthetics. The ball-and-stick representation collapses the complexity of an electronic molecular wave function into a single stick, which connects balls with varied radii. Both of these aesthetics are good at representing conformational wiggling and jiggling, but neither is particularly good at revealing reaction dynamics. In dS, we typically choose to represent molecules as an assortment of atoms, without showing the bonds, and simply show a time history of the nuclei as they move along their dynamical paths. This has the effect of revealing bonding structure, for the simple fact that molecules tend to undergo substantial vibrational motion in the direction of a bond. Certainly dS uses non-traditional molecular representations, and in that sense its aesthetics may be unfamiliar to those who are accustomed to the standard representations. We could easily implement the standard molecular representations in dS; however, I see this as an exciting area of research, and I believe that there are probably aesthetics which are better suited to representing dynamism than those which we tend to fall back on. One of the aims of dS has been to utilize aesthetics to engage non-specialists. This has opened up a range of new aesthetic territory, which is perhaps slightly less constrained than it would have been otherwise. Along these lines, I recently contributed to an MIT press volume entitled Molecular Aesthetics, which presented a range of articles related to the aesthetics of the invisible molecular world,[1] and may be of interest to those present at this discussion.

1 D. R. Glowacki, in *Molecular Aesthetics*, ed. P. Weibel and L. Fruk, MIT Press, Cambridge, MA, 2013, pp. 248–257.

Mr Kouyoumdjian asked: Given that one of your goals seems to be to target as wide an audience as possible, have you thought about porting a somewhat scaled-down version of your work to the Xbox One?

Dr Glowacki replied: Yes indeed, this is something we are thinking of and actively investigating at this very moment. One of the exciting possibilities to imagine is a scaled-back version of an MD code where people can find new protein configurations in their living room which could then be uploaded to some centralized server. One potential complication here is that Microsoft seems to be keeping Xbox One rather closer to its chest compared to Xbox 360, including the Kinect developer tools.

Dr Levieux remarked: During your presentation, you talked about gamification. In your experiment, did you provide the students with a clear set of goals or was it more like a contemplative experience? If not, do you think such an experience may be gamified as much as possible or that an intuitive, contemplative experience can be enough?

Dr Glowacki responded: Gamification is something that we have been considering in light of a few different observations: (1) crowd-sourced science platforms like Galaxy Zoo or Foldit have been successful; (2) depth sensors like the Kinect have become increasingly widespread items in the consumer market (Wikipedia reports over 24 million sold[1]); and (3) people engage with dS for a long time even if we do not set any tasks for them to accomplish. Taken together, these points suggest that the dS platform could be used for gamifying scientific problems related to dynamical path sampling in molecular systems.

1 http://en.wikipedia.org/wiki/Kinect

In the educational activities which we carried out with the students, we used dS to illustrate a number of chemical physics principles – *e.g.*, how energy transfer works, the relationship between attractive/repulsive forces and atomic motion, or how the vibrational motion of different molecules impacts their properties. We attempted to set simple tasks for the students to accomplish in order to reinforce these learning targets. One of the criticisms that we received in student feedback was that the extent of guidance in the tasks was too much, and that users wanted more freedom to experiment with the system to determine how things worked themselves. This seems to hint at the fact that the system's impact derives (at least in part) from the atmosphere that it creates – *i.e.*, what you have referred to as 'contemplative'. In less structured settings, the average time that users engage with dS is considerable – on the order of 10–20 min – which partially accounts for the interest it has received from arts institutions. Certainly, this contemplative experience is well suited to an arts setting. For gamifying science, the question is how to use the 'contemplative' appeal of dS to drive user engagement with scientific questions.

Dr Johnson asked: Technical question about Kinect: to reduce issues with occlusion, have you been able to successfully point the devices inward to triangulate multiple users?

Dr Glowacki replied: With any form of camera-based interaction strategy, occlusion is always an issue, and is best tackled by using configurations which involve multiple cameras.[1] We have indeed run successful experiments with three orthogonally-oriented depth sensors in order to build a real-time 3D mesh of users within a space and reduce occlusion. However, because our emphasis has been on portability and exploiting millions of depth sensors which are available in homes around the world, we have not generally adopted these multi-camera strategies. One of the nice things about the interactive algorithms which we have developed is that they work as well with a 2.5-dimensional depth sensor setup as they do with a fully 3D setup, so any strategy for reducing occlusion should be compatible with the human Hamiltonian algorithms that we have developed. A related research question that arises as one succeeds in reducing occlusion to accommodate multiple users is how best to render the simulation from the perspectives of the multiple viewers.

1 B. Petit, J.-D. Lesage, C. Menier, J. Allard, J.-S. Franco, B. Raffin, E. Boyer and F. Faure, in *International Journal of Digital Multimedia Broadcasting*, 2010, (2010), pp. 1–12.

Professor Brooks opened the discussion of the paper by Dr Reiher: What do you mean by "really feel" the forces on a molecule? Certainly you really feel something, but how are the forces your nervous system perceives related to those on the molecule. Do you computer according to Fechner's law that states that subjective sensation is proportional to the logarithm of the stimulus intensity?

Have you evidence beyond personal experience that haptic feedback helps docking?

Dr Reiher responded: I mean by "really feel" that the device mediates a response of a quantum system obtained from a quantum calculation.

In the Born–Oppenheimer approximation, this quantum force, which derives from the interaction of all electrons and nuclei, can be evaluated at any nuclear position. For a chosen nuclear position it is the driving force for the dynamics of this particular atomic nucleus (in the standard classical approximation for the dynamics of atomic nuclei). This quantum force is then upscaled to a macroscopic level to make it accessible to our tactile sense. Moreover, we may even damp this force in order to achieve an immersion in such a way that the features of the potential energy surface can be explored in a most convenient way (*e.g.*, abrupt large-scale changes in the force due to steep repulsive or deep attractive regions of the Born–Oppenheimer hypersurface are avoided in the exploration process). However, our scaling does not follow Fechner's law, but is instead solely driven by our objective to be able to explore the energy surface for mapping out a chemical reaction mechanism. Yet, we use an arctangent damping function to adjust to the force-resolution range of the haptic device, which bears similarities to a logarithmic function. One could easily introduce a logarithmic scaling to be in line with Fechner's law as the technical challenge in our scheme occurs at a different step, namely when we calculate the quantum (raw) force in real time.

The upscaling of the quantum force to our macroscopic level is performed in a spirit similar to that of building (macroscopic as well as virtual) ball-and-stick models of molecular structures (interestingly, these models also imply invoking the Born–Oppenheimer approximation[1,2]). As we have elaborated in our paper, for reactivity exploration it is mandatory to base the exploration on physical raw data (the quantum force that drives the nuclear motion) and to otherwise make sure that all relevant features needed for Eyring's absolute rate theory can be explored (*i.e.*, minima and first-order saddle points need to be detectable). With respect to your last question, I should emphasize that we do not study docking! We study chemical reactions.

In docking studies, conformational and entropy effects are absolutely decisive as many weak attractive contacts between a host and its guest are in operation (in addition to repulsive forces at short distances of host and guest atoms). By contrast, we form and break rather strong covalent bonds, in which the change of the Born–Oppenheimer electronic-energy surface yields the major contribution to the free energy change of each elementary reaction step. In most chemical reactions you break only one or two such bonds at a time. So, chemical reactivity is local and can be well controlled and manipulated by a haptic pointer device. Now, the main issue in chemical reactivity studies is to identify first-order saddle points of the surface that represent (high-energy) transition structures between reactant and product valleys. In order to explore uphill a high-dimensional Born–Oppenheimer electronic-energy hypersurface, we require a suitable descriptor. The force on an atomic nucleus is a well-suited descriptor. On a path uphill in electronic energy we may eventually find a saddle point and this stationary point is characterized by a vanishing force. So, in our case (and in contrast to docking studies) the force is carrying the essential information of the molecular manipulation. There is thus no doubt that one needs the force-feedback capability of the haptic device for such an exploration process.

You imply that the whole set-up might not be as convenient as any of the existing methods for quantum-mechanical reactivity explorations. If one person (that would be us) experiences the benefits of such an approach then this is already useful (*i.e.*, at least for us). For a general assertion of the usefulness of force-feedback in chemical reactivity studies, a thorough users' study would be required, which is not easy to conduct. Although we are convinced that our tool will be very helpful in quantum chemical reactivity studies, we lack objective evidence for this belief. Since we intend to turn our tool into a professional product, we are, however, confident that this product will find its users and thus, we currently consider the proof of the pudding to be in the eating.

1 E. Mátyus, J. Hutter, U. Müller-Herold, M. Reiher, On the emergence of molecular structure, *Phys. Rev. A*, 2011, **83**, 052512.
2 E. Mátyus, J. Hutter, U. Müller-Herold, M. Reiher, Extracting elements of molecular structure from the all-particle wave function, *J. Chem. Phys.*, 2011, **135**, 204302.

Dr Fowler opened a general discussion of Dr Glowacki's, Dr Reiher's and Mr Dreher's papers: Computational steering is more than applying forces to a simulation interactively using a haptic device (*e.g.* doi:10.1080/00107510310001605046). For example, one might wish to change the temperature, perhaps facilitating replica exchange simulations, or when something interesting is occurring increase the rate at which trajectory frames are written to disc. For educational purposes, one might even want to interactively alter the strength of individual harmonic bonds in a classical molecular dynamics simulation to show the effect.

My question comes in two parts: specifically, can ExaViz support steering of parameters other than forces? And generally, do the panel agree that steering in the manner I've described is useful?

Dr Reiher replied: Certainly, the haptic device can be used for more than applying forces. For instance, in the quantum chemical context we can envision the haptic device as a means to interactively shape a jacket potential that stabilizes a substructure relevant for a pre-defined chemical purpose[1] – pretty much in the same way as you would shape clay. In our specific case here, however, we were solely interested in applying forces as they counteract forces exerted by the potential energy on which we explore chemical mechanisms.

But I would like to clarify the point that you raised about the interactive altering of the strengths of individual harmonic bonds. This concept is only meaningful in a molecular mechanics setting in which you apply force fields. However, we carry out quantum chemical calculations which allow us to experience forces on nuclei when pulling them out of some bonding situation. These experienced forces mirror the 'real' quantum mechanical forces involved in bond breaking or bond forming processes. There are no classical spring models in a quantum description and thus there is no need to scale bond strengths. Instead, our tools allow you to experience the different bond strengths as evaluated by quantum mechanical calculations (in real time).

Professor Brooks noted that I should be careful when I say that one experiences the 'real' quantum forces on the atom. Surely, these forces are scaled to our macroscopic world and damped to increase the tactile perception, but they are derived from the actual quantum forces. In a sense, the tactile experience of these forces is as 'real' as the visual experience of a ball-and-stick model of a molecular structure. The experience is in both cases so 'natural' that it feels 'real'. The truth is as Dr Glowacki had pointed out: we just create images of the molecular world that allow us to conceptually work with certain aspects of it. Already assuming a molecular structure of atoms bound to form molecules is an abstraction that is hard to derive from quantum mechanical calculations which do not adopt the Born–Oppenheimer approximation.[2,3]

1 T. Weymuth, M. Reiher, Gradient-Driven Molecule Construction: An Inverse Approach Applied to the Design of Small-Molecule Fixating Catalysts, *Int. J. Quantum Chem.*, 2014, **114**, 838.
2 E. Mátyus, J. Hutter, U. Müller-Herold, M. Reiher, Extracting elements of molecular structure from the all-particle wave function, *J. Chem. Phys.*, 2011, **135**, 204302.
3 E. Mátyus, J. Hutter, U. Müller-Herold, M. Reiher, On the emergence of molecular structure, *Phys. Rev. A*, 2011, **83**, 052512.

Dr Glowacki responded: This is a great point. Every time we use a keyboard and mouse to set up a molecular simulation, we are in some sense 'interacting' with the simulation. When we visualize the simulation results, and then change our simulation based on those results, we are also 'interacting' with the simulation. With advances in computational architectures and interaction methodologies, the immediacy of our interaction is steadily increasing. Many of the papers in this discussion have focused on interactivity wherein users are actually manipulating the molecule. However, interactivity in molecular simulation can refer to a wide range of simulation features. For example, in the dS framework, users can interactively modify a wide range of simulation parameters – *e.g.*, the thermostat, the graphics rendering, and the strength with which a user's field couples into the MD simulation.

Dr Raffin commented: ExaViz is designed to be flexible enough to support a wide variety of steering scenarios. Applying a force is a classical steering scenario, but others can be developed. Basically we designed the application programming interface to be minimalist and limit intrusion in the simulation code. If I take the example of the temperature, you would have to add to the Gromacs simulation code a temperature input port. Next, still in the Gromacs simulation code, you need to add the code to retrieve the temperature value received at each iteration through this input port and apply where adequate in Gromacs code to change the temperature for the next iterations. You would also need to design a new GUI module (or extend an existing one) to add a knob or a simple text area, that the user uses to select the temperature he wants to apply to the running simulation. This module will need to send the new temperature through an output port. Then, the temperature output port needs to be connected to the temperature input port on the Gromacs side. Notice that the temperature is a global parameter, so it is required to check if Gromacs takes care of forwarding the temperature to all its processes (remember that Gromacs is running in parallel), or if you need to do it at the ExaViz level. If Gromacs takes care if it, you simply have to connect the GUI temperature output port to the Gromacs temperature input port. Otherwise ExaViz provides collective communication patterns to ensure the broadcast of a value: you need to use a broadcast tree to connect the GUI to each Gromacs process. This is done in the Python script configuring the application and does not require any modifications of the Gromacs code. By default, channels in ExaViz are FIFO. To avoid that, the simulation waits at each iteration that a new temperature is available, and you need to add what we call a greedy filter, that enables Gromacs to proceed if the FIFO channel is empty or to select the most recent temperature if several ones are present. One last detail of importance. A MD simulation is running very fast, while the user is comparatively slow. To make sure the temperature is applied at the right moment, *i.e.* iteration, you may need to add some extra code, for instance, to suspend the simulation while the user is selecting the new temperature to apply, or to transport with the temperature a target future iteration number when to start applying this new temperature. To summarise, I would say that ExaViz provides the basic tools to set up new steering scenarios but you still need to identify what makes sense to steer, how to inject the new parameter(s) in the parallel MD simulation, and what iteration paradigm (the HCI part) you need to associate to this steering scenario to make it practical for users.

Dr Baaden remarked: During discussion, Dr Reiher mentioned the necessity of a haptic device. I wonder whether the required functionality is mainly to have a sophisticated 3D input device with precise positioning, picking and eventually navigation in space. To what extent is the actual haptic capacity required and exploited?

Dr Reiher replied: The three-dimensional positioning and the additional three degrees of freedom that the input device offers are clearly useful. However, you refer to the (three-dimensional) force-feedback capability of the device and whether physical experience of the force is necessary for the exploration of chemical reactivity. Unfortunately, we cannot do without. In fact, it is the central ingredient. The reason for this is that the quantum mechanical force exerted on an atom (more precisely, on an atomic nucleus) in a molecular system by the surrounding nuclei and electrons (of the whole system) is our ONLY reactivity descriptor for the exploration of the Born–Oppenheimer potential energy surface. Imagine you want to explore reaction barriers starting from one reactant valley. It is clear that you need to push the reactive system energetically uphill. How do you know that you are pushing uphill towards reaching the barrier? How can you discriminate between different uphill paths? You might attempt a visualization solution of this problem, but the high dimensionality of the surface suggests to look for better solutions. And returning the force to the operator is the most intuitive way to know that you are working against a barrier. In this respect, force-feedback is absolutely essential.

Mr Stone asked: I was curious if you have considered evaluating the performance with TeraChem, a fast GPU-accelerated quantum chemistry package? One of the features we have added to VMD is the ability to interactively compute and animate molecular orbitals and electron densities, using one or several GPUs in parallel.[1,2] Would there be any value in a real-time display of this kind of information in an interactive reactivity simulation, or is a ball-and-stick graphical representation perfectly adequate? Are there other things you would want from an interactive visualization tool that you don't have for this type of simulation?

1 J. E. Stone, J. Saam, D. J. Hardy, K. L. Vandivort, W.-m. W. Hwu and K. Schulten, High Performance Computation and Interactive Display of Molecular Orbitals on GPUs and Multi-core CPUs, in *Proceedings of the 2nd Workshop on General-Purpose Processing on Graphics Processing Units, ACM International Conference Proceeding Series*, volume 383, pp. 9–18, 2009.
2 J. E. Stone, D. J. Hardy, J. Saam, K. L. Vandivort and K. Schulten, GPU-Accelerated Computation and Interactive Display of Molecular Orbitals, in *GPU Computing Gems*, ed. Wen-mei Hwu, 2011, ch. 1, pp. 5–18.

Dr Reiher answered: We currently do not have access to TeraChem nor to any other GPU-accelerated quantum chemistry code. However, we have already concluded that such an implementation would be most promising for real-time density functional based studies.[1] At the moment, our tight-binding density functional implementation[2] is so fast that we do not even need to consider its parallelization. Admittedly, this is an approximate density functional approach with undisputable semi-empirical character (in view of its basic assumptions), but we require from such an approach only that the qualitative features of the Born–Oppenheimer surface are reproduced – as elaborated in our paper.

Molecular properties can also be calculated in the interactive quantum-chemical structure optimization app of the SAMSON environment by Stephane Redon. SAMSON can also plot, for instance, the electron density on top of the visualized molecular structure. However, I have always been very reluctant with respect to such features for the simple reason that I have a hard time extracting useful information from looking at the electronic density or at any of the molecular orbitals. Clearly, seeing (localized) orbitals change upon structure optimization can be useful in view of the fact that this is the basis of correlation diagrams (like those by Woodward and Hoffmann,[3] for instance).

But thinking about it, I can imagine that some fields can carry very useful information. For instance, watching the spin density change in real time upon structure manipulation can be important if magnetic materials are the target of the exploration. Also the Laplacian field of the electron density can carry useful information as it extracts essential features from the otherwise rather structureless electron density (highlighted by Bader's Atoms in Molecules theory).

1 M. P. Haag, M. Reiher, Real-time Quantum Chemistry, *Int. J. Quantum Chem.*, 2013, **113**, 8–20.
2 M. P. Haag, A. C. Vaucher, M. Bosson, S. Redon, M. Reiher, Interactive Chemical Reactivity Exploration, *ChemPhysChem*, 2014, DOI: 10.1002/cphc.201402342.
3 R. B. Woodward, R. Hoffmann, *Angew. Chem., Int. Ed. Engl.*, 1969, **8**, 781–853.

Professor Olson remarked: You say that one could simulate chemical reactivity of systems of up to 100 atoms. It would seem that the dimensionality of the energy landscape would also be a factor in the amount of computation required. Is it always apparent what reaction coordinates are needed to be explored in these simulations?

Dr Reiher answered: Indeed, the effort for solving the electronic Schroedinger equation for one spatial arrangement of atomic nuclei scales with the number of atoms M and thus with the dimensionality of the surface $3M$. Although a more suitable scaling parameter would be the dimension of the one-particle space into which the molecular orbitals are expanded (*i.e.*, the number of basis functions), this number linearly scales with the number of atoms. For this reason, we had defined our primary target to be a reactive system of about 100 atoms for which we could show in our paper[1] that semi-empirical quantum-chemical calculations can already be carried out in real time without resorting to parallelized software of any kind (for a specialized unparallelized real-time tight-binding density functional implementation we obtained timings of less than 100 ms per electronic energy gradient calculation[2]). It can be expected that a parallelized code (especially when executed on GPUs) will allow us to eventually carry out full density functional gradient calculations in real time.

The reason why we restrict ourselves to about 100 atoms is the fact that typical chemical reactions are in general local. You usually break one or two bonds at a time, which involves a very limited number of atoms embedded in an environment of other atoms. Typical examples are catalytic processes at transition metal centers in synthetic catalysts or active sites of metalloenzymes. For such systems, structural models comprising on the order of 100 atoms are a standard choice for a reliable description of reaction mechanisms. Surely, these models may be further embedded into a molecular-mechanics force-field environment (modelling, for instance, a protein) or by a frozen-density embedding strategy (*e.g.*, in order to account for directed interaction of a protic solvent such as water). We are currently working on such embedding strategies which we extend to be flexible in a sense that they work like a magnifying glass that can be moved through a large molecular system. Looking through this magnifying glass automatically defines the quantum region, while the rest is treated by some approximate electrostatic model (*e.g.*, by a force-field approach). The boundary region is automatically taken care of in such a way that the operator may move this 'quantum lense' with the haptic pointer as he or she deems suitable.

The second part of your question actually was a decisive starting point for our developments. It is not at all clear what reaction coordinate needs to be explored. This general problem had led to a plethora of automated exploration methods. Our point is that exploration is simply more efficient if the operator can efficiently steer the exploration process. The current situation is like this: imagine you are standing in front of the lake of Zurich and want to cross the Swiss Alps to eventually arrive in Lugano in Ticino. While all existing automated methods would require you to carry a blindfold before you start the exploration process (*i.e.*, climbing across the Alps), we argue that a visual presentation of the structural changes in a reactive system and the tactile experience of the slopes removes this blindfold.

How then is our reaction coordinate obtained? Well, we manipulate the Cartesian coordinate of only a single atom. If this atom is moved by the haptic pointer device, then all other atoms in the reactive system experience a Hellmann–Feynman force calculated from the holistic quantum description of the entire electronic structure. These forces make atoms follow the moving atom (especially those atoms that one would call bound to the moving atom and then those bound to the latter and so forth). Depending on the speed of the exploration process either a whole subsystem of the reactive system may be (adiabatically) moved by dragging a single atom or the atom dissociates from its bonding partners in a quick move of the haptic pointer. Both situations are desirable for reactivity exploration as they allow the operator to explore the relevant configuration space. All nuclear configurations are then assigned an electronic energy and a low-energy path recorded by the operator (after a certain exploration period) is produced that can then be easily further optimized by standard techniques to generate a reliable minimum energy reaction path needed for kinetic modelling. Only then, has the true reaction coordinate been found. Thus, the reaction coordinate is the desired result and not the input to the exploration.

1 M. P. Haag, M. Reiher, Studying chemical reactivity in a virtual environment, *Faraday Discuss.*, 2014, (169), DOI: 10.1039/C4FD00021H.
2 M. P. Haag, A. C. Vaucher, M. Bosson, S. Redon, M. Reiher, Interactive Chemical Reactivity Exploration, *ChemPhysChem*, 2014, DOI: 10.1002/cphc.201402342.

Dr Brancale commented: In our experience with interactive docking, the combination of the haptic feedback with another form of real-time sensorial interaction (*e.g.* colouring the atoms of a molecule according to its specific binding score) greatly increases the overall experience of the user. However, these findings are based on anecdotal evidence.

Dr Glowacki responded: I think that Professor Brooks has made similar observations in previous work; in one user study (which I have cited elsewhere in this volume) he found that the haptic feedback in interactive molecular docking gave a factor of two increase in the ability of users to achieve certain outcomes. This is certainly a promising result, but it is by no means the end of the story – *i.e.*, further user studies are required to assess which combination of strategies is most efficient for allowing effective user interaction with interactive molecular simulation.

Dr Montes remarked: During the iGAM4ER 2013 competition that took place in the Cité des Sciences et de l'Industrie in Paris, there were teenagers that tried our Udock software. One of them who was 12 years old performed an accurate docking of Barnase and Barstar with our demo in a time range of 20 min, alternating the spaceship view and the classical view. His score was in the same range as the best I performed, knowing the solution.

Dr Glowacki responded: Indeed, it is observations of this sort that account for why crowd-sourced molecular simulations are capturing the imagination of research specialists. In many cases, non-specialists have shown the ability to quickly assimilate information about a given molecular system, and use their intuition to outperform both blind search algorithms as well as experts. These sorts of approaches are attractive insofar as they have the potential to both advance research as well as educate the public about research problems. I am curious whether any of the iGAM4ER results are published.

Dr Johnson commented: Researchers and content developers interested in substantial, long-term evaluation of their interactive projects should consider attending the Gordon Research Conference on Visualization in Science and Education, held every odd year. This conference invites communities "engaged in the production, study and application of visualizations to enhance and promote science" (http://www.grc.org/programs.aspx?year=2013&program=visualiz). It's a great place to meet and team up with collaborators who are educational evaluators, for example, who might want to do extended studies of the efficacy of your interfaces.

Mr Hall continued the discussion of the paper by Mr Dreher: For which systems have you used ExaViz to analyze multiple trajectories concurrently? Moreover, how do you deal with the presence of multiple trajectories in the ExaViz interface in terms of user interactions? For example, when trajectories are viewed concurrently, how do users select one trajectory and not another when selecting frames based on the line graph views?

Mr Dreher responded: The system was used on the FepA channel to study the number of water molecules present inside the channel. 5 concurrent trajectories were studied simultaneously with the same analysis as described in the paper. For now, the multiple trajectories feature is still at the state of "proof of concept". It is functional to make the same analysis on all the trajectories but the selection of frames is not available yet for multiple trajectories in the release. Regarding the interaction you are asking, you have the possibility to hide some curves like the ones you do not want to select which allow you to make a selection on a subset of the trajectories. We are currently working on the possibility to use directly these selections to create a true pipeline where a second analysis would run on the frames selected by the first analysis for instance.

Mr Stone remarked: I agree with many of the points made in this paper. The analysis of large scale parallel molecular dynamics simulations is becoming a significant bottleneck for many scientists.

I found the description of the "semi-automatic point of view extraction" feature very interesting. I would think that this feature would be very useful when working with complex models and with visualizations that involve a significant amount of occlusion. I was curious how many rays have to be cast into the molecular structure to determine the best viewing orientation for a particular component of the structure, and whether ray intersection test acceleration data structures are built on-the-fly just prior to ray casting.

The paper states that existing tools such as VMD and PyMol don't provide the ability to link analysis results such as plots, directly to the trajectory they are associated with, using interactive interfaces. I can't comment on other software packages, but VMD does provide interactive tools that closely integrate 3-D structure display, sequence analysis, trajectory analysis, and a number of plotting tools. The earliest examples of multi-modal analysis tools in VMD integrated sequence and structure analysis.[1] For trajectory analysis, VMD includes a "Timeline" plugin that allow users to interactively explore a trajectory with a direct link between the 3-D molecular structure display and the results of analysis calculations. The Timeline tool enables selection and highlighting of molecular structure associated with regions of interest in an analysis plot, with picking and selection in the analysis results causing the 3-D structure display to immediately display the associated trajectory frame and region of the structure associated with the result of interest. In addition to the Timeline tool, VMD also includes callback mechanisms that enable real-time plotting of interactive molecular dynamics simulations as they run, using any one of several plotting mechanisms. While the main VMD features described above have existed in various forms for roughly ten years, I should point out that they have only more recently been adapted to support much larger scale parallel analysis runs on petascale computers.[2,3]

1 E. Roberts, J. Eargle, D. Wright and Z. Luthey-Schulten, MultiSeq: Unifying sequence and structure data for evolutionary analysis, *BMC Bioinf.*, 2006, (7), 382.
2 J. E. Stone, B. Isralewitz, and K. Schulten, Early Experiences Scaling VMD Molecular Visualization and Analysis Jobs on Blue Waters, Extreme Scaling Workshop (XSW), pp. 43–50, 2013.
3 J. E. Stone, R. McGreevy, B. Isralewitz and K. Schulten, GPU-accelerated analysis and visualization of large structures solved by molecular dynamics flexible fitting, *Faraday Discuss.*, 2014, (169), DOI: 10.1039/c4fd00005f.

Dr Raffin replied: The number of rays cast into the molecular structure to determine the best point-of-view for a particular target mainly depends on the distance between the camera and the target, directly impacting to the size of the sphere from which the rays are cast. However, we found that using about 2000 rays can lead to quality results for a large range of camera distances. Obviously, one can improve the quality of the viewpoint by increasing the sampling rate. The acceleration data structures for the ray intersection tests are indeed built on-the-fly when the user selects his target and runs the computation. We are currently designing VMD and PyMol plugins to give the community access to this navigation feature.

Faraday Discussions

PAPER

Visual analysis for space–time aggregation of biomolecular simulations

Thomas Ertl,*[a] Michael Krone,[a] Stefan Kesselheim,[b]
Katrin Scharnowski,[a] Guido Reina[a] and Christian Holm[b]

Received 21st December 2013, Accepted 31st January 2014
DOI: 10.1039/c3fd00156c

Conducting a current through a nanopore allows for the analysis of molecules inside the pore because a current modulation caused by the electrostatic properties of the passing molecules can be measured. This mechanism shows great potential for DNA sequencing, as the four different nucleotide bases induce different current modulations. We present a visualisation approach to investigate this phenomenon in our simulations of DNA within a nanopore by combining state-of-the-art molecular visualisation with vector field illustration. By spatial and temporal aggregation of the ions transported through the pore, we construct a velocity field which exhibits the induced current modulations caused by ion flux. In our interactive analysis using parametrisable three-dimensional visualisations, we encountered regions where the ion motion unexpectedly opposes the direction of the applied electric field.

1 Introduction

Visualisation has become widely accepted for improving the understanding of complex data sets from real-world measurements as well as simulations. This is often referred to as 'generating insight'.[1] However, with the proliferation of data sets of rapidly increasing size, even the most concise and abstract visualisation techniques are facing scalability problems. Although this data scalability issue can be mitigated somewhat with algorithmic optimisation as well as with parallel hardware, the human employing a visualisation can hardly be scaled arbitrarily from a perceptual and cognitive point of view. These problems are analysed in detail by Thomas and Cook,[2] paving the way for a new approach that tightly integrates automated analysis steps with highly interactive visualisations. In this way, the human is put in the loop, steering the analysis process as insight is generated. This approach has been termed *visual analytics* and can be applied to generic abstract data from arbitrary domains. Keim *et al.*[3] established the *visual analytics mantra* as a simple guideline: "analyse first – show the important – zoom,

[a]*VISUS, University of Stuttgart, Germany. E-mail: ertl@vis.uni-stuttgart.de; kroneml@vis.uni-stuttgart.de; scharnkn@vis.uni-stuttgart.de*
[b]*ICP, University of Stuttgart, Germany. E-mail: kessel@icp.uni-stuttgart.de; holm@icp.uni-stuttgart.de*

filter and analyse further – details on demand". Lately, *visual analysis* is seen as a specific variant that applies these concepts to spatial data specifically prevalent in the life sciences. According to Kehrer *et al.*,[4] this requires a tight integration of the visual mapping, *i.e.* the representation of the data, visual interaction concepts and computational analysis.

We have already applied this concept successfully to different scenarios related to biophysics and biochemistry. One example is the time-dependent analysis of the interaction between polymer and solvent.[5] Here, the solvent in the relevant hydration shell is visualised, while the remaining solvent molecules can be filtered out in real time. Different properties of the solvent can be projected onto the solvent-excluded surface of the polymer, for example, the hydrogen bond loci, which are automatically computed for the data set. Additionally, the varying conformations over time can be accumulated to show the spatial extents covered during the whole simulation run and to illustrate the potential clustering of hydrogen bonds at the related surface area. Another example is the visualisation of solvent pathlines relevant to protein cavities.[6] The solvent movement is aggregated over time and can be clustered according to user-defined parameters, as well as filtered with respect to proximity to an area of interest. The resulting visualisation concisely depicts paths on which water can access, for example, the active site of a protein. In both cases, temporal or spatial aggregation is used to provide a meaningful visualisation that supports the analysis process. A further example in which aggregation is employed to give a clearer picture is the work of Lindow *et al.*[7] Here, a partially open channel in a protein is extracted and tracked over time. The extracted paths through the channel are aggregated over time. This results in an aggregated path that shows whether a small molecule would be able to traverse the channel during the simulation, despite the channel never being fully open.

In the work presented in this paper, we use aggregation to illustrate the behaviour of ions near DNA in a nanopore. Our visualisations allow researchers from the field of biophysics to interactively analyse their simulations, which is an important step in proposing or rejecting hypotheses about the characteristics of the investigated systems. Nanopores are isolated, nanometre-sized holes in thin membranes. When a voltage is applied between the two water compartments separated by the membrane, these pores turn into a sensitive biosensor. A small, but measurable electric current goes through the pore, and its magnitude depends largely on the conductance of the nanopore. The conductance of the nanopore interior changes whenever a biomolecule enters the pore. Detecting these modulations yields information about the molecule. The greater goal of nanopore scientists is to develop a read-out device for the sequence of DNA.[8,9] The electric conductance is caused by ions dissolved in the surrounding water. Typically, potassium chloride (KCl) is used as an electrolyte. The modulation of the current depends on the interaction of ions with the molecule that is studied. Understanding the origin of the current modulations is of crucial importance to interpret the observed electric current signal. In simulations presented before,[10] it was shown that a complex interplay of different interactions between DNA and ions, including water flow, constitute the observed current modulations. The complexity of the three-dimensional geometry, however, obscured the nature of the current modulation as well as the mechanisms creating it. Therefore, we aggregate the ions and their velocities spatially and temporally in order to analyse the ion flux in three dimensions.

We present an integrated visualisation application for the analysis of the aforementioned simulation of DNA within a nanopore. More specifically, we focus on the visual analysis of ion distribution and flux near the DNA. Our tool combines visualisation techniques for particle data and field data (3D scalar and vector fields). An analyst can choose between several methods to represent different characteristics of the data. The visualisations can be parametrised (*e.g.* using specific filters or thresholds) to reduce visual clutter and to refine data exploration. This allows experts from the field of biophysics to gain insight into their simulations and, thereby, supports visual discovery of interesting features. We employ GPU-accelerated algorithms for our application to maintain interactive frame rates, even during parameter adjustment. This facilitates the intuitive exploration of the data further, as it enables us to see the effect of a parameter on the visualisation immediately.

Publicly available programs for visualisation of molecular data sets like Visual Molecular Dynamics (VMD)[11] or Chimera[12] offer a lot of general functionality to illustrate and analyse biomolecules, but often lack some specific features to visualise other data types like vector fields. General purpose visualisation software like VTK/Paraview,[13] however, is usually not tailored to biomolecular data and does not provide sufficient functionality in this regard. Although all of the aforementioned frameworks can be extended using plugins, we opted for our own MegaMol[14] framework, as it supports efficient rapid prototyping of visualisations and is consistently designed for GPU-accelerated algorithms and fast rendering, which are crucial parts of our interactive visual analysis application.

2 Biophysical background

Fig. 1 shows a typical nanopore setup as used in experiments in the group of Dekker.[15,16] A small hole is drilled into a thin, free-standing silicon nitrite (SiN) membrane with a transmission electron microscope. The wafer is installed in a microfluidic setup such that the pore is the only connection between two water compartments. In one of the compartments, a double-stranded DNA is dissolved. Electrodes are inserted into both compartments and a voltage is applied between them. As the DNA is charged, molecules in the vicinity of the pore are dragged into

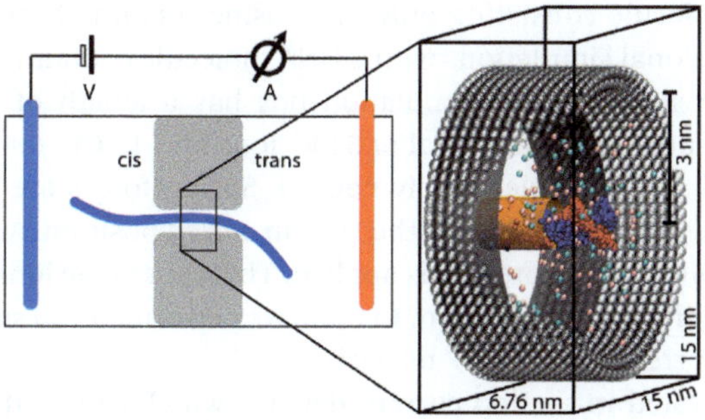

Fig. 1 Schematic setup of the simulated system. We concentrate on a small central region of a larger nanopore (right). The region includes a double-stranded DNA helix containing 20 base pairs surrounded by ions and water molecules (not shown here).

the pore and can be detected as they alter the electric current through the pore. Our study was intended to gain a quantitative understanding of the modulation of the current, following the experiments of Smeets *et al.*[15] They observed an interesting behaviour: if the concentration of the surrounding electrolyte is low, the presence of DNA increases the conductivity. In contrast, it is decreased for high electrolyte concentrations. This indicates a competition between different physical effects.

Our simulation approach to understanding the conduction is similar to that of Luan *et al.*[17] and sketched in Fig. 1. We concentrate on the central region of a longer nanopore and simulate a cylindrical pore with a DNA molecule. A DNA strand consisting of 20 base pairs is fixed at the centre of the box. Ions and water molecules are added to the system and an electric field is applied along the pore axis. Periodic boundary conditions along the axis assure that whenever a molecule leaves the simulation box on one side, it enters again on the other side. With this simulation setup, we sacrifice some realism in favour of a simplified analysis, but we still can find good agreement with experiments.

The total current is the product of the ion density and their velocity integrated over an arbitrary cross section. The magnitude of the current can hence be decomposed into two aspects: the number of ions that are present in the pore and the question of how easily they can move in the applied field. For the first aspect, we investigate the distribution of ions around a DNA molecule. Two competing mechanisms are present. First, the DNA occupies a certain volume in space, which is not available to ions and, therefore, the number of ions is reduced, compared to a DNA-free system. Second, the DNA is highly negatively charged and, therefore, surrounded by a cloud of positive K^+ ions with an extent of a few nanometres. The aspect of ion mobility also contains two contributions. The water movement is induced by the K^+ ions. This *electroosmotic flow* enhances the conductance as it facilitates the motion of the majority of ions. However, in the vicinity of the rough DNA surface, the ions are less mobile than in free solution, which reduces the conductance. The interplay of these four effects is responsible for the signal that is measured. Our simulations show excellent agreement with experiments regarding the change of conductivity with respect to the electrolyte concentration. We were able to show that all effects are of similar magnitude, but the surface friction effect is necessary for the observed current reduction.

The next challenge on the route towards a DNA sensor would be to reliably discriminate between different DNA sequences with this method. Originally, we used a DNA molecule consisting only of cytosine–guanine base pairs (CG). We performed additional simulations with a DNA molecule containing only adenine–thymine (AT) base pairs. The simulation box has a length of 6.76 nm in the direction of the pore axis and a total of 52 K^+ ions and 12 Cl^- ions were added so that the overall system is electrically neutral. Spring forces are added to the P-atoms of the DNA backbone to keep the DNA in a fixed position. An electric field of 0.2 V nm^{-1} parallel to the pore axis is applied. This causes the ions to move, which in turn also sets the water in motion. In total, we performed 100 simulations of 3.5 million steps, corresponding to 7 ns each.

We are interested in the ion current density, which can be decomposed into the product of the ion density and the mean local ion velocity. Since the DNA is fixed in space and can only vibrate weakly, the ion density can be calculated from aggregating the particle data on a uniform 3D grid in time. Every 0.4 picoseconds, corresponding to 200 simulation steps, the ion positions are written to disc for further analysis. The data aggregation is performed afterwards, using a trilinear interpolation scheme. By construction, the system is invariant under simultaneous translation along the axis by 3.38 Å and simultaneous rotation by 36 degrees. This invariance can be used to increase the statistical accuracy of the data. During aggregation, each ion is counted 20 times for each simulation time step, once for each of the 20 base pairs of the DNA. Every periodic instance of the ion is rotated and translated according to the orientation of the base pair. That is, we exploit not only the periodic boundary conditions of the simulation but also the periodicity of the DNA strand within the simulation box. The velocity is calculated by taking the ratio of flux density and ion density. For regions with low densities, the statistical accuracy of this quantity is very low, so that below a density threshold, the velocity is set to zero.

3 Visualisation techniques

We use a combination of established visualisation techniques to represent particle data, as well as scalar field data and vector field data. A general problem in dense 3D visualisation is visual clutter, especially when showing multiple features of a data set. Our application allows the analyst to use the different visualisation techniques selectively in order to remedy this issue. Furthermore, it offers specific filters for the individual techniques to reduce the amount of visualised data points and ranges. All our visualisations are fully interactive and use GPU-acceleration wherever possible in order to enable smooth, uninterrupted visual exploration of the data. In the following subsections, we summarise the employed visualisation methods and briefly describe their implementation.

3.1 Isosurface extraction

Isosurfaces, which extract features defined by a level set of a scalar function, are widely used to analyse three-dimensional, spatial scalar data. They are often visualised using either a mesh generated by a method like Marching Cubes[18] or direct volume rendering. In our application, we use GPU-based ray marching,[19] a direct volume rendering method that can be used to render a smooth isosurface. The basic idea of ray marching is to traverse the scene along a ray that starts at the

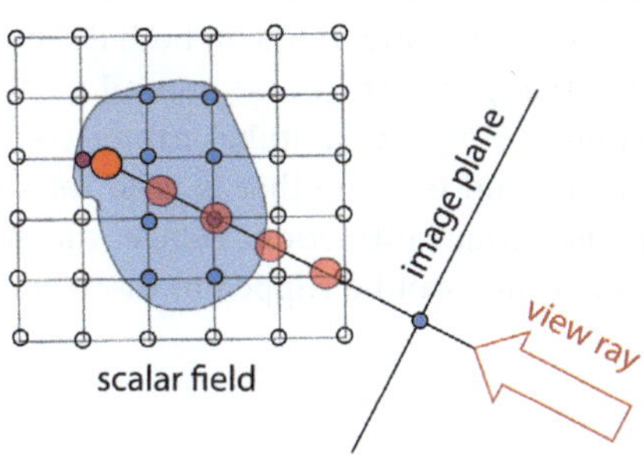

Fig. 2 Volume ray marching. For each pixel in the image plane, the scalar field is traversed along a view ray. Scalar values are sampled in a uniform interval using trilinear interpolation (red dots).

camera, thereby sampling the scalar field in a given interval (see Fig. 2). At each point, the scalar field is calculated from trilinear interpolation. If a given value, the isovalue, is crossed, the position of the isosurface along the ray is obtained from a linear interpolation. Isosurface ray marching yields high visual quality and runs very fast on current graphics hardware.[19] It has the additional inherent benefit that it renders semi-transparent surfaces correctly, whereas the triangles of mesh-based surfaces have to be sorted for correct transparency. In our OpenGL GLSL[20] implementation, we use multiple isovalues in order to capture the corresponding nested surfaces of the ion density. This allows us to encircle several density ranges at the same time (*cf.* Fig. 4). While transparency helps to see objects within the isosurface, it can also add to the visual clutter of the scene. Therefore, our application also offers the possibility of using clipping planes to partially remove individual isosurfaces.

3.2 Streamlines

Streamlines are often used in visualisation to represent flows, since they illustrate their dynamics intuitively. Given a time-varying vector field $v(x,t)$, a streamline $L(u)$ is a curve that is tangent to v at any fixed time t. Hence, streamlines can be seen as the solution to the initial value problem of an ordinary differential equation

$$\frac{\mathrm{d}L(u)}{\mathrm{d}u} = v(L(u), t), \tag{1}$$

where $L(0) = x_0$ is the location of the initial value or *seed point*. The most common method to numerically integrate streamlines is fourth order Runge–Kutta integration.

One of the biggest challenges in streamline integration is to choose the proper seed points. The goal is to render as few streamlines as possible in order to avoid visual clutter caused by occlusion and visual overload, while still representing all relevant characteristics of the underlying vector field. Various methods that address this problem have been presented. In our case, we simply seed the streamlines randomly within a certain range of isovalues on a user-defined plane. This is sufficient since it allows for the examination of the ion flux around the DNA. The user can also define the number of streamlines and adjust the number of iteration steps (*i.e.* the length of the streamlines). The streamlines are computed interactively on the GPU using CUDA. We use forward and backward integration in order to extend the streamline in both directions. A GLSL geometry shader implementation that generates a tube around a line is used to render the aforementioned streamtubes. The streamtubes can be colour-coded according to flux or velocity. The plane and isovalue that is used for streamline seeding can also be passed to the isosurface extraction. Hence, the streamlines are seeded inside of the isosurface, which can be clipped at the seeding plane.

3.3 Glyph rendering

Particles are typically visualised using a geometric object that represents the properties of the corresponding data point, so-called glyphs. To add further information, glyphs can be colour-coded according to a specific property of the underlying data, for example, coloured spheres representing atoms by element, charge or index of the polynucleotide chain.

The glyphs in our application are rendered using GPU-based ray casting,[21] as shown in Fig. 3. In principle, this method is analogous to ray tracing using only primary rays (*i.e.* no shadow rays, reflection rays *etc.*). A proxy geometry is rendered that covers the whole glyph with respect to the view and a fragment shader is used to compute the intersections between the view ray passing through every fragment and the glyph. If a ray does not hit the glyph, it is discarded, otherwise, the lighting at the hit point is computed based on the surface normal, the colour and the light properties. For objects with an implicit surface that can be described by a combination of few polynomial functions of low degree (*e.g.* spheres, cylinders, tori *etc.*), this technique is significantly faster than rendering classical triangle meshes.

The vector field describing the ion flux can also be rendered using glyphs. For each grid cell, we show an arrow glyph (formed by a cylinder and a cone) that points in the direction of the vector field. The size of the arrow can be scaled according to the ion density or flux at this point. Additionally, the arrow glyphs can be filtered by arrow length (*i.e.* ion flux) or ion density. Different colour mappings can be applied to show more information about the vector field at this point, *e.g.* density, flux, or direction. While the arrow glyphs are, in some ways, redundant with respect to the streamlines, they can give a more detailed representation of the field, while the streamlines show the general characteristics of the field more intuitively. The arrow glyphs, however, should be applied only sparingly, because they are prone to create occlusion and visual clutter.

4 Results & discussion

The visual analysis tool described above was applied to explore the simulation results for the nanopore conductance. It follows the visual analysis paradigm for exploratory data analysis and has led to these findings.

First we investigate features of the ion density. In the left panel of Fig. 4, we display the ion density and velocity, including the water velocity, in a one-dimensional projection of the data. The mean ion density is calculated from the trajectories by binning the data based on the radial distance. For both datasets, the ion density exhibits two peaks of different magnitude at around $r = 0.6$ nm and $r = 1.2$ nm. By comparison to the DNA structure, it is apparent that the

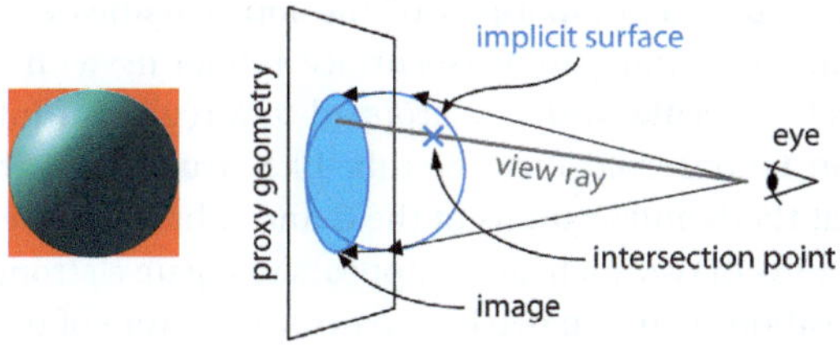

Fig. 3 Glyph ray casting for a sphere. Left: the final, pixel-accurate image of the sphere. Right: for each pixel of the proxy geometry, a view ray is intersected with the implicit surface of the sphere. If the ray hits the sphere, the colour of the corresponding pixel is determined based on the sphere colour and lighting (Phong shading).

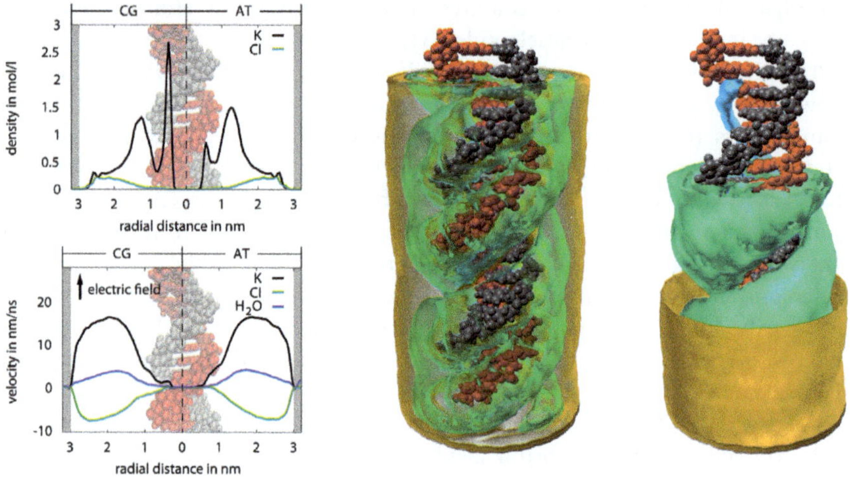

Fig. 4 Left: one-dimensional projection of the ion density (top) and the ion velocity (bottom) for both CG- and AT-DNA. The left parts of the diagrams show the curves for CG-DNA, while the right parts show the curves for AT-DNA. The DNA strand and the pore walls are given for reference. Right: two screenshots taken from our application showing three ion density isosurfaces for the CG-DNA molecule. The innermost, blue isosurface encapsulates ions inside the major groove of the DNA and strictly follows the helix shape. The outermost, yellow isosurface exhibits an almost perfectly circular cross section with a radius of ~2 nm, indicating that for larger distances, a rod-like model of DNA is absolutely adequate.

innermost peak corresponds to ions entering the grooves of the DNA molecule. The magnitude of these peaks is different for the CG- and the AT-DNA, which will be investigated later. At $r > 1.2$ nm, due to the charge of the DNA, a cloud of K^+ ions forms, which decays to around 0.3 mol l^{-1} at the pore boundary. This diffuse layer of counter ions is similar for both datasets. The velocities will be discussed below. In the rightmost panel of Fig. 4, we display three representative isosurfaces of the DNA density for the CG dataset. The blue surface corresponding to the isovalue of the highest density winds tube-like around the DNA helix inside the major groove of the DNA molecule. This structure corresponds to the first peak in the projected data. The intermediate, green isosurface has a complicated shape but is mostly cylindrical, with a gap in the major groove of the DNA. The third, yellow surface is almost indistinguishable from a cylinder, reflecting the fact that around $r > 1.5$ nm, almost all three-dimensional influence of the DNA double helix on the ion density has vanished.

In Fig. 5, we display a comparison of the ion densities closest to the DNA surface. We combine a transparent isosurface rendering with a stick-representation of the DNA molecule, which is cropped to a region of interest at integral base pairs. It can be observed that the tube-like structure of the CG dataset is located directly at the N and O atoms of the C and G bases in the major groove of the DNA. The vicinity of two such atoms appears to create a strongly attractive site for K^+ ions. At positions between two base pairs, the O atom of the next pair is also close, and this effect makes the tube structure continuous. For the AT-DNA, the situation is entirely different. There is no single attractive site in the groove, but the K^+ ions are close to the strongly negatively charged phosphate group of the backbone. There they can be in the vicinity of two oxygen atoms with negative

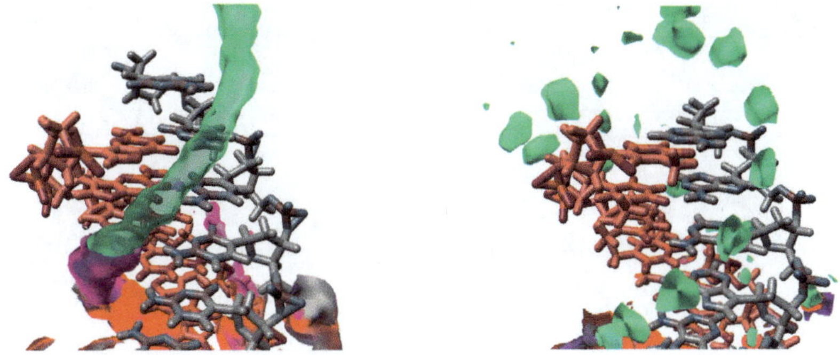

Fig. 5 Binding of K^+ ions to a CG-DNA (left) and an AT-DNA (right). For the CG-DNA, the innermost layer of ions binds very tightly to the major groove. The ions prefer close binding to the oxygen and nitrogen groups of the bases. In the case of AT-DNA, the major groove is far less attractive. Several binding sites can still be identified, most importantly the strongly negatively charged oxygen atoms on the backbone.

partial charges. No clear binding sites in the grooves can be found. This is reflected in the first peak in the upper diagram of Fig. 4 of the AT dataset being considerably lower than for the CG dataset.

Next, we focused on the vector field of ion velocities for the interesting regions with high ion density found in the last step. As described in Section 3.2, the streamlines (or -tubes) give a good impression of the general behaviour of the vector field. Fig. 6 shows a combination of streamtubes and arrow glyphs for regions with high ion density. Upon closer investigation of the velocity field for the CG data set, we noticed that there are regions where the majority of ions do not move in conformity with the electric current, as there are regions in the velocity field where the vectors point in the opposite direction. This effect

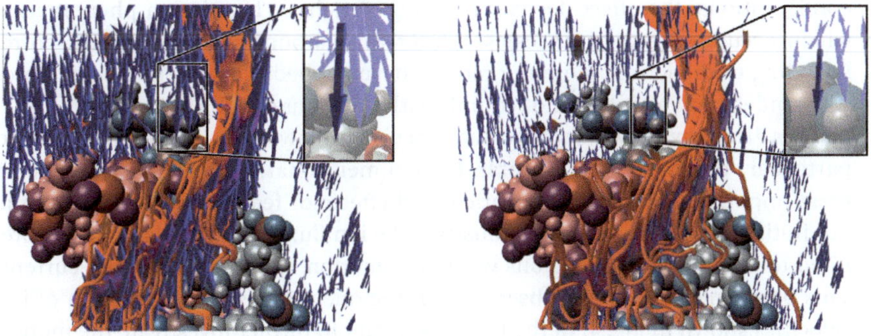

Fig. 6 Ion motion near the CG-DNA taken with two different sampling rates: 5 ps^{-1} (left) and 0.1 ps^{-1} (right). The streamlines are coloured according to the normal component of the local ion velocity. They give an overview of the vector field in interesting regions, while the arrows support a more detailed analysis of the velocity. On short time scales, a rapid, ballistic ion motion is apparent. If the trajectories are smoothed by using a lower sampling rate, the short-time motion is averaged out. On long time scales, the motion of the ions is largely suppressed, reflecting the friction effect near the surface. The current of the applied electrostatic field should pull all ions upwards; however, there are regions where the ion flow is directed downwards (see highlighted arrows in the inset cutouts). This effect is not due to short-time motion (left), as it is also observable on longer time scales (right).

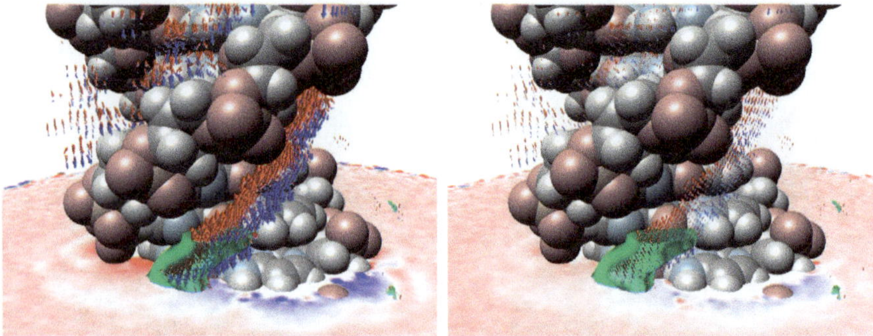

Fig. 7 Ion motion inside the major and minor groove of CG-DNA taken with two different sampling rates: 5 ps^{-1} (left) and 0.1 ps^{-1} (right). The green isosurface encircles the region with high ion density, which is following the major groove. In this region, ion motion is partially in compliance with the electric current (upwards, red arrow glyphs). There is, however, a distinct region where the direction of the ion motion opposes the direction of the current (blue arrow glyphs). The atoms of the DNA are colour-coded by their chemical element using the common CPK colouring scheme.

weakens if we use a larger temporal distance between subsequent snapshots to calculate the velocities (*cf.* Fig. 6).

We adapted our visualisations to investigate this phenomenon in more detail. Fig. 7 shows the arrow glyph visualisation of the ion velocity vector field inside the grooves (regions with high ion density, green isosurface). The arrows are colour-coded according to their direction: red arrows indicate velocity vectors that point in the direction of the electric current. This is the anticipated direction of ion motion. There is, however, a region in the major groove where the ions move in the opposite direction. Intuitively, this should not happen, since the force of the applied current is stronger than other competing effects (*e.g.* negatively charged atoms in the DNA). Again, by increasing the temporal distance between analysed configurations, the effect decreases (*cf.* Fig. 7 right). Thus, it is a phenomenon which occurs only on time-scales shorter than around 30 picoseconds, corresponding to the lowest sampling rate we have applied. This phenomenon is not fully understood as it does not comply with anticipated behaviour.

This discovery shows the importance of visual analysis of simulation results. In particular, parametrisable, flexible three-dimensional visualisations can be of great importance in finding unanticipated effects or features in the data. While our initial goal was to verify and illustrate the ion flux in the simulated nanopore, it led to the detection of regions where ion motion opposes the electric current. This phenomenon was not observable in the one-dimensional projection of ion velocities shown in Fig. 4. In the future, we want to further investigate the motion of the ions within the grooves more closely in order to determine their effect on the overall conductivity within the nanopore.

5 Future work

In the future, we want to extend our visualisation application in order to enable a more detailed analysis. As mentioned above, the regions within the grooves where the ion motion contradicts the electric current are of special interest. One possibility would be to apply the concept of solvent pathlines[6] to the trajectories of the ions. The paths of the ions, however, are likely to be much noisier than the water paths that Bidmon *et al.*[6] investigated. Therefore, we would need alternative methods for smoothing, filtering, and clustering. An example is the clustering of trajectories not only by their absolute spatial position but also by their position in relation to the nucleotide bases.

The visualisations used in this work to analyse the ion flux could also be applied to the water molecules in the simulations, since water molecules are brought into motion by the ions. Visualisations of the velocity of the ions and the velocity of the water molecules would show their mobility with respect to the pore and the DNA and to what extent they comply with each other.

Another interesting direction for future work would be to compare the all-atom simulations analysed in this paper with simulations using the Poisson–Boltzmann equation (implicit solvent and ions). Here, a specialized comparative visualisation could help to identify whether this would lead to differences in the resulting ion density and flux.

6 Summary & conclusions

In this paper, we presented a visualisation application for simulations of DNA within a nanopore. The density and motion of the ions is of special interest, since they play an important role in the resulting electric current. As this current depends on the molecule within the pore, it allows scientists to use the pore as a biosensor. The greater goal is to use nanopores as read-out devices for DNA sequencing. In our work, we investigated the current modulation caused by two different types of DNA, namely a DNA strand consisting of AT base pairs and one consisting of CG base pairs. Here, visual analysis turned out to be a driving force for making unexpected discoveries in the simulation data. Using our visualisations, we were able to identify regions in the CG-DNA where the ion motion contradicts the applied electric current. This might have an effect on the overall current through the nanopore, which would be important for the application of DNA sequencing. We also showed that dimensionality reduction, although helpful in many cases, can conceal interesting features that are only observable in 3D. Visual analysis is a powerful tool to generate insight. During our work, the close collaboration between experts from the fields of visualisation and biochemistry greatly facilitated the process, since it allowed us to create a visualisation tool that is tailored to a specific task but still flexible enough to support making novel discoveries. Adapting and refining the visualisations during the project was an important building block for the analysis leading to the results presented in this paper.

Acknowledgements

This work was partially funded by the German Research Foundation (DFG) as part of the Collaborative Research Centre SFB 716 (projects D.3, D.4 and C.5).

References

1 L. Rosenblum, R. Earnshaw, J. Encarnacao, H. Hagen, A. Kaufman, S. Klimenko, G. Nielson, F. Post and D. Thalmann, *Scientific Visualization: Advances and Challenges*, Academic Press, 1994.

2 *Illuminating the Path: The Research and Development Agenda for Visual Analytics*, ed. J. J. Thomas and K. A. Cook, 2005.
3 D. A. Keim, F. Mansmann, J. Schneidewind, J. Thomas and H. Ziegler, in *Visual Data Mining*, Springer, 2008, pp. 76–90.
4 J. Kehrer and H. Hauser, *IEEE Trans. Visualization Comput. Graphics*, 2013, **19**, 495–513.
5 B. Thomaß, J. Walter, M. Krone, H. Hasse and T. Ertl, *International Workshop on Vision, Modeling and Visualization*, 2011, pp. 301–308.
6 K. Bidmon, S. Grottel, F. Bs, J. Pleiss and T. Ertl, *Comput. Graphics Forum (Proceedings of EUROVIS 2008)*, 2008, **27**, 935–942.
7 N. Lindow, D. Baum, A.-N. Bondar and H.-C. Hege, *Proceedings of IEEE Symposium on Biological Data Visualization (biovis12)*, 2012, pp. 99–106.
8 J. J. Kasianowicz, E. Brandin, D. Branton and D. W. Deamer, *Proc. Natl. Acad. Sci. U. S. A.*, 1996, **93**, 13770–13773.
9 A. Meller, L. Nivon, E. Brandin, J. Golovchenko and D. Branton, *Proc. Natl. Acad. Sci. U. S. A.*, 2000, **97**, 1079–1084.
10 S. Kesselheim, W. Müller and C. Holm, *Physical Review Letters*, 2014, **112**, 018101.
11 W. Humphrey, A. Dalke and K. Schulten, *J. Mol. Graphics*, 1996, **14**, 33–38.
12 E. F. Pettersen, T. D. Goddard, C. C. Huang, G. S. Couch, D. M. Greenblatt, E. C. Meng and T. E. Ferrin, *J. Comput. Chem.*, 2004, **25**, 1605–1612.
13 W. J. Schroeder, K. M. Martin and W. E. Lorensen, *The Visualization Toolkit: An Object-Oriented Approach to 3-D Graphics*, Pearson, 1997.
14 MegaMol Project Website, http://www.vis.uni-stuttgart.de/megamol (last accessed: 2013-12-20).
15 R. M. M. Smeets, U. F. Keyser, D. Krapf, M.-Y. Wu, N. H. Dekker and C. Dekker, *Nano Lett.*, 2006, **6**, 89–95.
16 S. van Dorp, U. F. Keyser, N. H. Dekker, C. Dekker and S. G. Lemay, *Nat. Phys.*, 2009, **5**, 347–351.
17 B. Luan and A. Aksimentiev, *Phys. Rev. E: Stat., Nonlinear, Soft Matter Phys.*, 2008, **78**, 021912.
18 W. E. Lorensen and H. E. Cline, *Proceedings of ACM SIGGRAPH Computer Graphics and Interactive Techniques*, 1987, pp. 163–169.
19 M. Hadwiger, P. Ljung, C. R. Salama and T. Ropinski, *ACM SIGGRAPH Asia 2008 courses*, 2008, pp. 1–166.
20 R. J. Rost, B. Licea-Kane, D. Ginsburg, J. M. Kessenich, B. Lichtenbelt, H. Malan and M. Weiblen, *OpenGL Shading Language*, Addison-Wesley Longman, Amsterdam, 3rd edn, 2009.
21 G. Reina and T. Ertl, *Eurographics/IEEE VGTC Symposium on Visualization*, 2005, pp. 177–182.

Faraday Discussions

PAPER

Visualising intrinsic disorder and conformational variation in protein ensembles†

Julian Heinrich,*[ab] Michael Krone,[a] Seán I. O'Donoghue[bc] and Daniel Weiskopf[a]

Received 14th December 2013, Accepted 20th February 2014
DOI: 10.1039/c3fd00138e

Intrinsically disordered regions (IDRs) in proteins are still not well understood, but are increasingly recognised as important in key biological functions, as well as in diseases. IDRs often confound experimental structure determination—however, they are present in many of the available 3D structures, where they exhibit a wide range of conformations, from ill-defined and highly flexible to well-defined upon binding to partner molecules, or upon post-translational modifications. Analysing such large conformational variations across ensembles of 3D structures can be complex and difficult; our goal in this paper is to improve this situation by augmenting traditional approaches (molecular graphics and principal components) with methods from human–computer interaction and information visualisation, especially parallel coordinates. We present a new tool integrating these approaches, and demonstrate how it can dissect ensembles to reveal functional insights into conformational variation and intrinsic disorder.

1 Introduction

Over the past decade, the role of intrinsically disordered regions (IDRs) in proteins has been increasingly recognised as important, especially in eukaryotes. These regions are now known to play key roles in many biological functions, in regulatory control, and in many diseases.[1] The presence of IDRs in a protein is believed to often confound experimental structure determination, although these regions are present in many of the available 3D structures.[2] Some insights have been gained from examination of structures containing IDRs; in particular, it has become clear that IDRs can exhibit a wide range of structural conformations, from ill-defined and highly flexible to well-defined, upon binding of partner

[a]VISUS, University of Stuttgart, Germany. E-mail: {kroneml/weiskopf}@visus.uni-stuttgart.de
[b]CSIRO Computational Informatics, Sydney, Australia. E-mail: julian.heinrich@csiro.au
[c]Garvan Institute of Medical Research, Sydney, Australia. E-mail: sean@odonoghuelab.org

† Electronic Supplementary Information (ESI) available: http://bit.ly/mega-ensemble

molecules, or upon post-translational modifications.[1] Overall, however, many aspects of intrinsic disorder in proteins remain poorly understood.

Many structural studies of IDRs have used *homogeneous ensembles*, *i.e.*, ensembles composed of identical molecules that differ only in their 3D conformation. This includes ensembles derived from molecular dynamics (MD) simulations, where snapshots are taken at different time points. While MD can be powerful, it is often not feasible to compute sufficiently long trajectories to study key effects on IDRs, such as the binding of partner molecules. A second source of homogeneous ensembles that has been used to characterise IDRs are protein structures determined by nuclear magnetic resonance (NMR) studies.[3] These ensembles often exhibit large conformational variations that are widely believed to correlate with the dynamic behaviour of proteins in solution. However, there is good evidence that this belief may be wrong, and that variations observed in NMR ensembles derive primarily from a lack of data to describe the structure fully.[4] In contrast, when structures are derived from X-ray crystallography, any regions lacking sufficient data are simply removed, leaving apparent gaps in the polypeptide chain. A similar approach should probably be taken when using NMR ensembles to study IDRs: regions of the structure with little or no experimental data should often be removed from the analysis. When not removed, this may lead to overestimating the conformational variation of IDRs.

In this work, we focus on ensembles that are more heterogeneous, namely ensembles that contain all experimentally-determined structures that are judged to be significantly similar to one 'target' protein sequence, based on a template-based structure prediction method.[5] Currently, such ensembles are readily available for many proteins, often containing information on interactions with other proteins, DNA, RNA, or small molecules—such ensembles are likely to be of increasing significance for molecular biologists, as more structural data becomes available. Examining these ensembles can reveal a wealth of molecular detail on the range of conformations adopted with different binding partners, and can provide insight into IDRs, as these ensembles can capture ranges of conformations across different crystal packing environments, different experimental conditions, and across different molecular complexes. However, these ensembles can be quite complex, with sometimes hundreds, or even thousands of structures.

There are many methods to facilitate the analysis and visualisation of structural ensembles, one of the most widely used being principal components and related methods, which are typically used to find correlated motions[6] either in NMR ensembles or in crystal structures.[7] Such dimension reduction methods are useful for simplifying the resultant visualisations, thus aiding interpretation. However, it remains challenging to augment information about the spatial position of atoms, residues, or secondary structure elements with further attributes such as solvent accessibility, electrostatics, *etc.* Most methods developed to date focus either on homogeneous ensembles, or on ensembles of structural families,[8] which typically include a very diverse range of proteins. The ensembles considered in this work are an intermediate case, and there are few methods for using these ensembles to efficiently gain functional insight into IDRs or other aspects of conformational variation. A key problem with visualising such ensembles is that they are often highly cluttered, particularly in those regions that exhibit high flexibility.

Table 1 Residue attributes used in the parallel-coordinates view

Label	Description
2nd	Secondary structure state of residue determined by STRIDE.[12]
Contacts	Molecule in contact with current residue.
IUPRED	Predicted disorder score for current residue.[13]
Phi	The ϕ backbone angle for the current residue.
Psi	The ψ backbone angle for the current residue.
Position	Residue position in alignment to the target sequence.
RMSF*i*	Root mean square fluctuation of the Euclidean distance of the C_α atom position to the *i*-th eigenvector.
RMSF	Root mean square fluctuation of the Euclidean distance from the C_α atom to the mean structure.
SAS	Accessible surface area computed with the double cubic lattice method[14] *via* STRIDE.
Type	The amino acid type (mapped to arbitrary integer).
Chain	PDB[15] chain identifier (mapped to arbitrary integer).
ID	PDB identifier (mapped to arbitrary integer).
PCA*i*	3D coordinates of current structure projected along *i*-th principal component.[6]
RMSD	Root mean square deviation of current structure from the top-ranked structure.[16]

To address these challenges, we propose using parallel coordinates[9] in concert with traditional methods such as principal component analysis (PCA) and molecular graphics for the analysis of intrinsic disorder and conformational flexibility in heterogeneous protein ensembles. The use of parallel coordinates allows simultaneous visualisation of high-dimensional data—such as multiple attributes (see Table 1) from structure ensembles—and is particularly useful to facilitate exploring and finding patterns in the data. In this paper, we construct a multiple view setup[10] that allows residues selected in parallel coordinates to be directly highlighted in a 3D molecular graphics view *via* brushing-and-linking.[11]

2 Related work

The visualisation of homogeneous ensembles is particularly well supported by the VMD[17] molecular graphics tool, as well as other popular tools such as PyMOL[18] or Chimera.[19] Typically, all structures in an ensemble are visualised after being superimposed by minimising the root mean square deviation (RMSD) of the corresponding backbone atoms in the structure, typically using algorithms like those of Kabsch[16] or Coutsias *et al.*[20] However, the use of the superimposition approach quickly becomes limited for ensembles containing many structures or those exhibiting large conformational diversity. Therefore, in addition to the generic dimension reduction approaches mentioned above, a range of more tailored approaches have been developed to suit particular cases.

For visualising NMR ensembles, several specialised tools have been developed. One such tool is MOBI,[21] which compute a mobility score for the amino acid backbone, based on a combination of C_α interatomic distances and ϕ and ψ angles, then visualises the score using a colour-code mapped onto 3D structure representations. Another tool developed specifically for NMR ensembles is

MolMol,[22] which offers a 'sausage' visualisation, where a protein's backbone is represented by a tube of variable diameter, scaled according to the mobility of each amino acid.

For visualising MD simulations, specialist systems have been created for studying overall motions, including hierarchical, multiresolution trees,[23] as well as interactive linking between alternative visualisations (*e.g.*, DIVE[24]). In addition, methods have been developed for studying even more specialised cases, such as transient cavities[25] or molecular diffusion events.[26] Most MD methods make explicit use of temporal ordering, which is lacking in the ensembles studied in this work.

The above tools typically make use of a range of abstract visualisation methods. One of the first and most popular non-spatial visualisations in structural biology is the Ramachandran plot[27] for the investigation of the distribution of backbone torsion angles with respect to secondary structure elements. Other examples include hydropathy plots, RMSD plots, contact maps—for a recent review, see O'Donoghue *et al.*[28]

This work focuses on an abstract visualisation method—the parallel-coordinates plot[9]—that has not previously been applied to ensembles of molecular structures, or to intrinsic disorder. This method has been used to visualise high-dimensional data across various application domains, including bioinformatics[29] and systems biology,[30,31] where it has been shown to be useful for the analysis of regulatory networks or gene expression (see Heinrich and Weiskopf[32] for a recent survey).

We only found two previous reports using parallel coordinates with protein structure data. The first was from Luke[33] using parallel coordinates to visualise the conformation of the tetrapeptide Met-enkephalin using separate coordinate axes for each rotatable bond in the molecule, similar to the Ramachandran plot. However, this does not scale well as the number of axes increases with the protein size. The second application was from Becker,[34] which took a similar approach, but used only main-chain dihedral angles for conformational analysis of proteins. Becker recognised three major advantages of using parallel coordinates for conformational analysis: (i) multiple conformations can be displayed in the same plot, (ii) different types of axes can be mixed in a single plot, and (iii) dynamic clustering and filtering (hiding) can be conducted based on patterns emerging from the plot.

In this work, we further extend these approaches with a richer set of attributes, scalability, interactivity, multiple linked views,[10,11] and the integration of statistical methods in the analysis process.

3 Methods

For this work, a plugin to the MegaMol™ framework[35] was implemented to load sets of PDB[15] files, and compute a set of attributes to be used for the parallel-coordinates plot. The system was built using C++ and OpenGL and tested on a Windows workstation with an Intel Core i7, 6 GB RAM and an NVIDIA GeForce GTX 680 (4 GB VRAM).

3.1 Data preparation

We selected three well-studied human proteins where IDRs and conformational variation were believed to influence function (p53, RXR-α, and H2B). In each case, multiple experimentally determined structures are available—either for that sequence or highly similar sequences—that include a range of residues predicted to be disordered (determined using an IUPRED[13] score ≥ 0.5). In addition, the available structures include many cases with multiple partner molecules. We derived structural ensembles for each target protein sequence with the template-based structure prediction tool HHblits,[5] using it to find and align all PDB structures with a significantly similar sequence. We included only structures with an expected value of $<10^{-10}$, a threshold recommended to ensure that all structures are likely to have a similar fold to the target protein.[36,37]

The three resulting ensembles of PDB structures represent structural variations observed using different experimental methods (NMR, crystallography) across related proteins from several different organisms, and in the presence of a range of binding partners (*e.g.*, DNA or other proteins). For every structure, the HHblits output was used to produce an alignment between each residue in the ATOM records of the PDB file with a corresponding residue in the UniProt[38] sequence of the target protein.

Structures in the resulting ensembles were then clustered based on the region of the match to the full-length protein sequence. We selected one cluster for each sequence—corresponding to one sequence domain—that had a manageable yet sufficiently diverse set of PDB structures (from 47 to 78 cluster members). Structures in the cluster where ranked first by the number of identical residues to the full-length UniProt sequence; in the case of matches, PDB structures were then ranked by crystallographic resolution, with NMR structures ranked last. The clustering and ranking were performed using the Aquaria resource, currently in development at CSIRO and Garvan (http://aquaria.ws).

To prepare for visualisation and further analyses, each structure in an ensemble is superimposed onto the top-ranked structure using the Kabsch algorithm,[16] and the respective RMSD is recorded. For NMR structures consisting of multiple models, we used only the first model occurring in the PDB file. For each member of the ensemble, a set of additional attributes were computed. These attributes were selected to reveal different structural aspects that relate to both conformational variation and intrinsic disorder. The attributes are summarised in Table 1, and are further described below:

Secondary structure elements, backbone torsion angles, as well as the solvent accessible area per residue were computed *via* STRIDE.[12]

Intermolecular contacts were defined as follows: for each atom of each residue of the target protein, we searched within a distance of 5 Å—if any atoms were found within this distance belonging to another molecule in the PDB structure, this molecule was considered to be in contact with that protein residue.

For the ensemble, we computed a mean backbone structure by averaging the coordinates of superimposed C_α positions for all residues that HHblits matched to the residues in the query UniProt sequence. This mean structure was used to calculate a root mean square fluctuation (RMSF) for each C_α atom, providing a measure of local spatial variation at each residue.

In order to study correlated variations in structure, we applied PCA to each final ensemble using a standard approach[6] developed for analysing MD simulations. Here, a covariance matrix of atom coordinates is calculated and diagonalised to obtain the principal modes that describe most of the spatial variation within the ensemble. As all structures were superimposed prior to applying PCA, variations caused by the rotation and translation of a whole structure do not affect the computation. Again, only α-carbons are used. For the construction of the covariance matrix, only residues aligned to the target sequence in all members of the ensemble are considered (*i.e.*, gap residues were excluded).

3.2 Visualisation

We constructed a visualisation system using traditional 3D molecular graphics methods[28] to represent ensembles in a spatial context, in concert with a parallel-coordinates view to display additional multidimensional information about the same data set. The system allows users to select residues in parallel coordinates that exhibit certain attributes, with brushing-and-linking allowing the selection to be assessed in the spatial view.

The 3D view supports most commonly used molecular rendering modes, including ball-and-stick, stick, spacefilling (van-der-Waals), cartoon,[39] solvent excluded surface (SES),[40,41] and Gaussian surfaces[42] (see Fig. 1 for examples of a spline and stick rendering). Depending on the type of analysis to be conducted and the question to be answered, the standard practise in molecular graphics is to encode additional attributes (such as secondary structure, electrostatics, hydrophobicity, *etc.*) using colour or glyphs to be visualised together directly with the 3D structure in a spatial context. This approach works well for small numbers of attributes, but can become cumbersome for tasks that require consideration of many different attributes—the current standard practise is to enable switching between attributes, or to use multiple 3D visualisations. This can become tedious, especially for ensembles, which can impede the discovery of patterns in the data, such as relationships, clusters, dependencies, or outliers.

To facilitate the analysis of disorder in ensembles of structures, we augmented the traditional molecular graphics view with parallel coordinates,[9] which allow simultaneous visualisation of a large number of attributes across whole ensembles. In parallel coordinates, multidimensional data is represented by a set of axes arranged in parallel, as opposed to the orthogonal layout of axes in Cartesian coordinates. A data point in multidimensional space is then mapped to a poly-line (a set of line segments) in parallel coordinates, intersecting each axis at its respective coordinates. A point-line-duality between 2D Cartesian and parallel coordinates guarantees a unique mapping of patterns from a 2D scatterplot to a 2D parallel-coordinates plot and *vice-versa*. This allows us to incorporate well-known statistical plots such as the Ramachandran plot[27] into a parallel-coordinates system of protein ensembles. In addition, parallel coordinates allow us to visualise an arbitrary number of dimensions in a single plot, which can be useful to visually spot multidimensional outliers or clusters in the data and thus provide an analyst with information about protein ensembles that might be difficult or impossible to see using an isolated spatial view.

In our implementation, each poly-line represents a residue in one PDB structure, and each axis represents a residue attribute described in Table 1 (see also Fig. 1). In order to map categorical data to axes in parallel-coordinates, we cast non-numerical attributes (such as 'ID' or 'Type') to unique integers with no specific order. As a result of our residue-based representation, lines having an attribute in common will cross at the same point on that attribute axis; for example, all α-helical residues will cross the secondary structure axis at the same point. Axes are rendered as vertical

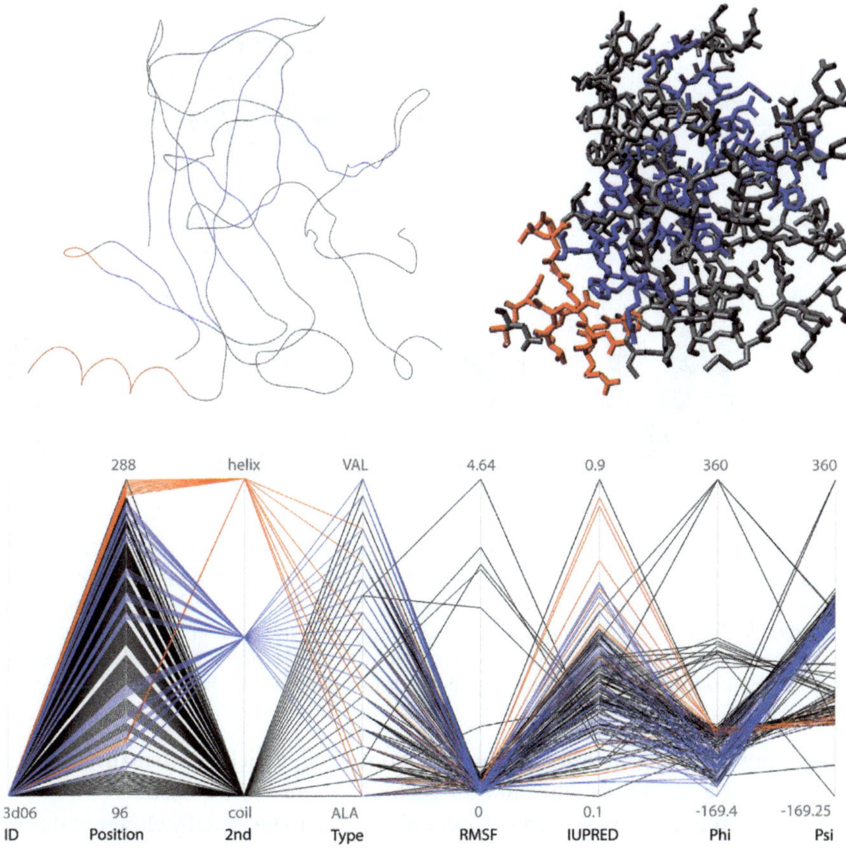

Fig. 1 Mapping residue attributes to poly-lines in parallel coordinates. This figure shows two representations (spline and stick, top) of a single structure from the PDB (3d06) and its representation in parallel coordinates (bottom). Each residue is represented as a poly-line (a set of line segments) crossing a set of axes, corresponding to attributes of the residue. Note that for some attributes (such as the PDB 'ID'), all lines from residues within the same structure will cross at the same point on the respective axis. In this view, the '2nd' axis (for secondary structure) was used to brush residues composing α-helices (red) and β-strands (blue).

lines with labels for the minimum and maximum of the respective dimension (note that we omit some labels in the figures for the sake of clarity). Since the order of axes is crucial for the determination of patterns in the data, our tool allows the order to be changed interactively.

As is typically done in parallel-coordinates views, our tool allows the user to select lines (representing residues), and to *brush* the selection with a user-defined colour. Selected residues are also immediately highlighted in the spatial view using the same colour. Furthermore, selection can be used to define a set of structures to be removed in both views (called *filtering*, as every structure that contains at least one selected residue is removed from the ensemble) or to hide lines in parallel coordinates; these simple but powerful features enable the user to interactively explore the ensemble based on attributes in parallel coordinates.

To further facilitate interactive exploration, we tailored our system for fast rendering. For the spatial view, we chose to represent the polypeptide backbone

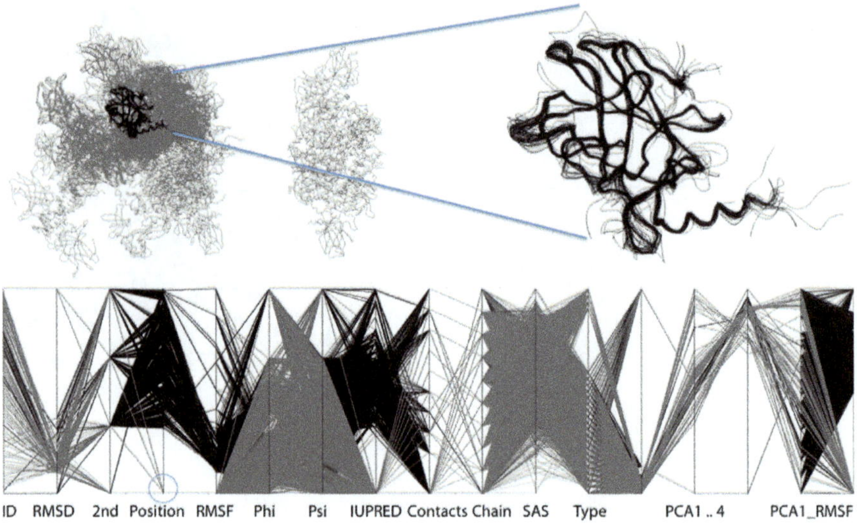

Fig. 2 P53 ensemble of 72 PDB structures, many containing partner proteins and DNA molecules (top left). Our system allows interactive dissection of the ensemble by hiding or revealing structures *via* selection of attributes from the parallel-coordinates view described in Table 1. The top right view was created from the original ensemble (top left) by a parallel-coordinates selection matching all PDB chains not aligned onto p53 (grey brush in the bottom plot), then filtering all structures that contain brushed residues.

using lines or splines (similar to the cartoon model), which was usually effective in providing a cogent visualisation for each of the ensembles used.

Initially, the ensembles used in this study tended to be visually cluttered due to the presence of multiple different molecules in various PDB files. With our tool, a user can easily focus on particular parts of the ensemble by selecting attributes from the parallel-coordinates view. For example, as a first step in our analyses, we used our tool to show only one chain in each PDB structure, namely the chain that aligns onto the target sequence (or the first such chain, in the case of oligomers)—see Fig. 2 (top).

We also designed our tool to automatically update the attributes of all parallel-coordinates axes whenever the user filters structures. For instance, this update process completely recalculates the PCA, based only on the currently visible structures and updates the 3D superposition using the top-ranked, non-hidden structure as the target structure.

4 Results

In this section, we tested our system by applying it to investigate the three protein ensembles described in Section 3.1. We show how our approach helped to gain insights into the relationship between intrinsic disorder and structural variation for these proteins.

4.1 Cellular tumour antigen P53

The ensemble for the human protein *cellular tumor antigen p53* consists of a set of 72 PDB structures that have been aligned to residue positions 94 to 295 of the

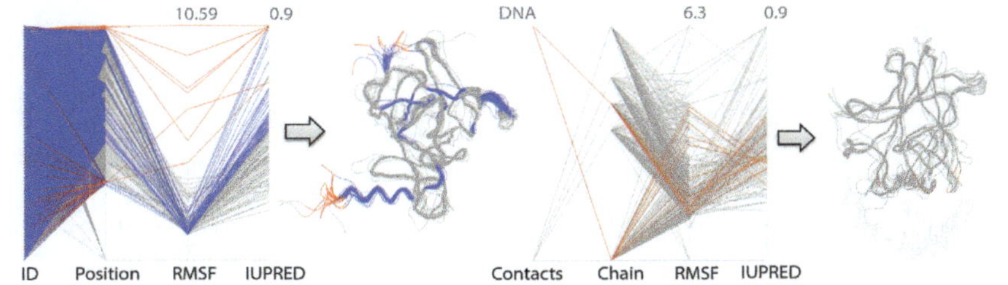

Fig. 3 Disorder and intermolecular contacts in the p53 ensemble. Left: Brushing was used to colour red all residues with visually outlying RMSF values (>3.8 Å)—all are predicted to be disordered (*i.e.*, have IUPRED score ≥0.5). However, many of the residues predicted to be disordered have low RMSF (blue). Right: Brushing in parallel coordinates allows users to focus on particular partner molecules. In this example, the 'Contacts' axis was used to highlight residues of p53 (red) in direct contact with DNA.

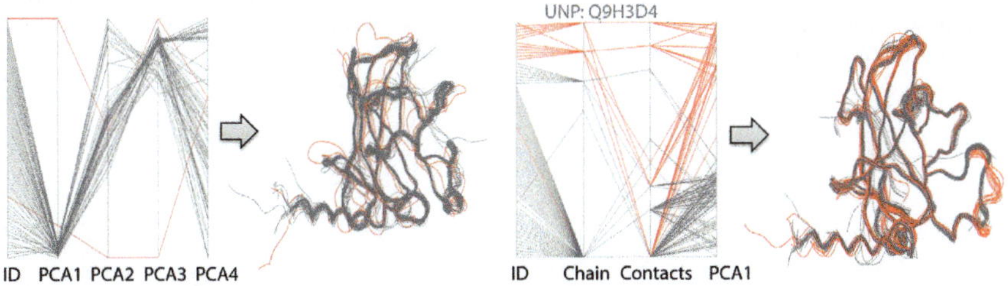

Fig. 4 Finding sub-states in the p53 ensemble. Left: From the set of 'PCA' axes, two prominent outliers in the ensemble are brushed red. Both structures show very different backbones from the ensemble all over the sequence. These outliers were removed for subsequent steps, causing all attributes of the parallel-coordinates plot to be recomputed automatically. Right: Of the remaining structures, selecting from the now updated 'PCA1' axis reveals another subset (red) with distinctly different structure—and somewhat higher apparent disorder—compared to the core ensemble (grey). From the 'Chain' axis, we see that this subset is composed of molecules p63 and p73 (Uniprot accessions Q9H3D4 and 015350), both close relatives of p53.

full-length sequence in UniProt (primary accession P04637). This region of p53 is known to bind DNA (*e.g.* 1TSR, 2AC0), as well as partner proteins, such as *p53 binding protein 1* (1GZH). Fig. 2 shows the initial view, with all PDB structures superimposed (top left) plus a view showing only PDB chains directly aligned onto p53.

From the 3D view, it seems that most structures form a rather rigid core, with two outlying regions of high conformational variation—one at the N-terminal α-helix and a second between residues 180 and 190. Applying our tool to this ensemble revealed that residues with very high observed disorder (RMSF) always had high predicted disorder (IUPRED), while the converse was not true (Fig. 3). Fig. 4 further illustrates how our tool can be used to successively dissect the p53 ensemble, for example by identifying and removing structures from divergent protein sequences, ultimately deriving a subset of highly similar structures that can be used, *e.g.*, to derive relationships between disorder and secondary structure (Fig. 5).

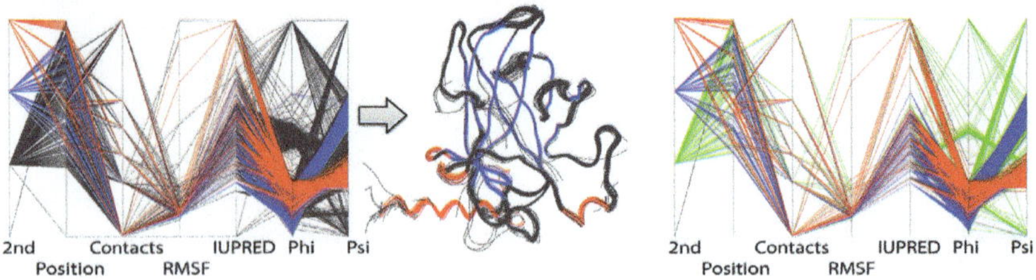

Fig. 5 Relationship between disorder and secondary structure in the p53 ensemble. Left: In this figure, the '2nd' axis (secondary structure) was used to brush α-helices (red) and β-strands (blue). The parallel-coordinates plot shows that helices in this ensemble are more likely to contain disordered residues than β strands, based on both IUPRED score and RMSF. The 'Phi' and 'Psi' axes show the expected configurations for both types of secondary structures, in accordance with the corresponding regions in the Ramachandran plot. Right: Brushing the '2nd' and 'Position' axes reveals residues that adopt different secondary structures across the ensemble. α-helices are shown in red, β-strands in blue, and coils in green. The plot indicates that only a small fraction of residues with ambiguous secondary structure have been predicted to be disordered by IUPRED. Among these, most are associated with low RMSFs. Next steps in the analysis might include filtering by 'Contacts' to investigate the source of the variation in secondary structure.

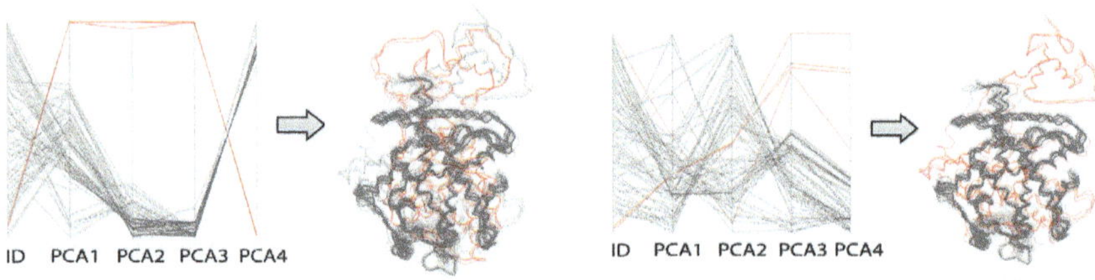

Fig. 6 Successive filtering of sub-states within the RXR-α ensemble. The pattern of lines across the 'PCA' axes clearly shows two distinct clusters (left). Brushing the small cluster (red) allows to remove the corresponding structures from the ensemble, and all attributes to be recomputed. The same procedure can be repeated until only a set of highly similar structures remains (right).

4.2 Retinoic acid receptor RXR-α

This example encompasses 78 structures that were aligned to residues 132 to 241 of human protein *retinoic acid receptor RXR-α* (P19793). The ensemble initially shows a high degree of variation; one large cluster of similar conformations can be seen, as well as two small clusters (Fig. 6). Using the principal component axes in the parallel-coordinates plot (in particular 'PCA3' and 'PCA4'), we were able to quickly identify and brush the smaller cluster. Fig. 6 shows two brushing and filtering steps used (from left to right) to filter out these sub-clusters; the remaining cluster, comprising most of the structures, was then examined for disordered regions. Fig. 7 compares residues with high predicted disorder (blue) *versus* those with high observed disorder (red). The blue selection includes many regions with very low observed disorder (*e.g.* the helices). Upon visual examination of the disordered region (bottom left in the 3D rendering in Fig. 7), we found a distinct cluster of structures that could be

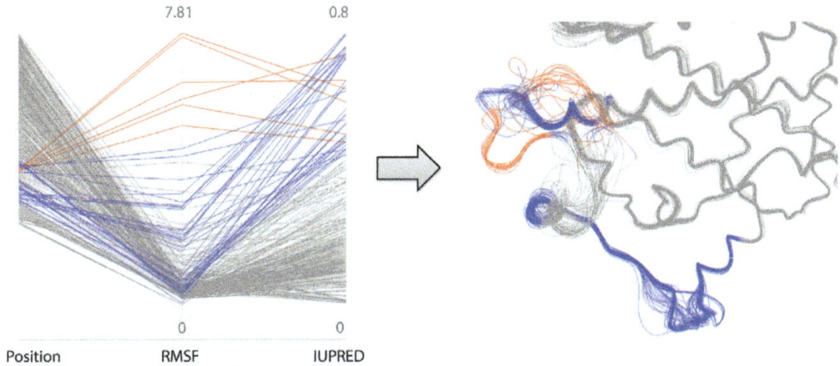

Fig. 7 Predicted *vs.* observed disorder for the RXR-α ensemble. Blue indicates residues predicted to be disordered (IUPRED score ≥0.5), while red indicates residues with visually outlying RMSF values (>5.2 Å). Note in the 3D structure the red colouring was rendered last, and hence conceals some residues coloured blue. The 'Position' axis shows the position of the selected amino acids in the sequence. As with p53, residues with high observed disorder (RMSF) tend to have high predicted disorder (IUPRED), however the converse trend is not as clear.

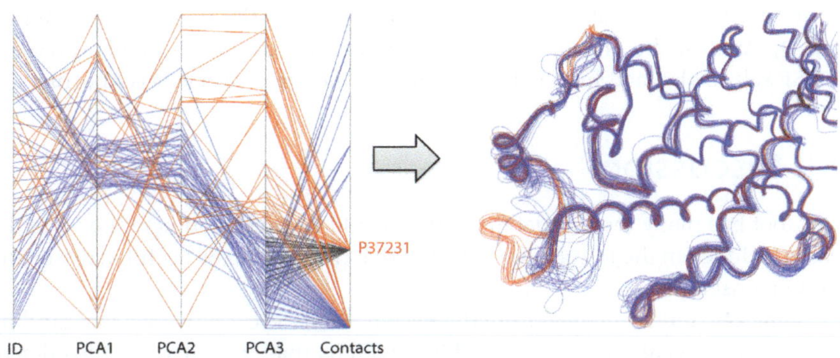

Fig. 8 Disordered region in the RXR-α ensemble. In the disordered region to the right, some structures form a relatively ordered subset, which has been selected (in red) by brushing the 'PCA3' axis. Looking at the 'Contacts' axis shows that all members of the ensemble that form the red cluster bind to the same protein (PPAR-γ), whereas none of the members that bind to another protein follows this conformation (blue selection). Thus, binding of PPAR-γ appears to induce order in this region.

highlighted using the third principal component (Fig. 8). Comparing the spatial views of Fig. 7 and 8 further shows that these structural differences are correlated with differences in other disordered regions of the ensemble. After adding the 'Contacts' axis to the parallel-coordinates plot, we can see that these structures are bound to the same partner protein (PPAR-γ, P37231, see rightmost axis in Fig. 8). Obviously, this conformation is not a requirement for binding PPAR-γ, as there are other structures in the ensemble that also bind to PPAR-γ (grey). However, this special conformation seems to prevent the binding of other possible binding partners (blue in selection in Fig. 8).

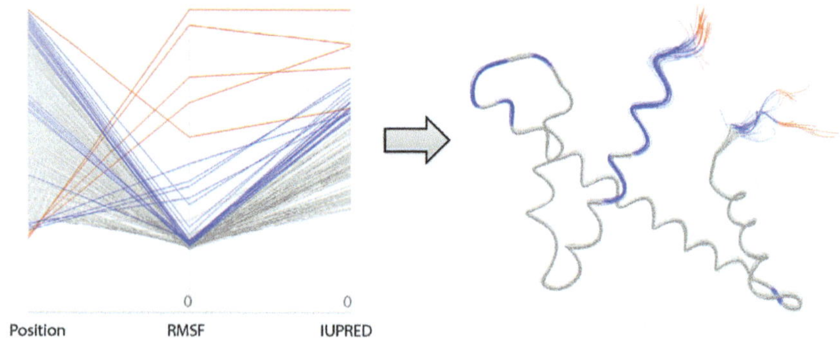

Fig. 9 Predicted *vs.* observed disorder for the H2B ensemble. Blue indicates residues predicted to be disordered (IUPRED score ≥0.5). Red indicates residues with visually outlying RMSF values (≥2.2 Å), *i.e.*, residues with high observed disorder. High observed disorder always corresponds to high IUPRED scores, but the converse is not true.

4.3 Histone H2B

This ensemble consists of 49 structures aligned to residues 30 to 127 of the human protein *histone H2B* (Q96A08). Overall, these structures are highly similar, with only two comparably small regions of disorder at the N- and C-termini (see Fig. 9). As with the previous two ensembles, all residues with highly divergent RMSF values also had high IUPRED scores, while the converse was not true.

5 Discussion

Our tool has many more capabilities than could be presented here, however the cases included in the Results demonstrate that the tool can be useful in dissecting protein ensembles.

Some key trends emerged from the Results. In some cases, we see that binding of partner molecules appears to stabilise regions that are otherwise disordered. This may explain the lack of observed disorder in many residues that were predicted to be disordered, especially for the H2B ensemble, in which all structures have been derived in complex with other histone proteins. The correlation of disorder with secondary structure observed in Fig. 5 is also interesting, and may merit further investigation using a larger set of ensembles.

However, the clearest trend to emerge was that in all three ensembles, all residues with high observed disorder (*i.e.*, outlying RMSF values) were predicted to be disordered (IUPRED score ≥0.5). Similarly, the converse was consistently not observed, *i.e.*, many residues predicted to be disordered had low RMSF. Like many other methods for predicting disorder, IUPRED is based purely on sequence, and measures the propensity of a sequence region to exhibit disorder, using only amino acid properties. Our results support the suggestion that IUPRED has high recall, but not high precision, for the task of predicting disorder in the heterogeneous ensembles used in this study. Stated another way, our results suggest that many of the residues predicted by IUPRED to be disordered are false positives. However, it is important to note that for the ensembles used in this study, the observed RMSF values may differ considerably from the 'true' disorder that occurs when these proteins are alone in solution, with no partner molecules—which is the state that IUPRED aims to predict. In contrast, the ensembles we used contained many structures with partner molecules, and almost all were derived from proteins in a crystalline state, not in solution. Nonetheless, such ensembles are a rich and detailed source of experimental data that we believe will be very useful in helping to improve our understanding of the functional roles and mechanisms of IDRs.

Overall, the results demonstrate that our tool makes it easy to explore interrelationships in these heterogeneous structural ensembles; in the near future, we intend to use our approach to look at a broader range of cases and, if these trends are maintained, to further test if these trends are statistically significant.

While we are using traditional molecular graphics techniques, the parallel-coordinates view greatly facilitates ensemble exploration by showing a large amount of additional information about the ensemble that otherwise would be hidden or difficult to see with a conventional molecular graphics approach. The combination with a spatial 3D view of the molecular structures further enables the analyst to cross-check selected patterns in a well-known environment. The use of parallel coordinates is powerful yet relatively easy to implement, and hence is a good candidate for inclusion in popular molecular graphics tools, such as VMD,[17] PyMol[18] or Chimera.[19] Such an inclusion would significantly extend the range of attributes, database support, and usability features compared to that currently available in our implementation, which is a research prototype.

There are some points that need further consideration when using parallel coordinates. One of the most criticised aspect is that patterns depend on the order of axes. There are several approaches to meet the challenge of finding a 'good' axis order: Some authors proposed using an automatic ordering based on various measures such as correlation coefficients or distance metrics (see Heinrich and Weiskopf[32] for an overview), others use manual, interactive reordering of axes (as we did for the system presented in this paper) or show all pairwise correlations in a matrix of parallel-coordinates systems.[43]

For very large ensembles and large structures, it may be useful to modify our approach, for example by adding another level of aggregation and compute statistics for whole chains or structures, instead of single residues. This would greatly reduce the number of graphical primitives that may occur, and so improve rendering speed. In part, this has been realised implicitly in our system for axes that depict information on a chain or structure basis (such as the 'Chain' or 'RMSD' axes).

In the future, we plan to extend our tool to achieve a tighter coupling between the parallel-coordinates plot and the 3D visualisation, for example by adding colour maps to an axis, thereby allowing users to select a parallel coordinate axis and to colour-code the 3D models according to the values on this axis (*e.g.* using a cool–warm shading). We also plan to add specialised protein ensemble representations (*e.g.* the 'sausage' visualisation used in MolMol[22]), and to add a range of further protein structure attributes—this will allow us to add further dimensions to the parallel coordinates and potentially find new patterns in the ensembles. Finally, we also plan to investigate the usefulness of this approach for analysing molecular dynamics simulations.

6 Conclusion

Adding views to a system that show different aspects of the data is a well-known and frequently practised approach for a wide range of applications. In this paper, we have shown that parallel coordinates can be a useful add-on to a molecular graphics environment. Applied to the rather complex use-case of analysing protein ensembles, the approach enabled us to dissect these complex datasets, and gain some insight into the correlation between observed and predicted disorder.

Acknowledgements

This work was partially funded by the German Research Foundation (DFG) as part of the Collaborative Research Centre SFB 716 (projects D.4 and D.5). The authors would like to thank Kenneth Sabir for proof reading the manuscript.

References

1 V. N. Uversky and A. K. Dunker, *Biochim. Biophys. Acta, Proteins Proteomics*, 2010, **1804**, 1231–1264.

2 M. Sickmeier, J. A. Hamilton, T. LeGall, V. Vacic, M. S. Cortese, A. Tantos, B. Szabo, P. Tompa, J. Chen, V. N. Uversky, Z. Obradovic and A. K. Dunker, *Nucleic Acids Res.*, 2007, **35**, D786–D793.

3 A. J. Martin, I. Walsh and S. C. Tosatto, *Bioinformatics*, 2010, **26**, 2916–2917.

4 W. Rieping, M. Habeck and M. Nilges, *Science*, 2005, **309**, 303–306.

5 M. Remmert, A. Biegert, A. Hauser and J. Söding, *Nat. Methods*, 2012, **9**, 173–175.

6 A. Amadei, A. B. Linssen and H. J. Berendsen, *Proteins: Struct., Funct., Genet.*, 1993, **17**, 412–425.

7 D. van Aalten, D. Conn, B. de Groot, H. Berendsen, J. Findlay and A. Amadei, *Biophys. J.*, 1997, **73**, 2891–2896.

8 U. Hensen, T. Meyer, J. Haas, R. Rex, G. Vriend and H. Grubmüller, *PLoS One*, 2012, **7**, e33931.

9 A. Inselberg, *Parallel Coordinates: Visual Multidimensional Geometry and Its Applications*, Springer, NY, USA, 2009.

10 J. Roberts, *Fifth International Conference on Coordinated and Multiple Views in Exploratory Visualization*, 2007, pp. 61–71.

11 M. A. Fisherkeller, J. H. Friedman and J. W. Tukey, *Proceedings of ACM Pacific*, 1975, pp. 140–145.

12 D. Frishman and P. Argos, *Proteins: Struct., Funct., Genet.*, 1995, **23**, 566–579.

13 Z. Dosztányi, V. Csizmók, P. Tompa and I. Simon, *J. Mol. Biol.*, 2005, **347**, 827–839.

14 F. Eisenhaber, P. Lijnzaad, P. Argos, C. Sander and M. Scharf, *J. Comput. Chem.*, 1995, **16**, 273–284.

15 H. M. Berman, J. Westbrook, Z. Feng, G. Gilliland, T. N. Bhat, H. Weissig, I. N. Shindyalov and P. E. Bourne, *Nucleic Acids Res.*, 2000, **28**, 235–242.

16 W. Kabsch, *Acta Crystallogr., Sect. A: Cryst. Phys., Diffr., Theor. Gen. Crystallogr.*, 1976, **32**, 922–923.

17 W. Humphrey, A. Dalke and K. Schulten, *J. Mol. Graphics*, 1996, **14**, 33–38.

18 W. L. DeLano, *The PyMOL Molecular Graphics System*, DeLano Scientific, Palo Alto, CA, USA, 2002.
19 E. F. Pettersen, T. D. Goddard, C. C. Huang, G. S. Couch, D. M. Greenblatt, E. C. Meng and T. E. Ferrin, *J. Comput. Chem.*, 2004, **25**, 1605–1612.
20 E. A. Coutsias, C. Seok and K. A. Dill, *J. Comput. Chem.*, 2004, **25**, 1849–1857.
21 A. J. M. Martin, I. Walsh and S. C. E. Tosatto, *Bioinformatics*, 2010, **26**, 2916–2917.
22 R. Koradi, M. Billeter and K. Wüthrich, *J. Mol. Graphics*, 1996, **14**, 51–55.
23 Y. Zhao, D. Stoffler and M. Sanner, *Bioinformatics*, 2006, **22**, 2768–2774.
24 D. Bromley, S. J. Rysavy, R. Su, R. D. Toofanny, T. Schmidlin and V. Daggett, *Bioinformatics*, 2014, **30**(4), 593–595.
25 M. Petřek, M. Otyepka, P. Banáš, P. Košinová, J. Koča and J. Damborský, *BMC Bioinformatics*, 2006, **7**, 316.
26 A. Spaar, C. Dammer, R. R. Gabdoulline, R. C. Wade and V. Helms, *Biophys. J.*, 2006, **90**, 1913–1924.
27 G. Ramachandran, C. Ramakrishnan and V. Sasisekharan, *J. Mol. Biol.*, 1963, **7**, 95–99.
28 S. I. O'Donoghue, D. S. Goodsell, A. S. Frangakis, F. Jossinet, R. A. Laskowski, M. Nilges, H. R. Saibil, A. Schafferhans, R. C. Wade, E. Westhof and A. J. Olson, *Nat. Methods*, 2010, **7**, S42–S55.
29 J. Dietzsch, J. Heinrich, K. Nieselt and D. Bartz, *Proceedings of the IEEE Symposium on Visual Analytics Science and Technology*, 2009, pp. 179–186.
30 J. Hasenauer, J. Heinrich, M. Doszczak, P. Scheurich, D. Weiskopf and F. Allgöwer, *EURASIP J. Bioinf. Syst. Biol.*, 2012, **2012**, 4–13.
31 O. Rübel, G. H. Weber, S. V. E. Keränen, C. C. Fowlkes, C. L. L. Hendriks, L. Simirenko, N. Y. Shah, M. B. Eisen, M. D. Biggin, H. Hagen, D. Sudar, J. Malik, D. W. Knowles and and B. Hamann, Proceedings of the Eurographics/IEEE-VGTC Symposium on Visualization, Eurographics Association, Lisbon, Portugal , 2006, pp. 203–210.
32 J. Heinrich and D. Weiskopf, *STAR Proceedings of Eurographics*, 2013, pp. 95–116.
33 B. T. Luke, *J. Chem. Inf. Model.*, 1993, **33**, 135–142.
34 O. M. Becker, *J. Comput. Chem.*, 1997, **18**, 1893–1902.
35 S. Grottel, G. Reina, C. Dachsbacher and T. Ertl, *Comput. Graphics Forum*, 2010, **29**, 953–962.
36 B. Wellmann, MSc thesis, Faculty for Informatics, Technical University of Munich, 2012.
37 M. Kalemanov, *BSc Honours thesis*, Faculty for Informatics, Technical University of Munich, 2012.
38 The UniProt Consortium, *Nucleic Acids Res.*, 2013, 41, D43–D47.
39 M. Krone, K. Bidmon and T. Ertl, *Proceedings of Theory and Practice of Computer Graphics*, 2008, pp. 115–122.
40 M. Krone, K. Bidmon and T. Ertl, *IEEE Trans. Visualization Comput. Graphics*, 2009, **15**, 1391–1398.
41 M. Krone, S. Grottel and T. Ertl, *Proceedings of IEEE Symposium on Biological Data Visualization*, 2011, pp. 17–22.
42 M. Krone, J. E. Stone, T. Ertl and K. Schulten, *EuroVis–Short Papers*, 2012, pp. 67–71.
43 J. Heinrich, J. Stasko and D. Weiskopf, *EuroVis–Short Papers*, 2012, pp. 37–41.

Faraday Discussions

PAPER

RiboVision suite for visualization and analysis of ribosomes†

Chad R. Bernier,‡[ab] Anton S. Petrov,‡[ab] Chris C. Waterbury,[ac] James Jett,[a] Fengbo Li,[a] Larry E. Freil,[a] Xiao Xiong,[a] Lan Wang,[a] Blacki L. R. Migliozzi,[a] Eli Hershkovits,[ab] Yuzhen Xue,[ad] Chiaolong Hsiao,[ab] Jessica C. Bowman,[ab] Stephen C. Harvey,[ac] Martha A. Grover,[ad] Zachary J. Wartell[e] and Loren Dean Williams*[ab]

Received 7th December 2013, Accepted 12th February 2014
DOI: 10.1039/c3fd00126a

RiboVision is a visualization and analysis tool for the simultaneous display of multiple layers of diverse information on primary (1D), secondary (2D), and three-dimensional (3D) structures of ribosomes. The ribosome is a macromolecular complex containing ribosomal RNA and ribosomal proteins and is a key component of life responsible for the synthesis of proteins in all living organisms. RiboVision is intended for rapid retrieval, analysis, filtering, and display of a variety of ribosomal data. Preloaded information includes 1D, 2D, and 3D structures augmented by base-pairing, base-stacking, and other molecular interactions. RiboVision is preloaded with rRNA secondary structures, rRNA domains and helical structures, phylogeny, crystallographic thermal factors, *etc.* RiboVision contains structures of ribosomal proteins and a database of their molecular interactions with rRNA. RiboVision contains preloaded structures and data for two bacterial ribosomes (*Thermus thermophilus* and *Escherichia coli*), one archaeal ribosome (*Haloarcula marismortui*), and three eukaryotic ribosomes (*Saccharomyces cerevisiae*, *Drosophila melanogaster*, and *Homo sapiens*). RiboVision revealed several major discrepancies between the 2D and 3D structures of the rRNAs of the small and large subunits (SSU and LSU). Revised structures mapped with a variety of data are available in RiboVision as well as in a public gallery (http://apollo.chemistry.gatech.edu/RibosomeGallery). RiboVision is designed to allow users to distill complex data quickly and to easily generate publication-quality images of data mapped onto secondary structures. Users can readily import and analyze their own data in the context of other work. This package allows users to import and map data from

[a]*Center for Ribosomal Origins and Evolution, Georgia Institute of Technology, Atlanta, Georgia 30332-0400, USA. E-mail: loren.williams@chemistry.gatech.edu*
[b]*School of Chemistry and Biochemistry, Georgia Institute of Technology, Atlanta, Georgia 30332-0400, USA*
[c]*School of Biology, Georgia Institute of Technology, Atlanta, Georgia 30332-0400, USA*
[d]*School of Chemical and Biomolecular Engineering, Georgia Institute of Technology, Atlanta, Georgia 30332-0400, USA*
[e]*Department of Computer Science, University of North Carolina at Charlotte, Charlotte, NC 28223-0001, USA*

† Electronic supplementary information (ESI) available. See DOI: 10.1039/c3fd00126a

‡ Joint first authors

CSV files directly onto 1D, 2D, and 3D levels of structure. RiboVision has features in rough analogy with web-based map services capable of seamlessly switching the type of data displayed and the resolution or magnification of the display. RiboVision is available at http://apollo.chemistry.gatech.edu/RiboVision.

Introduction

We present a tool devoted to the ribosome. This web-based portal, called RiboVision, is intended for rapid analysis, retrieval, filtering, and display of a variety of data types, simultaneously on three levels: primary (1D), secondary (2D), and three-dimensional (3D) structure, from ribosomes of six different species. RiboVision allows the mapping and display of new and preloaded data, swapping of data between species, and quick generation of publication quality images on any level of structure. RiboVision can maintain and display multiple layers of information, with controllable transparency. RiboVision allows users to import data from CSV files and map it directly onto all levels of structure. RiboVision has features in rough analogy with web-based map services capable of seamlessly switching the type of data displayed and the resolution or magnification of the display. RiboVision is available at http://apollo.chemistry.gatech.edu/RiboVision. We invite data for deposition, which would be made publically available.

The ribosome is a key component of life found in every living cell, and is responsible for the translation of genetic information from messenger RNA to coded protein.[1–3] The ribosome consists of several large ribosomal RNA molecules (rRNAs) and over fifty ribosomal proteins (rProteins). The ribosome contains over 250 000 atoms. Although eukaryotic ribosomes are larger and more sophisticated, many core functional regions of the ribosome are nearly identical in structure and sequence in all living organisms.[4–6] Comparison of ribosomal components reveals information about phylogeny and evolution.[7,8]

Ribosomes present special problems and opportunities for visualization and analysis. Ribosomes are exceedingly complex in terms of structure and function, highly abundant and widely distributed in biological systems, and are extremely dense in terms of available information. Easily accessible information on ribosomes includes (i) atomic positions of rRNA, rProtein, translation factors, associated ions, water molecules and antibiotics, obtained from X-ray and cryo-EM structures, (ii) molecular interactions between ribosomal components, inferred from 3D structures and from phylogeny, (iii) a very large and rapidly growing database of rRNA and rProtein sequences, (iv) phylogenetic relationships among various ribosomal components, (v) sites and types of modification of rRNA and rProteins, (vi) chemical mapping and reactivity data, (vii) functional and dynamical information, *etc.* Hundreds of secondary structures[9–15] and tens of thousands of primary structures (sequences) of ribosomes have been obtained.[16] High-resolution X-ray or cryo-EM 3D structures of ribosomal components are now available for species that represent the three major branches of the tree of life.[17–21] This portal addresses the challenge of making accessible integrated information on this key component of all biological systems. The basic features of RiboVision, such as selection of nucleotides, mapping the preloaded data onto 2D and 3D structures, and generating the figures, are intuitive. However, advanced features, such as mapping multiple sets of custom data, manual coloring of nucleotides and interactions, and asynchronous display of 2D and 3D structures, require reference to documentation available at http://apollo.chemistry.gatech.edu/RiboVision/Documentation.

A number of software packages have been designed to represent, edit, and manipulate secondary structures of RNA. This group includes RNA2DViewer/S2S,[22] RNAMovies,[23] PseudoViewer,[24] NavRNA,[25] RNAPLOT/Vienna,[26] RNAVIZ2,[27] XRNA,[28] and VARNA.[29] The functionalities of these viewers are summarized at http://varna.lri.fr/index.php?lang=en&page=links&css=varna. A specific secondary structure visualization tool has been recently developed for mapping structural and phylogenetic data onto 2D structures of ribosomal RNA [RNA2D-Map].[30] In addition, three-dimensional structures can be visualized using either stand-alone applications, *e.g.* PyMOL,[31] Chimera,[32] or VMD,[33] or java applets such as Jmol[34] and OpenAstexViewer.[35] Several RNA packages [*e.g.* Assemble2,[36] RNA2D3D[37]] provide integration of 2D and 3D visualization.

We believe RiboVision is made distinctive by our attempt to integrate a broad variety of data at all levels of structure into a single intuitive and publically accessible portal, allowing users to visualize and archive any type of data, and to generate editable publication-quality images. The RiboVision source code is fully open to the public and released under the MIT license. RiboVision is fully functional without the transfer of user data to remote servers, assuring confidentiality. RiboVision requires a modern browser with current Java. JSmol (to avoid Java) support is experimental, but causes degraded performance, so is not currently recommended.

Features

RiboVision contains preloaded ribosomal structures along with preprocessed information related to these structures. The left menu (Fig. 1a) controls the data display, data import, and output. The right side toolbar controls data visualization and website options. A detailed description of the toolbar is given in the website manual (which opens in a separate browser tab upon clicking on the RiboVision logo).

User interface

RiboVision's interface (Fig. 1) is based on modern web technology (jQueryUI) and design principles. Simple features should be easy to discover, while more advanced functionality is available and described in detail in the online documentation.

Included data

RiboVision contains preloaded structures and data for two bacterial ribosomes (*Thermus thermophilus* and *Escherichia coli*), one archaeal ribosome (*Haloarcula marismortui*), and three eukaryotic ribosomes (*Saccharomyces cerevisiae*, *Drosophila melanogaster* and *Homo sapiens*). For each species, both Large Subunits (LSUs) and Small Subunits (SSUs) are provided, except for the omission of the SSU of *H. marismortui*, for which no 3D structure is currently available. LSUs and SSUs are visualized separately and independently of each other.

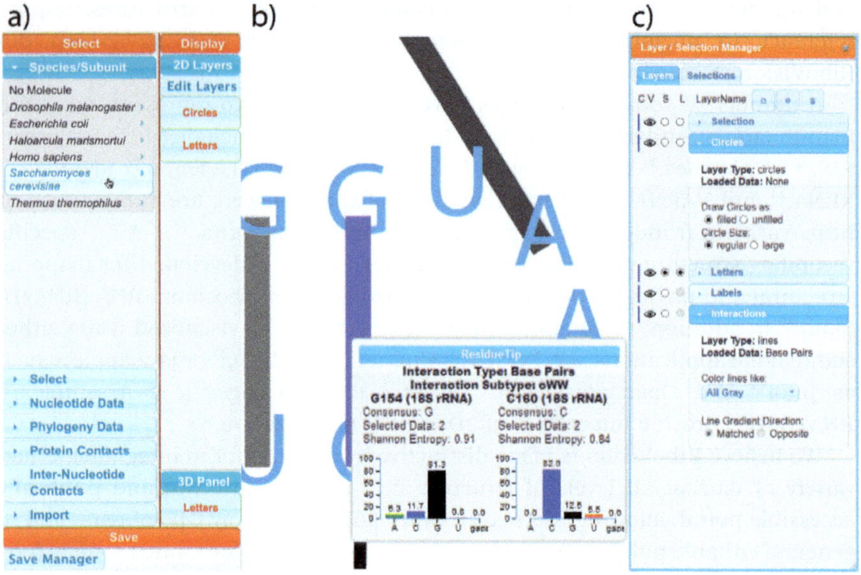

Fig. 1 Main Menu. The Species/Subunit menu offers selection of the LSU or SSU from six species. Nucleotide Selection provides options for the selection and display of specific fragments of rRNA. Nucleotide Data contains nucleotide-specific data from previous structural analyses of ribosomes. Phylogeny Data contains the Shannon entropies of each nucleotide. Protein Contacts allows users to map interactions between rProteins and rRNA. Inter-Nucleotide Contacts allows the users to display interactions between nucleotides by type. Import allows users to upload data for mapping onto each level of ribosomal structure. Display contains layer objects to which data can be loaded by dragging and dropping from the Data menus. Save allows export of figures, along with additional saving and exporting options. a) Species and subunit selection. b) ResidueTip. Hovering the mouse over data mapped on the 2D structure produces a ResidueTip, a pop-up box containing nucleotide-specific data. Hovering over an interaction line with the Alt key held down gives a ResidueTip with data on both nucleotides. c) Layer/Selection Manager. The Circles layer and the Interactions layer options panels are opened here, revealing advanced functionality.

Secondary structures

The 2D structures are composed of glyphs, one for each nucleotide. Each nucleotide glyph is a letter indicating the nucleotide type [A, C, G, U] centered on a circle. A user can map any desired data to the colors of the letters, to the colors of the circles and to lines linking the glyphs. The data are either preloaded or supplied by the user. Both traditional and corrected versions of secondary structures for SSU and LSU rRNAs are available.

3D structures

The coordinates of three-dimensional ribosomal structures were obtained from the Protein Data Bank (PDB). High resolution crystal structures were used for *E. coli*, PDB IDs: 4GD1 and 3R8S,[20] *T. thermophilus*, PDB IDs: 2J00 and 2J01,[19] *H. marismortui*, PDB ID: 1JJ2,[17] and *Saccharomyces cerevisiae*, PDB IDs: 3U5B,

3U5C, 3U5D, and 3U5E.[18] Cryo-EM structures were used for the three-dimensional coordinates of Metazoan species: *Drosophila melanogaster*, PDB IDs: 3J38, 3J3C, 3J39 and 3J3E, and *Homo sapiens*, PDB IDs: 3J3A, 3J3D, 3J3B, and 3J3F.[38]

Preloaded data

The data menus are populated with precomputed data sets. The default database currently contains nucleotide attributes that fall into three categories: Nucleotide Data, Phylogeny Data, and Protein Contacts.

Nucleotide Data allows glyph coloring and selection by (i) nucleotide number, (ii) helix or domain, (iii) radial distance from a geometric center [known as the onion partition],[4] (iv) crystallographic B-factor, and/or (v) proximity to a magnesium ion (within 2.4 Å, 2.6 Å, or 6.0 Å).

Protein Contacts allows glyph coloring and selection by molecular interactions with any desired subset of rProteins.

Phylogeny Data allows glyph coloring and selection by conservation statistics obtained from a preloaded multiple sequence alignment of rRNA sequences using the SILVA aligner.[16] The multiple sequence alignment was performed on a subset of 128 species including major phyla from all three branches of the tree of life as reported elsewhere.[39,40] Phylogeny data are represented by precomputed Shannon entropies (defined in a range of 0–2).

Inter-Nucleotide Contacts allows visualization of various types of molecular interactions between nucleotides. The basic types of interactions include *base-pairing*, *base-stacking*, *base–phosphate*, and *base–sugar*. Each of these interaction types is further sub-divided using the data and nomenclatures of the FR3D portal.[41] This menu also contains data related to RNA–RNA interactions mediated by proteins or magnesium ions, organized in *protein interactions* and *Magnesium2.4A* options. The interaction data between rRNA and rProteins can be limited by the user to any selected subset of proteins. Secondary structures do not explicitly reflect tertiary interactions, which are inferred from 3D structures. Tertiary interactions are mapped and visualized by lines on 2D structures according to their Inter-Nucleotide Contact attribute.

ResidueTip

ResidueTip is activated by hovering the mouse directly over a nucleotide glyph. Various information about that nucleotide is raised in the tip. The name of the selected nucleotide is given at the top, followed by the name of the consensus nucleotide. Next, the "Selected Data" is displayed. Selected Data is associated with that residue in the currently selected layer. The Shannon entropy and a nucleotide frequency bar chart are also included. This information is very useful to have at hand while exploring other mapped information. Holding down the Alt key activates the ResidueTip feature for interactions between glyphs. This function shows ResidueTips for both nucleotides, and elaborates on the type of interaction (Fig. 1b).

Layer/selection manager

Multiple layers of data, of various types, can be simultaneously displayed and independently manipulated. Basic manipulation of data is implemented in the

Main Menu (Select and Display panels). The advanced display and output are controlled by the Layer Manager (Fig. 1c) located in the toolbar. Sequence-related data is projected into the "Selected" Layer, controlled by the S column of radio buttons.

Each layer has a type that determines how it is mapped onto the 2D structure. The data can be used to color a nucleotide letter, to color a nucleotide circle, or to draw lines between nucleotide glyphs. The layer type can be seen by clicking and expanding the properties of a layer. The default layers are displayed in the Main Menu. In the "Letters" layer, data are represented by colors of the nucleotide letters. In the "Circles" layer, data are represented by the colors of the nucleotide circles. The users can create additional layers using the Layer Manager. User-added layers cannot be letter-type. The layers will also appear in the Display panel of the Main Menu. The ordering of the layers is controlled by dragging layer objects in the Display panel of the Main Menu or in the Layer Manager.

The color on the 3D structures can be made synchronous with any 2D layer by setting the "Linked" property (L column of radio buttons, Fig. 1c) for that layer. A layer can be temporarily turned off and on by clicking on the "V" property (the eye icon), or permanently deleted from the menu by clicking the Trash icon.

Coordinated nucleotide selection and coloring

Selection of nucleotides is coordinated across the 2D and 3D panels. The user can add nucleotides in the 2D panel view to the active selection group. Nucleotides can be selected individually or by using a click-and-drag selection box. A selected nucleotide is highlighed by a maroon circle. Multiple selection groups can be created and named using the Selections tab of the Layers/Selection Manager. The user can reactivate a given group and add/remove a nucleotide from the active selection group. In the 3D view, the selected nucleotides can be highlighted, or the user can choose to hide the non-selected nucleotides. Custom selections can be saved in a session in Save Manager.

Importing user data

The Import menu allows users to import their own data, such as conservation frequencies, SHAPE reactivity, footprinting data, *etc.* User data can be quickly and accurately mapped onto 1D, 2D, and 3D levels of structure. Data are imported as CSV files. User data does not go to the server; it is read and stored locally on the client. Pregenerated CSV input data templates are provided on the website. Differences in the numbering schemes between species require specific templates for each species, which are provided (see the detailed tutorial in the RiboVision manual for additional information.) Users can also import data from the CSV files generated by the Save Manager. These files may contain not only the species sequences, but also nucleotide and interaction data.

In the CSV template, the resNum column specifies the nucleotides in a format MoleculeName:NucleotideNumber(s) [*e.g.* 5S:35, 16S:(49–578), or 23S:271M]. Following convention, in some species NucleotideNumbers can consist of both digits and letter characters. The second column can have the heading DataCol, which contains numerical values corresponding to a desired property, or the heading ColorCol, with either hexadecimal color codes or supported color names

(1700+ colors, see manual). When ColorCol is not provided, numerical information from DataCol will be mapped to a rainbow scheme, mapping the minimum value to blue, and the maximum value to red. Alternatively, the user can supply their own color scheme.

Saving figures and work

A number of options are incorporated to allow facile production of publication quality figures. Using the Figures tab in the Save Manager, images in the 1D, 2D, and 3D panels can be individually exported in a number of file formats. Specifically, images from the 1D panel can be exported in SVG format. Images from the 2D panel can be exported as PDF, SVG, JPG, or PNG. For the vector formats (SVG and PDF), either the currently visible layer or all layers can be output. 3D images from the Jmol applet can be exported as JPG images, or can be exported to a PyMOL script. The program exports a ZIP file that contains the PyMOL script with a description of the current state and the necessary PDB files.

The Sequences & Data tab allows easy export of the actual data used in figure generation. The whole sequence is exported, along with subsets of sequences stored in the user selections. Additionally, any data used to create figures is also exported as a table.

Users can save their current work to disk for later retrieval using the Save/Restore Manager tool from the Tool Bar or by clicking the Session tab inside the Save Manager. All layers and selections can be saved. The default save location is the local browser LocalStorage cache. Restoring will reset the display including layers, selections, loaded data types, *etc.* but requires reloading of the user input data. This feature ensures user data privacy. Alternatively, users can save their work as a text file. This method allows users to restore the work session without an internet connection, or to move between browsers and computers.

Programming details

RiboVision is a cross-platform web app that integrates several advanced web solutions. The processing of data is done with jQuery, jQuery plugins, and JavaScript. The interface is based on the jQuery UI. The 1D drawing is done in SVG with the d3 library. The 2D structure drawing is done using the Canvas element. The 3D drawing is handled by the third party Java applet, Jmol. Our data are stored in a MySQL database and retrieved by a PHP server. The PHP server is also used to export file formatting and writing. Images are converted by ImageMagick. RiboVision functions, with certain limitations, on devices such as phones and tablets. RiboVision is under active development and still needs both client and server optimization. The code is available on https://github.com/RiboZones/RiboVision.

Examples

Here we demonstrate the basic functionality of RiboVision using examples. We highlight the major functions of RiboVision rather than describe the actual data used for the examples. A fully detailed description of features, as well as explanations of the methods used to generate preloaded data sets, is contained in the manual (http://apollo.chemistry.gatech.edu/RiboVision/Documentation).

Example 1. Mapping 1D data onto 2D and 3D structures

RiboVision allows users to map various data simultaneously onto all three levels of structure (1D, 2D, and 3D). In example 1 (shown in Fig. 2), we assign a color to each nucleotide glyph of the *E. coli* 23S rRNA. The color corresponds to the Shannon entropy precomputed for a set of species sampled over all phylogeny. This example, which in practice takes two steps and less than 30 s to accomplish, shows how to visualize, at nucleotide resolution, the degree of conservation of rRNA. The input data (as well as the data used in example 2) are a feature of RiboVision. The data contains the nucleotide number and a conservation score describing variability over a multiple sequence alignment using 128 representative species that sparsely represent the Woese tree of life. The 1D display (Fig. 2a) shows the phylogenetic Shannon entropy of each nucleotide, illustrating that rRNA nucleotides from 1400 to 1600 of the 23S rRNA of *E. coli* exhibit high variability. The 2D display (Fig. 2b) shows that the central loop of domain V and other non-duplex rRNA are generally conserved (dark blue); while double-stranded nucleotides are variable (red, paired nucleotides co-vary). The 3D panel (Fig. 2c) indicates that the central core of the LSU is highly conserved (blue) and the surface regions are variable (red).

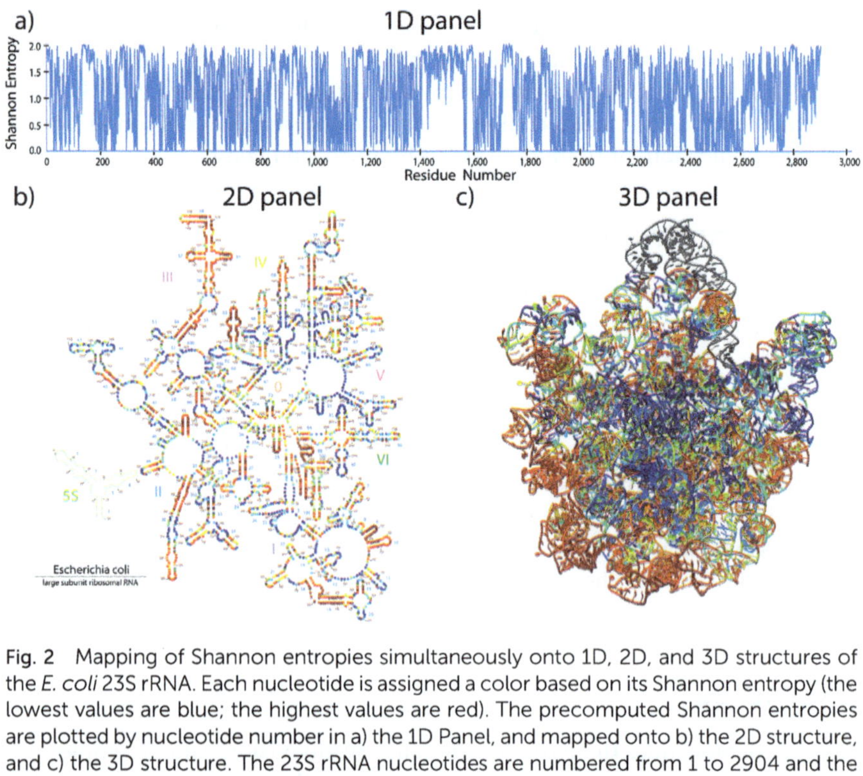

Fig. 2 Mapping of Shannon entropies simultaneously onto 1D, 2D, and 3D structures of the *E. coli* 23S rRNA. Each nucleotide is assigned a color based on its Shannon entropy (the lowest values are blue; the highest values are red). The precomputed Shannon entropies are plotted by nucleotide number in a) the 1D Panel, and mapped onto b) the 2D structure, and c) the 3D structure. The 23S rRNA nucleotides are numbered from 1 to 2904 and the 5S rRNA are numbered from 2905 to 3024 (Shannon entropies are not shown for 5S rRNA). Virtually any quantitative nucleotide-resolution data can be quickly mapped in this way.

To generate Fig. 2, select LSU rRNA of *E. coli* from the Species/Subunit section of the Main Menu (Fig. 1a). Drag Shannon Entropy from Phylogeny Data and drop it onto the Circles layer object in the Display panel of the Main Menu. In a similar fashion, essentially any type of data can be mapped simultaneously onto the 1D, 2D, and 3D structures. For example, RiboVision contains crystallographic data in the form of B-factors, which can be mapped in the same way. The preloaded database in RiboVision will grow over time.

Example 2. Mapping and visualizing the protein interactions in combination with other structural data

In this example, we illustrate a simultaneous mapping of the rRNA domain structure (from Nucleotide Data) and rRNA–rProtein molecular interactions (from Protein Contacts) onto the SSU rRNA of *T. thermophilus*. Fig. 3a depicts the 2D structure of the 16S rRNA of *T. thermophilus* with nucleotide letters colored by domain (domain 5′ is light blue, domain C is brown, domain 3′M is pink and domain 3′m is green). The same 2D structure contains the information about the

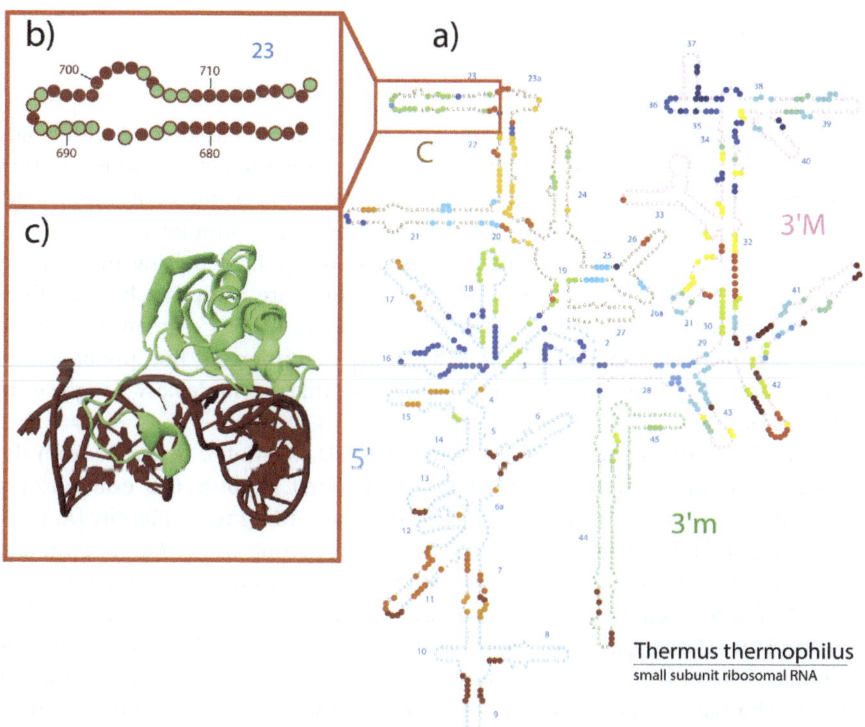

Fig. 3 Visualizing interactions between ribosomal proteins of the small subunit of *T. thermophilus* and the 16S rRNA using RiboVision. a) The nucleotides in the 2D structure of 16S rRNA are colored by domain (domain 5′ is light blue, domain C is brown, domain 3′M is pink, and domain 3′m is green), while nucleotides contacting ribosomal proteins are overlaid with colored circles; each protein is assigned a distinct color. Interactions of rProtein S11 (green) with helix 23 (maroon) of the SSU rRNA b) projected onto 2D structure and c) shown as an illustrated representation of the 3D structure.

molecular interactions of rProteins with rRNA marked with colored circles. Nucleotides within 3.4 Å of an rProtein are enclosed by circles that are color-coded by protein (S1, S2, *etc.*).

To generate Fig. 3, select SSU rRNA of *T. thermophilus* from the Species/Subunit section of the Main Menu. Drag Domains from Nucleotide Data and drop it onto the "Letters" layer object of the Display panel of the Main Menu. This action colors rRNA nucleotide letters by domain. Then select all proteins from the Protein Contacts Menu and drag the Protein Contacts object to the Circles Layer of the Display panel of the Main Menu. The name of the specific protein(s) interacting with a particular nucleotide can be seen in the ResidueTip window, which will automatically pop up upon hovering the cursor on top of a nucleotide glyph in the 2D structure.

Any given region of the ribosome can be represented in isolation and at high resolution. Fig. 3a shows isolated helix 23, and indicates that it interacts with rProtein S11 (light green circles). Fig. 3b contains a selected region of the 2D structure of helix 23. The nucleotides are selected using tools in the Select menu and marked by maroon circles. Nucleotides of helix 23 that are in contact with rProtein S11 are highlighted by green circles. The same selected fragment, along with rProtein S11, appears on the 3D structure (Fig. 3c).

Example 3. Visualizing user data and nucleotide interactions

RiboVision allows one to not only import and display data associated with single nucleotides, but also to visualize pairwise nucleotide interactions, drawing lines between them on the 2D structure. The interactions can be direct (*e.g.* base pairs or base stacking) or mediated (*e.g.* by proteins or magnesium ions).

In example 3, we demonstrate visualization of the base pairing interactions in the 18S rRNA of *S. cerevisiae*. The interactions are indicated by gray lines and assign a color to each nucleotide according to type: A – green, G – black, U – red, and C – blue (Fig. 4). The color definition is not preloaded to RiboVision but supplied by an external file (file example3.csv is given in the ESI†).

To achieve the display in Fig. 4, select the SSU rRNA of *S. cerevisiae* from the Species/Subunit section of the Main Menu. First, import the example3.csv containing the assigned colors of each residue according to its identity [A, G, U, C] (available in the ESI†) by opening the Import section of the Main Menu and selecting the file from a location on the local workstation. Then, drag the "User Data" object from the Import section of the Main Menu to the "Circles" layer object. Finally, choose the Base Pairs data set from the Inter-Nucleotide Contacts menu to visualize the Base Pairs interactions between rRNA nucleotides. The base pairs interaction data are automatically loaded into the Interactions layer of the Layer Manager (Fig. 1c). For this example, we further filtered the base pair interactions by choosing only a few sub-types. By default, the lines appear in gray. Colored lines are an advanced feature. Additional details about connected nucleotides and the interaction type are shown in the ResidueTip window upon hovering a mouse over a line while holding the ALT key (the key is required to distinguish selection of nucleotides from selection of lines).

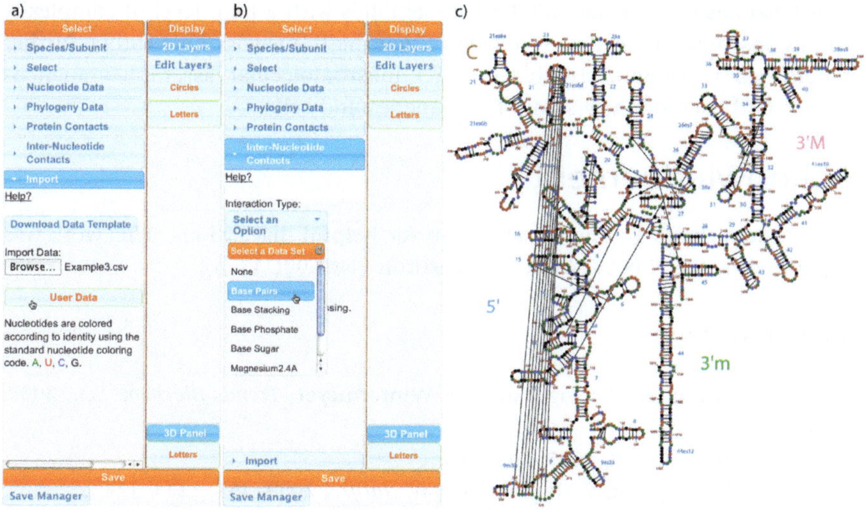

Fig. 4 A subset of base pair interactions in *S. cerevisiae* 18S rRNA visualized along with imported user data. a) The user data file "example3.csv" was loaded into RiboVision. A description of the file is visible under the "User Data" data object. b) Base Pairs was selected as the Interaction Type, from Inter-Nucleotide Contacts. Additional filtering was performed through selection of interaction sub-types. c) The resulting 2D display illustrates the standard nucleotide color code. Each nucleotide is assigned a color according to its identity. In addition, the gray lines connect nucleotides that form Watson–Crick/Watson–Crick interactions.

Conclusion and future perspectives

RiboVision was originally conceived as an in-house package for internal use in the Center for Ribosomal Origins and Evolution at Georgia Tech. We faced a bottleneck in visualization and analysis of a variety of data types on 1D, 2D, and 3D levels of structure, and in the rate of production of publication-quality figures. As the package matured, it was suggested that others might find it useful.

RiboVision has shown significant utility in our laboratories. This software enabled us to detect long-standing discrepancies between 2D and 3D structures of rRNAs. We have substantially revised the 2D structures of LSU rRNAs[40] to reflect a double-stranded region (helix 26a) formed between the left and right segments of the central loop in the traditional 2D structure. RiboVision enabled more modest revisions of the 2D structures of SSU rRNAs,[42] providing an accurate representation of the triple helical structure of the central pseudoknot and numerous non-canonical base pairs distributed throughout the 2D structure. We used RiboVision to create a public gallery of rRNA secondary structures, mapped with a variety of data (http://apollo.chemistry.gatech.edu/RibosomeGallery).

RiboVision is open source and is based on modern website technology. The version released here is focused primarily on manipulation and presentation at the 2D structural level with more limited options in 1D and 3D Panels. Visualization is limited to one subunit of one species at a time. The scope of the program will be extended in future versions.

Ribosomes are not the only large assemblies with a high level of complexity and extensive available data. We believe that the framework here can be further generalized to protein structures, other macromolecular assemblies such as viruses and bacteriophages, and to organelles and cells.

Acknowledgements

The authors thank Dr Hyman Hartman for helpful discussions. This work was supported by the NASA Astrobiology Institute (NNA09DA78A).

References

1 M. V. Rodnina, M. Beringer and W. Wintermeyer, *Trends Biochem. Sci.*, 2007, **32**, 20–26.
2 A. Yonath, *Mol. Cell*, 2005, **20**, 1–16.
3 T. M. Schmeing and V. Ramakrishnan, *Nature*, 2009, **461**, 1234–1242.
4 C. Hsiao, S. Mohan, B. K. Kalahar and L. D. Williams, *Mol. Biol. Evol.*, 2009, **26**, 2415–2425.
5 J. M. Harms, F. Schlunzen, P. Fucini, H. Bartels and A. Yonath, *BMC Biol.*, 2004, **2**, 4.
6 M. J. Belousoff, C. Davidovich, E. Zimmerman, Y. Caspi, I. Wekselman, L. Rozenszajn, T. Shapira, O. Sade-Falk, L. Taha, A. Bashan, M. S. Weiss and A. Yonath, *Biochem. Soc. Trans.*, 2010, **38**, 422–427.
7 G. E. Fox, *Cold Spring Harb. Perspect. Biol.*, 2010, **2**, a003483.
8 K. Bokov and S. V. Steinberg, *Nature*, 2009, **457**, 977–980.
9 C. R. Woese, L. J. Magrum, R. Gupta, R. B. Siegel, D. A. Stahl, J. Kop, N. Crawford, J. Brosius, R. Gutell, J. J. Hogan and H. F. Noller, *Nucleic Acids Res.*, 1980, **8**, 2275–2293.
10 H. F. Noller, J. Kop, V. Wheaton, J. Brosius, R. R. Gutell, A. M. Kopylov, F. Dohme, W. Herr, D. A. Stahl, R. Gupta and C. R. Woese, *Nucleic Acids Res.*, 1981, **9**, 6167–6189.
11 H. F. Noller, *Annu. Rev. Biochem.*, 1984, **53**, 119–162.
12 R. R. Gutell and G. E. Fox, *Nucleic Acids Res.*, 1988, **16**, R175–R269.
13 R. R. Gutell, N. Larsen and C. R. Woese, *Microbiol. Rev.*, 1994, **58**, 10–26.
14 J. J. Cannone, S. Subramanian, M. N. Schnare, J. R. Collett, L. M. D'Souza, Y. Du, B. Feng, N. Lin, L. V. Madabusi, K. M. Muller, N. Pande, Z. Shang, N. Yu and R. R. Gutell, *BMC Bioinformatics*, 2002, **3**, 2.
15 R. R. Gutell, B. Weiser, C. R. Woese and H. F. Noller, *Prog. Nucleic Acid Res. Mol. Biol.*, 1985, **32**, 155–216.
16 E. Pruesse, J. Peplies and F. O. Glockner, *Bioinformatics*, 2012, **28**, 1823–1829.
17 N. Ban, P. Nissen, J. Hansen, P. B. Moore and T. A. Steitz, *Science*, 2000, **289**, 905–920.
18 A. Ben-Shem, N. G. de Loubresse, S. Melnikov, L. Jenner, G. Yusupova and M. Yusupov, *Science*, 2011, **334**, 1524–1529.
19 M. M. Yusupov, G. Z. Yusupova, A. Baucom, K. Lieberman, T. N. Earnest, J. H. Cate and H. F. Noller, *Science*, 2001, **292**, 883–896.
20 J. A. Dunkle, L. Y. Wang, M. B. Feldman, A. Pulk, V. B. Chen, G. J. Kapral, J. Noeske, J. S. Richardson, S. C. Blanchard and J. H. D. Cate, *Science*, 2011, **332**, 981–984.

21 D. J. Taylor, B. Devkota, A. D. Huang, M. Topf, E. Narayanan, A. Sali, S. C. Harvey and J. Frank, *Structure*, 2009, **17**, 1591–1604.
22 F. Jossinet and E. Westhof, *Bioinformatics*, 2005, **21**, 3320–3321.
23 D. Evers and R. Giegerich, *Bioinformatics*, 1999, **15**, 32–37.
24 Y. Byun and K. Han, *Bioinformatics*, 2009, **25**, 1435–1437.
25 D. Auber, L. Nigay and G. Bailly, in *AVI 06: Proceedings of the International Conference on Advanced Visual Interfaces*, Italy, 2006, pp. 504–507.
26 I. L. Hofacker, *Nucleic Acids Res.*, 2003, **31**, 3429–3431.
27 P. De Rijk, J. Wuyts and R. De Wachter, *Bioinformatics*, 2003, **19**, 299–300.
28 http://rna.ucsc.edu/rnacenter/xrna/xrna.html.
29 K. Darty, A. Denise and Y. Ponty, *Bioinformatics*, 2009, **25**, 1974–1975.
30 X. Weijia, A. Wongsa, L. Jung, S. Lei, J. J. Cannone and R. R. Gutell, in *Proceedings of 2011 IEEE International Conference on Bioinformatics and Biomedicine*, Atlanta, GA, 2011, pp. 613–617.
31 *The PyMOL Molecular Graphics System*, Version 1.5.0.4, Schrödinger, LLC.
32 E. F. Pettersen, T. D. Goddard, C. C. Huang, G. S. Couch, D. M. Greenblatt, E. C. Meng and T. E. Ferrin, *J. Comput. Chem.*, 2004, **25**, 1605–1612.
33 W. Humphrey, A. Dalke and K. Schulten, *J. Mol. Graphics*, 1996, **14**, 33–38.
34 http://www.jmol.org/.
35 M. J. Hartshorn, *J. Comput.-Aided Mol. Des.*, 2002, **16**, 871–881.
36 F. Jossinet, T. E. Ludwig and E. Westhof, *Bioinformatics*, 2010, **26**, 2057–2059.
37 H. M. Martinez, J. V. Maizel and B. A. Shapiro, *J. Biomol. Struct. Dyn.*, 2008, **25**, 669–683.
38 A. M. Anger, J. P. Armache, O. Berninghausen, M. Habeck, M. Subklewe, D. N. Wilson and R. Beckmann, *Nature*, 2013, **497**, 80–85.
39 A. S. Petrov, C. R. Bernier, C. L. Hsiao, C. D. Okafor, E. Tannenbaum, J. Stern, E. Gaucher, D. Schneider, N. V. Hud, S. C. Harvey and L. D. Williams, *J. Phys. Chem. B*, 2012, **116**, 8113–8120.
40 A. S. Petrov, C. R. Bernier, E. Hershkovitz, Y. Xue, C. C. Waterbury, M. A. Grover, H. S. C., N. V. Hud, R. M. Wartell and L. D. Williams, *Nucleic Acids Res.*, 2013, **41**, 7522–7535.
41 M. Sarver, C. L. Zirbel, J. Stombaugh, A. Mokdad and N. B. Leontis, *J. Math. Biol.*, 2008, **56**, 215–252.
42 A. S. Petrov, C. R. Bernier, B. Gulen, C. C. Waterbury, E. Hershkovitz, C. Hsiao, S. C. Harvey, N. V. Hud, G. E. Fox, R. M. Wartell and L. D. Williams, *PLoS One*, 2014, **9**, e88222.

Faraday Discussions

PAPER

NRas slows the rate at which a model lipid bilayer phase separates†

Elizabeth Jefferys, Mark S. P. Sansom and Philip W. Fowler*

Received 12th December 2013, Accepted 26th February 2014
DOI: 10.1039/c3fd00131h

The Ras family of small membrane-associated GTP-ases are important components in many different cell signalling cascades. They are thought to cluster on the cell membrane through association with cholesterol-rich nanodomains. This process remains poorly understood. Here we test the effect of adding multiple copies of NRas, one of the canonical Ras proteins, to a three-component lipid bilayer that rapidly undergoes spinodal decomposition (*i.e.* unmixing), thereby creating ordered and disordered phases. Coarse-grained molecular dynamics simulations of a large bilayer containing 6000 lipids, with and without protein, are compared. NRas preferentially localises to the interface between the domains and slows the rate at which the domains grow. We infer that this doubly-lipidated cell signalling protein is reducing the line tension between the ordered and disordered regions. This analysis is facilitated by our use of techniques borrowed from image-processing. The conclusions above are contingent upon several assumptions, including the use of a model lipid with doubly unsaturated tails and the limited structural data available for the C-terminus of NRas, which is where the lipid anchors are found.

Introduction

The capacity of eukaryotic cells to sense signals and react is central to their ability to act in a cooperative way. Specific ligands are detected by a range of receptor proteins embedded in the cell membrane – these proteins then interact with other proteins to process, amplify and direct the signal to the appropriate location within the cell. Rather than being randomly distributed, these cell signalling proteins are localised to specific regions of the membrane and this is believed to be important for their function.[1] In 1997, Simons and Ikonen[2] proposed the "lipid-raft" hypothesis that identified small cholesterol-rich nanodomains as being the compartments into which cell signalling proteins partition. Although intuitively appealing, the lipid-raft hypothesis remains controversial,[3] in no small part because rafts have not been directly detected. For example, although there

Department of Biochemistry, University of Oxford, South Parks Rd, Oxford, OX1 3QU, UK. E-mail: philip.fowler@bioch.ox.ac.uk; Fax: +44 1865 613 201; Tel: +44 1865 613 200

† Electronic supplementary information (ESI) available: See DOI: 10.1039/c3fd00131h

have been many *in vitro* studies examining the ability of a lipid mixture to macroscopically phase separate into large ordered (L_o) and disordered (L_d) regions, this effect has never been directly observed *in vivo*. Little attention has been focussed on the other physical processes which also could be responsible for altering the local concentration of a protein: for example, the protein could interact with lipids, which could affect the dynamics of the bilayer. Also, the local curvature of a lipid bilayer could drive the preferential association of different proteins.

We shall focus here on NRas, a membrane-associated G-protein. Mutations of NRas, and the other members of the Ras family, are found to varying degrees in most human cancers,[4] underscoring their centrality in many different cell signalling cascades and hence their physiological importance. NRas comprises a G-domain and a hyper variable region (HVR, Fig. 1) that mediates its interactions with lipid bilayers. In common with the other members of the Ras family, the C-terminal cysteine (Cys-186) is covalently and irreversibly attached to an unsaturated lipid (a farnesyl moiety).[5] A second saturated lipid, usually a palmitoyl group, may be attached to Cys-181; this lipid modification can be reversed enzymatically. In this paper, we shall investigate using computer simulation how the protein itself affects the properties and behaviour of a phase-separating lipid bilayer and whether the segregation of NRas can be driven by the immiscibility of different lipid species.

Until recently, one could only simulate several hundred lipids for a few tens of nanoseconds, perhaps with a single protein, or protein complex (*e.g.* an ion

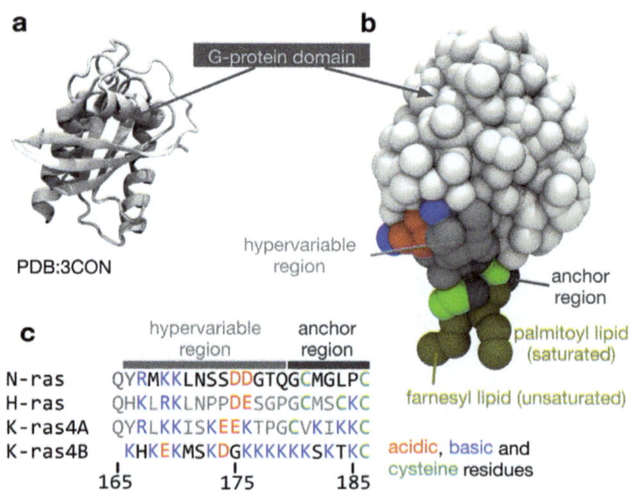

Fig. 1 The structure of NRas. (a) The structure of the G-protein domain of NRas with guanosine diphosphate (GDP) bound[6] at a resolution of 1.65 Å. Only residues 1 to 60 and 72 to 167 were resolved in the crystal structure. (b) The missing residues were modelled as described in the Methods and then converted to a coarse-grained MARTINI representation.[7] A palmitoyl lipid was attached to Cys-181 and a farnesyl lipid was attached to Cys-186 at the C-terminus. (c) A multiple sequence alignment between the canonical members of the Ras family highlighting the cysteines (green), to which lipids can be attached and the large variation in acidic (red) and basic (blue) residues. The same colouring scheme is used in the other panels of this figure.

channel), embedded, albeit with all atoms represented. With the advent of so-called 'coarse-grained' models for lipids and proteins and the continual growth in computational power, it has become possible in the past few years to simulate thousands of lipids for microseconds, a huge increase in capability in a short space of time. MARTINI is the most widely used coarse-grained forcefield[7] and it is parameterised to reproduce a range of experiment results for the lipids (such as the area per lipid) and the amino acids (such as the partition free energy). MARTINI has been used to simulate the phase separation of mixtures of lipids and cholesterol[8] and the behaviour of a range of proteins, including members of the Ras family, when inserted into phase-separating lipid mixtures.[9–11]

Unfortunately our ability to visualise and analyse such complex systems has not fully kept pace with these developments. Consider how one would determine if the lipid bilayer has separated into L_o and L_d domains and what their sizes are. Conventionally, spheres would be placed at the Cartesian coordinates of the different lipid atoms or coarse-grained beads, and an image rendered with a program such as VMD[12] or PyMol. This will show if phase separation is complete – quantification is more difficult.

Possible solutions come *via* the realisation that the vast majority of our existing visualisation and analysis tools are firmly rooted in the Cartesian paradigm and are not well suited to a study of this nature; alternative paradigms may be more appropriate (Fig. 2). The first of these borrows concepts from *fluid dynamics* and intuitively displays the velocity of particles using streamlines; this would help one study the diffusion and collective behaviour of lipids and, for example, if their motions correlate with embedded proteins. This is the subject of another paper at this Faraday Discussion meeting.[13] The second is to construct a *mathematical graph* where, for example, the nodes are the centres of mass of the proteins and an edge indicates that two proteins are interacting. The largest connected cluster and other useful statistics can be easily retrieved from such a

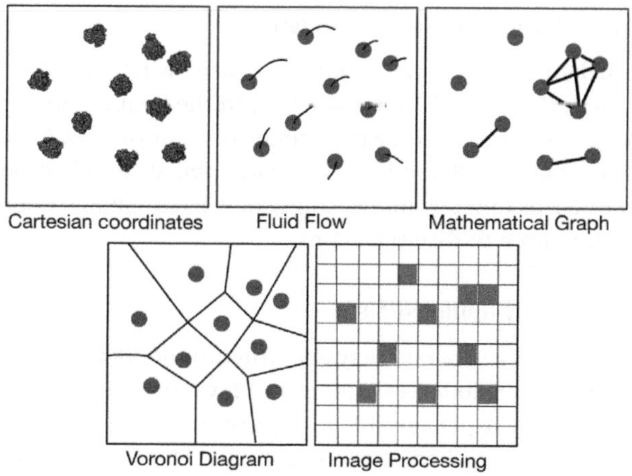

Fig. 2 Schematic illustration of several paradigms for visualisation and analysis. The first image is a top view of ten NRas proteins taken from one of the simulations. Van der Waals spheres have been placed at the centres of mass of all the coarse-grained beads and an image rendered using VMD.[12] All other images are illustrative.

graph. Alternatively, one can represent the surfaces of the lipid bilayer as either a *Voronoi tessellation* or a *Delaunay triangular mesh* (the latter is the dual graph of the former and hence they are related). Voronoi tessellation has been extensively used to measure the area per each lipid in a simulation[14–17] and has also been used to examine the phase separation of a mixed lipid bilayer.[18] Finally, one may borrow ideas from the field of *image processing* and represent the bilayer as a two-dimensional array of pixels whose values, for example, could indicate the density of a particular molecular species. This last paradigm is the best suited to our current problem and, as we shall see, standard image processing techniques and libraries will allow us to easily detect where the edges between different phases lie.

Methods

Setup of phase-separating lipid bilayer

A previously equilibrated patch of 1500 coarse-grained POPC lipids[19] was first tessellated to produce a square patch of 6000 lipids, 452 Å square. The sides of each patch fit perfectly together because the simulations are run with periodic boundary conditions. The lipids were then randomly 'mutated'[19] to achieve a 5 : 3 : 2 mixture of dipalmitoylphosphatidylcholine (DPPC, 751 lipids), dilinoleylphosphatidylcholine (DUPC, also known as DLiPC, 450 lipids) and cholesterol (299 lipids). A polyunsaturated lipid, such as DUPC, is required for phase separation using the MARTINI forcefield.[8,20] This mixture has also previously been shown to spontaneously phase separate in a reasonable time.[8,10] This patch was then solvated by the addition of 65 984 water beads, 2016 Na^+ beads and 2016 Cl^- beads, yielding an ionic strength of ~0.15 M and a total of 137 232 coarse-grained beads. An initial conformation was produced by minimising the energy of the system for 1000 steps using the steepest descents algorithm. Three 5 μs simulations of the lipid bilayer without any protein were run as controls using the GROMACS 4.6 molecular dynamics code[21] and the MARTINI v2.1 coarse-grained forcefield.[7]

Details of the molecular dynamics simulations

All non-bonded interactions were shifted to zero when the distance between beads was 12 Å. A timestep of 20 fs was used and the temperature was held at 310 K using a Berendsen thermostat with a time constant of 1 ps. The pressure was maintained at 1 bar by a Berendsen barostat, applied semi-isotropically, with a time constant of 1 ps and a compressibility of 3×10^{-4} bar.[22] Coordinates of all molecules were written to disc every 2 ns.

Construction of dually-lipidated NRas

As shown in Fig. 1, the available crystal structure of NRas[6] is incomplete. The missing surface-exposed loop (residues 60 to 72) was modelled using FREAD.[23] An initial structure of the hyper variable region (residues 167 to 179) was generated using Chimera[24] and attached to the G-domain. The resulting nearly-complete atomistic structure was converted to a coarse-grained MARTINI representation[7] and simulated for 0.1 μs in a 100 × 100 × 100 Å box containing 9565 water beads, 96 Na^+ beads and 90 Cl^- beads. Finally, a configuration of the anchor region (residues 180 to 186) was generated in Chimera[24] that satisfied a set of dihedral angles measured by solid state NMR of full-length NRas.[25] A harmonic bond with an equilibrium length of 5 Å and force constant 10 kJ mol^{-1} $Å^{-2}$ was also applied between the backbone beads of Cys181 and Leu184 to maintain the horseshoe-shape propensity that has been observed in numerous experimental studies.[25–27] The resulting, unlipidated anchor was simulated in a 60 × 60 × 60 Å box containing 1113 water beads, 20 Na^+ beads and 21 Cl^- beads to ensure the construct was stable.

All published studies that use the MARTINI coarse-grained forcefield to study how Ras proteins interact with lipid bilayers assume that farnesyl, a complex unsaturated lipid, can be represented as a linear sequence of 3 hydrophobic beads. We produced an improved set of MARTINI parameters for farnesyl, including the use of C4 beads to mimic the more polar nature of the double bonds, as described in the ESI.† This resulted in the coarse-grained beads being placed at the centre of mass of each isoprenyl unit, resulting in a naturally staggered geometry. Parameters and coordinates for palmitoyl were taken from the MARTINI parameters for DPPC.[28]

Farnesyl and palmitoyl were linked to Cys186 and Cys181 of the anchor *via* harmonic bonds of length 3.9 Å and force constant 50 kJ mol^{-1} $Å^{-2}$.[10] The equilibrium bond angles were 100 degrees with force constant 25 kJ mol^{-1} rad^{-2}, conforming to the carbon–sulphur–carbon (CSC) bond angle in thioethers. In order to model carboxymethylation, the Cys186 backbone particle was converted from type Qa with charge −1 to Na (uncharged). Comparison of the membrane association of peptides with and without this change demonstrated it to increase the membrane insertion depth of the C-terminus (data not shown), consistent with experimental data.[26] To assess the stability of the lipid-modified anchor, it was simulated alone in a 150-lipid DPPC bilayer for 100 ns. The dually lipidated anchor was attached to the C-terminus of the G-domain-linker construct using the usual parameters for backbone particle bonded interactions,[29] and the full-length model simulated for 100 ns in a 600-lipid DPPC bilayer with 6226 water molecules, 179 Na^+ and 174 Cl^- beads to assess stability of the construct and allow it to relax.

Insertion of NRas proteins into lipid bilayer

Since the model membrane is symmetric, we shall insert copies of NRas into both leaflets, rather than just one leaflet as would be the case in a physiological, asymmetric membrane. Ten copies of farnesylated and palmitoylated NRas were positioned so their lipid tails were inserted into each leaflet of the initial conformation of the phase-separating lipid bilayer, making twenty copies in each simulation. The proteins were arranged in a hexagonal pattern to optimise the packing of the protein. Rather than delete the lipids whose beads where within van der Waals distances of the lipidated proteins, we instead gradually "phased-in" over 1000 timesteps all the lipidated NRas proteins using the free energy module of GROMACS. A soft-core van der Waals potential[30] was used, with values of the soft-core parameter and power of 1.5 and 1.0, respectively. To provide room for the G-domains, a larger number of water beads (169 701) were added to increase the extent of the simulation unit cell along the *z* dimension, resulting in a total of 251 141 coarse-grained particles. Three 5 μs simulations were subsequently run of doubly-lipidated NRas. The simulation parameters are identical to those used for the control simulations of the lipid bilayers.

Analysis using image-processing techniques

We used Python 2.6 for all of the analysis. The coordinates and trajectories were read into MDAnalysis,[31] an open-source Python module that can read a wide range of coordinate and trajectory formats, including those produced by GROMACS. We also made extensive use of NumPy,[32] a Python module that creates *N*-dimensional array objects, and SciPy,[33] a library of functions that can act on NumPy arrays. Throughout, each pixel has dimensions 1 × 1 Å. First, two-dimensional NumPy arrays were created for each of the different species (DPPC, DUPC, cholesterol, protein) to cover the extent of the lipid bilayer. Since the initial configurations were all square and a semi-isotropic barostat was used, the lipid bilayers remained square, although their precise dimensions fluctuated during the simulations. Typically, the lipid bilayer measured ~400 Å to a side, hence the NumPy arrays measured ~400 × ~400 pixels. A pixel was set to unity if the coordinates of either the PO_4 bead (DPPC or DUPC), the ROH bead (cholesterol) or the centre of mass (protein) lay within its bounds (Fig. 3a). To give each species spatial extent (density), a Gaussian was then convolved with each NumPy array. We assumed that each lipid took up equal space in the leaflet and since there are 3000 lipids per leaflet and the bilayer is ~400 Å square, this equates to a Gaussian 4.1 Å wide. When two NRas proteins interact, their centres of mass are ~40 Å apart, hence, we chose the width of the Gaussian for the proteins to be 20 Å. To determine the extent of the L_o and L_d domains, we subtracted the density of the saturated lipid (DPPC) from that of the unsaturated lipid (DUPC, Fig. 3a). Next we created a mask which was unity everywhere this difference was greater than the mean, and zero everywhere else. This elegantly defined the L_o and L_d domains (Fig. 3a) and made subsequent analysis much more straightforward. Applying the Canny edge detection algorithm[34] with a smoothing Gaussian of width 2.0 Å accurately identified the edges between the two phases (Fig. 3a).

Finally, to analyse the topology of the surfaces of the two leaflets of the lipid bilayer, we create a NumPy array of the height (z) of the phosphate beads (Fig. 3c). Since the pixel density is much higher than the lipid density, this is initially sparse, however, we used a cubic interpolation routine (from SciPy) to construct an array that describes the topology of the surface of either leaflet. The difference between the upper and lower surface naturally gave us the thickness of the bilayer.

Results

The lipid bilayer rapidly forms L_o and L_d domains

Let us first examine how the 5 : 3 : 2 DPPC:DUPC:cholesterol bilayer behaves by examining the three 5 μs control simulations of the lipid bilayer (Fig. 4 and Fig. S2 & S3 in the ESI†). During the course of each simulation, the regions of predominantly saturated (L_o) and unsaturated (L_d) lipids increase in size (Fig. 4a). After ~3 μs, each leaflet has taken on a 'striped' appearance with (due to the periodic boundary conditions) one L_o stripe and one L_d stripe. The interface between the two phases remains very rough and there are small isolated islands of, for example, unsaturated lipids surrounded by saturated lipids. As one would expect, the cholesterol partitions primarily into the ordered L_o domain (Fig. S4 in the ESI†).

Adding the arrays of both leaflets shows that the formation of the L_o and L_d domains is highly correlated between the leaflets (Fig. 4b). The regions where

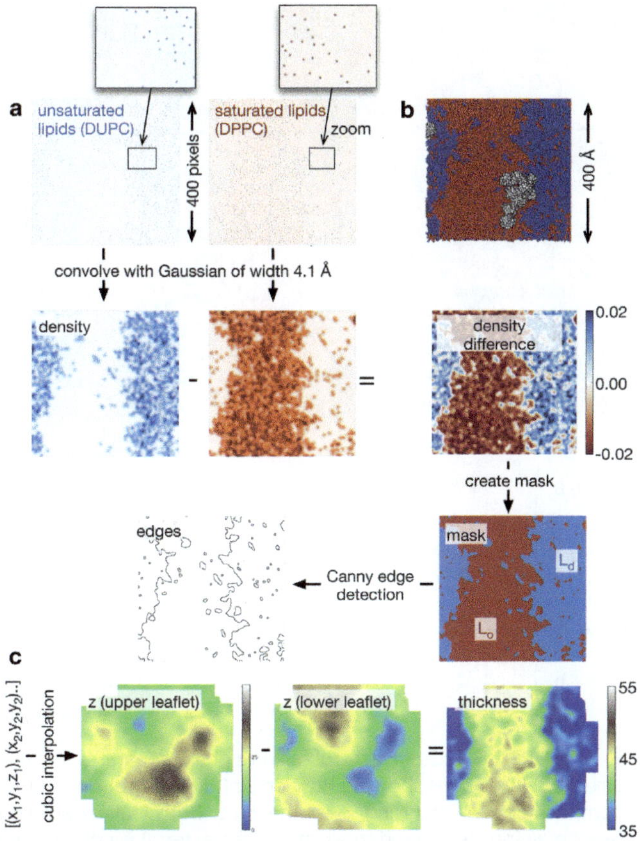

Fig. 3 Workflow showing how the image processing techniques are applied to each lipid bilayer. (a) The sparse arrays showing where the phosphate beads of the unsaturated and saturated lipid species are convolved with a Gaussian of width 4.1 Å. The pixels in the initial sparse arrays are very small, so a region in each array has been magnified. The difference in these two arrays identifies qualitatively the L_o and L_d regions. If we identify any pixel with a value > mean density as belonging to the L_d phase and the remainder the L_o phase, then we generate a mask. Applying the Canny edge detection algorithm identifies the interface between the two phases. (b) For comparison, an image rendered by VMD of the same data is shown. (c) We can create the surface of each leaflet by interpolating between the coordinates of the lipid phosphate beads. The difference in these arrays naturally gives us the thickness of the bilayer. This is coloured using a topographical palette: blues and greens represent smaller values of the thickness, yellows and browns, larger values of the thickness. For simplicity, we have not shown the proteins, but a similar procedure can be applied to the coordinates of their centres of mass to create arrays of their 'density'.

there is a domain mismatch between the upper and lower leaflets occur mainly at or near to the interfaces between the two phases. The proportion of the bilayer where there is a domain mismatch is initially 49% but falls gradually as the L_o and L_d domains grow until a plateau of ~15% appears to be reached after 4 μs (Fig. 4c) – we shall discuss this later. The other two control simulations behave in a similar manner (Fig. S2 & S3 in the ESI†).

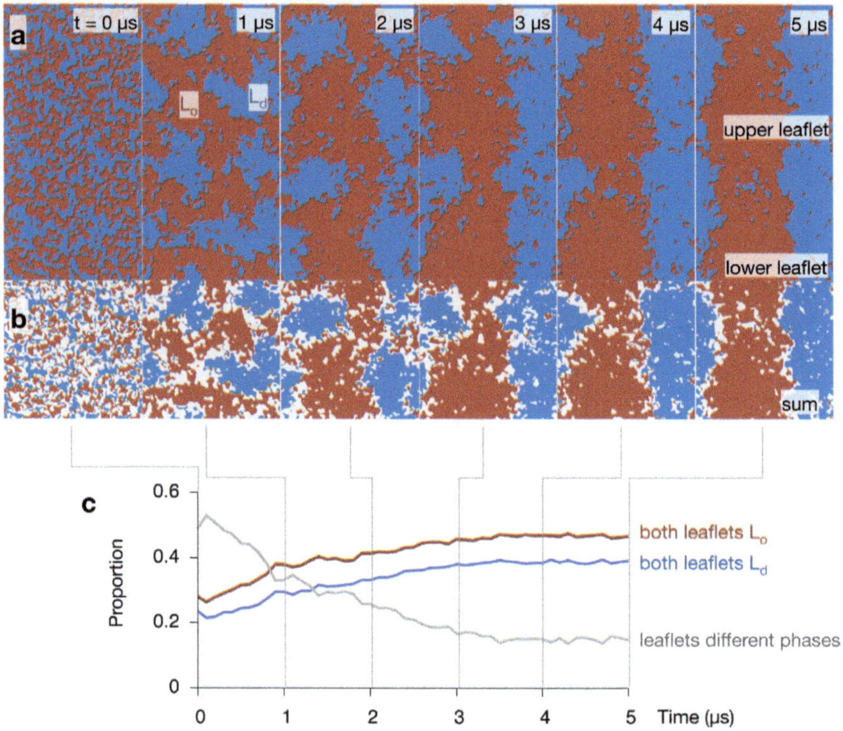

Fig. 4 The 5 : 3 : 2 DPPC:DUPC:cholesterol mixture phase separates and the two leaflets are highly correlated. (a) Images showing how one of the control simulations undergoes spinodal decomposition (*i.e.* demixing) over the 5 μs. Both leaflets are shown and the ordered phase containing DPPC (and most of the cholesterol) is coloured red, whilst the disordered phase containing DUPC is coloured blue. (b) The sum of the two arrays shows the correlation between the two leaflets. Regions where both leaflets are ordered or disordered are coloured red and blue, respectively. Any region where the leaflets are in different phases are coloured white. The majority of the differences between the leaflets is found at the interfaces. (c) Initially the phase of around half of the bilayer does not match the phase of the other leaflet. This proportion falls as the bilayer undergoes spinodal decomposition, whilst the proportion of the bilayer where the phases on both sides of the bilayer match increases.

The L_o domain is thicker than the L_d domain

The average thickness of the bilayer, measured between the phosphate beads, remains constant (~41 Å) during the formation of the L_o and L_d domains, however, this hides significant differences between the thicknesses of the two phases that emerge during the simulations (Fig. 5 and Fig. S5 & S6 in the ESI†). The average thickness of the bilayer where both leaflets are disordered decreases, reaching ~36 Å by the end of the simulation. At the same time, the average thickness where both leaflets are ordered increases, reaching ~45 Å. Examining the distribution of bilayer thicknesses clearly shows the bimodal behaviour (Fig. 5c) and further suggests that the L_o and L_d domains get thicker or thinner, respectively, as the simulations progress and they get larger.

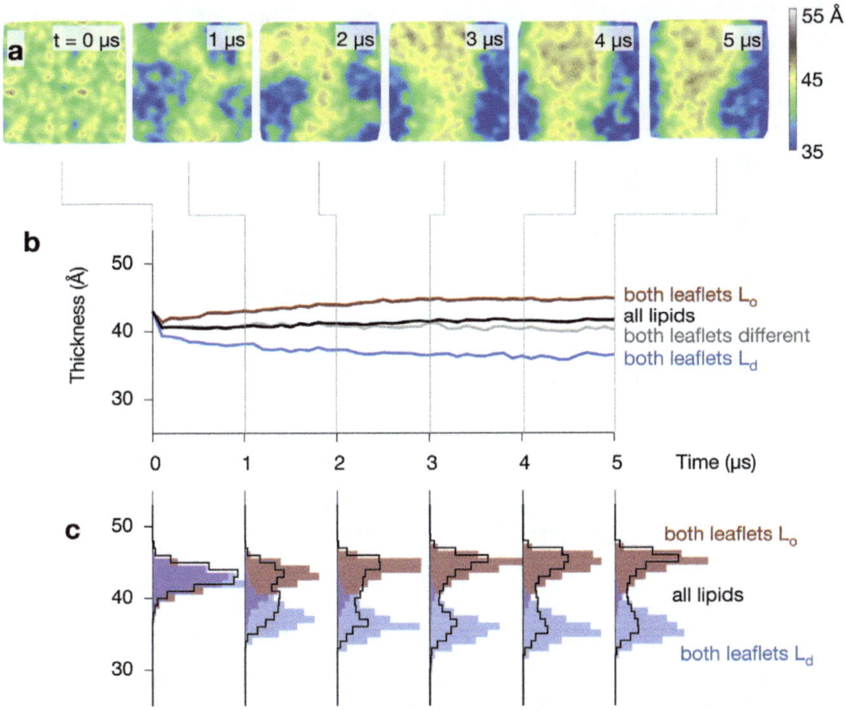

Fig. 5 The ordered phase is thicker than the disordered phase. (a) A series of images of the thickness of one of the control simulations, showing how the central L_o strip is thicker than the surrounding disordered regions (compare to Fig. 4). (b) The average thickness, measured between the phosphate beads, (in black) remains unchanged at ~41 Å during the formation of the L_o (red) and L_d (blue) domains, however, this hides the large difference in thickness between the two regions. (c) Examining the distributions shows how the complex behaviour develops.

NRas slows down the rate of spinodal decomposition

Since NRas has two lipid anchors, one saturated and one unsaturated, it is not clear what effect its presence in one of the two leaflets will have on the dynamics of the phase separation observed in purely lipid bilayers. Images taken from all three NRas simulations are shown in Fig. S7 in the ESI.† Comparing the upper leaflets of the three control simulations after 5 μs (Fig. 6a) to the upper leaflets of the three NRas simulations after 5 μs (Fig. 6b) is inconclusive; phase separation has occurred in both cases, although 'stripes' have not formed in the second NRas simulation, suggesting the process has been retarded. The total length of the interface between the L_d and L_o domains is a simple measure of the degree of phase separation. It rapidly decreases as the domains coarsen (Fig. 6c). However, when ten copies of NRas are added to each leaflet, we find that there are longer interfaces between the domains when NRas is present, compared to when it is not. A similar trend is observed when we plot the average radius of each domain against time (Fig. 6d); the average radius is reduced when NRas is added.

Whilst the proportion of the bilayer that has matched domains in both leaflets (Fig. 4c) appears to reach a plateau after 3–4 μs, the length of the interface and the

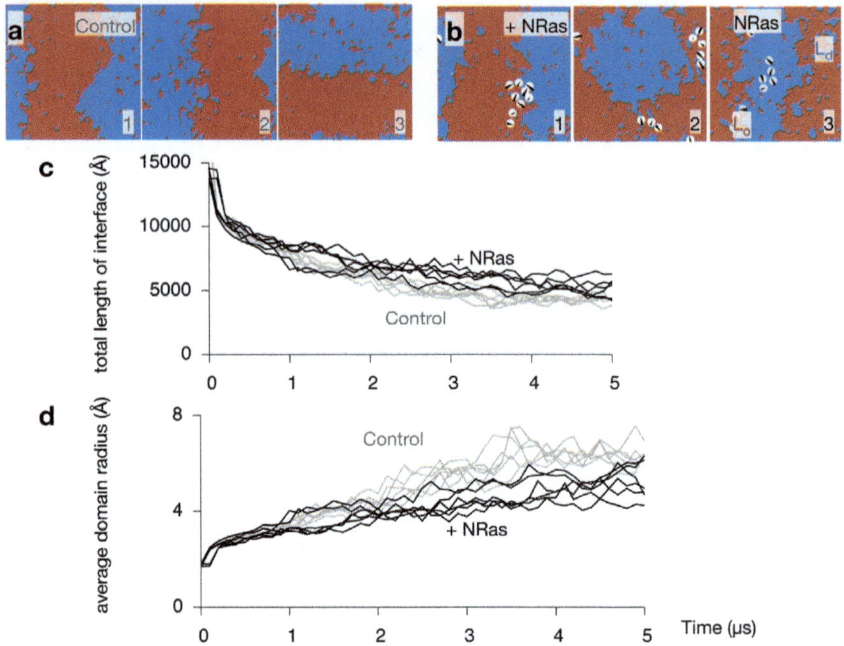

Fig. 6 NRas slows down the spinodal decomposition. Images of the upper leaflet at 5 μs of the (a) three control and (b) three NRas simulations. One can also see how the banding occurs stochastically along either the x or y axes. (c) The total length of the interface between the L_o and L_d phases decreases during the simulations. Both leaflets for all simulations are plotted. NRas appears to retard this effect (grey lines). (d) Plotting the average radius of a domain more clearly shows this effect.

average domain radius are still decreasing and increasing, respectively, after 5 μs. The theory of spinodal decomposition suggests that a system passes through several regimes, each of which follows a power law, $L(t) \sim t^{\alpha}$, where $L(t)$ is the characteristic length and α is a parameter that varies between regimes. This disparity suggests that perhaps the consolidation and coupling of the domains in the two leaflets is one regime, but the overall demixing has not yet reached completion, *i.e.* reached equilibrium. If we examine logarithm plots of the total length of the interface and the average domain radius (Fig. S8 in the ESI†), we see that there is a small change in gradient around 2–3 μs, which is consistent with the above suggestion. The evidence is not compelling, however, and confirmation requires much longer, and possibly larger, simulations.

Since we cannot be sure that the phase separation is complete, we cannot rule out the possibility that the lipidated protein is altering the thermodynamics of the process. If true, then the equilibrium conformation of the bilayer would be different in the presence of protein. We can conclude, however, that NRas is, at the very least, retarding the rate at which the L_o and L_d domains grow.

NRas localises to the interface between the L_o and L_d domains

Inspection of the upper leaflet of one of the NRas simulations suggests that the lipidated protein prefers the interface between the two domains (Fig. 7a).

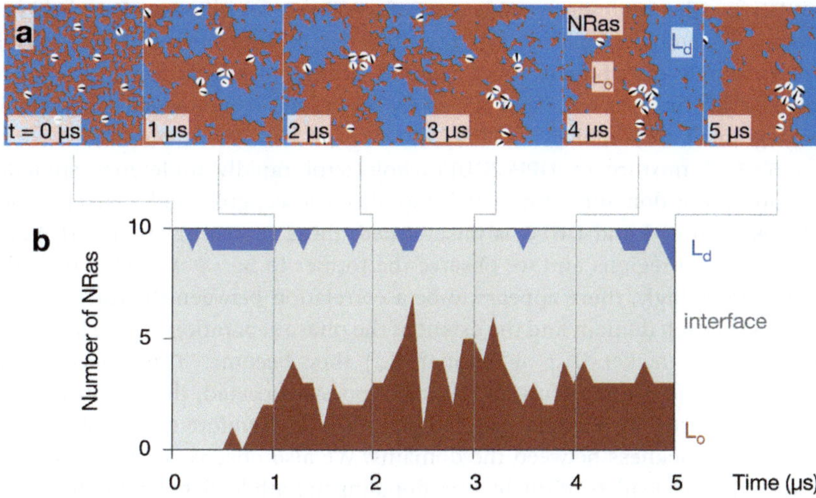

Fig. 7 NRas localises to the interface between the L_o and L_d domains. (a) A series of images of the upper leaflet of one of the control simulations showing the localisation of the ten NRas protein (shown as white circles). (b) A protein is defined as belonging to one of the two phases if the 20 Å footprint of the protein contains >80% of one phase or the other, otherwise it is categorised as interfacial.

Examining images of the other leaflet and the other simulations appears to confirm this (Fig. S7 in the ESI†). As mentioned in the Methods, the spatial extent of each protein is described by a Gaussian of width 20 Å. Examining the proportion of each domain under each of these Gaussians allows us to characterise whether each protein is sitting in the L_o domain (defined as >80% L_o), the L_d domain (>80% L_d) or at the interface (anything else). Since the lipids in the starting configuration are randomly mixed, 5/8th of the lipids under each protein are DPPC and hence all proteins are initially classified as interfacial (Fig. 7b). As the simulations progress and the domains coarsen, the number of proteins belonging to the L_o phase increases to ~3 out of the 10 proteins. Very few proteins (0 or 1) are found in the L_d phase with the remainder (5 or 6), which is the majority, sitting at the interface between the L_o and L_d domains. With one exception, the majority of the proteins partition to the interface in the other leaflets of the other simulations (Fig. S9 in the ESI†).

Discussion

Our simulations suggest that the cell-signalling protein NRas preferentially localises to the interface between L_d and L_o domains, which is perhaps not surprising since one of its two lipid anchors is saturated (palmitoyl), whilst the other is unsaturated (farnesyl). This is consistent with other simulation studies of NRas and other members of the Ras family that started from lipid bilayers that had already undergone phase separation.[10,11] We infer that NRas is reducing the line tension between the two immiscible lipid species and that this effect is responsible for reducing the rate of spinodal decomposition when NRas is present – this is the first time this effect has been directly observed using simulation. The preference NRas appears to have for the interface between domains will automatically increase the clustering of the protein compared to if it were randomly distributed. There may be additional effects causing NRas to segregate yet further, but these are likely too subtle to be detected using this coarse-grained methodology.

A 5 : 3 : 2 mixture of DPPC:DUPC:cholesterol rapidly undergoes spinodal decomposition; domains form which rapidly coalesce, grow and coarsen. Since DPPC is saturated and DUPC is unsaturated, these represent ordered (L_o) and disordered (L_d) regions and we observe the former to be ~9 Å thicker than the latter. Interestingly, there appears to be a correlation between the thickness (or thinness) of each domain and the extent of the phase separation *i.e.* the larger the domains, the thicker (L_o) or thinner (L_d) they become. This effect is not pronounced and requires further investigation. As expected, the majority of the cholesterol partitions into the ordered domain and therefore contributes to the disparity in thickness between the domains. We also observe a strong coupling between the two leaflets of the bilayer; domains in both leaflets tend to form in a synchronised manner such that, *e.g.*, L_o domains are found on top of L_o domains.

The introduction of coarse-grained approaches, combined with the continued increases in computational power, is permitting the study of ever larger patches of membrane. At 6000 lipids, this study is 3–4 times the size of previous studies.[10,11] This is important as, since the simulation unit cell is periodic, the dimensions of the bilayer constrain the maximum size of the coarsening lipid domains. Simulating what is currently considered a large patch of membrane has hence enabled us to study the process of spinodal decomposition in more detail than we would have otherwise been able to. A study of comparable size has also been recently published examining the aggregation of HRas.[35]

We have borrowed techniques from image-processing to analyse our simulations. This has not only simplified the analysis of the simulations but also enabled us to answer questions that would have been, at best, very challenging had we tried to use conventional Cartesian-based methods. For example, estimating the length of the interface between the L_o and L_d domains relies on our use of the Canny edge detection algorithm.[34] Once the edges have been found, however, measuring the length of the interface is a single line of Python code, since one just has to sum the elements in an array.

Central to our approach is the use of existing Python libraries and code, something that has been made possible by MDAnalysis,[31] a Python module that reads most molecular simulation coordinate formats. Once one can retrieve and store in Python the Cartesian coordinates of the different molecular species from a simulation, one has access to the full range of well-documented, optimised and powerful Python modules. Here we have made extensive use of NumPy[32] and SciPy[33] and also scikit-image for image processing. Furthermore, the *N*-dimensional array object that NumPy creates dramatically simplifies the code since many complex operations become single lines. The use of existing, optimised libraries meant that the Python code that performed all the analysis and produced all the images in Fig. 4–7 only required ~20 s per frame on a single CPU. Our approach is in contrast to the more traditional method of writing compiled standalone programs, which is time-consuming and more likely to introduce errors. These disadvantages will only grow as simulations and the questions we ask of them become ever more complex. Python is, of course, not the only possibility; MATLAB also has many similar libraries and tools.

We emphasise that all the visualisation and analysis paradigms listed in Fig. 2 are currently possible using Python. Mathematical graphs are straightforward to construct and interrogate using NetworkX,[36] another Python module, whilst Voronoi tesselations and Delaunay triangular meshes can be built using SciPy.[33]

Despite its prevalence in the field, DUPC, which has unsaturated bonds in both aliphatic tails, is not a physiological lipid. It was adopted after it was shown to be required for phase separation to occur within microseconds.[8,20] Unfortunately, this undermines the generality of the conclusions we draw, since it is possible that such rapid demixing as modelled here, does not occur *in vivo*. There is at present, however, no other way of modelling the formation of L_o and L_d phases using a coarse-grained approach that retains some chemical specificity of the lipids and the proteins. Our simulations therefore assume that this type of behaviour occurs in cells. The simulations, of course, could be longer, however 5 μs of a bilayer of 6000 lipids is already at the cutting-edge of what can be achieved, and it is preferable to have repeats of each simulation rather than a single, very long simulation. The use of a simulation box of finite size places an upper limit on the length of any features observed. This, when combined with the temporal resolution of the simulations, naturally places limits on the phenomena that can be modelled. We are, of course, assuming that all relevant physiological effects occur on the 10 nm length scale and 5 μs timescale.

As we have described, the available crystal structure of NRas is incomplete.[6] Although we carefully constrained a model of the remainder of the protein using other available biophysical data, it is possible that our coarse-grained model of NRas is inaccurate. In particular, the relative separation of orientation of the two cysteine residues that are lipidated might be different. Fortunately, these two residues are only five amino acids apart so they must be in proximity to one another, regardless of the exact structure.

Our analysis also assumes that we can represent a lipid bilayer by a two-dimensional image, which is true only if the bilayer does not undulate significantly. Once this occurs, the link is broken between the Cartesian coordinates of the simulation unit cell and the local coordinate frames on the surface of the lipid bilayer, making analysis difficult. Fortunately, our bilayers do not significantly undulate, but this assumption limits the applicability of some of our methods. Finally, it is likely that there are subtle effects that are not described by the MARTINI coarse-grained forcefield. It is also possible that some of the effects we have observed are also artefacts, for example, more work on the observed clustering of NRas is required.

In future work, we will analyse in more detail the observed phase separation. The theory of spinodal decomposition posits that phase separation can be divided into several regimes, each of which conforms to a dynamic scaling hypothesis, *i.e.* its characteristic length, $L(t)$, has a simple power law time dependence, $L(t) \sim t^{\alpha}$, where α is a parameter that varies between regimes. The curvature of the lipid bilayer has also been suggested to play a role in the segregation of Ras family proteins. Several methods to measure curvature have been recently introduced[17,18] but these are at a nascent stage and require refining before they can be used.

Acknowledgements

This work was supported by the Wellcome Trust through grant 092970/Z/10/Z and the BBSRC. We would also like to thank the EU PRACE network, through projects 2010PA1752 and 2012061115, and Oxford Advanced Research Computing for access to high performance computing.

References

1 F. Spira, N. S. Mueller, G. Beck, P. von Olshausen, J. Beig and R. Wedlich-Söldner, *Nat. Cell Biol.*, 2012, **14**, 640–8.
2 K. Simons and E. Ikonen, *Nature*, 1997, **387**, 569–72.
3 S. Munro, *Cell*, 2003, **115**, 377–88.
4 E. Castellano and E. Santos, *Genes Cancer*, 2011, **2**, 216–31.
5 J. F. Hancock, A. I. Magee, J. E. Childs, J. Marshall, T. Ridgeway and M. Hill, *Cell*, 1989, **57**, 1167–1177.
6 L. Nedyalkova, Y. Tong, W. Tempel, L. Shen, P. Loppnau, C. H. Arrowsmith, A. M. Edwards, C. Bountra, J. Weigelt, A. Bochkarev and H. Park, *Structural Genomics Consortium*, 2008, DOI: 10.2210/pdb3con/pdb.
7 S. J. Marrink and D. P. Tieleman, *Chem. Soc. Rev.*, 2013, **42**, 6801–22.
8 J. Domaski, S. J. Marrink and L. V. Schäfer, *Biochim. Biophys. Acta, Biomembr.*, 2012, **1818**, 984–94.
9 L. V. Schäfer, D. H. de Jong, A. Holt, A. J. Rzepiela, A. H. de Vries, B. Poolman, J. A. Killian and S. J. Marrink, *Proc. Natl. Acad. Sci. U. S. A.*, 2011, **108**, 1343–8.
10 D. H. de Jong, C. A. Lopez and S. J. Marrink, *Faraday Discuss.*, 2013, **161**, 347.
11 L. Janosi, Z. Li, J. F. Hancock and A. A. Gorfe, *Proc. Natl. Acad. Sci. U. S. A.*, 2012, **109**, 8097–8102.
12 W. Humphrey, A. Dalke and K. Schulten, *J. Mol. Graphics*, 1996, **14**, 33–38.
13 M. Chavent, T. Reddy, J. Goose, A. C. E. Dahl, J. E. Stone, B. Jobard and M. S. P. Sansom, *Faraday Discuss.*, 2014, DOI: 10.1039/c3fd00145h.
14 S. J. Marrink, J. Risselada and A. E. Mark, *Chem. Phys. Lipids*, 2005, **135**, 223–44.
15 K. Murzyn, T. Róg and M. Pasenkiewicz-Gierula, *Biophys. J.*, 2005, **88**, 1091–103.
16 G. Lukat, J. Krüger and B. Sommer, *J. Chem. Inf. Model.*, 2013, **53**, 2908–25.
17 V. Gapsys, B. L. de Groot and R. Briones, *J. Comput.-Aided Mol. Des.*, 2013, **27**, 845–858.
18 S. Baoukina, E. Mendez-Villuendas and D. P. Tieleman, *J. Am. Chem. Soc.*, 2012, **134**, 17543–53.
19 H. Koldsø, D. Shorthouse, J. Helie and M. S. P. Sansom, (submitted), 2014.
20 R. S. Davis, P. B. Sunil Kumar, M. M. Sperotto and M. Laradji, *J. Phys. Chem. B*, 2013, **117**, 4072–4080.
21 S. Pronk, S. Páll, R. Schulz, P. Larsson, P. Bjelkmar, R. Apostolov, M. R. Shirts, J. C. Smith, P. M. Kasson, D. van der Spoel, B. Hess and E. Lindahl, *Bioinformatics*, 2013, **29**, 845–854.
22 H. J. C. Berendsen, J. P. M. Postma, W. F. van Gunsteren, A. DiNola and J. R. Haak, *J. Chem. Phys.*, 1984, **81**, 3684–3690.
23 Y. Choi and C. M. Deane, *Proteins: Struct., Funct., Bioinf.*, 2010, **78**, 1431–1440.

24 E. F. Pettersen, T. D. Goddard, C. C. Huang, G. S. Couch, D. M. Greenblatt, E. C. Meng and T. E. Ferrin, *J. Comput. Chem.*, 2004, **25**, 1605–1612.

25 G. Reuther, K.-T. Tan, A. Vogel, C. Nowak, K. Arnold, J. Kuhlmann, H. Waldmann and D. Huster, *J. Am. Chem. Soc.*, 2006, **128**, 13840–6.

26 L. Brunsveld, H. Waldmann and D. Huster, *Biochim. Biophys. Acta, Biomembr.*, 2009, **1788**, 273–288.

27 A. Vogel, G. Reuther, M. B. Roark, K.-T. Tan, H. Waldmann, S. E. Feller and D. Huster, *Biochim. Biophys. Acta, Biomembr.*, 2010, **1798**, 275–85.

28 S. J. Marrink, H. J. Risselada, S. Yefimov, D. P. Tieleman and A. H. de Vries, *J. Phys. Chem. B*, 2007, **111**, 7812–24.

29 L. Monticelli, S. K. Kandasamy, X. Periole, R. G. Larson, D. P. Tieleman and S.-J. Marrink, *J. Chem. Theory Comput.*, 2008, **4**, 819–834.

30 T. C. Beutler, A. E. Mark, R. C. van Schaik, P. R. Gerber and W. F. van Gunsteren, *Chem. Phys. Lett.*, 1994, **222**, 529–539.

31 N. Michaud-Agrawal, E. J. Denning, T. B. Woolf and O. Beckstein, *J. Comput. Chem.*, 2011, **32**, 2319–2327.

32 T. E. Oliphant, *Guide to NumPy*, Brigham Young University, Provo, UT, 2006.

33 E. Jones, T. E. Oliphant, P. Peterson and Others, *SciPy: Open Source Scientific Tools for Python.*, 2001, http://www.scipy.org/.

34 J. Canny, *IEEE Trans. Pattern Anal. Mach. Intell.*, 1986, **8**, 679–98.

35 H. Li and A. A. Gorfe, *PLoS One*, 2013, **8**, e71018.

36 A. A. Hagberg, D. A. Schult and P. J. Swart, *Proceedings of the 7th Python in Science Conference (SciPy2008)*, Pasadena, CA USA, 2008, pp. 11–15.

Faraday Discussions

PAPER

Density based visualization for molecular simulation

Dmitri Rozmanov,* Svetlana Baoukina and D. Peter Tieleman

Received 6th December 2013, Accepted 31st January 2014
DOI: 10.1039/c3fd00124e

Molecular visualization of structural information obtained from computer simulations is an important part of research work flow. A good visualization technique should be capable of eliminating redundant information and highlight important effects clarifying the key phenomena in the system. Current methods of presenting structural data are mostly limited to variants of the traditional ball-and-stick representation. This approach becomes less attractive when very large biological systems are simulated at microsecond timescales, and is less effective when coarse-grained models are used. Real time rendering of such large systems becomes a difficult task; the amount of information in one single frame of a simulation trajectory is enormous given the large number of particles; at the same time, each structure contains information about one configurational point of the system and no information about statistical weight of this specific configuration. In this paper we report a novel visualization technique based on spatial particle densities. The atomic densities are sampled on a high resolution 3-dimensional grid along a relatively short molecular dynamics trajectory using hundreds of configurations. The density information is then analyzed and visualized using the open-source ParaView software. The performance and capability of the method are demonstrated on two large systems simulated with the MARTINI coarse-grained force field: a lipid nanoparticle for delivering siRNA molecules and monolayers with a complex composition under conditions that induce monolayer collapse.

1 Introduction

Molecular visualization of structural information obtained from computer simulations is an important part of research work flow. A good visualization technique should be capable of eliminating unnecessary redundant information and highlight important effects clarifying the key phenomena for the researcher as well as for the audience.

Current methods of presenting molecular structural information are generally limited to variants of the traditional "ball-and-stick" representation or various kinds of molecular surfaces. This approach, while good for simple chemical

Department of Biological Sciences and Centre for Molecular Simulation, University of Calgary, 2500 University Drive NW, Calgary, Alberta T2N 1N4, Canada. E-mail: rozmanov@gmail.com; tieleman@ucalgary.ca

systems, becomes less attractive when very large biological systems, such as lipid nano-particles (LNP) for drug delivery,[1] are simulated at microsecond timescales. In such simulations the structural aspects of the entire system are of interest and the visualization cannot be reduced to a detailed representation of a relatively small active site as is the case in some ribosome simulations,[2] for example. Real time rendering of such large systems becomes a difficult task even on high end workstations; the amount of information in one single frame of a molecular dynamics trajectory is enormous at full atomic details. In addition, in coarse-grained simulations the features of interest typically are not at the level of individual atoms, and visualization methods that are based on the detailed chemical structure are by default not necessarily the best choice.

There have been previous attempts to improve the visualization quality and perception for structural data. For biologically relevant molecules, for example, there are several molecular representations which greatly help with interpretation of protein structures, such as various cartoon-based representations implemented in the VMD[3] molecular visualization software. Such representations reduce the amount of redundant chemical information and highlight the features of the protein secondary structure relevant to its biological function. A limitation of this and similar methods is that they are only applicable to proteins.

A more general approach is to use average configurations for the visualization of ordered systems with low mobility.[4,5] The method helps to remove structural variation of the visualized structures due to thermal vibrational motion and highlights the underlying ordered structure. At the same time, the method is poorly suited for disordered systems with significant molecular mobility at the time scale of the averaging and may result in non-physically looking structures. Mobility information is not a part of the average configuration but can be partially communicated by appropriate particle coloring.[5]

In this paper we report a novel visualization technique based on spatial atomic densities. The method can be considered a natural extension and combination of the average configuration[4,5] and the spatial distribution function[6] approaches for molecular visualization. The power of this method is demonstrated on two large-scale test systems simulated using the MARTINI coarse-grained force field[7,8] on a length-scale of *ca.* 50 nm. Software tools and utilities necessary for application of this technique, as well as the workflow are also reported in this paper.

The rest of the paper is organized as follows. The following section 2 details the method and the workflow required to apply the proposed technique. An application of the method to a case study of the structural features and dynamics of a drug delivery lipid nano-particle will be given in section 3.1. The section 3.2 will discuss the visualization and analysis of the collapse of the lipid monolayers at the air–water interface. Finally, our conclusions will be summarized in section 4.

2 The method

2.1 Property grids

The method is based on visualization of spatial information sampled from molecular simulations and mapped on a high-resolution 3-dimensional property grid for some specific time frame of interest. Such a grid represents a 3-dimensional field and can contain average information about some property of the

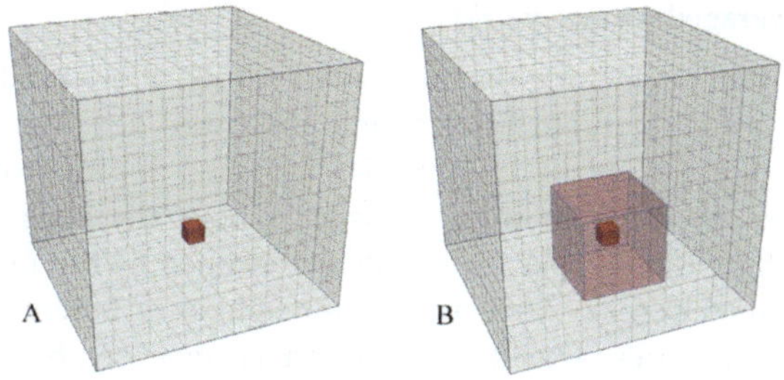

Fig. 1 (A) A simulation box is mapped to a 3D grid; Properties of the particles found in a volume element (red) are sampled and averaged to produce the grid value for that volume element using a large number of the simulation trajectory snapshots for a specific time interval. (B) 3D grid of a locally averaged property. Values of the cells are obtained by applying an 3-dimensional moving average to the original grid. The values inside the local 3D domain (red transparent box) are averaged and assigned to the central cell (opaque red cell). The local domains are subjected to the periodic boundary conditions.

system in each volume element for the duration of the sampling time frame [see Fig. 1 (A)]. Therefore, the averaging is performed for a spatial location rather than for a specific particle. Generally, such grids can contain various types of data and represent scalar, vector, or tensor property fields such as, for example, density, potential energy, mobility, order parameter, linear momenta, force, or stress tensor. Each property of interest is sampled on a separate grid using a large number of snapshots of the system. This paper mainly focuses on the density as the sampled property of the system for illustration purposes but much of the power of the approach is based on the ability to combine density maps with arbitrary other grids, such as the densities of different entities or an average energy grid, for example, and do calculations on them.

In this method the resolution of the 3D grid should be fine enough to resolve the smallest structural feature of interest; in atomistic simulations the grid spacing can be 0.5 Å so that individual atoms can be resolved. For coarse grained (CG) simulations using the MARTINI[7] force field the grid spacing can be in the range 1.5–2.0 Å to resolve the larger particles (beads) representing several atoms. The time frame for sampling should be short enough to prevent averaging of the temporal features of interest, such as structural elements or dynamical or configurational heterogeneities.[9–11] At the same time the sampling period must be sufficiently long to allow for representative sampling of the accessible phase space of the system. The 3D grids obtained using this approach are essentially 3-dimensional histograms with the number of bins significantly larger (by 1–2 orders of magnitude) than the number of samples (particles) in a single configurational snapshot of the system. Hence, in order to generate smooth and representative density grids the number of sampled configurations should be large, typically, several hundreds or even thousands, depending on the grid resolution, number of particles in the system, and the desired quality of the density grid.

The property grids then can be visualized directly using appropriate visualization tools or manipulated to produce more complex fields, such as cross-correlation densities or local average property grids.[12]

2.2 Locally averaged property grids

Property grids generated using the approach outlined above produce continuous property fields which contain information about the density distribution of a specific set of particles in the system for a specified duration of time. At short sampling times the diffusive mass transfer is normally insignificant and the accessible configurational space is mostly determined by the molecular rattling motion determined by the nearest molecular neighbours. Essentially, under these conditions, the system's density distribution is an ensemble of weakly overlapping single particle density distributions.

When coexisting phases in the system are of interest such short time density grids may not show a sufficiently clear difference in the density (or any other property) between these phases. Following a definition of a phase as a uniform continuous domain of matter with similar properties, a phase property is a collective quantity determined by many molecules comprising the phase. An obvious solution then is to provide additional averaging of the density grid at the characteristic length scale for the phases of interest. In practice, application of a 3-dimensional moving average over such a characteristic local domain around each grid cell is a convenient method to achieve such local averaging as shown in Fig. 1(B). The resulting property grid has the same dimensions and resolution as the original grid but its cell values are averaged over spatial domains local to each of the cells. The size of the domain necessary to provide the desired averaging can be estimated from the particle density and a typical number of molecules required to distinguish the phases.

2.3 Visualization workflow

To generate density grids suitable for the visualization approach described in this paper a number of steps are required:

■ The starting point is a single initial configuration of the system to be visualized. The configuration has to contain all required information so that a new simulation can be initiated from it to generate enough additional configurations for the averaging procedure, if the original trajectory was not saved at a high enough frequency.

■ The atomic densities are sampled on a high resolution grid in this method. The grid resolution should be appropriately selected. Typically each atom or particle is expected to occupy the volume of at least $3 \times 3 \times 3$ grid cells. Many configurational snapshots are necessary for a sufficient density sampling on such a high resolution grid. Normally, hundreds or thousands of configurations are needed to generate smooth density maps. A good rule for a uniform system is to have the number of particles times the number of configurations at least ten times the number of grid cells.

■ The simulation length is chosen so that it is short enough to avoid smearing of the structural features of interest due to diffusion, but sufficiently long to provide adequate sampling of the system's accessible volume. It can be 10 ps for an atomistic simulation of liquid water and 10 ns for a coarse grained simulation of a lipid membrane. By increasing the sampling time one can shift the focus of attention from short time behaviour in the system to processes occurring at longer time scales.

■ Once the number of required configurations and the length of the simulation for grid sampling is known, the simulation is performed and the configurations are saved to a trajectory data file. The GROMACS simulation package was used in this work, and the configurations were saved as an XTC trajectory file.[13]

■ The saved trajectory information is then used to sample the particle locations for density grid generation. For each density of interest a separate density grid is generated, so they can be analyzed and visualized independently. In this work for each grid to be sampled a particle index was generated and added to a GROMACS index file. The trajectory and the index files were used by software developed in our group to generate and save the density grids.[14] Such high resolution grids carry a large amount of information, of the order of hundreds of megabytes per grid, so the data is compressed before it is written to disk to reduce the amount of storage space required.

■ Once the grids are generated they can be used for numerical analyses or manipulated to produce new grids, such as cross density products.

■ For visualization, normally, the grids have to be converted to an appropriate format first. In this work, ParaView,[15] an open source multi-platform data analysis and visualization application, was used to generate most of the figures. The grids, therefore, were converted to the XML base VTI format, native to the VTK visualization library[16] ParaView is based on, using software developed in our group.[14]

If an animation is to be generated using this density base approach, then all these steps have to be repeated for each frame of the animation.

3 Results and discussion

3.1 Structural features of a lipid nano-particle

To demonstrate an application of the proposed density based technique, a simulation of a lipid nano-particle (LNP) in an aqueous environment will be discussed in this section. The LNP studied in this work is designed for therapeutic siRNA drug delivery to a target cell to induce specific gene silencing following the RNA interference pathway.[17] In addition to water and the siRNA drug, the particle is mainly composed of cholesterol and a specially designed ionizable cationic lipid, as well as a small amount of a structural phospholipid. A protective shell of a polyethylene glycol (PEG) grafted lipid is required for the particles to prevent aggregation and to protect the particles from being degraded by the immune system during systemic delivery.[1,17] A more detailed understanding of the structure and the roles of all the components of the particle is required for designing more efficient and precisely targeted drug delivery systems. We have previously built a model of this particular particle and described initial simulations, as well as characterized its structure using traditional visualization and analysis techniques,[1] but the simulations presented here are new and start from a different procedure to make the initial model.

The simulations were performed in a 57.6^3 nm^3 cubic simulation box under periodic boundary conditions using the GROMACS simulation package[13] and the coarse grained MARTINI force field.[7,8] The system was composed of 15984 molecules of 2,2-dilinoleyl-4-(2-dimethylaminoethyl)-[1,3]-dioxolane (DLin-KC2-DMA),[1] an ionizable cationic lipid, 15984 cholesterol (CHOL) molecules,[7] 3996 molecules of 1,2-distearoyl-sn-glycerophosphocholine (DSPC),[7] a phospholipid, 3996 PEG-lipid molecules (PEG-C2),[1] and 1179533 water particles. The siRNA in this simulation was represented by 216 12 base pair double stranded RNA MARTINI molecules, in the current model indistinguishable from DNA,[18] and 11232 chloride anion particles were added to the system to neutralize the system. In the MARTINI force field the coarse graining is implemented by representing 4 heavy atoms, typically, by a single effective particle.[7] In total the simulation system contained 1714349 MARTINI particles, and represented a system of approximately 20 million atoms.

The particle was self-assembled from a mixture of the DLin-KC2-DMA, DSPC, cholesterol, and RNA molecules randomly distributed inside a sphere 25.5 nm in radius. The initial configuration for the LNP simulations was built using the bio.b-gen simulation box generator developed in our group.[19] The PEG-lipid molecules were placed randomly around the sphere in a spherical shell between radii 26.0 and 26.5 nm. 130950 water particles were placed inside a sphere of radius 27.0 nm and 1059815 water particles were added to the rest of the system to achieve the desired level of lipid hydration and to maintain a uniform density throughout the system. After that, 11232 randomly selected water particles were replaced by Cl^- ions to neutralize the system. The system was propagated for 200 ns at a temperature of 320 K using a 10 fs time step, during which the nano-particle self-assembled and equilibrated. The spacing for the density grids in this study was slightly less than 2 Å with grid dimensions of $300 \times 300 \times 300$ cells. The grid resolution was a compromise between the captured structural details and the computational effort of visualization: a typical MARTINI bead has an effective diameter of 4.7 Å and can be resolved at this grid spacing; at the same time a grid of 27,000,000 cells still can be manipulated on current hardware. The grids were sampled using 1000 configurational snapshots of the system during a 100 ps simulation.

3.1.1 General structure of the nano-particle. The general view of the lipid nano-particle simulated in this work is shown in Fig. 2. A protective shell formed by the PEG-lipid wrapping the particle can be seen in Fig. 2(A) as well as a number of RNA molecules present at the external surface. A cross-sectional view in Fig. 2(B) reveals the internal structure of the particle.[1] Several water filled cavities

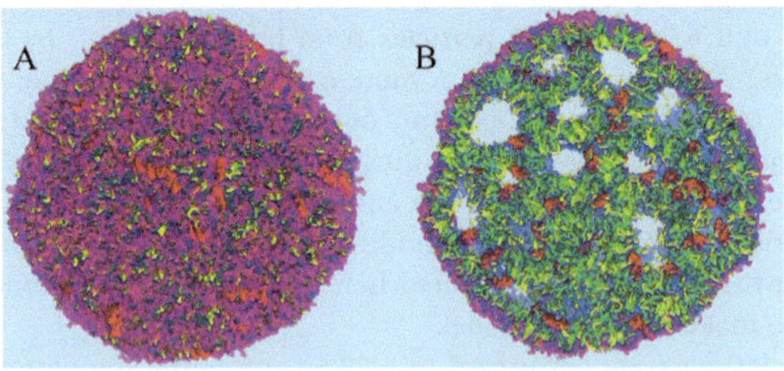

Fig. 2 An overview of the lipid nano-particle studied in this work. (A) The external view of the particle. (B) A cross-sectional view represented by a 3.5 nm slab cut from the middle of the particles. The molecular densities of the components are shown by a set of semi-transparent iso-surfaces: magenta, the protective PEG-lipid (PEG-C2), red, the therapeutic siRNA (RNA), green, cholesterol (CHOL), yellow, the phospholipid (DSPC), and blue, the ionizable cationic lipid (DLin-KC2-DMA). All iso-surfaces were generated using 5 nm^{-3} density values.

are visible inside the particle separated by a bilayer-like lipid membranes. The RNA molecules are clearly seen inside the lipid membranes and appear to be uniformly distributed. The PEG-lipid density is mostly limited to the external surface but a few PEG-lipid molecules can be seen inside the particle.

The density of water (solvent) shown in Fig. 3(A) provides a representation of the external interface of the particle with the surrounding water. At the 100 ps time scale the diffusive contribution to molecular motion is small, so that the high resolution density grid contains most of the molecular level details cluttering the representation. By averaging the density grid locally the amount of detail is reduced, so that the general shape of the particle can be seen more clearly, see Fig. 3(B). This representation of the particle external surfaces reveals these details much more clearly when compared with a more conventional particle view shown in Fig. 2(A). Fig. 3(C) shows the same water density iso-surface as Fig. 3(B) but with a semi-transparent surface. Such a representation shows the water cavities inside the particle, their distribution, and sizes. Apparently, water entrapped inside the particle during the particle construction process aggregated into compact water pockets. Significant variation of the water pocket sizes can also be noticed. A search for continuous domains in the averaged grid revealed 69 water cavities inside the particle. The plot of the estimated cavity diameters in Fig. 4 shows a two-tier cavity organization, with the larger cavities being in the 6–10 nm diameter range and the smaller cavities in the 0.5–2 nm range. The total volume of the internal cavities is 8172 nm^3 and their total estimated area is 6638 nm^2. The domain of the solvent density lower than the threshold value corresponds to the rest of the particle and had a volume of 50475 nm^3. The sum of these two volumes gives the total particle volume which can be used to estimate the particle radius of 24.1 nm. Thus, the external area of the LNP assuming its spherical shape can be approximated as 7300 nm^2, and the total area of the interface with the solvent of approximately 14000 nm^2. This analysis shows that the areas of the external and internal interfaces of the particle are similar, with 52 and 48% respective contributions to the total area.

The distribution and structural ordering of the main lipid components of the nano-particle can be evaluated from their particle densities shown in Fig. 5. The

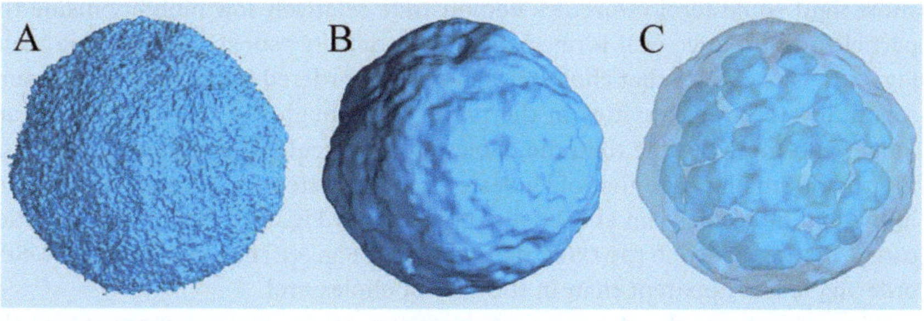

Fig. 3 Iso-surfaces of the solvent density in the lipid nano-particle system at 5 nm^{-3} density values. (A) The original density shows too many molecular details, obscuring the overall topology of the particle surface. (B) A locally averaged water density using 9^3 cell averaging ($\approx 1.7^3$ nm^3) shows the particle surface topology more clearly. (C) A semi-transparent representation of the same local-average density iso-surface showing water cavities inside the particle.

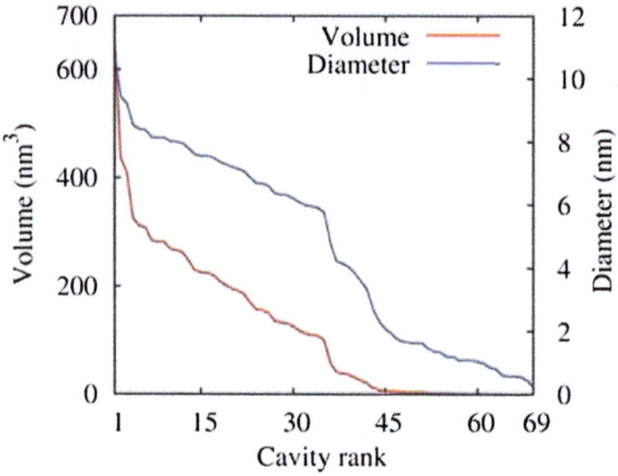

Fig. 4 Volumes (red curve) of the 69 water cavities identified inside the lipid nano-particle shown in Fig. 3(C) sorted by size. The corresponding cavity diameters are also shown (blue curve), assuming a spherical shape for the cavities. The threshold density value for the density domain analysis was 5 nm^{-3}.

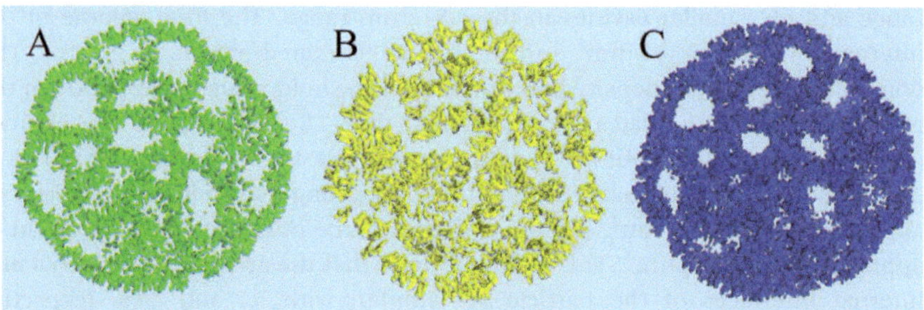

Fig. 5 Molecular densities of the components represented by density iso-surfaces: (A) Cholesterol, (B) the DSPC lipid, (C) the DLin-KC2-DMA cationic lipid. Shown in a 3.5 nm cross-sectional slab using 5 nm^{-3} density iso-values.

short rigid cholesterol molecules demonstrate relatively low mobility inside the particle and produce well formed distinctive density "sticks" at this time scale, Fig. 5(A). One can see that cholesterol is strongly ordered in the lipid membranes separating the water cavities in the particle. Although the DSPC molecules are longer and more flexible than the cholesterol molecules, the amount of DSPC is four times smaller than that of cholesterol in the system, so that the densities of individual molecules can still be distinguished in Fig. 5(B). Some water cavities can be distinguished in the DSPC density plot in Fig. 5(B) but, generally, the DSPC ordering is less apparent than in the case of cholesterol.

Fig. 5(C) shows the density representation of the main component of the particle, the DLin-KC2-DMA cationic lipid. Its molar fraction in the particle is the same as that of cholesterol, but the DLin-KC2-DMA molecules are longer and more flexible, so that its density in Fig. 5(C) is fused together and the ordering around the cavities is significantly more difficult to see. To address the question

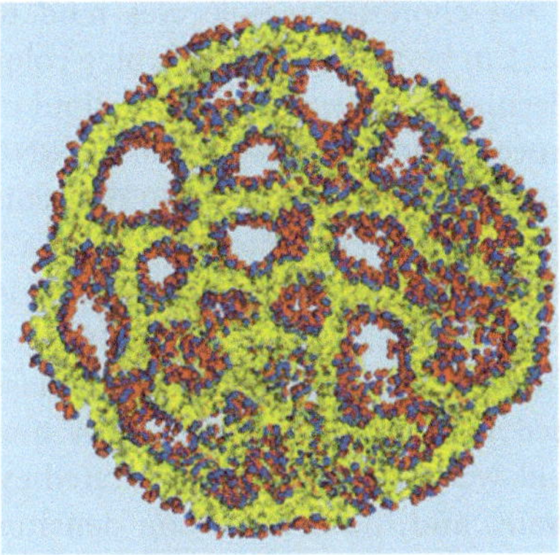

Fig. 6 The density of the DLin-KC2-DMA cationic lipid in the nano-particle. The densities of the three parts of the lipid molecules were sampled on separate grids and plotted as separate iso-surfaces through 3 nm^{-3} densities: the hydrophilic head groups (red), linker groups (blue), and the hydrophobic tail groups (yellow).

about DLin-KC2-DMA orientational ordering in the particle the densities of its hydrophilic head, linker, and hydrophobic tail groups were sampled separately and plotted in Fig. 6. This representation clearly demonstrates strong ordering of the cationic lipid head groups around the water pockets.

A more careful examination of the component densities in Fig. 5 shows that the water cavity openings in the cholesterol density seem larger than in either the DSPC and DLin-KC2-DMA density plots. This observation suggests that the cholesterol molecules must reside deeper in the lipid membranes separating the water pockets and are not present at the lipid–water interface. To observe the effect more clearly the iso-surface of DLin-KC2-DMA density shown in Fig. 6(C) was colored by the cholesterol density and plotted in Fig. 7. In this representation

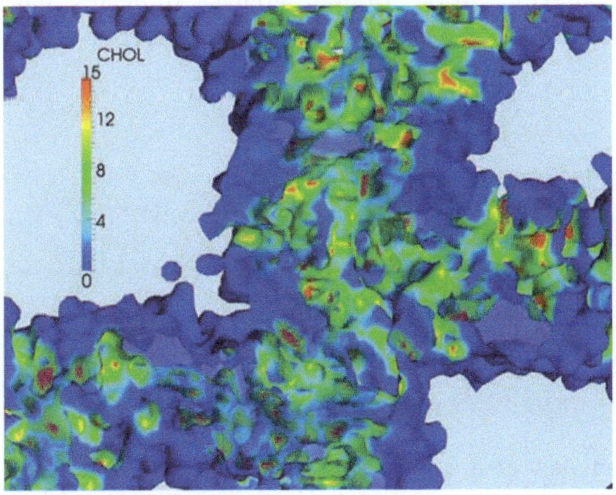

Fig. 7 A fragment of the DLin-KC2-DMA density iso-surface from Fig. 5(C) colored by the cholesterol density. The volumes with the DLin-KC2-DMA density above the surface iso-value are shown by a haze of the same coloring. The DLin-KC2-DMA iso-density value is 3 nm^{-3}, the units for the color scale of the cholesterol density are nm^{-3}.

it becomes apparent that cholesterol molecules tend to be absent from the interfaces with water, as can be seen by the dark blue color of the cavity surfaces; all the cholesterol molecules are deeper inside the lipid content of the particle.

3.1.2 Structural aspects of RNA encapsulation. No RNA molecules are present inside the water cavities (Fig. 2B), but some RNA molecules are located at lipid–water interfaces, at the lipid side of the interface. Fig. 8(A) provides a clearer view of RNA located at the lipid–water interface. The interface is mainly formed by the DSPC and DLin-KC2-DMA lipids. An RNA molecule is seen adsorbed at the interface with most of the molecule being in the lipid phase. Further examination of the system indicated that all encapsulated RNA molecules are located at internal or external lipid–water interfaces as illustrated by Fig. 9.

An iso-surface of water and the chloride ion densities, as well as the RNA density provide a different representation of the same water cavity in Fig. 8(B). Apparently, the water pockets carry a negative charge spread at the interface due to the chloride ions entrapped inside the cavities. The ions are adsorbed at the interface from the water side. The lipid side of the interface is mainly formed by the cationic, DLin-KC2-DMA, and zwitterionic, DSPC, lipids and is expected to be positively charged, explaining this attraction. Interestingly, the RNA molecules carry a negative charge of −22 and despite this fact they still prefer to adsorb onto the negatively charged water cavities.

It has been previously noted that the vicinity of the RNA molecules in the nanoparticle contains approximately fourteen times more DLin-KC2-DMA lipid molecules than DSPC molecules.[1] This ratio was estimated from RNA–lipid radial distribution functions and was significantly higher than the 4 : 1 molar ratio of these lipids in the particle composition. This observation indicated a higher affinity of the DLin-KC2-DMA lipid for RNA molecules in comparison with the DSPC lipid. In this work the number of RNA–lipid contacts were visualized directly using spatial density correlations between the RNA molecules and the DSPC and DLin-KC2-DMA lipids (Fig. 10). The number of RNA-DLin-KC2-DMA (blue) contacts is higher than RNA-DSPC contacts (yellow); their ratio is visually also higher than the 4 : 1 ratio expected from the composition, in qualitative agreement with previous work.[1] Domain analysis indicates that the ratio of the number of RNA-DLin-KC2-DMA contacts to the number of RNA-DSPC contacts is

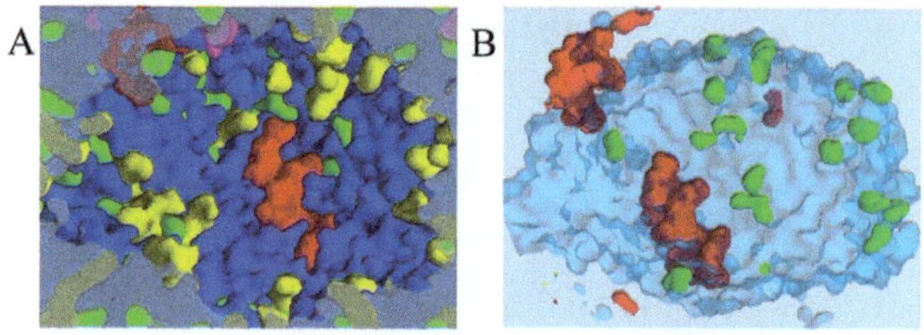

Fig. 8 A detailed view of a water cavity. (A) The interface surface. The component densities are shown as 5 nm^{-3} iso-surfaces: red, RNA, green, CHOL, yellow, DSPC, and blue, DLin-KC2-DMA. The cutting plane is shown by a semi-transparent grey color. (B) The same cavity represented by the iso-surfaces of water (semi-transparent blue, 5 nm^{-3}), chloride ion (green, 1 nm^{-3}), and RNA (red, 5 nm^{-3}) densities. Shown from a slightly different angle.

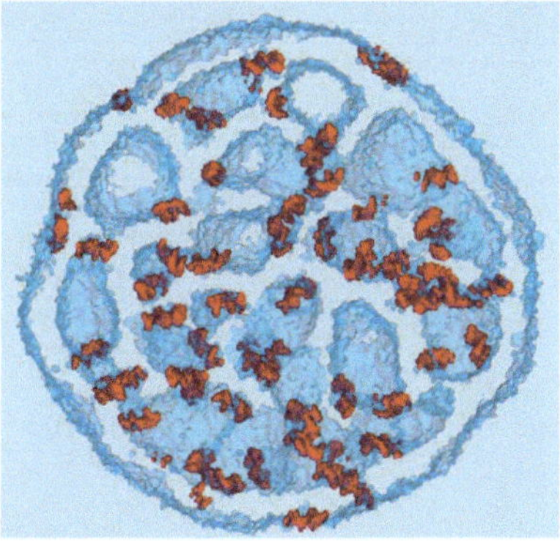

Fig. 9 RNA molecules adsorbed on water interfaces. The RNA molecules are shown as an iso-surface of the RNA density (red), the water interfaces are represented by the iso-surface of water density (semi-transparent blue). Shown in a 3.5 nm cross-sectional slab using 5 nm^{-3} iso-densities.

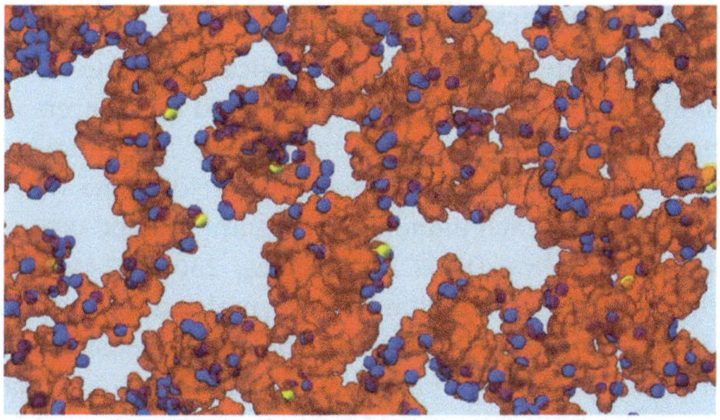

Fig. 10 The spatial correlations or contacts of the RNA molecules with the DLin-KC2-DMA and DSPC lipids inside the nano-particle shown as the iso-surfaces of the product of the RNA and the lipid density fields. The RNA density is represented by the red iso-surface at 5 nm^{-3}; the DLin-KC2-DMA-RNA density product field is shown in blue, and the DSPC-RNA product field is shown in yellow. Both the product field surfaces are plotted using 0.001 nm^{-6} values.

approximately 40 : 1, or 10 times higher than the molar ratio of these lipids. Apparently, the RNA molecules are mostly surrounded by the cationic DLin-KC2-DMA lipid and DSPC is forced from these regions; under such circumstances DSPC is mostly expected to be present at water interfaces rather than in the vicinity of the RNA molecules. This observation is illustrated by Fig. 11 where an RNA molecule is shown bound to the surface of a water cavity. The RNA molecule lays flat on the water surface and there is practically no DSPC density in its immediate vicinity.

The actual role of the DSPC lipid in formation of lipid nano-particles is not yet fully understood, but it is known that some amount of DSPC is required for nano-particle formulation to have *in vivo* drug activity.[20] Based on the observations made in this work we can speculate that the DSPC lipid is responsible for formation of more stable interfaces between the lipid content and water.

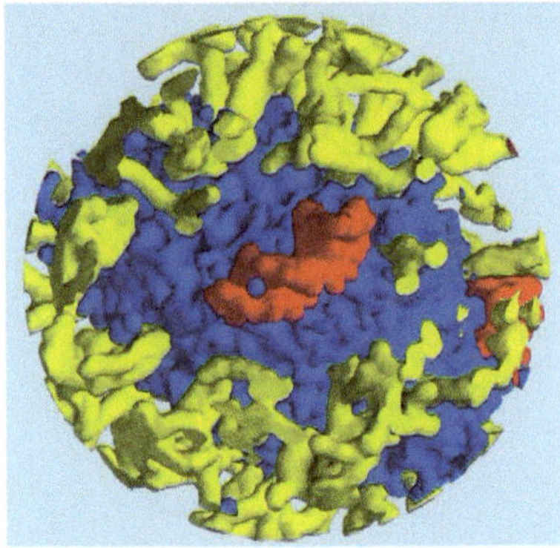

Fig. 11 A cut-away representation showing an RNA (red) molecule on the surface of a water pocket inside the lipid particle. The water interface is shown as the iso-surface of water density (blue). The yellow density iso-surface represents the DSPC lipid molecules. The surfaces were plotted using 5 nm^{-3} iso-densities.

3.2 Structure and dynamics of lipid monolayers with phase coexistence

The second case study demonstrating the application of the density based technique for visualization is a simulation study of lipid monolayers at the air–water interface. The simulations were performed using the GROMACS simulation package[13] and the MARTINI coarse grained force field.[7,8] The system used in this study is shown in Fig. 12. Each monolayer was composed of 3840 1,2-dipalmitoyl-sn-glycero-3-phosphocholine (DPPC) molecules, 2304 1,2-(*cis*,*cis*-9,12-octadecadienoyl)-sn-glycero-3-phosphocholine (DOPC) molecules, and 3072 cholesterol (CHOL) molecules. Monolayers with these specific compositions in simulations separate into liquid-ordered (Lo) and liquid-disordered (Ld) phases at surface tensions in the range from 0 to 30 mM m^{-1} at a temperature of 290 K.[21]

The system set-up comprised a water slab in a vacuum with two symmetric monolayers at the two vapor–water interfaces (see Fig. 12(A)). Partial collapse of

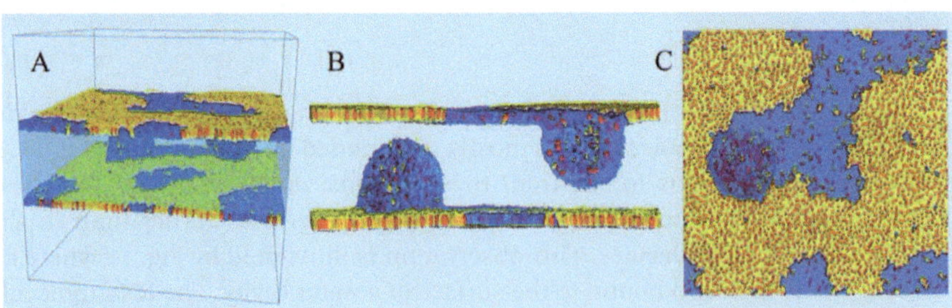

Fig. 12 The system for simulations of lipid monolayers. (A) A general view of the simulation box and two monolayers. The lipid components of the monolayers are shown by the iso-surfaces of their atomic densities sampled over 1 ns and using the density value of 5 nm^{-3}: DPPC (yellow), DOPC (blue), and cholesterol (red). The water slab is between the monolayers in the middle of the box, shown by a semi-transparent blue haze. Above and below the monolayers is the vapor phase. (B) A side view of the system demonstrating lipid semi-vesicles caused by the monolayers' collapse. Water is not shown. (C) A top view of the upper monolayer from the hydrophilic side.

the monolayer was induced by applying a near-zero (<5 mN m^{-1}) surface tension using a surface-tension coupling scheme, after which the monolayer compression was stopped by switching to constant volume simulation. This led to monolayer folding[22] into bilayer discs in water; the bilayer discs bent and formed semi-vesicles attached to the monolayers (and filled with water), shown in Fig. 12(B) and (C). This resulted in a complex 3D structure with the lipid components heterogeneously distributed between the monolayer at the interface and the bilayers in water. This structure is directly relevant for lung surfactant, a mixture of lipids with a small amount of proteins lining the gas exchange interface in the lungs. Lung surfactant monolayer at the air–water interface is connected to the bilayer reservoirs in water; this connectivity provides low surface tensions on interface compression and expansion, which is necessary for breathing.[23–25] The phase distribution in the monolayers and their properties are the main focus of this paper. The results reported in this section are a part of a larger study of lipid monolayers (Baoukina *et al.*, in preparation).

The simulation box was 61.7 × 61.7 × 60.0 nm^3 (with periodic boundary conditions); the spacing for the density grids in this study was slightly less than 2 Å with dimensions of 315 × 315 × 305 grid cells. The grids were sampled using two different time frames, 1 ns and 10 ns, with 1000 configurations in each time frame.

3.2.1 Lipid dynamics in the monolayers. The monolayers contain two coexisting phases: the Lo phase is rich in DPPC and CHOL, and the Ld phase mainly contains the unsaturated DOPC lipid. Such a phase separation can be clearly seen in the top view of the monolayer in Fig. 12(C).

The cholesterol ordering in the DPPC (yellow) lipid domain can be clearly seen; the red dots of cholesterol can also be seen in the blue disordered (DOPC) phase. While cholesterol ordering is very different in the two phases, the actual cholesterol mobility for the 1 ns sampling time appears similar as can be judged from the size of the accessible cholesterol volume. A cross-sectional view of the same cholesterol density in Fig. 13(A) confirms this similarity: the cholesterol density distributions in the Ld phase are comparable to those in the Lo phase on the 1 ns time scale. However, the cholesterol density on the 10 ns time scale, shown in Fig. 13(B), demonstrates a significant difference in mobility between the two phases; the more mobile cholesterol molecules in the Ld phase produce diffuse

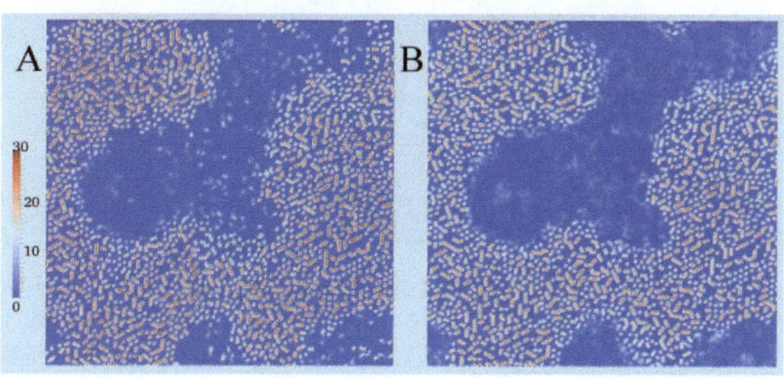

Fig. 13 The atomic density of cholesterol shown on a plane cutting the system along the middle of the top monolayer. (A) 1 ns sampling time. (B) 10 ns sampling time. The units for the color scale are nm^{-3}.

delocalized density clouds at this time scale, while the character of the density of less mobile cholesterol molecules in the Lo phase has not changed. A similar insight can be gained from the total lipid density plots shown in Fig. 14, where the Lo phase maintains the molecular details on a 10 ns scale while in the Ld phase these details are completely smeared out by the molecular motion. Thus, by employing different time scales for density grid generation it is possible to visualize the difference in molecular mobility in two lipid phases.

3.2.2 **Properties of the lipid phases.** As was already mentioned in Section 2.2, identification of a chemical phase or distinguishing between several phases in the system often requires a local averaging to account for the collective nature of the phase behaviour. The graph in Fig. 15(A) shows the probability distributions of the density values in the original density grid of the DPPC lipid as well as in two locally averaged grids. The monotonous shape of the original density probability distribution is typical for a molecular rattling motion, when all the molecules are vibrating around their positions determined by the nearest neighbours without actual diffusion. Under these conditions, there is no significant overlap between the accessible volumes of neighbouring molecules and the total probability

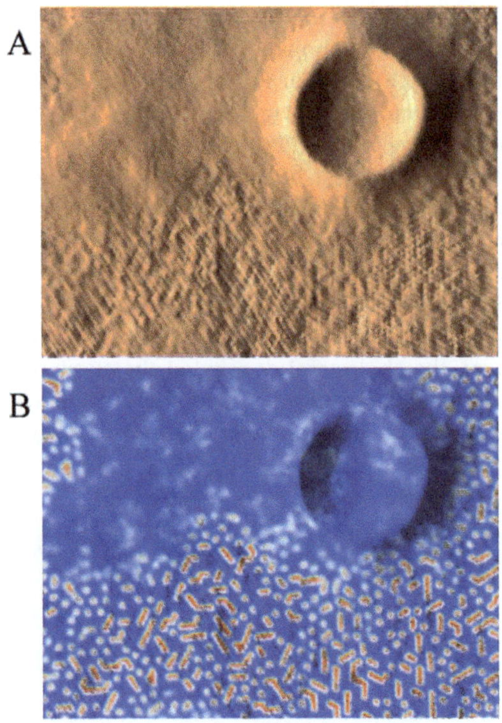

Fig. 14 A close-up of the total lipid density 3 nm^{-3} iso-surface sampled over a 10 ns trajectory. The hydrophobic (lipid tail) side of the top monolayer is shown. The circular formation in the top right part is the spot where the lipid semi-vesicle is attached to the other side of the monolayer. (A) A shaded representation demonstrating the difference between Lo and Ld phases. Crystal lattice-like formations are seen in the Lo phase while the surface of the Ld phase appears smooth due to the diffusive averaging. (B) The same fragment of the iso-surface colored by the cholesterol density, for comparative purpose.

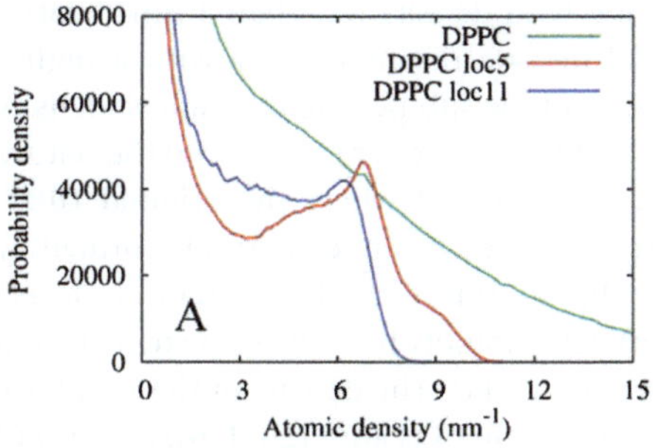

Fig. 15 Probability distributions of the grid density values in the lipid monolayers from a 10 ns simulation. (A) For the DPPC lipid. The original density (DPPC) is shown as well as two locally averaged densities using 5 × 5 × 5 cell (loc5) and 11 × 11 × 11 cell (loc11) averaging volumes. (B) Corresponding probability distributions of the atomic density of the DOPC lipid.

distribution is similar to that of a single molecule. When accessible volumes overlap, several molecules contribute to the property value of the overlap region, which results in an enhanced probability for that average value. Such overlapping regions have similar properties and the density probability shows a peak at the density value specific for this phase.

The graphs in Fig. 15(A) show negligible overlap in the original DPPC density grid, and the presence of a distinct phase in the grid averaged over 11^3 cell local volumes ($\approx 2.2^3$ nm^3), with an average density of approximately 6.5 nm^{-3}. The grid averaged over smaller local regions of 5^3 cells ($\approx 1.0^3$ nm^3), shows a more complex shape with at least two regions of enhanced density probability, in the ranges of 4–8 and 8–11 nm^{-3}. These distributions can be explained by the presence of small clusters of another phase within the Lo phase, with a characteristic size of 1.0–2.2 nm. Fig. 16 shows the isosurfaces for the DPPC density corresponding to these density values.

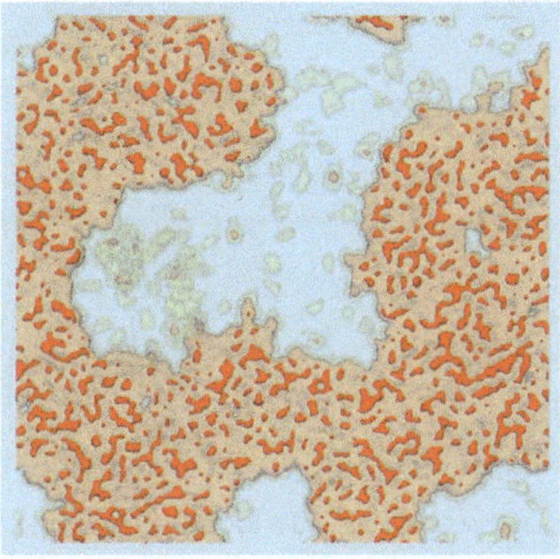

Fig. 16 Density iso-surfaces of the DPPC lipid demonstrating DPPC domains of different locally averaged densities: 1 nm^{-3} (semi-transparent yellow), 4 nm^{-3} (semi-transparent orange), 8 nm^{-3} (solid red). Top monolayer, 10 ns sampling time.

The part of the probability distribution with density less than 4 nm^{-3} corresponds to interface regions (and the Ld phase); the main peak with a higher density corresponds to the bulk Lo phase; and the sub-peak of the highest density is due to the DPPC clusters inside the bulk phase. A better insight into the nature of the DPPC clusters can be obtained from Fig. 17 where the original DPPC density is shown together with the cluster surfaces. These clusters are formed by several DPPC molecules, with no other lipids. Interestingly, clusters of closely packed and ordered DPPC molecules with low mobility have the same density as a liquid condensed (LC) phase formed by pure DPPC. The density of the LC phase was calculated for lipid monolayers with coexisting LC and liquid-expanded (LE) phases (results not shown). Due to their small size, we believe the clusters represent compositional fluctuations in the vicinity to the transition to the LC phase, rather than a bulk LC phase. The monolayers at the conditions in the simulations are likely close to a three phase coexistence region, similar to the gel-Lo-Ld coexistence region in the ternary phase diagrams of DPPC : DOPC : cholesterol mixtures forming lipid bilayers.[26,27]

To contrast the differences between the immobile Lo phase and the mobile Ld phase, Fig. 15(B) shows analogous density distributions for the DOPC lipid prevalent in the Ld phase. Apparently, at this timescale, the natural diffusion in the phase provides sufficient averaging, so that the presence of a bulk phase can be seen in the probability distribution of the original density. Local averaging of the density grids does not produce new evidence of any additional phases enriched in DOPC.

4 Conclusions

A density based visualization technique has been applied to the results of molecular dynamics simulation of two different systems in the framework of the MARTINI force field. It has been shown that very large systems can be visualized because the density based approach to visualization does not depend on the

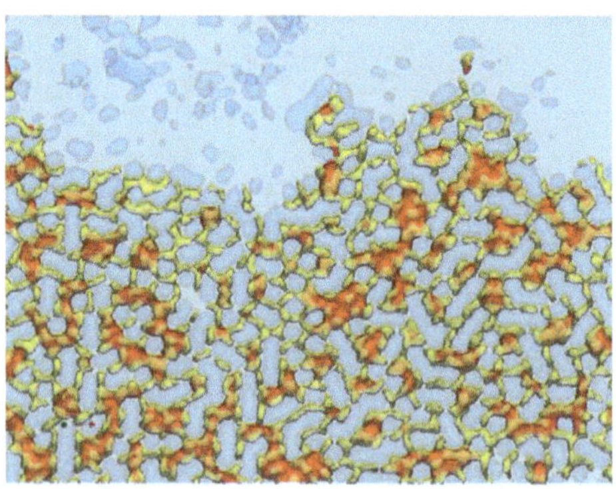

Fig. 17 The iso-surfaces of the original and a locally averaged (over 5^3 cells) DPPC density. The original density is shown by the yellow semi-transparent surface through 9 nm^{-3} density values. The locally averaged density is shown by the red semi-transparent surface with an iso-value of 8 nm^{-3}. The semi-transparent blue surface represents the cholesterol density with an iso-value of 3 nm^{-3}.

number of particles but on the density grid resolution. The grid resolution can be adjusted according to available resources.

Density fields contain information about short term statistics of particle motion and accessible volume but still retain the main structural features of the system when an appropriate time scale is used. This aspect of generated visual representations is similar to the statistical nature of most experimental methods of structural analysis. Many experimental methods produce signals proportional to the volumes of phases with specific properties and averaged over time, such as X-ray diffraction for example. The volumes of the domains with specific densities can be related to the intensities of such experimental signals. The lipid density surface shown in Fig. 14(A) shows a striking resemblance with images produced in atomic force or scanning tunneling microscopy experiments. This similarity of the proposed visualization approach should simplify making connections between the results from simulations with the results obtained in experimental studies, the most crucial step of any simulation study. In addition, moving the data from the discrete space of molecular or atomic properties to the domain of continuous property fields offers new opportunities for more advanced structural analysis.

The density based approach to visualization and analysis of the simulation results reported in this paper have assisted us in revealing structural features of a drug delivery nano-particle as well as gaining new insights into the structure and dynamics of lipid monolayers at the air–water interface. The continuous nature of the density fields allowed for the structural correlational analysis of the RNA–lipid distribution, quantitative analysis of the water cavities and the lipid–water interfacial areas, and the clear visualization of the cholesterol distribution relative to the cationic lipid inside a lipid nano-particle in Section 3.1. The DPPC clusters in the lipid monolayers in Section 3.2 were identified after the probability distributions of the locally averaged grid density values were plotted and then used for targeted density visualization. The approach for data analysis and visualization proposed in this work undoubtedly offers a number of unique options which can potentially benefit any modern molecular dynamics simulation study.

Acknowledgements

This work was supported by the Natural Sciences and Engineering Research Council (Canada). Simulations were carried out on Compute Canada facilities. DPT is an Alberta Innovates Health Solutions Scientist and Alberta Innovates Technology Futures Strategic Chair in (Bio)Molecular Simulation.

References

1 A. K. K. Leung, I. M. Hafez, S. Baoukina, N. M. Belliveau, I. V. Zhigaltsev, E. Afshinmanesh, D. P. Tieleman, C. L. Hansen, M. J. Hope and P. R. Cullis, Lipid Nanoparticles Containing siRNA Synthesized by Microfluidic Mixing Exhibit an Electron-Dense Nanostructured Core, *J. Phys. Chem. C*, 2012, **116**, 18440–18450.

2 L. V. Bock, C. Blau, G. F. Schroeder, I. I. Davydov, N. Fisher, H. Stark, M. V. Rodina, A. C. Vaiana and H. Grubmueller, Energy barriers and driving

forces in *t*RNA translocation through the ribosome, *Nat. Struct. Mol. Biol.*, 2013, **20**(12), 1390–1397.
3 W. Humphrey, A. Dalke and K. Schulten, VMD – Visual Molecular Dynamics, *J. Mol. Graphics*, 1996, **14**, 33–38.
4 M. S. G. Razul, J. G. Hendry and P. G. Kusalik, Mechanisms of heterogeneous crystal growth in atomic systems: Insights from computer simulations, *J. Chem. Phys.*, 2005, **123**, 204722.
5 J. Vatamanu and P. G. Kusalik, Molecular insights into the heterogeneous crystal growth of sI methane hydrate, *J. Phys. Chem. B*, 2006, **110**, 15896–15907.
6 I. M. Svishchev and P. G. Kusalik, Crystallization of Liquid Water in a Molecular Dynamics Simulation, *Phys. Rev. Lett.*, 1994, **73**(7), 975–978.
7 S. J. Marrink, H. J. Risselada, S. Yefimov, D. P. Tieleman and A. H. de Vries, The MARTINI Force Field: Coarse Grained Model for Biomolecular Simulations, *J. Phys. Chem. B*, 2007, **111**, 7812–7824.
8 S. J. Marrink and D. P. Tieleman, Perspective on the Martini model, *Chem. Soc. Rev.*, 2013, **42**, 6801–6822.
9 A. Widmer-Cooper, P. Harrowell and H. Fynewever, How reproducible are dynamic heterogeneities in a supercooled liquid?, *Phys. Rev. Lett.*, 2004, **93**(13), 135701.
10 A. Widmer-Cooper and P. Harrowell, On the relationship between structure and dynamics in a supercooled liquid, *J. Phys.: Condens. Matter*, 2005, **17**, S4025–S4034.
11 G. S. Matharoo, M. S. G. Razul and P. H. Poole, Structural and dynamical heterogeneity in a glass-forming liquids, *Phys. Rev. E: Stat., Nonlinear, Soft Matter Phys.*, 2006, **74**, 050502.
12 D. Rozmanov and P. G. Kusalik, Isocofigurational molecular dynamics study of the kinetics of ice crystal growth, *Phys. Chem. Chem. Phys.*, 2012, **14**, 13010–13018.
13 B. Hess, C. Kutzner, D. van der Spoel and E. Lindahl, GROMACS 4: Algorithms for Highly Efficient Load-Balances, and Scalable Molecular Simulations, *J. Chem. Theory Comput.*, 2008, **4**, 435–447.
14 "Grid3D Toolbox for molecular visualization." on-line: http://sourceforge.net/projects/grid3d, Jan 2014.
15 A. Henderson, "*ParaView Guide, A Parallel Visualization Application*," Kitware Inc., 2007.
16 W. Shroeder, K. Martin, and B. Lorensen, *The Visualization Toolkit* (4th edn), Kitware Inc., 2006.
17 L. Huang and Y. Liu, *In vivo* delivery of RNAi with lipid-based nanoparticles, *Annu. Rev. Biomed. Eng.*, 2011, **13**, 507–530.
18 S. Khalid, P. J. Bond, J. Holyoake, R. W. Hawtin and M. S. P. Sansom, DNA and lipid bilayers: self-assembly and insertion, *J. R. Soc. Interface*, 2008, **5**, 241–250.
19 "Bio.B-Gen, a simulation box generator for biological simulations." on-line: http://sourceforge.net/projects/biobgen, Jan 2014.
20 K. T. Love, K. P. Mahon, C. G. Levins, K. A. Whitehead, W. Querbes, J. R. Dorkin, J. Qin, W. Cantley, L. L. Qin, T. Racie, M. Frank-Kamenetsky, K. N. Yip, R. Alvarez, D. W. Y. Sah, A. de Fougerolles, K. Fitzgerald, V. Koteliansky, A. Akinc, R. Langer and D. G. Anderson, Lipid-like materials for low-dose in vivo gene silencing, *Proc. Natl. Acad. Sci. U. S. A.*, 2010, **107**(5), 1864–1869.

21 S. Baoukina, E. Mendez-Villuendas and D. P. Tieleman, Molecular View of Phase Coexistence in Lipid Monolayers, *J. Am. Chem. Soc.*, 2012, **134**, 17543–17553.

22 S. Baoukina, L. Monticelli, H. J. Risselada, S. J. Marrink and D. P. Tieleman, The molecular mechanism of lipid monolayer collapse, *Proc. Natl. Acad. Sci. U. S. A.*, 2008, **105**, 10803–10808.

23 Y. Y. Zuo, R. A. W. Veldhuizen, A. W. Newmann, N. O. Petersen and F. Possmayer, Current perspectives in pulmonary surfactant – Inhibition, enhancement and evaluation, *Biochim. Biophys. Acta, Biomembr.*, 2008, **1778**, 1947–1977.

24 B. Piknova, V. Schram and S. B. Hall, Pulmonary surfactant: phase behaviour and function, *Curr. Opin. Struct. Biol.*, 2002, **12**, 487–494.

25 J. Perez-Gil and T. E. Weaver, Pulmonary Surfactant Pathophysiology: Current Models and Open Questions, *Physiology*, 2010, **25**, 132–141.

26 J. H. Davis, J. J. Clair and J. Juhasz, Phase Equilibria in DOPC/DPPC-d(62)/ Cholesterol Mixtures, *Biophys. J.*, 2009, **96**, 521–539.

27 P. Uppamoochikkal, S. Tristram-Nagle and J. F. Nagle, Orientation of Tie-Lines in the Phase Diagram of DOPC/DPPC/Cholesterol Model Biomembranes, *Langmuir*, 2010, **26**, 17363–17368.

ROYAL SOCIETY OF CHEMISTRY

DISCUSSIONS

Advanced visualization and visual analytics: general discussion

DOI: 10.1039/c4fd90020k

Dr Procter opened a general discussion of Professor Dr Ertl's, Dr O'Donoghue's and Dr Petrov's papers: All of your systems were developed to provide new insight to specialists from specific simulations and/or data. Please briefly explain the kinds of user evaluation processes you applied (if any!) and any additional advice concerning the role of user evaluation in the creation and optimisation of these systems.

Professor Dr Ertl replied: We performed this work in very close collaboration with visualisation specialists and physicists. The visualisation techniques and their parameters were selected as the result of direct and personal interaction. The users, in this case the physicists, were put in the loop of visualisation development. Beyond that, a broader user evaluation was not performed.

Dr Petrov responded: RiboVision has been under development for two years. During this time, we have been taking into account comments regarding features or interface/user-friendliness from local users and from outside members of the community. At the initial stages of development, RiboVision was used and tested by a number of people from the RiboEvo center at Georgia Tech, who were using it for their own projects, presentations, figure preparation *etc.* During the later stages, we expanded the pool of users and added additional features. This process is ongoing. We also received comments from external reviewers; we redesigned the RiboVision interface to be more user-friendly based on these comments. Following the advice of colleagues, to facilitate learning, we also created the video tutorials and placed them on YouTube (https://www.youtube.com/watch?v=yroCBIyB9Y0). Finally, in a response to the experiences of our colleagues and our own experience as users of RiboVision, we are optimizing the code to make RiboVison faster and more efficient.

Dr O'Donoghue replied: The work described in our paper was a research prototype only. Although we do provide access to the complete working code and data, our goal at this point was not to create a usable system for others. We were first concerned with evaluating whether the proposed visualisation approach had any merit, and whether it would enable new insight. Regarding advice for creating usable systems, we believe one of the key barriers here is the prevailing concept which artificially separates 'developers' and 'users' – ideally, we would always develop systems in close collaboration with people who use them. So, 'users' are really 'collaborators' or 'co-developers'. Another concept that can be harmful is that of the 'domain expert' – when developers talks about 'domain experts', it could suggest they are developing software to performs functions they don't understand.

Professor Brooks commented: We need to develop the relationship between the visualisation and application communities, occasionally attending each other's conferences. Computer scientists are not in fact scientists, but toolsmiths; often building tools to help scientists. In my experience, it has always been a mistake to start with a vision for a tool (except the most general), and then hunt for a problem. What has worked for us is to start with a real user problem and a real user. Something useful to "somebody" is much more apt to be useful to many people than something not yet useful to anybody.

Professor Dr Ertl replied: I agree and want to stress again that the Faraday Discussion is such an event which fosters collaboration. Actually, I think that more core visualization researchers should have attended and I advertised the event at a recent Dagstuhl seminar, reminding visualization colleagues that we should not wait until application partners come to us but that we need to actively reach out. While I do not have a problem being a toolsmith, I consider it a bit dangerous if we concede that computer scientists are in fact not scientists. Effective collaboration needs partners on the same level of respective appreciation. This is difficult if the "real" scientists think of us as their programmers.

Professor Brooks commented: As toolsmiths, we operate on a "no money changes hands" basis: we raise our funding, our collaborators raise theirs.

I've always found that it takes 20-25% of my interactive graphics team's effort to do routine, non-research things for our collaborators' needs. E.g., making nice pictures for their publications in their journals, and videos for their conferences.

Such service is cheap at that cost to us: really good collaborators—great scientists and great to work with—are very precious. They stimulate and guide strong computer science on our part.

Professor Dr Ertl responded: I fully agree with the toolsmith perspective. However, in most cases we do not get away with 20% service. Since tools are constantly improving, users' expectations rise about what the minimum functionality a tool needs to have before they consider using it. This makes it more and more tedious to work on interesting aspects of research prototype while trying to catch up with standard functions. Nevertheless, for our group it works well with an increasing application base.

Professor Hirst opened the discussion of the paper by Dr O'Donoghue: The parallel coordinate representation seems very useful. What consideration is given to the order in which the coordinates are presented? For example, one could imagine identifying a maximally informative order, which might make strong correlations more readily apparent.

Dr O'Donoghue responded: There are many ways to precompute the order of axes (see ref. 32, Section 6.2 for an overview), all with a different purpose and optimality criterion that usually depend on the type of data under investigation and the analysis goals. In our system, the order of axes can be easily changed. We believe that this is the most effective strategy in an explorative analysis setting: to empower the user to experiment and find particular axes orderings that they judge to be the most useful. Once a good optimality criterion is found for the axis order in protein ensembles, it might be useful to incorporate it into the system.

Mr Stone asked: In reading the paper, I was curious how labels are dealt with for the lines on the interior of the parallel coordinates view. How does a user see the labels associated with the points on the interior when they want to know their associated values?

Mr Krone replied: If the user hovers over one of the polylines in the Parallel Coordinates plot with the mouse, the polyline is highlighted and all labels for this polyline are displayed. The labels then show the associated value for each axis. Using this simple user interaction, we remedy the problem of occlusion.

Dr Zoppè opened the discussion of the paper by Professor Dr Ertl: The observation of MD simulation is very difficult, especially for those that do not work with it everyday. What I find most problematic is the very fast vibration of atoms, which really hurts the sight of (some) observers. The computer graphics community has developed many ways to make it easier for the eye to look at a fast moving object, by introducing motion blur, or using other softening techniques; however, it seems that the MD people are reluctant to show animations made softer in some ways. In my opinion, if we could make the movies more pleasant to look at, we might look at them for longer times, and possibly get from the animations more information that becomes evident only after we have spent some time watching it.

A long time ago, an experiment was made with school children that were shown animations of water molecules. They took several minutes before they could 'see' that the molecules were 'searching for each other' or 'escaping each other'. More complex objects might require a longer observation time, but it is difficult to spend time looking at something that is not pleasant to look at.

Professor Dr Ertl replied: We agree that vibration of atoms can be disturbing. Moreover, in simulation data, there are discontinuities due to the temporal discretization. We agree with you that this can make it hard to follow certain atoms and might distract the analyst. For the application presented in our paper, we used a temporal aggregation, so we don't have this problem. In the case where the user wants to see an animation of the data, smoothly interpolating between time steps can be a solution. Here, it is important to note that linear interpolation of the atom positions can lead to wrong binding lengths; therefore, we think that spherical linear interpolation is a more promising approach. However, if the elapsed time between two snapshots is too long, there is no way to faithfully represent the path of an atom. Using methods from computer graphics like motion blur, as you proposed, could remedy this issue as it could be used to show the uncertainty of the interpolated data. A user study that examines the benefit of such a method would be necessary and might help to overcome the reluctance of the expert users.

Mr Hall remarked: In Section 6, it is stated "We also showed that dimensionality reduction, although helpful in many cases, can conceal interesting features that are only observable in 3D." This is very true, and the chemistry community is aware of this. It is important to note that the computational chemistry community has represented scalar fields for 3D space using 3D representations, *e.g.* spatial distribution functions, for over 20 years. As an example of representing such fields using 3D isosurfaces, consider P. G. Kusalik and I. M. Svishchev, *Science*, 1994, **265**, 1219. Therefore, the concept of representing the spatially 3D scalar fields resulting from simulations using 3D isosurfaces, *e.g.* as in Fig.4 and Fig. 5 of the paper, is not a novel idea for chemical simulations, though the authors give the impression that it is novel in their paper.

Mr Krone responded: You are, of course, right that 3D scalar field representations of molecular data are not a novel contribution. Our statement was with respect to the special use case, where 2D or even 1D aggregations and projections of the data are commonly used for analysis (see *e.g.* Fig. 4, left, in our paper). With such simple visualizations, it is hard or even impossible to see three-dimensional features in the data. We think the contribution is not a technical one but lies in the application we designed and implemented together with our project partners and the discoveries they made using the application.

Dr Baaden commented: The proposed space-time aggregation procedure combines results derived from simulation snapshots into a uniform spatial grid. Having a reliable origin and fitting of frames might be important in this context. The chosen DNA-within-a-nanopore example seems to be particularly well suited for such an averaging as it is bounded in space and the structure does not move very much. Is this a special case, or can it easily be expanded even to more flexible molecules moving freely in space, in which case I guess some fitting is required and some limits may be imposed on the timescale that can be averaged over such that the structure does not deviate too much?

Professor Dr Ertl replied: The investigations so far have been performed on a fixed rigid double stranded DNA structure. Extending our investigations to flexible molecules would be highly desirable, since it is of high practical importance and experiments with the much more flexible single-stranded DNA are abundant. Unfortunately a simple extension of the proposed method to characterise the ion behaviour around such a molecule indeed requires further conceptual work.

Mr Krone added: The data sets we were investigating are, indeed, very well suited for the method since they can be fitted easily. In the future, we want to extend our method to DNA that is more flexible. We strongly believe that a flexible fitting that uses the molecular structure of the DNA for such a case is possible. This should allow faithful aggregation over long periods of simulation time and still give good results. However, the location of the base pair within the nanopore would probably also have to be considered in this case.

Dr Zoppè opened the discussion of the paper by Dr O'Donoghue: The idea of disorder comes from a description derived from the practice of X-ray crystallography. I consider that the name 'disordered protein' or domain is a very unfortunate choice, because it gives a wrong idea. What we really think is that the 'disordered' parts are instead parts that can take up different conformations (even in the crystal condition, which freezes most parts of the protein). Being the most flexible, they might well be the most interesting, or most active and significant, just like, for example, the hands of humans that are the 'most disordered' and are indeed the tools that allow us to do so many different things.

Dr O'Donoghue replied: There is growing evidence that, in eukaryotic proteins at least, intrinsically disordered regions do indeed play key functional roles. It seems that many of these regions are involved in functional regulation, particularly *via* post-translational modification (see http://nar.oxfordjournals.org/content/40/21/10628.full and http://pubs.acs.org/doi/abs/10.1021/cr400695p). We are not aware of evidence that these sites are often directly associated with binding to ligands or other proteins. Therefore, to continue the human analogy, these regions may often be sensors, like ears and eyes, rather than hands. There is growing recognition in the community of the fundamental importance of these regions – this is reflected by the steady rise of papers addressing this topic. Our paper describes a new way to examine the direct, structural evidence on disorder, and we hope this may contribute to improving the understanding of this complex and poorly understood phenomenon.

Dr Hall asked: Your manuscript describes the application of parallel coordinates to plot the analysis of an ensemble of structures with no intrinsic ordering. Can you plot dynamic properties (*e.g.* from a simulation) using parallel coordinates?

Dr O'Donoghue replied: Yes, absolutely this would be feasible and likely useful. We alluded to this possibility in the introduction of our paper. The standard approach to visualising dynamic properties in parallel coordinates is by adding one axis for time and choosing an appropriate sampling for that dimension. Grottel *et al.* recently proposed to apply interpolation for the visualisation of multivariate trajectories in parallel coordinates, which we think is a very interesting approach that is directly applicable, *e.g.*, to MD simulations of protein structures. For an overview of other ways to visualise dynamic properties in parallel coordinates, see Section 6.5 in ref. 32 of the paper.

Dr Baaden queried: Would it make sense to (optionally) add a 3rd dimension to the 2D polylines in parallel coordinates (*e.g.* bottom of Fig. 1) such that the lines would traverse not a vertical axis in 2 dimensions but rather a two-dimensional plane spanned by two attribute axes in 3 dimensions. As I understand it, the plots illustrate correlations between adjacent axes, hence up to 3 interrelated attributes. Once a first round of correlations is determined between adjacent coordinates on the plot, this dimension extension may allow the unravelling of complex correlations between 4 to 6 attributes in adjacent planes that are difficult to spot in the 2D view.

Dr O'Donoghue answered: In the data visualization community it is usually believed to be best to avoid the use of 3 dimensions for abstract data, such as the parallel coordinate data we are displaying in this work. The general thinking is that 3D visualization should be used only when necessary, *i.e.*, only when visualizing 3D spatial data. Even for 3D spatial data, it is often advantageous to reduce to two dimensions *via* principal component analysis or other such methods. In 3D, occlusion and distortion effects often obscure relationships and require interaction to overcome. Thus, our thinking is that adding a 3rd dimension would undermine some of the advantages of the parallel coordinate visualization.

Mr Hall addressed Mr Krone and Dr O'Donoghue : Did you consider letting people select polylines in the parallel coordinate view based on the slopes of polylines between adjacent coordinates? The concept of using the slope to select polylines in parallel coordinate views was put forth in the paper H. Hauser, F. Ledermann and H. Doleisch in *IEEE Symposium on Information Visualization 2002* (*InfoVis '02*), Boston, USA, 2002, pp. 127–130. In the future, how do you plan to extend the brushing and filtering of the polylines in the parallel coordinate view?

Mr Krone replied: More sophisticated interaction techniques for the parallel coordinates like the selection based on slope that you mentioned might be very helpful and we might consider adding such techniques in the future. Currently, we use just a simple, range-based brushing, which is probably the easiest method for novice users who are not familiar with more sophisticated interaction for parallel coordinates.

Dr O'Donoghue replied: The technique you mentioned is called angular brushing (*i.e.*, selection based on slope) – it is a powerful selection technique for parallel coordinates, as it allows the user to select all lines of the same slope between any two axes. This essentially translates to selecting data points with a positive linear correlation ("ideal points", see Fig. 4 and 10 in ref. 32 of the paper) and thus can also be achieved by inverting one of the axes and selecting all lines that intersect at a point. We will therefore consider both axis inversion and angular brushing as valuable extensions to future versions of our system. Currently, however, we use just a simple, range-based brushing, which is probably the easiest method for novice users who are not so familiar with parallel coordinates.

Dr Reiher addressed Professor Dr Ertl: In quantum chemistry, the amount of data worth reporting has always been quite limited in the past (because the molecular sizes have been comparatively small, while the computational methods are well defined in such a way that calculations are, in general, very well reproducible). When we laid out the Real-time Quantum Chemistry approach[1] we realized that the rate of data production will become so huge that it is not at all clear how much of the final results are to be reported in a scientific publication (suddenly, the number of stable molecular structures and transition states simply explodes for a large molecular system and, for instance in catalytic reactions, these can all be important data). Clearly, in the field of classical molecular dynamics this situation has been part of the game for decades. As a visualization and simulation expert with a bird's eye view on simulation in the natural sciences

I would like to ask you about your opinion on how such huge amounts of data should be prepared for publication, *i.e.* selected and reported? How can one guarantee reproducibility of such a work – especially if carried out with specialized non-standard tools? How do the scientific conclusions in such a study need to be supported in order to be reproducible or verifiable in, say, 20 years time?

1 M. P. Haag, M. Reiher, Real-time Quantum Chemistry, *Int. J. Quantum Chem.*, 2013, **113**, 8–20.

Professor Dr Ertl answered: Simulation of biological or soft matter systems is challenging because energy and entropy are typically well balanced. Therefore single states of the system are not of importance if weighted by their probability. Typical Molecular Dynamics Algorithms ensure that every configuration is visited with a probability according to its Boltzmann weight. The applied mechanisms are called thermostat. Common Molecular Dynamics programs, such as the employed tool GROMACS, allow for a bit-wise reproducibility of trajectories, and therefore allow for reproducibility. Writing a parallel software based on random number that fulfil this prerequisite is, however, challenging. All physical conclusions are drawn from a statistical analysis of whole trajectories. Therefore the statistical analysis, in our case the aggregation mechanism, must be reported as well. A thorough analysis of statistical errors is required, as it sets the reference frame, to which accuracy the data is claimed to be "correct". As individual states of the system are of little interest, their documentation is typically of little direct value. Therefore the question posed is much less problematic than in fields like catalysis, where individual states can carry much more value.

Dr Sommer opened the discussion of the paper by Dr Petrov: During the discussion of your work, you have compared your RiboVision to Google Maps®. Of course it is a good idea to make the usability of our tools comparable to Google Maps, but the comparison to your 2D maps is a little bit misleading. Because Google Maps basically just copies the two-dimensional road layouts to their two-dimensional maps. In case of RiboVision, this is more complicated, because you have to compare the 3D representation of the ribosome with the 2D map of the secondary structure. So this task is more demanding. *Question*: Is there a way to automatize the secondary structure generation process for such a complex structure as the ribosome?

Dr Petrov replied: The automated production of rRNA secondary structures is theoretically possible. The process would probably require significant manual intervention. There are a number of applications that automatically generate predicted secondary structures of large RNAs. These automated methods provide arbitrary layouts of RNA secondary structural elements, and it is unlikely they would be accepted by the ribosome community. Secondary structures of the LSU and SSU ribosomal RNA of *E. coli* were originally laid out manually (Woese *et al.*, *Nucleic Acids Res.*, 1980, **8**, 2275; H. F. Noller, J. Kop, V. Wheaton, J. Brosius, R. R. Gutell, A. M. Kopylov, F. Dohme, W. Herr, D. A. Stahl, R. Gupta and C. R. Woese, *Nucleic Acids Res.*, 1981, **9**, 6167) in an artistic, aesthetic and informative manner. Over the next several decades, secondary structures for other prokaryotic species were generated using the *E. coli* secondary structure as a template using the XRNA program from the Noller lab (http://rna.ucsc.edu/rnacenter/xrna). However, eukaryotic rRNAs are larger, containing expansion segments that cannot be fully accommodated by the E. coli template (Cannone *et al.*, *BMC Bioinformatics*, 2002, **3**, 2). The expansion segments of large eukaryotic ribosomes must be detached from the template and are placed at arbitrary positions into available space.

We have recently proposed a family of new rRNA secondary structures that can be used as templates for all existing species, as far as we know. We have constructed secondary structures for a subset of bacterial, archaeal, and eukaryotic species ranging from bacteria to human (Petrov *et al.*, *Nucleic Acids Res.*, 2013, **41**, 7522; Petrov *et al.*, *PLoS ONE*, 2014, **9**, e88222). In these templates, we unified as much as possible the layouts between bacterial/archaeal species and eukaryotic rRNAs. We integrated many of the eukaryotic expansion segments into the secondary structures. We have also corrected some inconsistencies in the secondary structures that existed in the original layouts. The new layouts were essentially done manually but inherited some features of the original layouts. The modified secondary structures can serve as templates to generate diagrams of other closely related species.

However, to generate a new secondary structure from a template, the secondary interactions (*i.e.* a set of all base pairs) must be known. The most reliable sources of this data are three dimensional structures. However, only a handful of them are available. Alternatively, secondary structures can be predicted using the methods of co-variational analysis, or using folding algorithms. Although these methods work well for bacterial/archaeal rRNAs, the full secondary structures of eukaryotic expansion segments, which are highly variable and idiosyncratic, cannot be predicted with confidence. At this point we consider that developing an automated generation of the secondary structures is beyond the scope of RiboVision. Nevertheless, in the future we hope to develop an interface, which would allow the users to import a pre-generated secondary structure into RiboVision.

Dr Baaden asked: Is it possible to introduce additional forms of data related to structure and dynamics into the RiboVision suite? I am particularly thinking about cryo-EM-derived conformations and conformational transitions, as well as molecular dynamics simulation trajectories. Dynamics and conformational changes may represent an interesting new dimension of data.

Dr Petrov answered: Potentially any information that can be quantified and associated in a data structure with nucleotide positions can be readily integrated into RiboVision. RiboVision is currently using Jmol as an applet to visualize three dimensional conformations. Jmol does support visualization of density maps, but this feature has not been implemented in RiboVision yet. As crystal structures of ribosomes trapped in different conformations are available for a number of species, we are planning to develop a multi structure support option, which would include visualization as well as analysis of conformational changes features. Currently RiboVision contains a set of preloaded structures. We plan to extend RiboVision to allow users to upload a desired conformation from the MD simulations and to compare it with a preloaded structure. However, we are not planning to expand RiboVision to directly visualize MD simulations, in part due to limitations of Jmol. Once an alternative Java script-based molecular visualization applet becomes available, we may reconsider.

Following the meeting, a group of delegates participated in additional discussion on the topic of **WebMolVis3D: Web-based Molecular 3D Viewers and Formats for large(r) Structures**. This meeting has been summarised below by Dr Björn Sommer:

Introduction In the evening of the 8th May, a special meeting was initialized by Dr Sean O'Donoghue in the context of the Faraday Discussion 169, with the aim to talk about the future of web visualization in 3D. Shortly, I want to sum up the current problems which led to our meeting. A standard example of molecular web-compatible visualization is Jmol, which is especially used to visualize PDB files. It is also used for the official PDB website to interactively explore protein structures. Jmol is implemented in Java and was therefore usually used as a Java applet. *Via* Javascript it was easily and interactively integrated into a website. But thanks to the new security requirements of Oracle's Java, this is no longer possible in a user-convenient way. If an actual Java version is not installed, the applet will not run. And in case Java is up-to-date, then a number of warning messages have to be confirmed by the user before the applet starts. In addition, the developer is forced to obtain a valid software-signing certificate to authenticate the applet. But these problems apply for all Java developers in general. For 3D visualization, other particular issues have to be taken into account. In recent years, Java 3D was used by many projects, because it was quite intuitive for the software developer with moderate graphic programming knowledge. But as of today, there are two major drawbacks: for applets and applications, additional OS-dependent libraries are required, and second, the development of the original Java 3D project has been stopped. Still, the actual Java 3D libraries are compatible to most actual systems (Windows, Linux, Mac OS X), but in the near future it might not be supported anymore by new OS releases.

Discussion Therefore, actual developments have to be taken into account to identify new perspectives for web 3D development. But instead of simultaneously developing many different approaches, it would be a good idea if all biology-related visualization experts join their efforts to create a community-driven initiative uniting the basic ideas. During the meeting, the following ideas were collected.

1. WebGL should be used for 3D visualization in browsers, because it provides a direct access to OpenGL. However, it has to be mentioned that it is not possible to access the complete functionality of native OpenGL. Most browsers use Google's graphic system ANGLE, which may be deactivated in some browsers such as Firefox to directly access WebGL. However, this functionality is not straightforward for standard users.
2. JavaScript should be the language of choice to interact with the WebGL visualization in the browser; because it is already used by many existing approaches, its current implementation is quite fast (*e.g.* in Firefox) and it is compatible with all mainstream browsers.
3. Because JavaScript should be used, the appropriate data exchange format should be JSON (JavaScript Object Notation). A PDB importer and exporter will be required for JSON. The new JSON-based format should overcome the drawbacks and limitations of the PDB format and moreover meet the requirements of all participants.
4. An alternative way of visualizing especially huge structures in browsers could be provided by a server-client architecture: a server is used to compute the visualization in the background and to deliver images *via* the Internet to the client computer.
5. An initial meeting for all potential participants of a WebGL initiative for Bio Visualization should be prepared. Two ideas exist: a) a separate two weeks workshop, where all

attendees try to develop a framework which meets the requirements of all involved projects, and b) an initial meeting in context of the VIZBI 2015 conference in Boston. In both cases, only invited attendees should be addressed. A good idea would be the involvement of experts of related fields such as the SBML initiative (*e.g.* Michael Hucka).

6. Different funding opportunities for a larger international initiative have been discussed, such as EU proposals for Horizon 2020, a COST proposal, NIH or NCBI. Different biology-related application cases could be combined for this purpose.

Participants The Participants of this initial meeting were: Marc Baaden (Paris, France), Graham Johnson (San Francisco, USA), Michael Krone (Stuttgart, Germany), Séan O'Donoghue (Sydney, Australia), Arthur Olson (San Francisco, USA), Anton S. Petrov (Georgia, USA), Björn Sommer (Bielefeld, Germany), John Stone (Illinois, USA), Jens Thomas (Liverpool, UK), Jim Zheng (Houston, USA).

Dr Baaden responded: Molecular visualization for the web is an important topic and WebGL is certainly a very promising approach. For some applications, the fact that it is a rather low level approach might represent an inconvenience. For that purpose, I would strongly advocate to also discuss more abstracted solutions that will in many cases carry out the job to translate a scene to some standard such as WebGL. Of course this typically comes at some cost, such as lower performance. On the other hand, it may come with free benefits such as being able to export to specific mobile platforms (iOS, Android,..). So far, we have experimented with three approaches: using the Processing framework (processing.org) for simple molecular graphics, using the Unity3D game engine[1] or interacting with the SpiderGL framework.[2,3] Simple Unity scenes may also be exported directly to HTML5.[4] Such *meta*-frameworks make it surprisingly simple to put together not too demanding web-based molecular viewers.

1 Z. Lv *et al.*, Game on, Science – how video game technology may help biologists tackle visualization challenges, *PLoS ONE*, 2013, **8**, e57990.
2 M. Callieri *et al.*, Visualization methods for molecular studies on the web platform, The Web3D 2010 Conference, 22-24 July 2010, Los Angeles, California (http://www.scivis.it/images/stories/PDFarticle/spidermol.pdf).
3 http://www.baaden.ibpc.fr/pub/blt2/jbn11.html
4 https://github.com/drojdjou/J3D/wiki/Unity-exporter-tutorial

Dr O'Donoghue answered: Many thanks for documenting this discussion. I would like to elaborate on the point raised about the current security issues that block Java applets, giving some background on why this is such a concern, and suggesting how the JavaScript efforts can hopefully avoid some of these problems. Our group maintains a service for the academic community that provides a concise visual summary of all 3D structures significantly similar to any protein (*e.g.*, http://aquaria.ws/P04637/). The core of this service is a free, open source Java3D applet (http://srs3d.org/) that has required over 6 man-years of development. In recent years, however, most browsers have added increasingly heavy-handed security restrictions on Java applets to prevent malicious attacks. Unfortunately, as with the proverb about throwing out the baby with the bath-water, these restrictions now effectively block completely legitimate Java applets, including many such as ours that have been developed by and for the scientific community. Currently, the most obvious response is to cut our losses and start over, completely re-implementing these scientific web applications from scratch in JavaScript, while taking advantage of WebGL to use GPU-accelerated 3D graphics. For many reasons, I think this is the right response – however, technologies that access the GPU raise their own security issues. Hence, I wish to raise the concern that we may potentially face a similar issue in the future as we do now for Java applets. I'm proposing that the nascent JavaScript-based molecular graphics initiatives combine forces and work with standards bodies such as the Khronos group to hopefully ensure that the work we are now beginning is not squandered again in the future.

Mr Stone commented: In regards to the discussion of Java based molecular viewers, we developed a simple Java based viewer and experienced many of the same issues that others have raised here. One concern I have about solutions that are focused on Java- or browser-based JavaScript molecular viewing technologies is that they tend to only address the 3-D rasterization part of the problem, and they don't appear to provide a good solution for computationally demanding visualization tasks. Many of the advanced visualization features in programs like VMD are based on rapid computation using multi-core CPUs and GPUs, *e.g.* for computation and display of secondary structure, electrostatic fields, and molecular surfaces. I would be concerned about developing new molecular viewers using any technology that doesn't adequately address both the graphics and computational needs of advanced molecular visualization.

Mr Stone opened the discussion of the paper by Dr Fowler: Do you think that the 2D image processing approach taken in the paper could be extended to 3D by measuring surface area associated with each region of density within the simulation volume? From the perspective of processing efficiency, would there have been any advantage in doing the density map averaging calculation inside of VMD, using the "VolMap" analysis tool or the associated lower level analysis commands, or is this a relatively inexpensive task in terms of I/O and computation?

Dr Fowler replied: Yes, the approach could, and should, be extended to 3D as the reliance on the bilayer approximating a 2D surface is a weakness in our approach. This is especially important as more complex lipid structures, such as viruses or highly-curved membranes, are now being studied. At present, the python-based approach is not computationally expensive, in part, we assume, due to the optimisation of the NumPy and SciPy libraries. For example, processing a single trajectory to produce all the data and images necessary for Fig. 4–6 took around 20 s per frame on a single 2008 Intel Xeon CPU core. Most of this would have been computation. That said, it is not difficult to imagine situations where the analysis does become expensive (such as doing it in 3D), in which using the GPU-accelerated VolMap tool would be necessary.

Mr Stone addressed Dr Fowler and Dr Rozmanov: The VMD "QuickSurf" surface algorithm begins by computing a density map very rapidly using GPUs or multi-core CPUs.[1] The present algorithm generates a density map that is the result of only a single trajectory frame, but it is straightforward to extend this to allow temporal averaging over a range of trajectory frames near to the "current" frame, and to implement a wider variety of density functions than are currently implemented. Can you think of other potential extensions to QuickSurf along these lines that would be beneficial for these types of visualizations and analyses?

By extending the built-in features of VMD to encompass more of the features described in these papers, these capabilities could become available to a much larger number of researchers, including those users that don't have the necessary skills to develop their own tools as has been done in these papers.

1 M. Krone, J. E. Stone, T. Ertl, and K. Schulten, Fast Visualization of Gaussian Density Surfaces for Molecular Dynamics and Particle System Trajectories, *EuroVis - Short Papers*, 2012, 67–71.

Dr Fowler responded: There are two key components here to the analysis: (1) generating the "image" (if 2D) or "density map" (if 3D) of the atoms or beads of interest in the simulation and (2) analysing that data, perhaps using image processing tools or other methods. The fast, GPU- or multi-CPU accelerated VolMap tool within VMD could help accelerate the first step, but the difficulty is in interfacing with appropriate libraries to do the second. The most obvious step would be to have Tk/Tcl command line access to the VolMap tool and being able to write out the data to disc in an appropriate format. The user could then read it into *e.g.* python and perform the required analysis. A longer term solution would be to make VMD, or the analysis parts, such as VolMap, available as, for example, a python module, which would streamline such an analysis process.

Dr Rozmanov responded: If VMD could natively work with densities and conveniently manipulate them this would be a great help. Note, when an object is visualized, in VMD for example, very often it needs to be quantified and manipulated; currently, there are not many options for this that I am aware of. It would be nice if there was a clear way to not only show the density isosurfaces in VMD but also save them in a well documented format so they could be analyzed outside VMD. It would be useful if VMD could save the 3D density it computes in a compact documented form. Doing computation on the density grids, such as computation of a product of two or more grids is a good way to visualize spacial correlations between species, as it is shown in Fig. 10 of the paper. I am offering the "g3d" format as a grid container, which is a part of the Grid3D Toolbox (http://sourceforge.net/projects/grid3d).

Dr Glowacki addressed Dr Fowler: In your presentation, you seemed to suggest that your lipid investigations were limited by Cartesian representations of molecular dynamics and structure, but it seems to me that the image processing methods you've used are also Cartesian – *i.e.*, they are representations of 2D Cartesian space (x,y) in a sort of coarse-grained pixel format. Is it really a case that you were limited by Cartesian representations, and not rather that you were simply limited by the standard analysis approaches? Furthermore, I wonder whether you've thought about any other areas where image processing approaches could usefully be applied to molecular analysis?

Dr Fowler replied: I should clarify what I mean by Cartesian: here I take it to mean plotting and analysing the centres of atoms (or coarse-grained beads) in 3D Cartesian space. You are correct that I was limited by standard analysis approaches, but I argue that this is because the vast majority of analysis tools in the field of biomolecular simulation are based on manipulating the 3D Cartesian coordinates of the atoms/beads (for example, computing a radial distribution function). Once one starts viewing the simulation differently, as an image for example, then some tasks that were previously difficult (like measuring the length of the interface) become easier.

I fully expect other techniques from image processing to be used to analyse and visualise molecular simulations, in part as the number of atoms or beads in the simulations grows (10^6 to 10^7 is common) thereby becoming more complex not only just through size but also by being able to model more complex phenomena. One particularly pressing example is how do we analyse the curvature of an undulating lipid bilayer in a simulation? There are approaches from computer vision (Delaunay triangular meshes, Bézier patches) that could be applied but we are only at the start of this process.

Dr Glowacki commented: With regards to Fig. 6c in your paper, you have claimed that this makes the case that NRas has a statistically significant effect on spinodal decomposition. But there are appears to be a lot of noise in the graphs. On the surface, it's not clear to me if the results are actually distinct within the noise. Is it possible to quote error bars for the spinodal decomposition rates to make this claim more quantitative? Is it possible that some of the error might arise from the edge detection algorithm itself? I wonder whether you have tried any other edge detection algorithms – *e.g.*, gradient following, or whether you think that is something worth doing.

Dr Fowler answered: You are correct that we did not test the significance of this result, either by calculating error bars or by some other approach. If we assume that each simulation contributes two independent measurements of the total length of the interface (since there are two leaflets to the bilayer), then there are two sets of six data points evolving with time. Fig. 1 here shows the p-value of these datasets having the same expected values, calculated using the independent

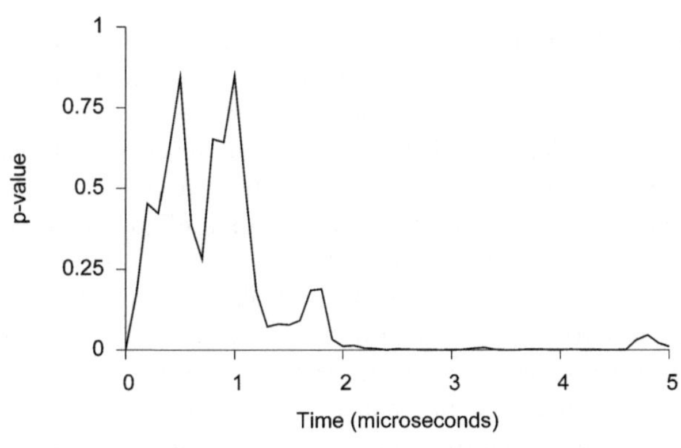

Fig. 1 NRas significantly slows the rate at which our model ternary lipids bilayer phase separates. The p-value shown here is the probability, according to the independent two-sample Student's t-test, that the two sets of interface lengths plotted in Fig. 6c of our paper (DOI: 10.1039/c3fd00131h) are drawn from the same distribution. After c. 2 μs this probability is very small, and hence the difference is statistically significant.

two-sample Student's *t*-test. This shows that once the simulations have run for 2 microseconds, it is highly likely that NRas is retarding the rate at which the bilayer phase separates. We did not try any other edge detection algorithms but we did try altering the width of the Gaussian used in the process, but there was no change in the final result.

Dr Hall queried: Voronoi tessellation (mentioned in your paper) could also be used to calculate the edges between domains, and has been used for this purpose in membrane simulations by a number of different groups in the past. What are the key advantages of this method over Voronoi tessellation? Additionally, have you benchmarked the pixel-based technique against the traditional approach?

Dr Fowler responded: Yes, Voronoi tessellation could be used to divide the leaflet into a series of "cells", each of which would hold a single lipid, and therefore domains could be identified and the total length of the interface measured, just as we have done. I would expect that this method would suffer from higher noise due to the discrete nature of the tessellation, which might have made detecting the effect we saw more difficult. Additional advantages of our approach are that (1) it is straightforward to produce meshes (variation in height over the surface) for each leaflet and thereby analyse the variation in height (Fig. 5 in the paper) and (2) one can more easily examine where the other components, such as protein or cholesterol, are. It is not obvious how one would do that in a consistent way using Voronoi tessellation. We haven't benchmarked the traditional approach as I have not been able to measure the interface using traditional coordinate-based methods.

Dr Zoppè commented: Your model shows that lipid mixing is slowed by the presence of NRas, but it also shows that NRas aggregates. Is there any experimental evidence for this behavior of NRas or of any other membrane protein? If the proteins in the model behave in a non-physical way, why should one believe that the lipids behave in the model like they do *in vivo*?

Dr Fowler responded: Yes, there is some aggregation of the ten NRas proteins in each leaflet of the bilayer. This appears to occur through the soluble G-domain. There is no direct experimental evidence of this behaviour, although it is known that Ras proteins do cluster. This behaviour is, however, usually ascribed to the proposed preference of Ras proteins for cholesterol-rich domains in the cell membrane (the "lipid raft" hypothesis) rather than oligomerisation. One of the main strengths of the MARTINI force field that we have used is its modelling of lipid bilayers so we do not have any concerns about the behaviour of the lipids. That said, we did have to use an unphysiological lipid, DUPC, to see phase separation within the microsecond timescale. In future work we are using a truncated form of NRas that has had the G-domain removed. This mimics a common experimental construct and not only should avoid any possible artifactual clustering but also accelerate the diffusion of the protein in the lipid bilayer.

Dr Sommer remarked: If, for example, approaches such as Voronoi tessellation are used for the visualization and analysis of the area per lipid in a membrane, the results strongly depend on the chosen key atoms. In case of APL@Voro (ref. 16 in the paper), we used a specific key atom selection to visualize phase separation. *Question 1*: Does your approach also depend on the selection of specific key atoms? *Question 2*: Or, more precisely: did you test different key atom selections with your approach for specific application cases? *Remark*: It might be interesting to compare the results of your methodology with the ones of APL@Voro, because both approaches use Gromacs with the MARTINI force field in the corresponding publications. So it might be relatively easy to identify pros and cons of both approaches.

Dr Fowler replied: Throughout we identified the position of each lipid using its MARTINI coarse-grained phosphate bead. We did not test the effect of, for example, using the headgroup bead. One could also, of course, use instead the centre of mass of the lipid. It would, of course, be interesting to compare the two approaches as it is not clear which method will work best in any given situation.

Mr Krone opened the discussion of the paper by Dmitri Rozmanov: You recommend to use a grid resolution of 0.5 Å to capture phenomena at an atomic level of detail. On the other hand, you argue that the density based (volumetric) representation is good for visualizing very large data since a coarser grid can be used. However, this obviously leads to reduced detail. Do you think that a coarser grid could lead to the loss of very localized phenomena and, therefore, to wrong assumptions about the data due to this information loss?

Dr Rozmanov responded: Yes, you are right. If very localized phenomena is important and the grid resolution is not fine enough then the interpretation of the results may be affected. This is true not only for the spatial resolution but for temporal resolution as well. The sampling is done over some specific time assuming that the lifetime of important structural features is longer than the sampling time. If sampling time is too long then the features may be time averaged and this may also affect the interpretation. To overcome this potential problem, a good understanding of the system is required. From the technical side, a specific part of the system can be re-sampled on a higher resolution grid if it is known to be of specific interest. This, however, is not implemented in the toolbox we are providing.

Mr Krone commented: You said that you are using ParaView for isosurface extraction (marching cubes). I think that you could benefit from a specialized implementation that would allow you to process your density grid only partially. This would allow you to handle larger data sets at higher resolution. A similar scheme was used for the QuickSurf representation in VMD (details can be found in: Krone *et al.*, Fast Visualization of Gaussian Density Surfaces for Molecular Dynamics and Particle System Trajectories, *EuroVis - Short Papers*, 2012, 67–71).

Dr Rozmanov answered: We use ParaView for visualizing the system and exploring it. Using isosurfaces is a convenient way to visualize the 3-dimensional density grids we have. To this end, we are not concerned with a specific implementation of a surface extraction algorithm provided that it works. Generally, we found that ParaView implementation is adequate for most of our needs. If it can be implemented more efficiently that would be nice and desirable, of course. When we want isosurfaces for later analysis, we have our own non-interactive surface extracting tool, which can save triangulated surfaces for us using the marching cube algorithm. The speed is no concern in this case as it is not time critical.

Mr Stone commented: I agree that density based visualizations of simulation trajectories are useful, and that many of the same points made in this paper drove development of particular features in VMD. In particular, it is worth mentioning that the "QuickSurf" surface visualization feature of VMD[1] is one example where we regularly use a density-based visualization scheme to allow interactive display of very large complexes that are difficult to handle with other techniques.

VMD includes several trajectory density averaging capabilities as part of the "VolMap" analysis tool and associated volumetric processing functions that were developed later for a variety of closely related purposes.[2–5] The VolMap tool in VMD would appear to implement a few of the property/density schemes mentioned in this paper, but the paper also mentions that a large number of other properties may be computed similarly. I was wondering if there is a particular motivation to compute the property grids within the simulation engine itself, or if this could just as easily be done in VMD or a similar visualization or analysis tool post-simulation? Are any of the properties dependent on data that is only available within the simulation tool, or could they just as easily be computed later from a simulation trajectory file? Are there any algorithmic efficiencies that are possible due to implementation within a simulation code that would not be available otherwise?

Does the approach described in the paper take advantage of the sparsity in cases where much of the volume might be "empty"? When implementing QuickSurf in VMD,[1] we found memory capacity to be a significant challenge, so I was wondering if you had any comments on whether memory capacity limits were a problem in this work?

1 M. Krone, J. E. Stone, T. Ertl and K. Schulten, Fast Visualization of Gaussian Density Surfaces for Molecular Dynamics and Particle System Trajectories, *EuroVis - Short Papers*, 2012, 67–71.
2 J. Cohen, K. Kim, P. King, M. Seibert and K. Schulten, Finding gas diffusion pathways in proteins: Application to O_2 and H_2 transport in CpI [FeFe]-hydrogenase and the role of packing defects, *Structure*, 2005, **13**, 1321–1329.
3 J. Cohen and K. Schulten, O_2 migration pathways are not conserved across proteins of a similar fold, *Biophys. J.*, 2007, **93**, 3591–3600.
4 B. J. Johnson, J. Cohen, R. W. Welford, A. R. Pearson, K. Schulten, J. P. Klinman and C. M. Wilmot, Exploring molecular oxygen pathways in Hanseluna Polymorpha copper-containing amine oxidase, *J. Biol. Chem.*, 2007, **282**, 17767–17776.
5 J. E. Knapp, R. Pahl, J. Cohen, J. C. Nichols, K. Schulten, Q. H. Gibson, V. Srajer, and W. E. Royer Jr., Ligand migration and cavities within scapharca dimeric HbI: Studies by time-resolved crystallography, Xe binding, and computational analysis, *Structure*, 2009, **17**, 1494–1504.

Dr Rozmanov responded: Yes, there is motivation to compute properties on the fly during simulation run time. In fact, the initial implementation of property grids was done on the fly in an unpublished simulations code MDIce that is used in Prof. P. G. Kusalik's group for ice–water interface detection (see Fig. 5 in ref. 1). The method depends on atomic (molecular) properties or contributions so they can be sampled on a spatial grid. This kind of information is rarely available for atoms besides locations, momenta, and possibly forces. But other properties, such as atomic potential energies, stress tensor contributions, and mobility can be accessed from within the simulation engine during run time. In addition, on the fly generation avoids saving huge trajectory files and saves only necessary property grids. Such functionality not only would allow accessing additional properties but also would eliminate much of the grid generation effort, which currently is significant.

The grid generation facility in the MDIce code mentioned did take advantage of the grid sparsity. It was possible because, in that work,[1] water molecules did not move much and multiple grid cells were never visited during a single sampling time window, 2–10 ps. However, once the time window becomes longer than the molecular diffusion time, the benefits of sparse mapping diminish and at some point, say 100 ps, there will be no advantage of using sparse grid storage. At higher temperatures, this crossover in storage efficiency will happen even at shorter times.

Memory capacity can be an issue when working with multiple density grids. The lipid nanoparticle system described in this paper was analyzed on a 300x300x300 grid, which requires *ca.* 200 MB of memory for a single grid of a scalar property, such as density. Loading and analyzing a data file containing 20–30 density grids in ParaView (http://paraview.org) may be a challenge on a desktop machine with 16 GB of memory, but increasing the available memory to 32 or 64 GB normally solves this problem.

1 D. Rozmanov and P. G. Kusalik, *PCCP*, 2012, **14**, 13010–13018.

Dr Glowacki remarked: I found your density-based visualization methods very interesting – in particular their ability to reveal time-dependent phenomena in stationary snapshots, which continues to be a problem. You mentioned in your talk that you have applied these methods to a range of observables, including energy. For the last few years, I have been active in carrying out non-equilibrium MD simulations to investigate the time-dependent diffusion of energetic "hot-spots" in complex molecular systems.[1] On the heels of this work, I have very recently begun examining the information content of molecular trajectory snapshots when one simply includes atomic 'trails' derived from feedback of previous rendering buffers. Such snapshots effectively illustrate the time history of a trajectory, as shown in Fig. 2. For gas phase scattering and energy transfer processes, these sorts of figures clearly provide information on the extent of rovibrational energy transfer that occurs in a gas phase collision. I was wondering whether you have undertaken any strategies for visualizing energy transfer processes, in either condensed phases or in the gas phase.

1 D. R. Glowacki, A. J. Orr-Ewing and J. N. Harvey, *J. Chem. Phys.*, 2011, **134**, 214508; D. R. Glowacki, R. A. Rose, S. J. Greaves, A. J. Orr-Ewing and J. N. Harvey, *Nat. Chem.*, 2011, **3**, 850–855; S. J. Greaves, R. A. Rose, T. A. A. Oliver, D. R. Glowacki, M. N. R. Ashfold, J. N. Harvey, I. P. Clark, G. M. Greetham, A. W. Parker, M. Towrie and A. J. Orr-Ewing, *Science*, 2011, **331**, 1423–1426.

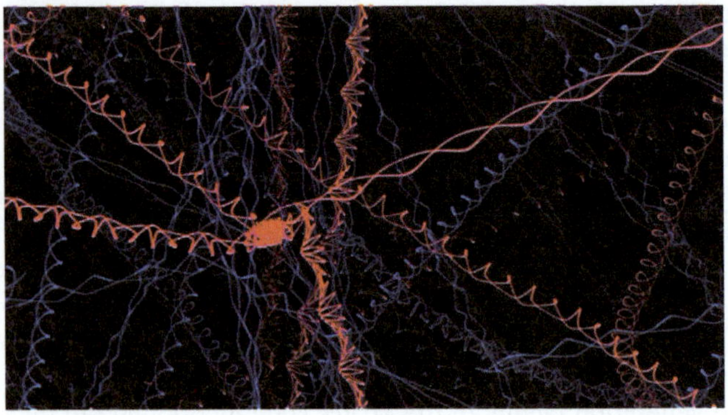

Fig. 2 Time history of a semiclassical collision trajectory between two diatomic molecules.

Dr Rozmanov replied: This density based method is better suited for condensed systems, where much better density sampling can be achieved in a relatively short time when compared with a gas phase. The proposed approach is very general and, in principle, can sample any property of the system. Such property fields or property density fields can be visualized to reveal spatial variations in the property in the system for the sampling time. Fig. 3 here shows such a 3D field (profile) of the potential energy in a 2-phase ice–water system at 10 ps time scale. Fig. 4 and 5 provide different representations of the potential energy on a 2D slice through this profile. The middle liquid part of the system demonstrates significant variation of the potential energy at this time scale. Animation of such a profile along the time line of the energy transfer event may be a good way to visualize it.

Finally, it has to be noted in regard to this visualization technique that it tries to move from discrete single atom data to a statistically averaged continuous

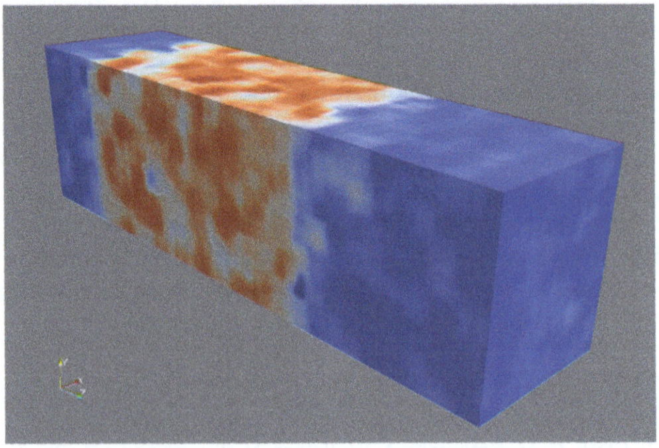

Fig. 3 3D field (profile) of the potential energy in a 2-phase ice–water system at 10 ps time scale.

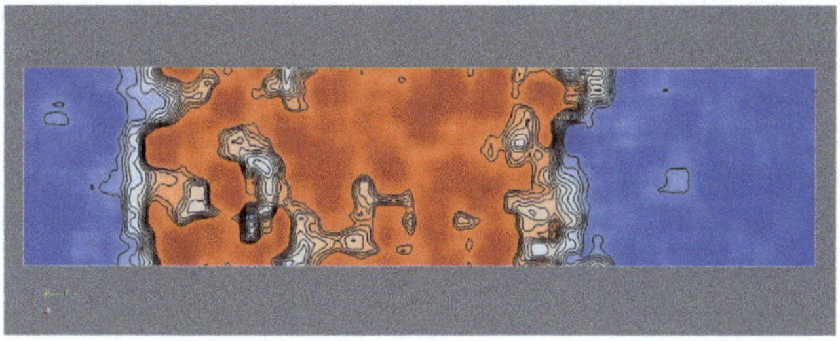

Fig. 4 A contour plot representation of the potential energy on a 2D slice through the profile in Fig. 3.

Fig. 5 A carpet plot representation of the potential energy on a 2D slice through the profile in Fig. 3.

density field. This implies that the events of interest have to be statistically significant for the system and localized in some region of space of the system. This requirement may make visualization of gas phase events difficult.

Dr Baaden commented: I wonder whether trajectory data requires specific fitting for the density to be meaningful and how such a fitting would affect the visualization. This might be an issue when attempting to sample densities over longer timescales, where the underlying structures may move significantly such that an average density only hardly captures the model. An example could be movement, diffusion or rearrangement of internal cavities such as those shown in Fig. 2B of the paper. E.g. when interested in following a specific cavity, would the same result be obtained without fitting the trajectory, with a global fit on the whole lipid nano-particle and with a locally restricted fit on the vicinity of the cavity of interest?

Dr Rozmanov responded: We stress in the paper that the sampling time is very important for the method and, in fact, is a parameter that should be estimated beforehand. The sampling time must be shorter than the lifetime of the features of interest; in this case, than the diffusion time of the water cavities. This might seem as a limitation initially, but instead we see it as an additional power. The time scale is another degree of freedom in visual representation and data analysis.

The system does look differently on different time scales. It is the same as different experimental structural methods give different information about the same system depending on their characteristic time scales. Looking at the system using different time scales may give additional insight into its behavior. A good example for this is Fig. 13 in the paper, where the comparison of the cholesterol density sampled over 1 and 10 ns time windows clearly shows the onset of diffusive motion at the longer time scale.

Faraday Discussions

PAPER

GPU-accelerated analysis and visualization of large structures solved by molecular dynamics flexible fitting

John E. Stone,*[a] Ryan McGreevy,[a] Barry Isralewitz[a] and Klaus Schulten*[b]

Received 1st February 2014, Accepted 17th March 2014
DOI: 10.1039/c4fd00005f

Hybrid structure fitting methods combine data from cryo-electron microscopy and X-ray crystallography with molecular dynamics simulations for the determination of all-atom structures of large biomolecular complexes. Evaluating the quality-of-fit obtained from hybrid fitting is computationally demanding, particularly in the context of a multiplicity of structural conformations that must be evaluated. Existing tools for quality-of-fit analysis and visualization have previously targeted small structures and are too slow to be used interactively for large biomolecular complexes of particular interest today such as viruses or for long molecular dynamics trajectories as they arise in protein folding. We present new data-parallel and GPU-accelerated algorithms for rapid interactive computation of quality-of-fit metrics linking all-atom structures and molecular dynamics trajectories to experimentally-determined density maps obtained from cryo-electron microscopy or X-ray crystallography. We evaluate the performance and accuracy of the new quality-of-fit analysis algorithms *vis-à-vis* existing tools, examine algorithm performance on GPU-accelerated desktop workstations and supercomputers, and describe new visualization techniques for results of hybrid structure fitting methods.

1 Introduction

Molecular dynamics simulations provide researchers with a "computational microscope", a powerful tool that provides dynamic views of cellular processes with atomic detail and nanosecond temporal resolution that can not be achieved through experimental methods alone. Petascale supercomputers extend the reach of the computational microscope to very large biomolecular systems of particular interest to public health, *e.g.*, in the case of viruses such as HIV.[1]

Molecular dynamics simulations depend on the availability of all-atom molecular structures, but it can be extremely difficult to obtain all-atom atomic structures for large biomolecular complexes through traditional approaches. The most widely used method for acquiring structures of biomolecules is X-ray

[a]*Beckman Institute, University of Illinois, 405 N. Mathews Ave, Urbana, IL, USA*
[b]*Department of Physics, University of Illinois, 1110 W. Green, Urbana, IL, USA*

crystallography. However, crystallization of large biomolecules and macromolecular complexes can be challenging. Instead, cryo-electron microscopy (cryo-EM) single-particle reconstruction is becoming a central technique for structure determination of large biological complexes. Cryo-EM does not require the difficult crystallization step and allows the structure to be imaged in solution, more closely reproducing physiological conditions. Although cryo-EM yields sub-nanometer resolution data, sometimes approaching atomic resolution,[2–9] crystallographic structures are generally still used for interpreting the cryo-EM data. This approach requires combining data from different imaging modalities using techniques known as hybrid fitting methods. Many hybrid fitting methods that combine X-ray crystallography structures and cryo-EM density for structure determination have been developed in recent years. Some of these methods use rigid-fragment fitting,[10,11] while others such as Rosetta,[12] DireX,[13] Gorgon,[5] and FRODA[14] perform flexible fitting which allows conformational changes to better shape the structure to the density. Some approaches include the use of low-frequency normal modes,[15] deformable elastic networks,[13] and cross correlation[16] or least-squares difference between experimental and simulated maps[17] to drive the structure into the cryo-EM density. There are also fitting methods that use a Monte Carlo-based approach,[18] while others such as our own use molecular dynamics.[19]

Our method, Molecular Dynamics Flexible Fitting[19,20] (MDFF), matches a crystal structure to a cryo-EM density potential energy function during an MD simulation. The additive modification of the potential energy function is defined on a 3-D grid and incorporated into an MD simulation using the **gridForces** feature of NAMD.[21,22] Forces are computed from the added potential and applied to each atom depending on its position on the grid using an interpolation scheme. The computed forces push the atoms toward areas of higher density within the EM map. Restraints imposed during the simulation help preserve the secondary structure, stereochemical correctness,[23] and symmetry[24] of the proteins. MDFF has proven to be successful, as evidenced by its many applications solving structural models for the ribosome,[25–33] photosynthetic proteins,[34,35] and the first all-atom structure of the HIV capsid.[1]

An important step in any hybrid fitting method is the evaluation of the quality-of-fit of the final structure to the cryo-EM density. One of the most common scoring methods is the cross correlation coefficient between the experimental density map and a simulated density map. Other scoring functions exist, such as the Laplacian-filtered cross correlation or an envelope score which determines the amount of density filled with atoms, and uses different approximations resulting in varying levels of accuracy.[36] Approximations are useful because they enable the fast computation necessary for interactive visualization and analysis of fittings, particularly for large structures (>1 million atoms) or long time-scale (microsecond range) simulations. Some scoring functions, including cross correlation, require the calculation of a simulated synthetic density map from the atomic structure which can then be compared to the experimental map. The traditional way of accomplishing this employs techniques created for X-ray crystallography which represent the density contribution of each atom as a spherical Gaussian function.[37,38]

Below, we describe new data-parallel CPU and GPU algorithms for the rapid computation of simulated density maps, density difference maps, cross correlation, and spatially localized cross correlation maps. The new algorithms serve both analytical and visualization uses and achieve performance levels that enable their effective use on large-size biomolecular complexes such as viruses, and for the detailed analysis of fitting results in the case of long MDFF trajectories. We illustrate the use of the new analysis features in several visualization scenarios and in the context of new MDFF quality-of-fit analysis features running on both desktop workstations and supercomputers.

2 Related work

The original sequential CPU implementation of density simulation for MDFF was based on a method described earlier,[39] which involves interpolating atomic weights onto a grid and convolving the grid with a Gaussian filter whose standard deviation is assumed to be half the target resolution. One problem with this implementation is that the spreading of atoms which occurs when interpolating the weights onto a grid is not accounted for when applying the Gaussian filter. The Situs[40] program **pdb2vol**[41] takes a similar approach to creating synthetic densities, but it corrects for the lattice smoothing by subtracting the lattice projection mean-square deviation from the Gaussian kernel variance. Situs also allows the user to select kernel types other than Gaussian, such as triangular or hard sphere. For the Situs-generated density maps created for the comparisons reported in this paper, a Gaussian kernel with lattice smoothing correction and mass-weighted atoms was used (Situs version 2.7.1 obtained from http://situs.biomachina.org/). Chimera[42] produces synthetic density maps with the *molmap* command by describing each atom as a 3-D Gaussian distribution of width proportional to the map resolution and amplitude proportional to the atomic number; the command is based on the **pdb2mrc** program of the EMAN package.[43] All Chimera calculations reported in this paper employed the *measure correlation* command with default parameters, except for aboveThreshold set to false. The resulting correlation about mean values were used as the point of comparison (Chimera version 1.8.1 obtained from https://www.cgl.ucsf.edu/chimera/).

All of these methods are similar in that they use spherical Gaussian functions to represent the density contribution of each atom, a technique created for X-ray crystallography.[37,38] One shortcoming of this approach is that it does not incorporate the blurring required to account for the attenuation of a high resolution signal.[17] Approximations for the effect of truncation were derived for X-ray crystallography,[37] but not for EM applications. One method for overcoming this limitation is to create the map from simulated densities of each atom calculated by a resolution-limited Fourier transform of the atomic scattering factor.[44] This is the approach taken in DireX[13] which also uses an *ad hoc* blurring method to account for the instrumental signal attenuation at high resolution. Another program, **RSRef**, implements a similar method which does not rely on any *ad hoc* assumptions.[17]

We compared densities produced by our new GPU-based algorithms and by Chimera against those generated from DireX for the ribose binding protein (1URP). The cross correlations for a 3 Å and 8 Å map were 0.938 and 0.966, respectively, for our method, and 0.900 and 0.965 for Chimera. These correlation values are, as expected, lower than correlation values for densities created by Situs (Table 1), though the VMD and Chimera values are still within 5% of each other.

Table 1 Comparison of cross correlation results for the new GPU and CPU algorithms in VMD, for Chimera, and for the original sequential VMD algorithm, on densities simulated using the Situs[40] program pdb2vol.[41] These tests show the agreement between the densities generated by each of the algorithms and account for any discrepancies in the comparison of subsequent cross correlations on experimental data

	RHDV	1URP	1URP
Resolution(Å)	6.5	8	3
VMD-GPU-CUDA cc	0.998	0.999	0.989
VMD-CPU-SSE cc	0.996	0.997	0.991
Chimera cc	0.784	0.989	0.922
VMD-CPU-SEQ cc	0.997	0.999	0.950

These differences may be more important for fitting methods which, similar to **RSRef**, directly minimize the least-squares difference between the simulated and experimental density maps. Conversely, MDFF derives forces from the experimental map alone and, thus, uses the cross-correlation only in subsequent analysis steps. The more accurate density synthesis methods which utilize Fourier transforms require much more computation than the Gaussian approximations. Future work could experiment with hybrid approaches that borrow aspects from both methods to improve accuracy without sacrificing the speed of computation. The goal of the method described here is fast computation suited to visualization and analysis, a use which is amenable to approximation.

3 Hybrid fitting process

In many hybrid methods, the initial structure must first be docked into the EM density before further fitting can proceed. This process is generally an expedient way of rigidly docking the entire structure, or several independent rigid pieces, into the density so that they are correctly oriented within the map. To achieve this with MDFF, rigid-body docking is performed using the Situs[40] program **colores**.[45] Once the structure has been docked to the density, it is often useful at this stage to calculate the quality-of-fit. In particular, independent local correlations of each domain or secondary structure are useful in order to visualize which areas of the structure are initially poorly fit to the map. This quickly identifies areas that may be of greatest concern during the subsequent fitting simulation.

The next step in the fitting process is the determination of the finer-grained movements and any large scale deformations that are required to improve the fit to the reference map. For MDFF, this is the point where molecular dynamics simulations are used to flexibly guide the structure into the density. The VMD[46] plugin **mdff** is used to generate a potential energy function U_{EM} defined on a 3-D grid derived from the cryo-EM density map

$$U_{\mathrm{EM}}(\boldsymbol{R}) = \sum_j w_j V_{\mathrm{EM}}(\boldsymbol{r}_j), \tag{1}$$

where

$$V_{\mathrm{EM}}(\boldsymbol{r}) = \begin{cases} \xi \left[1 - \dfrac{\Phi(\boldsymbol{r}) - \Phi_{\mathrm{thr}}}{\Phi_{\max} - \Phi_{\mathrm{thr}}}\right] & \text{if } \Phi(\boldsymbol{r}) \geq \Phi_{\mathrm{thr}}, \\ \xi & \text{if } \Phi(\boldsymbol{r}) < \Phi_{\mathrm{thr}}. \end{cases} \tag{2}$$

Here $\Phi(\boldsymbol{r})$ is the Coulomb potential; it and its maximum value Φ_{max} are obtained from cryo-EM. In eqn (2) a threshold Φ_{thr} is used to clamp all $\Phi(\boldsymbol{r})$ values which are lower than Φ_{thr}, effectively removing the solvent contribution from the map and creating a flat potential for those regions. This threshold Φ_{thr} is chosen to be at or above the value at which a histogram of the EM density contains a large peak corresponding to the solvent. The scaling factor ξ uniformly adjusts the strength of the influence of the cryo-EM map on the molecular system. In addition, $V_{EM}(\boldsymbol{r})$ has a weight w_j for each atom j present at position $\boldsymbol{r}_j$. Generally, w_j is set to the atomic mass which follows from the roughly linear correspondence between the mass of atoms and their density in a cryo-EM map.

Forces computed from the potential defined in eqn (1) push the structure into regions of high density. This push is often divided into a multi-stage process where, in the various stages, parts of the atomic structure can be completely decoupled from the potential, or have the magnitude of the force varied by changing the scaling factor ξ or weights w_j. In the case of the ribosome,[25–33] protein and RNA are often each fitted separately while the other is constrained. In such a multistage scheme, it is beneficial to calculate the quality-of-fit at each step separately for each component of the molecular complex, enabling the identification of problematic regions.

For many systems, once the entire structure has been fitted, a single correlation value for the entire protein will be published, giving a coarse overall evaluation of the model. Because this value may be used for comparison across different methods or studies, an accurate, reproducible, and comparable quantity must be used. However, it is often useful to calculate the correlation at a finer decomposition (Fig. 1), for example per-residue. Such an analysis could resolve which parts of the structure are fitting well and which parts might require additional fitting or investigation. Additionally, it can be beneficial to determine how local correlations change, and hopefully improve, over the course of the fitting. A visualization of a fine-grained quality-of-fit over the entire simulation can provide the researcher with insight into the progression of the fitting process and the dynamics of each part of the structure. Computing the simulated densities for many different regions and their subsequent correlations for each step of the fitting can be extremely demanding computationally, especially for large structures or long fitting simulations. Practical application would require efficient algorithms permitting fast analysis turnaround.

The VMD TimeLine trajectory analysis plugin[47] makes use of the new data-parallel CPU and GPU quality-of-fit algorithms to provide an interactive visualization of the entire fitting process. TimeLine is designed to help identify and assess trajectory events by performing analysis calculations for each component of a molecular system, and for every frame of a simulation trajectory. The heatmap-style 2-D matrix TimeLine plot provides a "whole-trajectory" and a "whole-structure" view of the calculated property. Fig. 1c shows an example plot: the time dimension is displayed horizontally, the structure-component dimension is displayed vertically, and the property of interest – here, density map cross correlation score – is presented by colored rectangles. The TimeLine plot is connected interactively to the VMD 3-D structure display. When the user "scrubs" the mouse cursor around features on the plot, TimeLine highlights the associated 3-D molecular structures and displays their configurations and motions at the times of the corresponding events. A researcher can identify structure transitions,

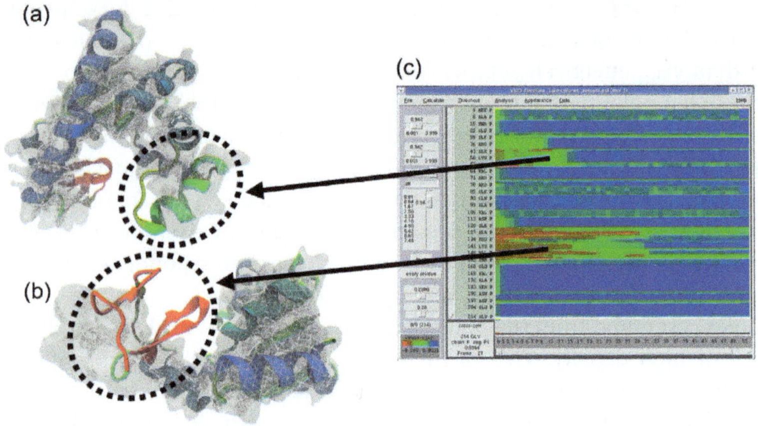

Fig. 1 Per-secondary-structure-component local cross correlation of adenylate kinase. MDFF was run on an initial conformation of adenylate kinase (PDB 1AKE) in order to fit it to a simulated density at 5 Å resolution from another conformation (PDB 4AKE). Panels (a) and (b) show a point during the simulation before the structure has been completely fit, colored by per-secondary-structure-component local cross correlation. Darker blue indicates the areas of best fit, while red indicates poorly fitting regions. The target density is drawn as a transparent gray isosurface with a complementary wireframe representation. Panel (a) shows a helix partly in the density; the helix is colored green to indicate a mediocre cross correlation indicative of a relatively close, but not perfect, fit to the density. Panel (b) shows a beta sheet almost completely outside of the reference density; the beta sheet is colored red to indicate a poor cross correlation. (c) VMD TimeLine analysis of the cross correlation over the entire simulation. Note how regions of poor fit (red) turn green and blue as the fit improves during the course of the simulation.

trends, and sequences in a trajectory more quickly by using a single zoomable, interactive 2-D TimeLine plot than by examining a complex 3-D structure over many trajectory frames. When calculating density map cross correlations, different types of trajectory information can be obtained by dividing the system into analysis components of different sizes: fine division among individual residues or contiguous secondary structures (as in Fig. 1c) allows detailed structural assessment, while coarser division to fractions of protein segments or even entire protein segments provides information about larger structural elements, particularly for larger systems.

If quality-of-fit analysis can be performed fast enough for interactive updates to keep pace with a running simulation, the correlations can be visualized by the user in real time. Such a real-time analysis can enable the user, along with interactive molecular dynamics (IMD),[48] to monitor fitting progress and manually manipulate parts of the structure. The user could move the structure in an attempt to improve the correlation while being provided with immediate feedback about the resulting effects, with trajectory and local trend context provided by an updating TimeLine view. This technique could make the fitting of difficult structures a much easier process while providing detailed analysis at every step. Additionally, this method could help fit small regions which normally take a long time to fit, often well after most of the structure has long since converged. Interactive MDFF has been performed in the past for studies on the ribosome,[29] but real-time quality-of-fit analysis could provide for better fits and easier simulations.

4 Methods for evaluating quality of fit

In MDFF, the simulated density is compared to the experimental map using Pearson's correlation coefficient, given by

$$\rho_{SE} = \frac{\langle (S - \langle S \rangle)(E - \langle E \rangle) \rangle}{\sigma_S \sigma_E}, \tag{3}$$

where $\langle S \rangle$ and $\langle E \rangle$ are the average voxel values of the simulated and experimental density maps, respectively, and σ_S and σ_E are their standard deviations.[49] The result is a normalized ($\rho_{SE} \in [-1, 1]$) indicator of the quality-of-fit of a given structure in the experimental density. The cross correlation defined in eqn (3) is traditionally used as a global measure that includes all voxels in the density maps, or as a local correlation[11] where only the volume surrounding the molecule is considered. The global correlation is affected by the size of the volume and includes contributions not only from the molecule, but also the solvent. Global correlations can report artificially high correlation values when considering large volumes that have been arbitrarily padded. Accordingly, local correlations tend to offer a more useful and unbiased characteristic. In MDFF, local correlations are calculated by specifying a threshold stated in units of standard deviations from the mean value of the simulated density. All voxels containing density values lower than this threshold are excluded from the cross correlation calculation. With higher density values naturally occurring closer to the structure, the density being used for the calculation is thus contained within the molecular envelope.

One impediment to the creation of a fast algorithm for computation of the cross correlation is the need for knowledge of the mean value of each map for both the covariance and standard deviations. A simple two-pass algorithm would first iterate through the maps and compute the mean, followed by a second iteration which would compute the covariance and standard deviation. In the interest of speed and efficiency, a single-pass algorithm is desirable, where only one iteration over the densities would be required. For a single-pass version of the standard deviation calculation, the equation can be rewritten

$$\begin{aligned} \sigma_S^2 &= \frac{1}{N}\sum_{i=1}^{N}(x_i - \overline{x})^2 = \frac{1}{N}\sum_{i=1}^{N}(x_i^2 - 2\overline{x}x_i + \overline{x}^2) \\ &= \frac{1}{N}\sum_{i=1}^{N} x_i^2 - \frac{2\overline{x}}{N}\sum_{i=1}^{N} x_i + \frac{N\overline{x}^2}{N} = \left(\frac{1}{N}\sum_{i=1}^{N} x_i^2\right) - 2\overline{x}^2 + \overline{x}^2 \end{aligned} \tag{4}$$

or

$$\sigma_S^2 = \left(\frac{1}{N}\sum_{i=1}^{N} x_i^2\right) - \overline{x}^2. \tag{5}$$

This expression permits one to compute the sum and the sum of squares of the voxels in each map in a single pass, with the standard deviation being calculated afterwards. This technique can be extended similarly to create a single-pass formulation of covariance. Unfortunately, when the single-pass equations are implemented in floating point arithmetic, they are susceptible to floating point truncation and cancellation error, which can affect the accuracy of the results. Floating point error can be minimized by shifting the means of the reference and simulated maps to zero, and by ensuring that the simulated density maps closely approximate the standard deviation of a reference map, thereby reducing the magnitudes of accumulated squared quantities. Further improvements to floating point precision can be achieved by performing summations in a hierarchical fashion, by using double-precision arithmetic, or through the use of precision enhancement techniques such as Kahan's compensated summation, native-pair arithmetic, and others.[50–52] In our tests, we have confirmed the presence of floating point error, but even without implementing all of the amelioration techniques described above, we have concluded that the errors have a negligible effect on the final correlation results for the test cases we have run to date. We believe this to be the direct result of a well-conditioned input scenario due to the limited dynamic range of the density values contained within both experimental and simulated density maps.

5 Data-parallel cross-correlation algorithms

We have devised high-performance data-parallel algorithms for computing simulated density maps, and comparing them with reference density maps using multi-core CPUs with vector instructions and massively parallel GPUs. Each of the methods we have devised for evaluating the quality-of-fit between an all-atom structure and a reference experimental density map requires the computation of a simulated density map from the all-atom structure. Once computed, the simulated density map is then compared with the experimental reference map using voxel-by-voxel subtraction, Pearson correlation, sum of absolute differences, and similar methods. In all of the methods we have evaluated thus far, the generation of the simulated density map is by far the most computationally demanding algorithm step. Since the simulated density value at a given point in space can be computed independently from the values of all other points, the simulated density map computation is well-suited to data-parallel implementations using CPU vector instructions such as the Intel x86 SSE and AVX instructions, or the ARM NEON instructions. Similarly, the computation can take advantage of GPU-accelerated algorithms that employ tens of thousands of threads running on thousands of arithmetic units. Once the simulated density map has been computed, it can be compared with the experimental reference map using a variety of methods.

A key difference between different density map comparison techniques lies in their respective memory capacity requirements and, in the case of GPU implementations, the host-GPU memory transfer characteristics of each algorithm. It is a straightforward task to implement quality-of-fit algorithms using multi-pass approaches that compute densities, means, and covariances in distinct steps, but these approaches lose performance relative to implementations that combine all steps into a single-pass algorithm. By combining all algorithm steps into a single-pass approach, intermediate values can be stored in fast on-chip registers, shared memory, and caches, that are typically two orders of magnitude faster than accesses to off-chip DRAM. When possible, we employ single-pass cross correlation algorithms, where the simulated density values are computed only ephemerally in CPU or GPU on-chip registers without ever storing the computed density values to off-chip DRAM memory. This approach avoids performance limitations that arise from a roughly factor of 40 performance gap between peak GPU arithmetic throughput (*e.g.*, 2–4.5 TFLOPS) and GPU DRAM memory bandwidth (*e.g.*, 200–300 GB/s).

We have previously developed fast algorithms for calculating and visualizing molecular orbitals present in quantum chemistry calculations,[53] and for the display of molecular surfaces, biomolecular complexes, and cellular structures.[54–56] These algorithms each require rapid evaluation of Gaussian radial basis functions, which is also necessary for computation of simulated density maps used in computing cross correlations. In the present work, the simulated density map is computed from a linear combination of Gaussian radial basis functions, with a per-atom weighting factor based on atomic radii. The simulated density map can be computed in a single-pass, for single-coefficient atomic weighting factors, or in multiple passes when creating temporally averaged density maps, and to support the use of more sophisticated schemes that use radial basis functions with multiple Gaussian terms or other basis functions. The reference density is tested for exclusion, and the simulated densities are compared with an optional user-defined threshold criterion to determine if they should contribute to the cross correlation. Reference voxels assigned with the IEEE floating point special value NaN (not-a-number), are excluded from consideration. Similarly, simulated density values that fall below an optional user-defined density threshold are also excluded from contributing to the cross correlation. All voxels for which both the reference and simulated densities pass the exclusion and threshold tests contribute then to the partial sums for the overall cross correlation. When all densities have been processed, a parallel reduction is performed to compute the final cross correlation from all partial sums that were previously computed.

The new single-pass cross correlation algorithm for GPUs is based on the NVIDIA CUDA programming toolkit,[57] and is referred to hereafter as VMD-GPU-CUDA. The GPU cross correlation algorithm uses a parallel decomposition outlined in Fig. 2. Density contributions from neighboring atoms within the cutoff distance are summed into thread-local registers. The cutoff distance is chosen such that density contributions at or sufficiently near zero are discarded, thereby ensuring that the algorithm achieves linear time complexity. Each thread produces four density values by looping over consecutive densities along the z-axis. Each thread computes its own local contributions to the per-tile cross correlation partial sums, for its four densities according to the voxel exclusion and thresholding criteria described above. Each CUDA thread block performs a local intra-block parallel reduction among all threads, summing each of the thread-local partial sums into a single set of cross-correlation partial sums associated with that block's $8 \times 8 \times 8$ tile. Once the intra-block parallel reduction is complete, the first thread of the thread block computes a tile-local cross correlation, and stores both the tile-local cross correlation as well as the per-tile cross correlation partial sums to GPU global memory, in preparation for a final inter-block global parallel reduction that yields the overall cross correlation result. One important performance-critical detail about our implementation is that we employ special counters and atomic memory update instructions to allow a single

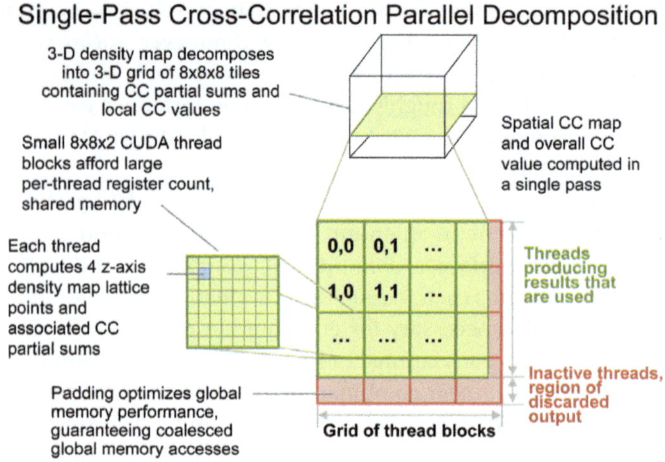

Fig. 2 Parallel decomposition for the single-pass GPU-accelerated cross correlation algorithm. The simulated density map is decomposed into 8 × 8 × 8 tiles which are processed by individual CUDA thread blocks. Each thread block is 8 × 8 × 2 threads in size. Each thread computes 4 consecutive densities along the *z*-axis, thereby enabling reuse of atom data in on-chip registers and increasing arithmetic throughput. Each CUDA thread block performs thresholding and computes partial sums for the overall cross correlation, as well as the local cross correlation value for its 8 × 8 × 8 tile.

GPU kernel invocation to perform the entire cross correlation calculation. Each thread block increments a global counter using an *atomic-add* machine instruction as it completes its tile-local work. All but the last thread block to complete simply exit once they have finished their tile-local work. The last thread block to complete its tile-local work performs the final inter-block parallel reduction of all partial sums and stores the overall cross correlation result to GPU global memory.

The new cross-correlation algorithm for multi-core CPUs employs CPU vector instructions and multithreading. The new CPU algorithm is hereafter referred to as VMD-CPU-SSE. The CPU algorithm decomposes the density map into planar slices which are dynamically assigned to a fixed pool of CPU worker threads by a dynamic load balancer built into VMD.[47] Each CPU worker thread runs a loop that accumulates densities from each atom onto the region surrounding the atom up to a defined cutoff limit where the Gaussian is assumed to have decayed to zero. Mainstream multi-core CPUs lack dedicated machine instructions for calculation of exponentials; as a result, application software must either use general purpose math library routines or incorporate custom functions for this purpose. Since the sign of the exponential function's input domain is always negative for the density map synthesis algorithm we have developed, we use a high-performance Taylor series approximation of the exponential function. Our specialized negative-domain exponential approximation is hand-written in assembly language intrinsics for Intel x86 SSE 4-way vector instructions, based on a similar function we originally developed for molecular orbital display.[53] We have improved the performance of the new exponential approximation beyond what we had previously achieved by premultiplying input values with constants to convert to base-*e*, by eliminating conditional tests that check for input domain cases that cannot occur in our density map algorithm, and by guaranteeing multiple-of-four data alignment. The end result of these optimizations is that the multi-core CPU density map algorithm can be competitive with the performance obtained with GPUs, which have machine instructions for computing exponentials due to their heavy use in computer graphics lighting algorithms.

6 Test cases and performance results

In order to test the new data-parallel GPU and CPU cross correlation algorithms, nine test cases were chosen from structures of varying size with density maps at resolutions between 3 Å and 8 Å (Fig. 3, Table 1, Table 2). Six of the structures are final results of MDFF simulations with their corresponding experimental cryo-EM maps, while the remaining three are structures with simulated density maps using the Situs[40] program *pdb2vol*.[41] The largest structure is the rabbit hemorrhagic disease virus (RHDV) capsid solved with a 6.5 Å resolution density map.[59] Structures of aquaporin, GroEL, GroEL-GroES complex in the ATP-bound state, Mm-cpn in the closed state, and lidless Mm-cpn in the open state, were taken from the results of a cryo-EM modeling challenge.[60] The final structure, ribose binding protein (PDB: 1URP) was used to simulate density maps at 3 Å and 8 Å using Situs, as a means of comparing to density maps generated from another program (Table 1). The size of the virus capsid and the difficult-to-fit nature of the modeling challenge structures both pose problems to researchers using hybrid fitting methods. Our data-parallel GPU and multi-core CPU cross correlation

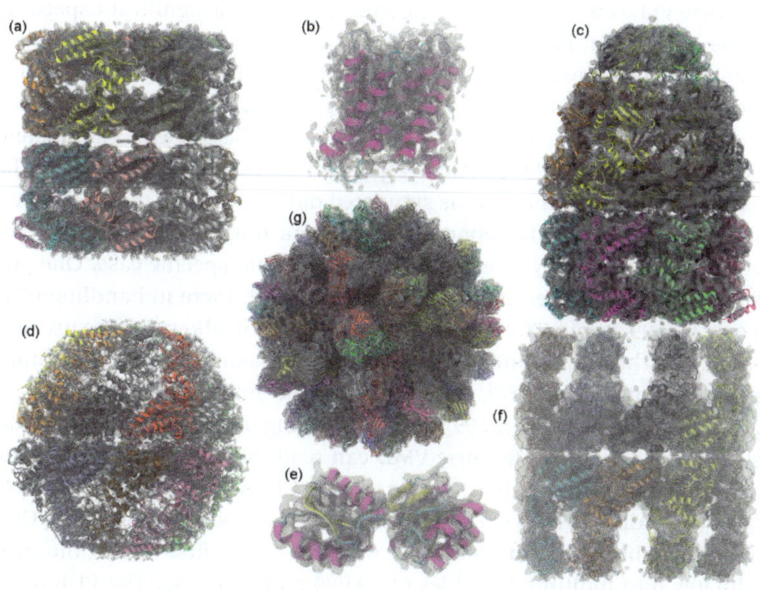

Fig. 3 All test systems fitted within their experimental density, or in the case of 1URP, a synthetic one. (a) GroEL (PDB 3E76); (b) aquaporin-0 (PDB 3M9I); (c) GroEL-GroES complex in the ATP-bound state (PDB 2C7D); (d) Mm-cpn in the closed state (PDB 3LOS); (e) D-ribose-binding protein (PDB 1URP); (f) lidless Mm-cpn in the open state; PDB 3LOS was used as a template to construct a homology model using Modeler;[58] (g) rabbit hemorrhagic disease virus capsid (RHDV) (PDB 3J1P).

Table 2 Comparison of cross correlation algorithm results, runtimes, and speedups for the new GPU and CPU algorithms in VMD, for Chimera, and for the original sequential VMD algorithm. All speedups are normalized using Chimera runtime as the baseline

	RHDV	Mm-cpn open	GroEL	GroEL-GroES	Mm-cpn closed	Aquaporin
Resolution(Å)	6.5	8	4	7.7	4.3	3
# atoms	702K	61K	54K	59K	64K	1.6K
# voxels	111M	14M	8M	7M	7M	2.6M
VMD-GPU-CUDA cc	0.744	0.857	0.769	0.886	0.658	0.754
VMD-CPU-SSE cc	0.736	0.866	0.839	0.892	0.712	0.771
Chimera cc	0.687	0.837	0.727	0.867	0.635	0.739
VMD-CPU-SEQ cc	0.743	0.869	0.846	0.898	0.728	0.653
VMD-GPU-CUDA	0.458 s	0.06 s	0.034 s	0.038 s	0.028 s	0.007 s
	34.6×	25.7×	36.8×	35.0×	57.5×	55.7×
VMD-CPU-SSE	0.779 s	0.085 s	0.159 s	0.077 s	0.179 s	0.033 s
	20.3×	18.1×	7.9×	17.3×	9.0×	11.8×
Chimera	15.86 s	1.54 s	1.25 s	1.33 s	1.61 s	0.39 s
	1.0×	1.0×	1.0×	1.0×	1.0×	1.0×
VMD-CPU-SEQ	62.89 s	2.9 s	1.57 s	2.49 s	2.13 s	0.04 s
	0.25×	0.53×	0.79×	0.53×	0.75×	9.7×

methods were compared to the previous sequential algorithm in VMD, as well as to Chimera,[42] a program widely used by cryo-EM researchers.

The new VMD cross correlation algorithms provide a significant speedup over both the old sequential CPU code in VMD as well as Chimera. The algorithms provide correlation results that agree with Chimera to within 5% for most cases. In the case where the two have the greatest divergence (7.92% difference), the virus capsid, our method results in a 0.998 cross correlation with a simulated map produced by Situs, while Chimera reports a correlation of 0.7842 (Table 1). Because of this discrepancy on the simulated data, we believe that the correlation result for the large capsid reported by Chimera might be flawed, or that it is otherwise not necessarily directly comparable in that specific case. One point of possible error is the difference between VMD and Chimera in handling PDB files with over 99 999 atoms. The original PDB specification allows for the use of only 5 columns for the atom index, restricting the maximum officially supported number of atoms to 99 999. To overcome this limitation, VMD uses alpha-numeric atom indices beyond 99 999 when writing out a PDB file, while Chimera uses additional columns. Because VMD can read these large structure PDB files written in Chimera format but not *vice versa*, a Chimera format PDB was used in all cases (including RHDV) where it was required, for both the VMD and Chimera calculations. Chimera format PDB files were created by first writing the structure coordinates into multiple PDB files of 99 999 atoms or less. The PDB files were then read into Chimera, followed by using the "copy/combine" feature to combine every PDB into a single structure which could be saved to a single PDB file. An additional overlooked complication regarding the difference in PDB file formats could be the cause of the discrepancy in correlation values obtained for RHDV.

The new feature for calculating spatial cross correlation maps provides a rapid means of visualizing 3-D correlations and resolving areas of poor fit, as seen in Fig. 4. These spatial correlation maps are especially useful for large complexes, for example virus capsids, because only a single density calculation is required to obtain localized information. By using the VolumeSlice representation in VMD to view 3-D textured slice planes of the correlation map in the *x*-, *y*-, or *z*-axis, the map can be quickly scanned for regions of low correlation. Structure components located in regions of poor correlation can be identified by selecting them or coloring them based on local cross correlation values (Fig. 5).

All single-node workstation class tests were run on a workstation containing two Intel Xeon E5-2687W processors each consisting of 8 physical processor cores and 16 hardware threads, for a total of up to 32 threads of execution in hardware. The workstation was configured with 256 GB of memory, and an NVIDIA Quadro K6000 GPU with 12 GB of on-board memory. All software was run on CentOS Linux version 5.9, and NVIDIA driver version 331.22. VMD was compiled using Intel C/C++ version 12.0.2, and CUDA toolkit version 4.0. Tests involving Situs

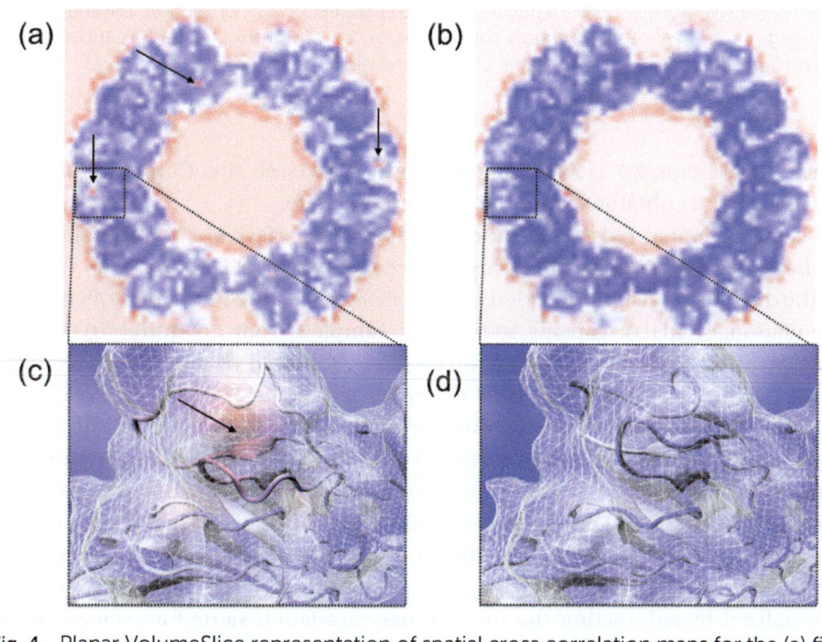

Fig. 4 Planar VolumeSlice representation of spatial cross correlation maps for the (a) first and (b) last frames of an MDFF simulation of the rabbit hemorrhagic disease virus capsid. These maps visualize local cross correlation of 8 × 8 × 8 voxel regions, with red indicating poorest correlation and dark blue indicating best correlation. Regions of poorest correlation lie on the exterior and interior faces of the capsid, as would be expected. Poor-fitting regions within the subunits of the structure are seen as red spots and are marked with arrows. Panels (c) and (d) show closer views of one of these poor-fitting regions, before and after fitting, respectively. The structure is shown inside the experimental density map and is colored according to the corresponding spatial cross correlation volume around that region, to further visualize which parts of the structure have a poor fit. Panel (c) clearly shows a poorly fit region, with parts of the structure entirely outside of the density; the poor fit has been fixed by MDFF as is evident in (d).

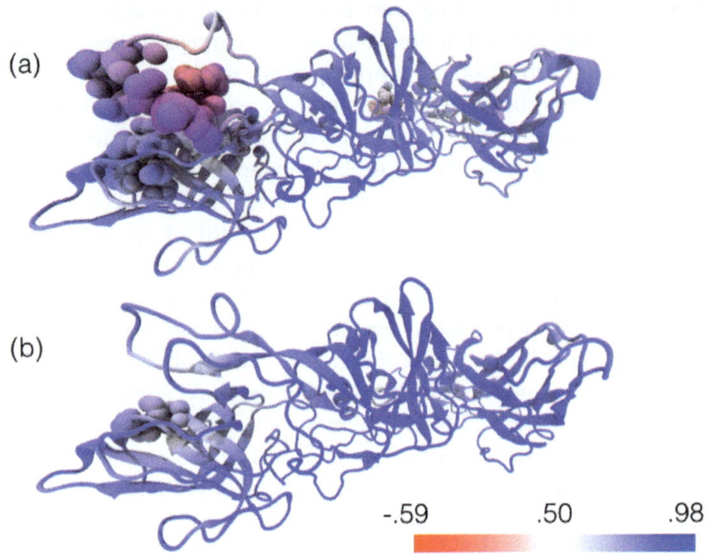

Fig. 5 One subunit of the rabbit hemorrhagic disease virus capsid before (a) and after (b) MDFF. The structure is colored according to the spatial cross correlation volume around that region. A VDW representation shows part of the structure selected as those atoms which lie inside voxels of the spatial cross correlation volume with values <0.1.

used Situs version 2.7.1, as obtained from the Situs web site. Chimera tests used version 1.8.1 as obtained from the Chimera web site.

The accelerated cross correlation method makes feasible both detailed and global trajectory assessments of large molecular systems. Fig. 6 shows an example of the diagnostic insight provided by a TimeLine view of the initial part of an MDFF simulation of RHDV capsid: 100 frames sampled from an initial 10 000-step energy minimization with MDFF forces active, followed by 400 frames sampled from the initial 40 ps of the MDFF dynamics trajectory. The structure is divided into 180 components: the 180 protein segments, each with an identical residue sequence, that make up the symmetrical virus capsid. In Fig. 6a, the pronounced pattern of horizontal stripes indicates a periodic segment-to-segment variation of cross correlation values, following differences in the capsid's symmetrical segment arrangement, as seen in Fig. 6b. Fig. 6c shows relative changes observed for the individual capsid segments during the trajectory: the trajectory values are normalized by subtracting the initial cross correlation value for each segment. About half of the cross-correlation changes seen in Fig. 6c occur during the initial minimization to frame 100, with much of the remaining improvement occuring quickly during the first picoseconds of the MDFF simulation, from frames 100 to 125, and minor fluctuations ceasing for the majority of segments by frame 350. Several segments can be seen to have relatively stable correlation values during both MDFF minimization and MDFF dynamics; a researcher studying the system might investigate these segments further.

Detailed analysis of very large systems with long trajectories can require a large number of individual calculations, and impose a heavy I/O burden on a computer system. We implemented a parallel TimeLine cross correlation analysis using the

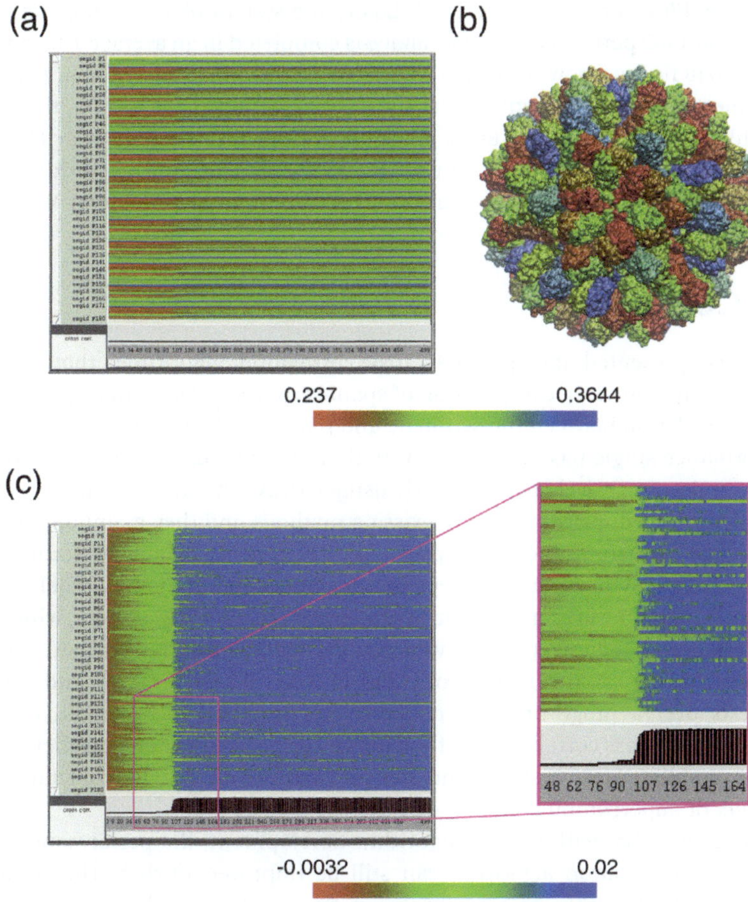

Fig. 6 Cross correlation coefficients in a rabbit hemorrhagic disease virus (RHDV) MDFF trajectory. Panel (a) contains a TimeLine plot of the 500-frame RHDV trajectory discussed in the text, displaying the per-segment cross correlation coefficients. Panel (b) shows the RHDV structure colored by the values in the first frame (left-most column) of panel (a). Panel (c) contains a TimeLine plot of the same trajectory as in panel (a), displaying the change from initial value of per-segment cross correlation, namely, the per-segment values in panel (a) after subtracting the value for each segment in the first frame. The bar graph at the bottom of the plot shows (in black and coral bar height) the proportion of segments at each timestep with a change in cross correlation greater than +0.01.

new GPU cross correlation algorithm, for use on clusters or supercomputers. The new TimeLine analysis calls VMD parallel scripting primitives[47] to perform the needed calculation, synchronization, and data reduction operations. A 10 000 frame trajectory of the 1.38 million atom RHDV capsid structure was examined. For cross correlation analysis, the capsid was divided into 720 structural components; each of the capsid's 180 525-residue segments were divided into four nearly-equal contiguous segments. The calculation was run on the NSF Blue Waters petascale supercomputer, using the XK7 hybrid GPU-accelerated compute nodes. Each XK7 node contains an AMD Opteron 6276 CPU with 32 GB of RAM and an NVIDIA Tesla K20X GPU accelerator with 6 GB of RAM. Structure and trajectory files were striped over 160 Lustre file system object storage targets to ensure best I/O performance.[47] The analysis completed in an average 19.5 minutes using 2048 nodes, showing a 1035× speedup over a projected 336 hour (14.0 day) runtime using a single node. Using the Table 2 performance results, this would roughly correspond to a 35 800× speedup over an estimated 485 day (1.3 year) runtime for Chimera running on a single node and a 142 100× speedup over an estimated 1924 day (5.3 year) runtime for the original sequential VMD CPU algorithm running on a single node.

7 Conclusions

We have presented data-parallel cross correlation algorithms that combine density map synthesis, computation of spatial cross correlation maps, and overall cross correlation between the synthetic and reference density maps, using a high performance single-pass approach on multi-core CPUs using SSE vector instructions, and on massively parallel GPUs using CUDA. The new algorithms achieve significant speedups compared to existing methods and they maintain adequate numerical precision. The single-pass algorithms achieve high performance by fusing independent algorithm steps into a single computational kernel, thereby maximizing the ratio of arithmetic operations *vs.* memory references, which is of greatest overall benefit to massively parallel GPU hardware. We have demonstrated the use of the new algorithms in several visualization and analysis contexts over a range of problem sizes, and we have incorporated the high performance cross correlation method into a TimeLine multi-node parallel analysis script enabling large systems and long trajectories to be analyzed rapidly on clusters or supercomputers.

We expect that with further algorithm development, the performance of the new cross correlation algorithm can still be improved further. The simulated density map algorithm presently operates on a uniform grid that encompasses the bounding volume of an atom selection. The use of the uniform grid decomposition for the density map calculation can result in some unnecessary computation for "empty" regions of the volume that are not associated with selected atoms, since they must still participate in parallel reduction operations that yield the final correlation results. The use of the uniform grid approach provides high performance in many common cases, but it is clear that as cryo-EM imaging techniques improve and both the resolution and size of resolved biomolecular complexes increase, there will be a need for a density map algorithm that performs well for atom selections that yield sparse regions of computation. There are also several opportunities for further performance improvements in the case of trajectory analyses, such as those performed with TimeLine, by eliminating host-GPU memory copies associated with the reference map for all but the first timestep, and by eliminating computations that are common to multiple atom selections in analyses that compute cross correlations for a large number of atom selections for each trajectory frame.[47]

New data-parallel *shuffle* instructions available on recent NVIDIA GPUs could be used to accelerate the intra-block and inter-block parallel reductions that take place within the GPU algorithm. The most recent Intel and AMD processors have added new 8-element AVX and AVX2 vector instructions, offering up to a 2× performance gain over SSE instructions. The greatly increased performance levels achieved by the new algorithms create many new opportunities for interactive analyses and visualizations and for interactive molecular dynamics simulations[48] using the MDFF method.

Acknowledgements

This research is part of the Blue Waters sustained-petascale computing project supported by NSF award OCI 07-25070, and "The Computational Microscope" NSF PRAC award. Blue Waters is a joint effort of the University of Illinois at Urbana-Champaign and its National Center for Supercomputing Applications (NCSA). The authors thank Yanxin Liu for providing simulation trajectories. The authors wish to acknowledge support of the CUDA Center of Excellence at the University of Illinois and NIH funding through grants 9P41GM104601 and 5R01GM098243-02.

References

1 G. Zhao, J. R. Perilla, E. L. Yufenyuy, X. Meng, B. Chen, J. Ning, J. Ahn, A. M. Gronenborn, K. Schulten, C. Aiken and P. Zhang, *Nature*, 2013, **497**, 643–646.

2 J. Zhang, M. L. Baker, G. F. Schröder, N. R. Douglas, S. Reissmann, J. Jakana, M. Dougherty, C. J. Fu, M. Levitt, S. J. Ludtke, J. Frydman and W. Chiu, *Nature*, 2010, **463**, 379–383.

3 S. J. Ludtke, M. L. Baker, D.-H. Chen, J.-L. Song, D. T. Chuang and W. Chiu, *Structure*, 2008, **16**, 441–448.

4 X. Zhang, E. Settembre, C. Xu, P. R. Dormitzer, R. Bellamy, S. C. Harrison and N. Grigorieff, *Proc. Natl. Acad. Sci. U. S. A.*, 2008, **105**, 1867–1872.

5 M. L. Baker, J. Zhang, S. J. Ludtke and W. Chiu, *Nat. Protoc.*, 2010, **5**, 1697–1708.

6 C. F. Hryc, D.-H. Chen and W. Chiu, *Curr. Opin. Virol.*, 2011, **1**, 110–117.

7 R. Zhang, C. F. Hryc, Y. Cong, X. Liu, J. Jakana, R. Gorchakov, M. L. Baker, S. C. Weaver and W. Chiu, *EMBO J.*, 2011, **30**, 3854–3863.

8 S. Maki-Yonekura, K. Yonekura and K. Namba, *Nat. Struct. Mol. Biol.*, 2010, **17**, 417–422.

9 L. Cheng, J. Zhu, W. H. Hui, Z. Zhang, B. Honig, Q. Fang and Z.-H. Zhou, *J. Mol. Biol.*, 2010, **397**, 852–863.

10 F. Fabiola and M. S. Chapman, *Structure*, 2005, **13**, 389–400.

11 A. M. Roseman, *Acta Crystallogr., Sect. D: Biol. Crystallogr.*, 2000, **56**, 1332–1340.

12 R. Das and D. Baker, *Annu. Rev. Biochem.*, 2008, **77**, 363–382.

13 G. F. Schröder, A. T. Brunger and M. Levitt, *Structure*, 2007, **15**, 1630–1641.

14 C. C. Jolley, S. A. Wells, P. Fromme and M. F. Thorpe, *Biophys. J.*, 2008, **94**, 1613–1621.

15 F. Tama, O. Miyashita and C. L. Brooks III, *J. Struct. Biol.*, 2004, **147**, 315–326.

16 M. Orzechowski and F. Tama, *Biophys. J.*, 2008, **95**, 5692–5705.

17 M. S. Chapman, A. Trzynka and B. K. Chapman, *J. Struct. Biol.*, 2013, **182**, 10–21.

18 M. Topf, K. Lasker, B. Webb, H. Wolfson, W. Chiu and A. Sali, *Structure*, 2008, **16**, 295–307.

19 L. G. Trabuco, E. Villa, K. Mitra, J. Frank and K. Schulten, *Structure*, 2008, **16**, 673–683.
20 L. G. Trabuco, E. Villa, E. Schreiner, C. B. Harrison and K. Schulten, *Methods*, 2009, **49**, 174–180.
21 J. C. Phillips, R. Braun, W. Wang, J. Gumbart, E. Tajkhorshid, E. Villa, C. Chipot, R. D. Skeel, L. Kale and K. Schulten, *J. Comput. Chem.*, 2005, **26**, 1781–1802.
22 D. Wells, V. Abramkina and A. Aksimentiev, *J. Chem. Phys.*, 2007, **127**, 125101.
23 E. Schreiner, L. G. Trabuco, P. L. Freddolino and K. Schulten, *BMC Bioinformatics*, 2011, **12**, 190.
24 K.-Y. Chan, J. Gumbart, R. McGreevy, J. M. Watermeyer, B. T. Sewell and K. Schulten, *Structure*, 2011, **19**, 1211–1218.
25 E. Villa, J. Sengupta, L. G. Trabuco, J. LeBarron, W. T. Baxter, T. R. Shaikh, R. A. Grassucci, P. Nissen, M. Ehrenberg, K. Schulten and J. Frank, *Proc. Natl. Acad. Sci. U. S. A.*, 2009, **106**, 1063–1068.
26 J. Gumbart, L. G. Trabuco, E. Schreiner, E. Villa and K. Schulten, *Structure*, 2009, **17**, 1453–1464.
27 T. Becker, S. Bhushan, A. Jarasch, J.-P. Armache, S. Funes, F. Jossinet, J. Gumbart, T. Mielke, O. Berninghausen, K. Schulten, E. Westhof, R. Gilmore, E. C. Mandon and R. Beckmann, *Science*, 2009, **326**, 1369–1373.
28 B. Seidelt, C. A. Innis, D. N. Wilson, M. Gartmann, J.-P. Armache, E. Villa, L. G. Trabuco, T. Becker, T. Mielke, K. Schulten, T. A. Steitz and R. Beckmann, *Science*, 2009, **326**, 1412–1415.
29 L. G. Trabuco, E. Schreiner, J. Eargle, P. Cornish, T. Ha, Z. Luthey-Schulten and K. Schulten, *J. Mol. Biol.*, 2010, **402**, 741–760.
30 J. Gumbart, E. Schreiner, L. G. Trabuco, K.-Y. Chan and K. Schulten, *Molecular Machines in Biology*, Cambridge University Press, 2011, ch. 8, pp. 142–157.
31 J. Frauenfeld, J. Gumbart, E. O. van der Sluis, S. Funes, M. Gartmann, B. Beatrix, T. Mielke, O. Berninghausen, T. Becker, K. Schulten and R. Beckmann, *Nat. Struct. Mol. Biol.*, 2011, **18**, 614–621.
32 X. Agirrezabala, E. Schreiner, L. G. Trabuco, J. Lei, R. F. Ortiz-Meoz, K. Schulten, R. Green and J. Frank, *EMBO J.*, 2011, **30**, 1497–1507.
33 W. Li, L. G. Trabuco, K. Schulten and J. Frank, *Proteins: Struct., Funct., Bioinf.*, 2011, **79**, 1478–1486.
34 J. Hsin, J. Gumbart, L. G. Trabuco, E. Villa, P. Qian, C. N. Hunter and K. Schulten, *Biophys. J.*, 2009, **97**, 321–329.
35 M. K. Sener, J. Hsin, L. G. Trabuco, E. Villa, P. Qian, C. N. Hunter and K. Schulten, *Chem. Phys.*, 2009, **357**, 188–197.
36 D. Vasishtan and M. Topf, *J. Struct. Biol.*, 2011, **174**, 333–343.
37 R. Diamond, *Acta Crystallogr., Sect. A: Cryst. Phys., Diffr.,Theor. Gen. Crystallogr.*, 1971, **27**, 436–452.
38 T. A. Jones and L. Liljas, *Acta Crystallogr., Sect. A: Found. Crystallogr.*, 1984, **40**, 50–57.
39 P. L. Stewart, S. D. Fuller and R. M. Burnett, *EMBO J.*, 1993, **12**, 2589–2599.
40 W. Wriggers, *Biophys. Rev.*, 2010, **2**, 21–27.
41 W. Wriggers, *Acta Crystallogr., Sect. D: Biol. Crystallogr.*, 2012, **68**, 344–351.
42 E. F. Pettersen, T. D. Goddard, C. C. Huang, G. S. Couch, D. M. Greenblatt, E. C. Meng and T. E. Ferrin, *J. Comput. Chem.*, 2004, **25**, 1605–1612.

43 S. J. Ludtke, P. R. Baldwin and W. Chiu, *J. Struct. Biol.*, 1999, **128**, 82–97.
44 M. S. Chapman, *Acta Crystallogr., Sect. A: Found. Crystallogr.*, 1995, **51**, 69–80.
45 P. Chacón and W. Wriggers, *J. Mol. Biol.*, 2002, **317**, 375–384.
46 W. Humphrey, A. Dalke and K. Schulten, *J. Mol. Graphics*, 1996, **14**, 33–38.
47 J. E. Stone, B. Isralewitz and K. Schulten, *Proceedings of the XSEDE Extreme Scaling Workshop (XSW)*, 2013, pp. 43–50.
48 J. E. Stone, J. Gullingsrud, P. Grayson and K. Schulten, *2001 ACM Symposium on Interactive 3D Graphics*, New York, 2001, pp. 191–194.
49 J. Frank, *Three-dimensional electron microscopy of macromolecular assemblies*, Oxford University Press, New York, 2006.
50 W. Kahan, *Commun. ACM*, 1965, **8**, 40.
51 Y. He and C. H. Q. Ding, *ICS '00: Proceedings of the 14th international conference on Supercomputing*, New York, NY, USA, 2000, pp. 225–234.
52 D. H. Bailey, *Comput. Sci. Eng.*, 2005, **07**, 54–61.
53 J. E. Stone, J. Saam, D. J. Hardy, K. L. Vandivort, W. W. Hwu and K. Schulten, *Proceedings of the 2nd Workshop on General-Purpose Processing on Graphics Processing Units*, ACM International Conference Proceeding Series, New York, NY, USA, 2009, pp. 9–18.
54 M. Krone, J. E. Stone, T. Ertl and K. Schulten, *EuroVis - Short Papers 2012*, 2012, pp. 67–71.
55 E. Roberts, J. E. Stone and Z. Luthey-Schulten, *J. Comput. Chem.*, 2013, **34**, 245–255.
56 J. E. Stone, K. L. Vandivort and K. Schulten, *Proceedings of the 8th International Workshop on Ultrascale Visualization*, New York, NY, USA, 2013, pp. 6:1–6:8.
57 J. Nickolls, I. Buck, M. Garland and K. Skadron, *Queue*, 2008, **6**, 40–53.
58 A. Sali and T. L. Blundell, *J. Mol. Biol.*, 1993, **234**, 779.
59 X. Wang, F. Xu, J. Liu, B. Gao, Y. Liu, Y. Zhai, J. Ma, K. Zhang, T. S. Baker, K. Schulten, D. Zheng, H. Pang and F. Sun, *PLoS Pathog.*, 2013, **9**, e1003132.
60 K.-Y. Chan, L. G. Trabuco, E. Schreiner and K. Schulten, *Biopolymers*, 2012, **97**, 678–686.

Faraday Discussions

PAPER

Visualising and controlling the flow in biomolecular systems at and between multiple scales: from atoms to hydrodynamics at different locations in time and space

Evgen Pavlov,*[a] Makoto Taiji,[c] Arturs Scukins,[a] Anton Markesteijn,[b] Sergey Karabasov[b] and Dmitry Nerukh[a]

Received 24th December 2013, Accepted 31st January 2014
DOI: 10.1039/c3fd00159h

A novel framework for modelling biomolecular systems at multiple scales in space and time simultaneously is described. The atomistic molecular dynamics representation is smoothly connected with a statistical continuum hydrodynamics description. The system behaves correctly at the limits of pure molecular dynamics (hydrodynamics) and at the intermediate regimes when the atoms move partly as atomistic particles, and at the same time follow the hydrodynamic flows. The corresponding contributions are controlled by a parameter, which is defined as an arbitrary function of space and time, thus, allowing an effective separation of the atomistic 'core' and continuum 'environment'. To fill the scale gap between the atomistic and the continuum representations our special purpose computer for molecular dynamics, MDGRAPE-4, as well as GPU-based computing were used for developing the framework. These hardware developments also include interactive molecular dynamics simulations that allow intervention of the modelling through force-feedback devices.

1 Introduction

The entirety of physical processes in complex systems can be represented as a hierarchical structure. The processes in chemical system, macromolecules, or biological objects proceed simultaneously at different temporal and spatial scales. Recently attempts have been made in building integrated models that allow carrying out computer experiments simultaneously at several scales.[1–3] This

[a]Non-linearity and Complexity Research Group, Aston University, Birmingham, B4 7ET, UK. E-mail: e.pavlov@aston.ac.uk

[b]School of Engineering & Materials Science, Queen Mary, University of London, Mile End Road, London E1 4NS, UK. E-mail: s.karabasov@qmul.ac.uk

[c]Computational Biology Research Core, RIKEN QBiC (Quantitative Biology Center), 1-6-5 Minatojima-minamimachi, Chuo, Kobe, Hyogo 650-0047, Japan. E-mail: taiji@gsc.riken.jp

is especially important for biological objects, starting from protein molecules and ending with human organs.[4,5] This kind of model contains and exchanges information between the components at different scales providing a more complete picture and uncovering a qualitatively new phenomena.[6,7]

While statistical mechanics methods allow the description of the system as a whole using average values, the direct implementation of models with different levels of hierarchy demands understanding the physical behaviour of materials at different levels. For example, in solid state physics interatomic interaction models[8] describe the type of bond formation and the structure of the material. On the other hand, the method of finite elements, which completely ignores internal atomistic structure, describes it as a continuum model.[9]

An example of such models applied to the physics of liquids is the Landau–Lifshitz fluctuating hydrodynamics.[10] This model maps collective dynamics of the continuum phase to the atomistic level. The hydrodynamic Landau–Lifshitz equations (LL–FH) are classical continuum fluid dynamics equations with added stochastic flows. They include the statistical model of temperature fluctuations around the local equilibrium. The fluctuations satisfy the equipartition theorem.[11] Combining this model with the method of molecular dynamics (MD), which represents the atomistic approach (everything is determined through interatomic interactions), allows construction of an hierarchical model. It is now possible to take into account the atomistic character of the movements of the molecules and the continuum approach for stochastic flows in a fluid at the same time.[12–14] The application of this approach to the macromolecule–water system is promising for studying the properties of proteins and their environment.

Historically one of the first attempts to unify the continuum description of liquids with their atomistic representation was the Langevin equation used for the descriptions of the particle motion in continuum:

$$\frac{1}{m}f(t) = \frac{\mathrm{d}u}{\mathrm{d}t} + \gamma u, \tag{1}$$

where u is the velocity of the particle of mass m moving under the influence of the force f and γ is the friction coefficient. The force is assumed to be random with correlations determined by the expression where it is defined as a white noise:

$$\langle f_i(t) f_j(t')\rangle = 2D\delta_{ij}\delta(t - t'), \tag{2}$$

where

$$D = m\gamma k_{\mathrm{B}}T. \tag{3}$$

The spectral density[15] of the force is

$$\frac{1}{3}\langle f^2(t)\rangle_\omega = 2D. \tag{4}$$

It shows that D is a diffusion coefficient and at the same time it is the noise intensity in the Langevin equation. The eqn (2,4) express the fluctuation–dissipation relationship.[15]

In the hydrodynamic limit the simultaneous fluctuations at different locations in space are not correlated and considered as white noise. To determine the fluctuations of the local stresses and heat fluxes in the case of continuous hydrodynamic environment Landau and Lifshitz suggested a modification of the classical Navier–Stokes equations in which the random variables are present. As the development of Langevin's idea (1,2), the stochastic terms are determined based on the fluctuation–dissipation theorem,[11] which provides the balance between the fluctuations in the system and its dissipative properties. When numerically solving the fluctuation hydrodynamics equations it is important to keep this balance. If it is violated the system can either develop instability (with the dominance of stochastic flows) or the value of the fluctuations can be too small (under the dominance of dissipative flows in the system).

For a correct transition between the MD (atomistic) and the hydrodynamic (continuous) descriptions the correspondence of the statistics of fluctuations in both representations is required. This is computationally demanding because some characteristics of the liquid are particularly slow converging (such as, for example, the isothermal compressibility calculated using the derivative of the pressure with respect to volume). This is especially true at the stage of framework development as the equations at all levels have to be solved at the same time for benchmarking. To overcome the computational difficulties two ways are possible. The first path uses inexpensive gaming platforms, the second one implements supercomputers. We have used both in our work while developing our approach.

The widespread use of computer games has led to the rapid evolution of graphics processors, the development of which is faster than the development of conventional CPUs. The main complication for molecular modelling applications is the need in software coding that requires special knowledge. Nevertheless, the trend is such that, possibly, the graphics processors will soon be one of the most attractive alternatives for cheap and relatively high performance computing (HPC). Our results reported below show high efficiency of GPU based computation in solving LL–FH equations.

The development of specialised petaflops platforms is more difficult, but it has a potential of fundamental breakthroughs in niche simulations such as molecular modelling as they can achieve timescales infeasible for any other hardware. The current trend in peta- (and exa-) flops HPC can be divided into two classes, Fig. 1. The first uses general purpose CPUs and achieves high performance by utilising

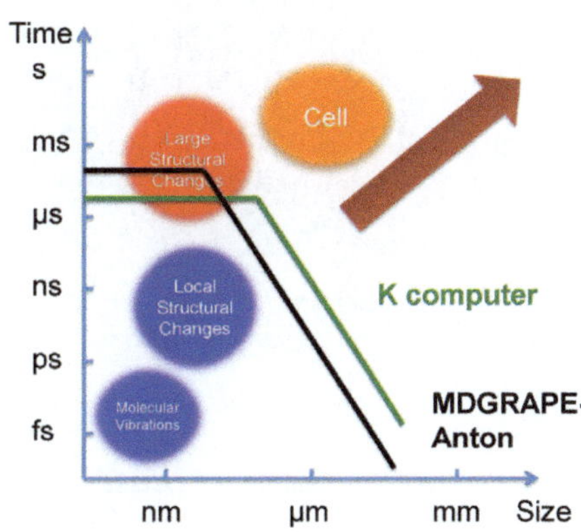

Fig. 1 Time and space scales of biomolecular processes and the limits of MD simulation feasibility.

a very large number of computing nodes. An example of such a machine is the K-computer built in RIKEN. The second class is built around specialised accelerators where the most time consuming molecular computation is performed on the integrated chip, for example, the MDGRAPE or ANTON computers, Fig. 1. We have used both types of supercomputers for generating MD data necessary in our hybrid method development.

In this publication we report on the research of an international team in the framework of a G8 project 'Using next generation computers and algorithms for modelling the dynamics of large biomolecular systems' that includes the theoretical, computational, and engineering development of a framework for hybrid multiscale atomistic/hydrodynamic modelling of liquid solutions. We first describe the main idea and theoretical implementation of the approach. Then, the results on the method's numerical implementation as well as GPU and high performance specialised computer MDGRAPE are provided.

2 Multiscale hybrid method

2.1 The general idea

Our model describing both hydrodynamic (continuous) and molecular dynamic (discrete) components is based on the model of a miscible two-phase fluid. The computational domain is a double periodic box which is covered by a uniform Eulerian grid and filled with Lagrangian particles. At some regions, HD and MD, the liquid is described by purely hydrodynamic and purely Newtonian equations of motion respectively. In the hybrid domain the fluid consists of two "phases", Fig. 2. Phase 1 is a continuum phase with volume fraction $s = \frac{V_1}{V}$, where V_1 is the volume of phase 1 and V is the total volume of the domain. Phase 2 is

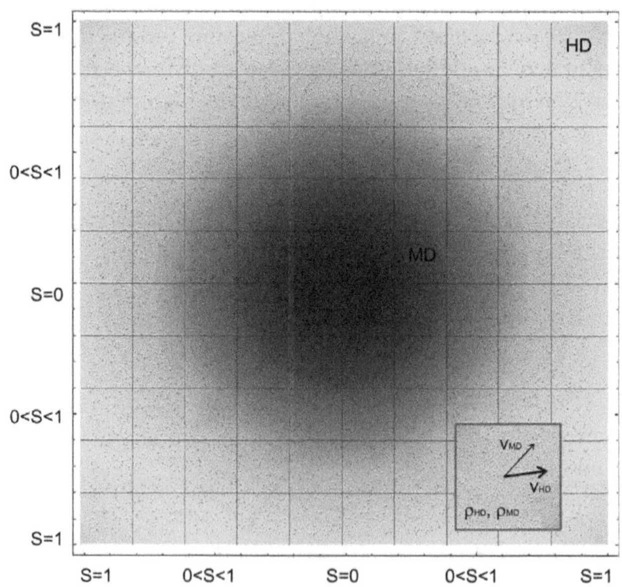

Fig. 2 Hybrid hydrodynamics – molecular dynamics model and the hydrodynamic cell (control volume).

a hydrodynamic phase that incorporates molecular particles. Its volume fraction is $(1-s)$ and it is characterised by N, the number of particles per elementary volume dV. The phase fraction $s = s(x)$ is a function of space coordinates x such that $s = 1$ in the HD domain, $s = 0$ in the MD domain, and it smoothly varies, $0 < s(x) < 1$, in the hybrid domain, Fig. 2. For each hydrodynamic cell the velocity u and the density ρ of the hydrodynamic phase are defined. The averaged value of the atomistic velocities and densities are also defined for each cell, u_p and ρ_p, Fig. 2.

The equations of motion are derived from the equations of conservation of total mass and momentum in each elementary Eulerian volume (control volume). The coupling between HD and MD substances is established in the HMD region by introducing the birth and death source terms for each substance. For determining the densities and velocities of the continuum phase the system of Landau–Lifshitz equations is solved:

$$\begin{aligned}
&\frac{\partial \rho}{\partial t} + \nabla(\rho \vec{u}) = 0, \\
&\frac{\partial \rho_i}{\partial t} + \nabla(\rho u_i \vec{u}) = \nabla_j\left(\tilde{\Pi}_{ij} + \Pi_{ij}\right), \\
&\frac{\partial \rho E}{\partial t} + \nabla(\rho E \vec{u}) = \nabla_j\left(\left(\Pi_{ij} + \tilde{\Pi}_{ij}\right)\cdot u_i\right) + \nabla\left(\vec{q} + \tilde{\vec{q}}\right),
\end{aligned} \tag{5}$$

where the stress tensor consists of a deterministic part

$$\Pi_{ij} = -(p - \xi\nabla\cdot\vec{u})\delta_{ij} + \eta(\partial_i u_j + \partial_j u_i - 2d^{-1}\nabla\vec{u}\cdot\delta_{ij}) \tag{6}$$

and a stochastic part, a random Gaussian matrix with zero mean and the covariance given by the formula

$$\begin{aligned}
&\left\langle \tilde{\Pi}_{ij}(r_1, t_1)\cdot\tilde{\Pi}_{kl}(r_2, t_2)\right\rangle \\
&= 2k_B T\left[\eta(\delta_{ij}\delta_{ik} + \delta_{ik}\delta_{jl}) + \left(\eta_V - \frac{2}{3}\eta\right)\times\right. \\
&\left.\times\ \delta_{ij}\delta_{jk}\right]\delta(r_1 - r_2)\delta(t_1 - t_2).
\end{aligned} \tag{7}$$

The stochastic stress tensor can be written explicitly[11] as

$$\tilde{\Pi}_{ij} = \sqrt{\frac{2k_B T}{\delta t \delta V}}\left(\sqrt{2}\sqrt{\eta}\cdot \boldsymbol{G}^s_{ij} + \sqrt{d}\sqrt{\xi}\ \frac{tr[\boldsymbol{G}]}{d}\boldsymbol{E}_{ij}\right), \tag{8}$$

where $\boldsymbol{G}$ is the Gaussian random matrix with zero mean and covariance $\langle \boldsymbol{G}_{ij}\boldsymbol{G}_{kl}\rangle = \delta_{ik}\delta_{jl}$, $\boldsymbol{G}^s_{ij} = \dfrac{\boldsymbol{G}_{ij} + \boldsymbol{G}^T_{ij}}{2} - \dfrac{tr[\boldsymbol{G}]}{d}\mathbf{E}_{ij}$ are symmetric random matrices with zero trace, $\boldsymbol{E}_{ij}$ is the identity matrix, $tr[\boldsymbol{G}]$ is the trace of the matrix. This form of correlation follows from the equipartition theorem, which relates the thermal fluctuations to temperature.[11]

The averaged heat flux is

$$q\alpha = \kappa\cdot\partial\alpha T, \tag{9}$$

where κ is the heat conductivity coefficient, ξ and η are the shear and bulk viscosities, d is the dimensionality of the system. An additional stochastic heat flux also has zero mean and the covariance flux components are defined by

$$\langle \tilde{q}_i(r_1, t_1)\cdot\tilde{q}_j(r_2, t_2)\rangle = 2k_B\lambda T^2\delta_{ij}\beta\delta(r_1 - r_2)\delta(t_1 - t_2)$$

or, explicitly,

$$\tilde{q}_i = \sqrt{\frac{2k_B\lambda T^2}{\delta t\delta V}}\,\mathbf{G}_i. \tag{10}$$

In our case the stochastic stress tensor and the stochastic heat fluxes are independent. For this reason in the following the heat flows are not considered.

The solution provides the correspondence between the averaged MD velocities of the particles in the cell, the fluctuating hydrodynamic flows, and the temperature. Our method provides the equations that give matching values of the particle velocities and the fluid velocities.

2.2 Main governing equations of the method

We use the following notations: x_i is the Cartesian coordinate component, u_i is the velocity component, ρ is the density, m is the mass, F is the force per unit Eulerian volume $V = h,^3$ N is the number of particles per unit volume, and subindex p stands for particle (MD model). Assume a summation over any index repeated twice. For simplicity here we assume that space and time is uniform everywhere in the MD and HD regions of the box. The mass conservation for the HD substance of density ρ that occupies the partial volume $0 < s < 1$ is

$$\frac{\partial(s\rho)}{\partial t} + \frac{\partial(u_i s\rho)}{\partial x_i} = J, \tag{11}$$

with an equivalent to the conservation law:

$$s(\rho(t+\tau) - \rho(t)) + \sum_p \left(s\int_t^{t+\tau}\rho u_i \mathrm{d}t\right)\mathrm{d}S_\beta/V = -\int_t^{t+\tau} J\mathrm{d}t, \tag{12}$$

where J and ρ represent the quantities averaged over the control volume, that is the Eulerian "bin" for MD of volume $V = V_{bin}$, area S and β is a coordinate direction. J is the HD substance birth rate due to the coupling with MD particles in the HD zone. A similar equation is formed for the MD substance that occupies the partial volume $(1 - s)$:

$$\frac{\partial}{\partial t}\left((1-s)\sum_p \rho_\mathrm{p}\right) + \frac{\partial}{\partial x_i}\left((1-s)\sum_p \rho_\mathrm{p} u_{ip}\right) = -J, \tag{13}$$

where $\rho_\mathrm{p} = m_\mathrm{p}/V$ is the density of the MD particles. Again, this differential equation is just a convenient form for the equation, which in practice is solved in the conservation form

$$(1-s)\left(\sum_p \rho_p(t+\tau) - \sum_p \rho(t)\right) + \sum_\alpha \left((1-s)\int_t^{t+\tau}\sum_p \rho_p u_{\beta_p}\,\mathrm{d}t\right)\mathrm{d}S_\beta\big/V = -\int_t^{t+\tau} J\,\mathrm{d}t, \tag{14}$$

where the flux divergence can be found from the MD statistic

$$\sum\left((1-s)\int_t^{t+\tau}\sum_p \rho_p u_{\beta_p}\mathrm{d}t\right)\mathrm{d}S_\beta/V$$
$$=\sum_\beta\left((1-s)\left(\sum_{\beta-\text{particles-out}}\rho_p-\sum_{\beta-\text{particles-in}}\rho_p\right)\right). \tag{15}$$

By adding the same term $\frac{\partial}{\partial t}((1-s)\rho)+\frac{\partial}{\partial x_i}((1-s)\rho u_i)$ to both sides of equation of HD phase (11), it is rearranged to

$$\frac{\partial}{\partial t}p+\frac{\partial}{\partial x_i}(u_i\rho)=J+$$
$$+\frac{\partial}{\partial t}((1-s)\rho)+\frac{\partial}{\partial x_i}((1-s)\rho u_i). \tag{16}$$

The mass birth/death function J entering the balance laws is not arbitrary but has to be determined in accordance with the modification of the MD equations. It is assumed that in the HD zone the evolution of the coordinate x_{pi} of each particle is determined by its MD velocity obtained from the Newton law that takes into account the two-phase force coupling, u_{ip}^{Newton} plus a correction. The correction is needed to constrain the velocity of the particles to the HD value in the pure HD limit $s\rightarrow 1$:

$$\frac{\mathrm{d}x_{pi}}{\mathrm{d}t}=u_{pi}+s(u_i-u_{ip})+$$
$$s(1-s)\alpha(x)\frac{\partial}{\partial x_i}\left(\tilde{\rho}-\sum_p\rho_p\right)/\rho_p/N(t), \tag{17}$$
$$\tilde{\rho}=s\rho+(1-s)\sum_p\rho_p.$$

Eqn (17) is equivalent to using a modified force potential in the usual MD equations (although the explicit formula for the modified potential that produces the corrected velocity field is not possible to obtain explicitly from the pair potential).

The momentum equation for the HD substance follows from the conservation law:

$$\frac{\partial(s\rho u_i)}{\partial t}+\frac{\partial(u_j u_i s\rho)}{\partial x_j}=sF_i+J_2, \tag{18}$$

where J_2 is the MD interaction force.

A similar conservation law for the MD particles per volume is:

$$\frac{\partial\left((1-s)\sum_p\rho_{pi}u_{pi}\right)}{\partial t}+\frac{\partial\left((1-s)\sum_p\rho u_{pi}u_{pj}\right)}{\partial x_j}=$$
$$=(1-s)\sum_p F_{pi}-J_2, \tag{19}$$

where $F_{pi} = F_{pi}^{\mathrm{MD}} = \dfrac{\partial V_p}{\partial x_i}$, V_{P}-interparticle pair potential. In the above the HD force is computed in accordance with the Landau–Lifshitz FH model:

$$F_i = -\frac{\partial(\Pi_{ij} + \tilde{\Pi}_{ij})}{\partial x_i}. \tag{20}$$

Similar to the mass equation, the momentum birth/death function J_2 entering the balance laws is not arbitrary but has to be determined in accordance with the modification of the MD equations, achieved by generalising the second Newton law:

$$\frac{\mathrm{d}u_{pj}^{\mathrm{Newton}}}{\mathrm{d}t} = (1-s)F_{pj}/\rho$$

$$+\frac{\partial}{\partial x_i}\left(s(1-s)\alpha(x,t)\sum_p u_{pj}/N(t)\frac{\partial}{\partial x_i}\left(\tilde{\rho} - \sum_p \rho_p\right)\right)/\rho_p/N(t)-$$

$$-\frac{\partial}{\partial x_i}\left(s\beta(x,t)\frac{\partial}{\partial x_i}\left(\tilde{\rho}\tilde{u}_j - \sum_p \rho_p u_{pj}\right)\right)/\rho_p/N(t), \tag{21}$$

where the density is $\tilde{\rho} = s\rho + (1-s)\sum\rho_p$, and the velocity is $\tilde{u}_i = [s\rho u_i + (1-s)\sum_p \rho_p u_{pi}]/\tilde{\rho}$, and $\alpha(x,t) = \alpha(x)$, $\beta(x,t) = \beta(x)$, for $N(t) > 0$, otherwise $\beta(x,t) = 0$.

3 Results

3.1 Results on the hybrid method

For computer simulations a system of 40k two-dimensional particles that interact *via* the Lennard-Jones potential was chosen. The simulation area was a square with periodic boundary conditions.[16] Temperature and density corresponded to water density at 122.4 K. The region was divided into 100 hydrodynamic cells,

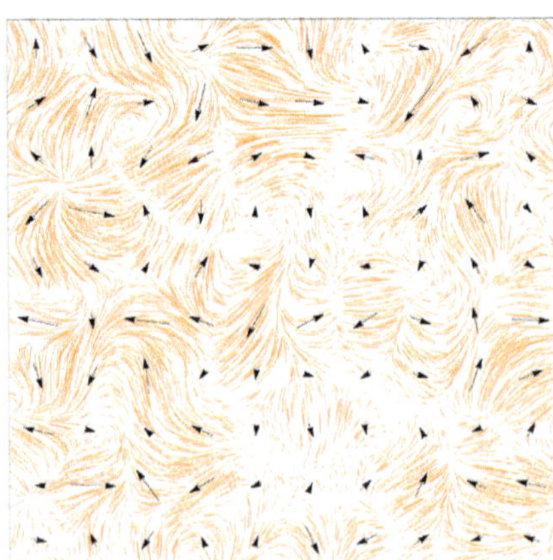

Fig. 3 Velocity fields of MD (grey) and hydrodynamic (black) phases for $s = 0.6$.

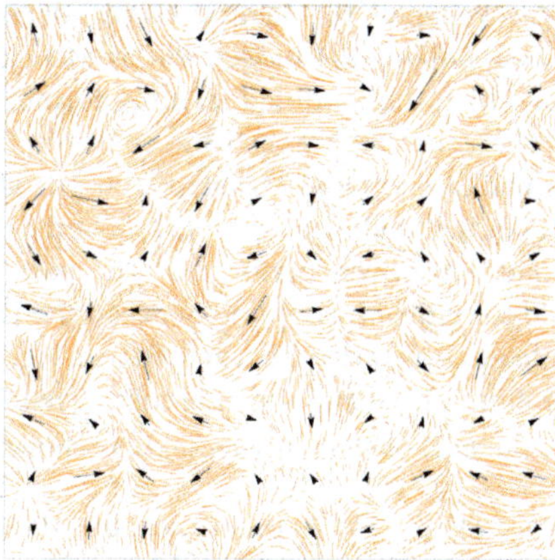

Fig. 4 Same as in Fig. 3 for $s = 0.8$.

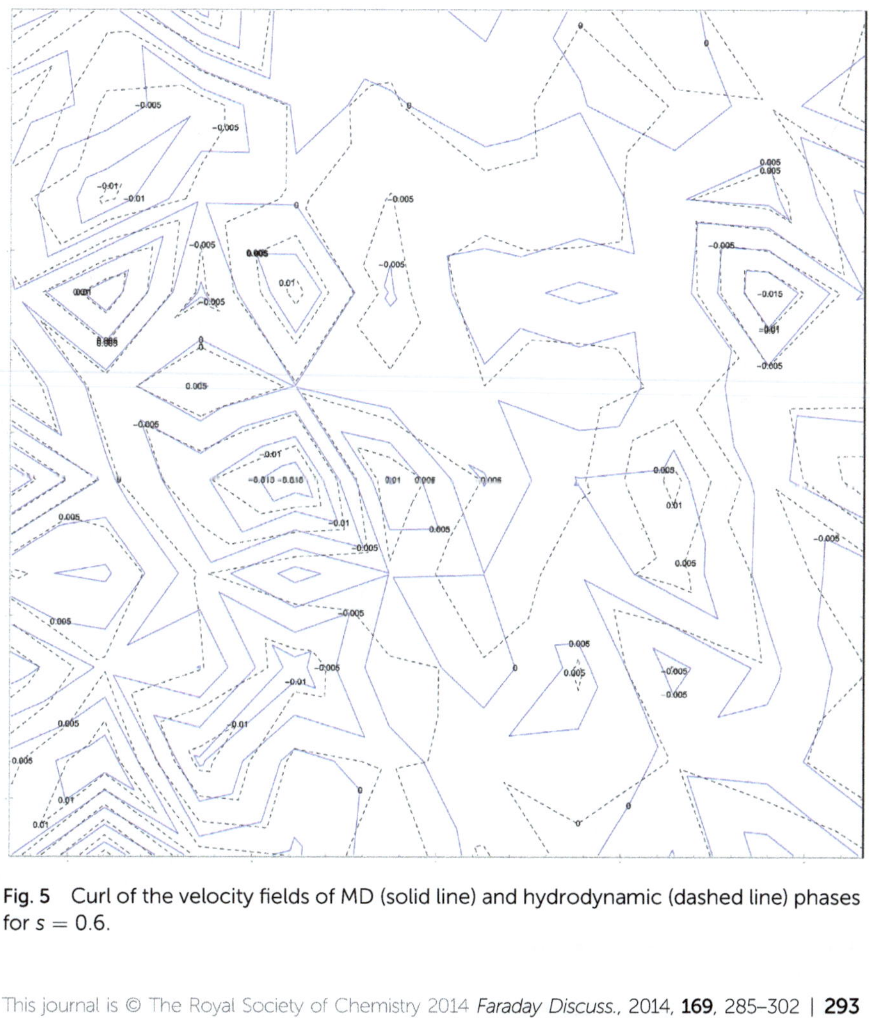

Fig. 5 Curl of the velocity fields of MD (solid line) and hydrodynamic (dashed line) phases for $s = 0.6$.

each of which had the value of the hydrodynamic density and temperature derived from the Landau–Lifshitz FH model. Purely MD calculations were used for equilibration, after which the modified leapfrog formulae (21) and (17) were used for integration. The coupling between the continuum phase (FH) and the MD phase was performed at every 100 MD step (100 MD steps is equal to 1 FH step). For several values of the model parameters s, α, and β the simulations of the hybrid LL-FH/MD model have been performed. The parameters α and β were chosen equal to 10 000 for efficient coupling between the phases.

The coupling parameter s was varied from zero to one. Fig. 3 and 4 show a typical behaviour of the continuum phase and the molecular dynamics phase velocities u_i and $\sum_p u_{pi}/N(t)$. Also, the curl of the velocity fields is shown in Fig. 5 and 6. The density fields are shown in Fig. 7 and 8. In all cases the MD and FH phases follow each other. The best correspondence is achieved for $s = 0.8$, as expected, since for this case the MD particles act as almost passive traces that follow the hydrodynamic flows, whereas for $s = 0.6$ discrepancies between the fields are present.

The solutions for the two phases are driven to each other and the standard deviation of the main variables converge to the reference values of the pure MD solution, which is the same as that of the pure LL–FH solution. It should be

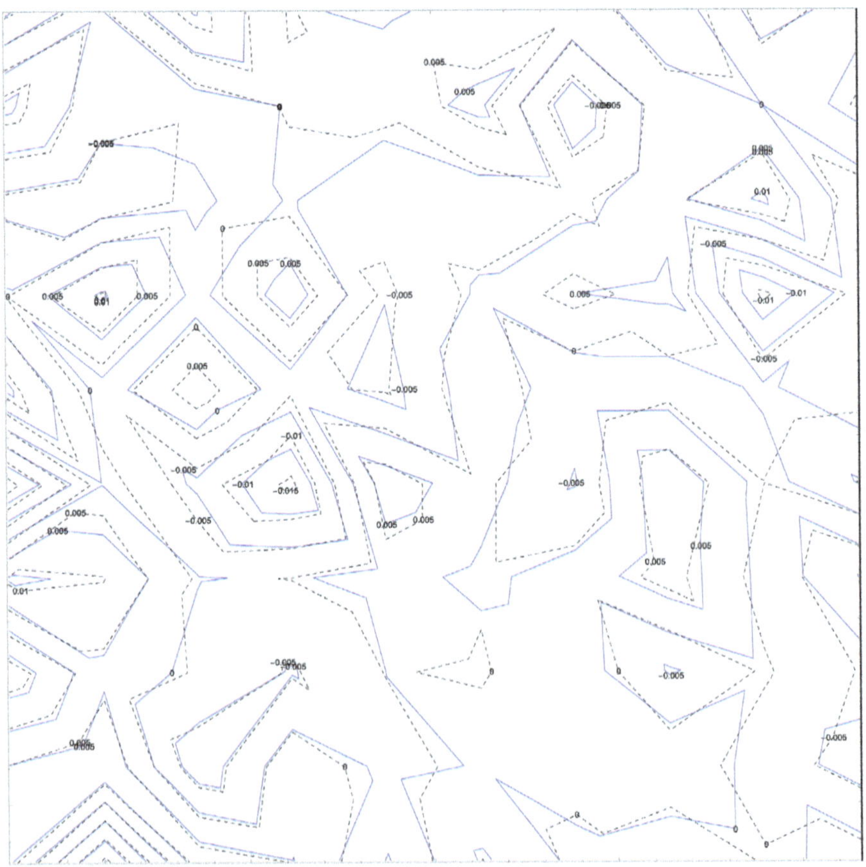

Fig. 6 Same as in Fig. 5 for $s = 0.8$.

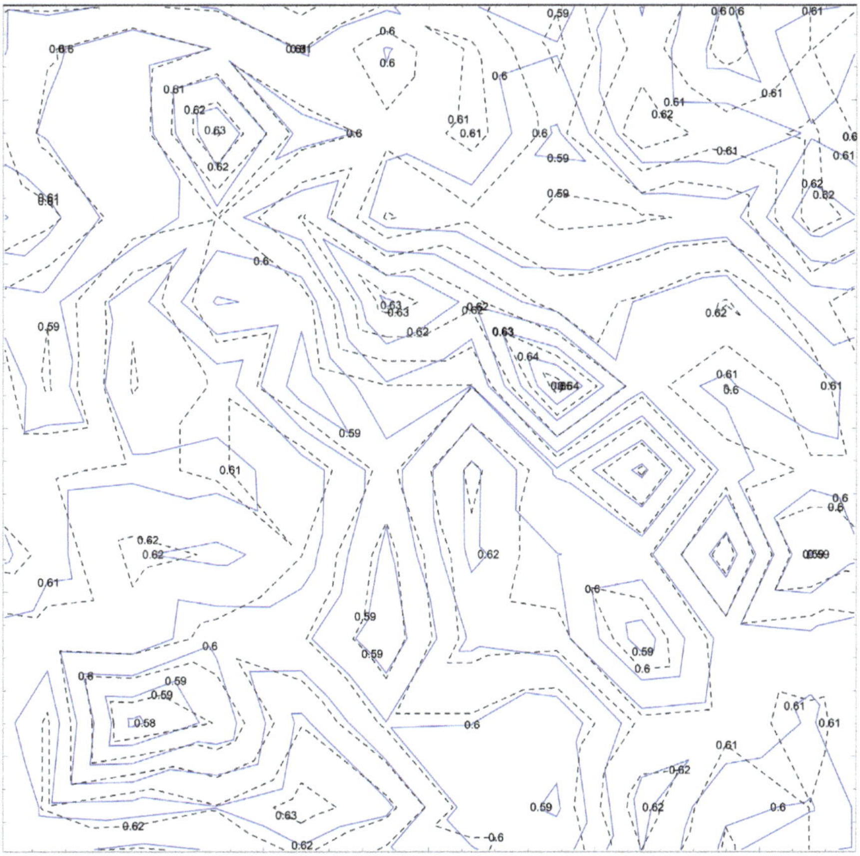

Fig. 7 Density fields of MD (solid line) and hydrodynamic (dashed line) phases for $s = 0.6$.

stressed that comparisons of the field values of the hybrid method with those of pure MD (FH) are meaningless because the solutions are stochastic (chaotic) and particular realisations of the fields are different. A meaningful comparison is possible in the statistics of the fluctuations of the field values, which do match as described above.

3.2 High performance molecular modelling: MDGRAPE

In multiscale simulations the computational speed is limited by high-resolution parts due to the large differences in time and space scales. The basic time step of an all-atom MD simulation is 1-2 fs, while those of hydrodynamics is of the order of hundreds of picoseconds. Thus, we need to fill the gap of several orders of magnitude in time scale. We have done this by using HPC.

The general CPU based supercomputers, such as the K-computer, are most effective for simulating large molecular systems comprising millions of atoms. In this case the scaling of the computation speed with the number of CPUs is very good, Fig. 9. In contrast, the specialised computers allow faster calculations for smaller systems, of the size of several thousands of atoms, Fig. 1. We have developed such special-purpose computer systems for MD simulations named

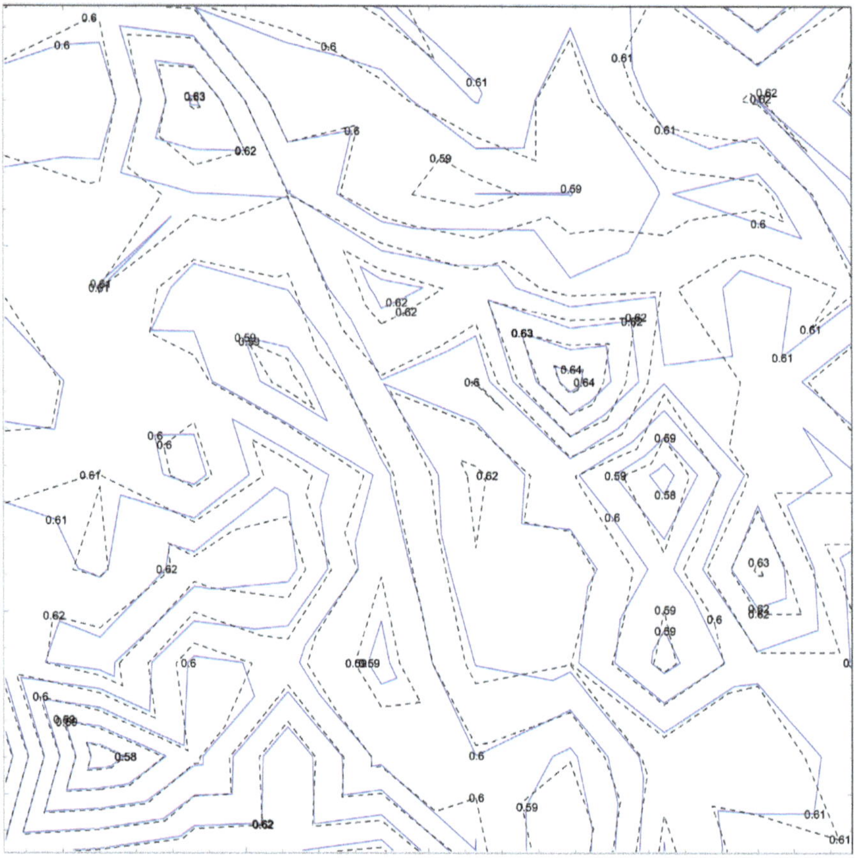

Fig. 8 Same as in Fig. 7 for $s = 0.8$.

MDGRAPE,[17] Fig. 10. The MDGRAPE computers were accelerators of non-bonded force calculations.

Recently, the ANTON machine, which is also a special-purpose computer for MD simulations, has been developed by D. E. Shaw research.[18] They integrated the non-bonded force accelerators, general-purpose processor cores, memories, and network interfaces in a single System-on-Chip (SoC). The integration minimizes the latency between these computing elements. As a result, it enables 100 times better performance for small-scale systems than normal PC clusters and the high performance of the full system can be achieved for systems as small as 10K atoms. Thus, it enables 100 ps MD simulation per second for these systems.

The ability of enabling fast simulations of small systems is especially useful for protein MD simulations as well as multiscale simulations. For this purpose we are currently developing the fourth-generation machine, MDGRAPE-4, Fig. 11. The MDGRAPE-4 also aims at achieving fast simulations of small systems. It is also based on the SoC technology, similar to ANTON. The target performance is 100 ps simulation per second for a system with 100K atoms. Currently, we have almost finished the evaluation of the system board. We plan to finish the system in 2014

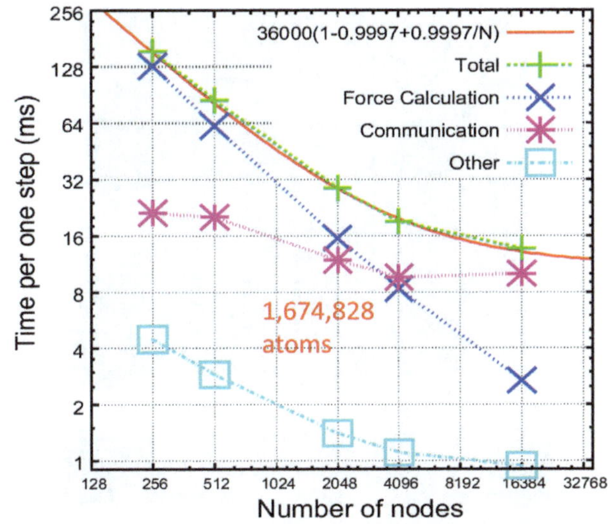

Fig. 9 Scaling with the number of computation nodes for MD simulations using the K-computer at RIKEN.

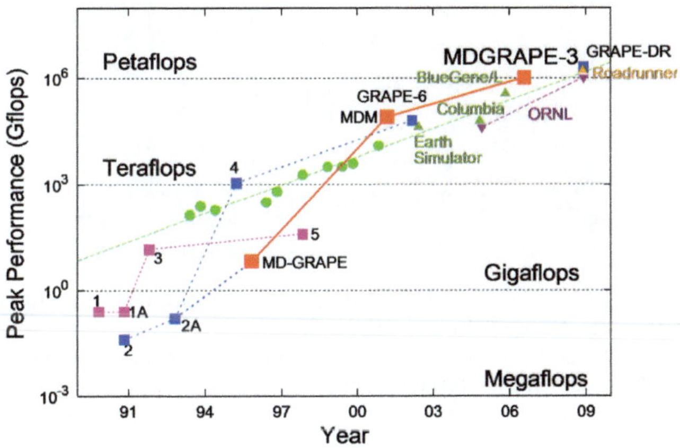

Fig. 10 The development of the GRAPE family accelerators (Gordon Bell Prizes won in 1995, 1996, 1999, 2000 (double), 2001, 2003, and 2006).

when the software development will be finalised. The latter will be based on GROMACS and contain a C library set for accessing the GP core.

The calculation speed is critical for visual interaction with the simulated systems since the time of data processing in a human brain and a visual system is of the order of 10 ms. Thus, to enable real-time interactive simulations, we need the performance of more than 30 frames per second and more than 300 simulation steps per second. We have developed such a system using the previous generation machine MDGRAPE-3, Fig. 12. In this system, only a few layers of water molecules can be treated due to the performance limitation. A typical speed is ≈15 frames per

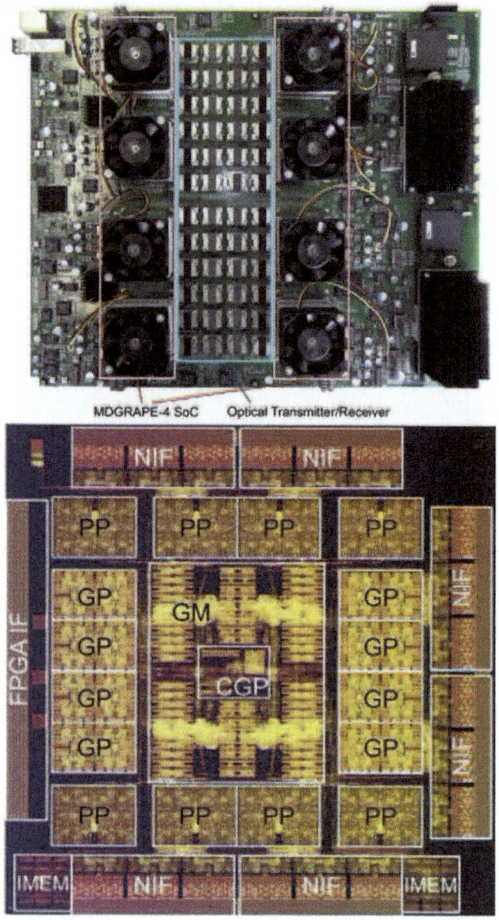

Fig. 11 The MDGRAPE-4 system board and System-on-Chip. The board has 8 SoCs and 48 optical transmitter/receiver for networking. The SoC has 8 nonbond force calculation modules (PP), 8 general-purpose processor modules (GP) each of which has 8 cores, memories (GM), and network interfaces (NIF).

second for a $\approx$ 150 residue globular protein. The multiscale simulation coupling MD on the MDGRAPE-4 and FH on conventional CPU/GPU machines will enable smoother and more precise visual simulations of solvated biomolecules.

3.3 GPU results

Another branch of HPC implementation of our framework uses GPU based calculations. Because of the nature of stochastic simulations, it is the statistical behaviour of the LL–NS equations which is relevant for the model rather than an instantaneous solution. For statistical convergence of low frequencies typical of the hydrodynamics fluctuations, one needs a sufficiently long simulation time. For example, the solution of the stochastic diffusion problem converges as

$$\frac{1}{\sqrt{N_t}}$$

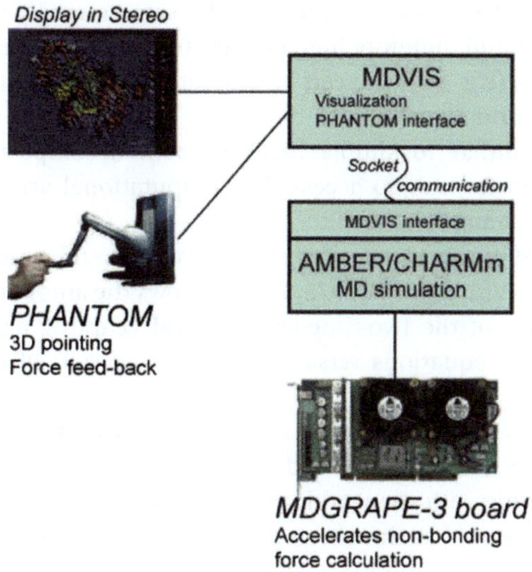

Fig. 12 Interactive MD simulation system using MDGRAPE-3. User can intervene simulations using the 3D force-feedbacked joystick.

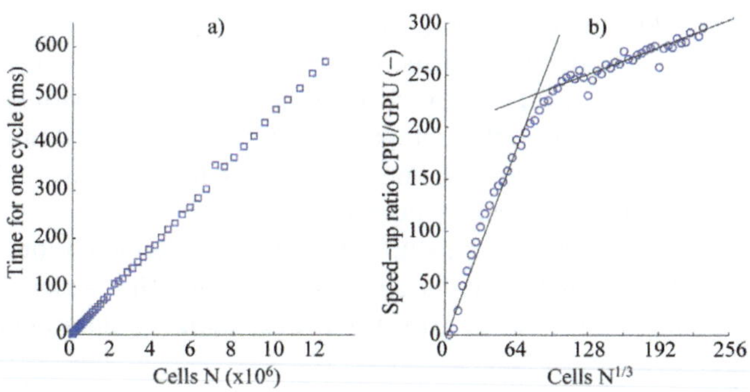

Fig. 13 (a) The amount of time one cycle takes on the GPU *versus* the number of cells (N). (b) The ratio between the time it takes to execute one cycle on the CPU *versus* the time on the GPU *versus* the number of cells in one direction of the cubed domain.

where N_t is the number of time steps of the simulation, which is typically of the order of several millions to get the error in fluctuations down to less than a few percent.

For three-dimensional problems, this leads to considerable computational times in case it is run in a serial manner based on a single computational unit (CPU). On the other hand, the LL–NS Fluctuating Hydrodynamics equations are well suited for an implementation in NVidia's CUDA. The latter allows one to off-load the computations on to the Graphical Processing Units (GPU) with performing massively parallel computations based on hundreds of computational threads.

The basic idea behind computations on the GPU, using the terminology of CUDA, are based on the fact that certain bits of code (kernels) can be executed simultaneously in several blocks each containing hundreds of threads. The key for optimisation is to limit the request for the same memory space, resulting in memory collisions and therefore high latency. In the case of the LL–NS equations, each block of cells (*e.g.* 16 × 16) can be evaluated independently, where only the single layer of boundary cells need to request memory space shared with other blocks. This is similar to simple block domain decomposition. Besides the different approach on how to access the computational arrays, the (C++) code itself needs little change.

The GPU used in the current simulations of liquid argon at equilibrium conditions is the NVidia Tesla C2075. Fig. 13 shows the amount of time it takes to compute one cycle of the two-time-level Central Leapfrog algorithm for three-dimensional LL–NS equations *versus* the number of cells of the computational domain.

The dependency of the amount of workload increases linearly with the amount of cells, *i.e.* a linear scaling if workload balance is achieved. It is also interesting to compare this scaling with the workload based on a single CPU. Fig. 13(b) shows how the GPU computation times compare to the computational times based on a single core of an Intel Xeon E5-2609 processor. There are two "regimes" on the CPU *versus* GPU comparison plot.

In the first part of the curve 13 corresponding to relatively small job sizes, the curve rises steeply indicating a sharp growth of efficiency when only a fraction of the GPU threads are fully utilised and their competition for shared memory is low. After all when GPU threads are fully engaged, the rate of the curve growth stagnates. The latter is likely to be caused by the fact that the system is reaching its limit of available RAM (GPU dedicated 5GB *versus* CPU shared with system 6GB). Nevertheless, as follows from this comparison graph, that the GPU can do the task up to 300 times faster compared to one CPU core, or at best about 75 times faster in comparison with the four-core CPU processor.

Because of the significant speedup due to GPU implementation, the LL–NS fluctuating hydrodynamics simulations performed for this article took a few hours rather than a week on the grids up to 10 million computational cells.

4 Conclusions

A theoretical framework that allows a correct hybridisation of the atomistic (molecular dynamics) and continuum (hydrodynamics) representations of a liquid system is described. The velocity and density fields of such hybrid systems are calculated which demonstrate that the atoms can follow the continuous flows without violating the conservation laws. Importantly, the framework does not require any artificial repulsive walls or forces that separate the representations. Instead, the contribution of each representation is governed by an arbitrary function of space and time. This allows the definition of different domains, which can also move, containing different degrees of 'atomisation'. This can be used in simulating, for example, a protein surrounded by water that gradually changes its representation from purely atomistic in the vicinity of the protein to purely continuous far from it.

For implementing the framework, a special purpose accelerator, MDGRAPE, has been used. The latest version of it, MDGRAPE-4, can achieve the 100 ps per second performance for a 100K atoms system. The details of the system are described and compared to other high performance computer systems used for molecular modelling.

An implementation of the continuum fluctuating hydrodynamics solver using NVidia Tesla C2075 GPU is described. A very effective speed up in computation is demonstrated.

Our current research is centred around the applications of the described framework and its HPC implementation to water–protein realistic systems. We expect the first results of this work to be published by spring 2014.

Acknowledgements

This work has been supported in the framework of the G8 Research Councils Initiative on Multilateral Research Funding (Engineering and Physical Sciences Research Council Grant No. EP/J004308/1).

References

1 P. Espanol, *Europhys. Lett.*, 1997, **40**, 631.
2 N. Pasquale, D. Marchisio and P. Carbone, *J. Chem. Phys.*, 2012, **137**, 164111–164119.
3 S. Fritsch, S. Poblete, C. Junghans, G. Ciccotti, L. D. Site and K. Kremer, *Phys. Rev. Lett.*, 2012, **108**, 170602.
4 D. Nerukh and S. Karabasov, *J. Phys. Chem. Lett.*, 2013, **4**, 815–819.
5 H. Frauenfelder, G. Chen, J. Berendzen, P. Fenimore, H. Jansson, B. McMahon, I. Stroe, J. Swenson and R. Young, *Proc. Natl. Acad. Sci. U. S. A.*, 2009, **106**, 5129–5134.
6 *Multiscale Modeling and Simulation of Composite Materials and Structures*, ed. W. Young, H. David and R. Talreja, Springer, Chichester, 6th edn, 2007.
7 P. Ajayan, L. Schadler and P. Braun, *Nanocomposite Science and Technology*, Wiley, New York, 3rd edn, 2006.
8 W. Harrison, *Pseudopotentials in the theory of metals*, W. A. Benjamin, New York, 1966.
9 G. Dhatt, E. Lefrancois and G. Touzot, *Finite Element Method*, Wiley, New York, USA, 2012.
10 G. K. Batchelor, *An Introduction to Fluid Dynamics*, Cambridge University Press, Cambridge, England, 1967.
11 L. Landau and E. Lifshitz, *Statistical Physics*, Pergamon, Oxford, 1980.
12 A. Asproulis, M. Kalweit and D. Drikakis, *Adv. Eng. Software*, 2012, **46**, 85–92.
13 G. Fabritiis, R. Delgado-Buscalioni and P. Coveney, *Phys. Rev. Lett.*, 2006, **97**, 134501.
14 G. Fabritiis, M. Serrano, R. Delgado-Buscalioni and P. Coveney, *Phys. Rev. E: Stat., Nonlinear, Soft Matter Phys.*, 2007, **75**, 026307.
15 Y. Klimontovich, *Usp. Fiz. Nauk*, 1987, **151**, 309–332.
16 D. Rapaport, *The Art of Molecular Dynamics Simulation*, Cambridge University Press, 2004.
17 J. Makino and M. Taiji, *Scientific Simulations with Special-Purpose Computers*, John Wiley & Sons, Tokyo, Japan, 1997.
18 D. E. Shaw, M. M. Deneroff, R. O. Dror, J. S. Kuskin, R. H. Larson, J. K. Salmon, C. Young, B. Batson, K. J. Bowers, J. C. Chao, M. P. Eastwood, J. Gagliardo,

J. P. Grossman, C. R. Ho, D. J. Ierardi, I. Kolossváry, J. L. Klepeis, T. Layman, C. McLeavey, M. A. Moraes, R. Mueller, E. C. Priest, Y. Shan, J. Spengler, M. Theobald, B. Towles and S. C. Wang, *Commun. ACM*, 2008, **51**, 91–97.

Faraday Discussions

PAPER

Allosteric pathway identification through network analysis: from molecular dynamics simulations to interactive 2D and 3D graphs†

Ariane Allain,[a] Isaure Chauvot de Beauchêne,‡[a] Florent Langenfeld,[a] Yann Guarracino,[a] Elodie Laine§[a] and Luba Tchertanov*[ab]

Received 27th February 2014, Accepted 3rd March 2014
DOI: 10.1039/c4fd00024b

Allostery is a universal phenomenon that couples the information induced by a local perturbation (effector) in a protein to spatially distant regulated sites. Such an event can be described in terms of a large scale transmission of information (communication) through a dynamic coupling between structurally rigid (minimally frustrated) and plastic (locally frustrated) clusters of residues. To elaborate a rational description of allosteric coupling, we propose an original approach – MOdular NETwork Analysis (MONETA) – based on the analysis of inter-residue dynamical correlations to localize the propagation of both structural and dynamical effects of a perturbation throughout a protein structure. MONETA uses inter-residue cross-correlations and commute times computed from molecular dynamics simulations and a topological description of a protein to build a modular network representation composed of clusters of residues (*dynamic segments*) linked together by chains of residues (*communication pathways*). MONETA provides a brand new direct and simple visualization of protein allosteric communication. A GEPHI module implemented in the MONETA package allows the generation of 2D graphs of the communication network. An interactive PyMOL plugin permits drawing of the *communication pathways* between chosen protein fragments or residues on a 3D representation. MONETA is a powerful tool for on-the-fly display of communication networks in proteins. We applied MONETA for the analysis of *communication pathways* (i) between the main regulatory fragments of receptors tyrosine kinases (RTKs), KIT and CSF-1R, in the native and mutated states and (ii) in proteins STAT5 (STAT5a and STAT5b) in the phosphorylated and the unphosphorylated

[a]*Bioinformatics, Molecular Dynamics & Modeling (BiMoDyM), Laboratoire de Biologie et Pharmacologie Appliquée (LBPA UMR8113 CNRS), École Normale Supérieure de Cachan, 61 avenue du Président Wilson, 94235 Cachan, France. E-mail: Luba.Tchertanov@lbpa.ens-cachan.fr*

[b]*Centre de Mathématiques et de Leurs Applications (CMLA), Ecole Normale Supérieure de Cachan, 61 avenue du Président Wilson, 94235 Cachan, France*

† Electronic supplementary information (ESI) available: See DOI: 10.1039/c4fd00024b

‡ Present address: Physik-Department (T38), Technische Universität München, James-Franck-Str. 1, 85748 Garching, Germany

§ Present address: Biologie Computationnelle et Quantitative, UMR 7238 CNRS-Université Pierre et Marie Curie, 15, Rue de l'Ecole de Médecine, 75006 Paris, France forms. The description of the physical support for allosteric coupling by MONETA allowed a comparison of the mechanisms of (a) constitutive activation induced by equivalent mutations in two RTKs and (b) allosteric regulation in the activated and non-activated STAT5 proteins. Our theoretical prediction based on results obtained with MONETA was validated for KIT by *in vitro* experiments. MONETA is a versatile analytical and visualization tool entirely devoted to the understanding of the functioning/malfunctioning of allosteric regulation in proteins – a crucial basis to guide the discovery of next-generation allosteric drugs.

1. Introduction

Data and computational tools

With the permanently increasing computational means and the rapid development of efficient algorithms, the biomolecular modeling field has entered a new era. The exponential growth of data produced by molecular dynamics (MD) simulations calls for the development of tools able to handle and analyze these large amounts of data by reducing their dimensionality to extract the relevant and pertinent information. Such tools should enable visualization of this information to unveil its biological implications and help in generating new hypotheses. This issue is one of a multi-scale nature. On a more systemic level, cellular spatial organization, signaling and communication also occur at different scales – from distances of some angstroms in binary protein complexes and assemblies to global micrometer distances across and between cells. Recently, a new view of cellular spatial organization was proposed[1] in which cell signaling is described in terms of dynamic allosteric interactions within and among distinct, spatially organized transient clusters. The clusters vary over time and space and their lengths range from nanometers to micrometers. A key challenge is to understand the interplay across these multiple scales, link it to the physicochemical basis of the conformational behavior of single molecules and ultimately relate it to cellular function. These conceptions have emphasized the necessity for the development of computing tools for a rational understanding of allosteric regulation and its manifestations at different levels.

The last few decades have faced a similar issue with the explosion of the amount of genomics and protein–protein interactions data related to cell metabolism and regulation. As a response, many performance tools were developed to integrate the data in an accurate, complete and ergonomic way. Apart from genome-wide association studies (GWAS) were developed graphical representations of gene regulatory networks (*TranscriptomeBrowser* 3.0[2]), of networks of functionally related genes (*PainNetworks*[3]), of protein–protein interaction networks in bacteria,[4] and even multi-scale and multi-object graphs of organ–disease–drug causal relationship networks.[5]

Increasing interest in the allosteric phenomenon stimulated the development of tools capable of analyzing and visualizing protein communication at the atomic level. Some of them, *e.g.* RINalyzer,[6] AlloPathFinder[7] and others[8–10] build an interaction network from the native contacts observed in the crystallographic structure of a protein. SPACER[11] was developed using theoretical concepts based on the thermodynamic view of allostery. xPyder[12] and similar studies[13–15] exploit molecular dynamics-derived dynamical correlations, while AD-ENM[16] and PARS[17] analyze structural communication routes based on highly simplified physical models (elastic network models or normal mode analysis). Piazza and colleagues used off-lattice and non-linear network protein models to localize modes of non-linear origin (*discrete breathers*) that accumulate energy on few sites in response to perturbations even occurring at distant locations.[18–20] Methodological efforts have also been directed toward the identification of evolutionarily conserved networks of residues thought to mediate allosteric communication.[7,21]

The development of tools for an accurate and complete description of protein allosteric communication requires a deep discerning of the allosteric phenomenon and associated regulation mechanisms.

Allostery and associated regulation mechanisms

Allosteric regulation is an ubiquitous mechanism of intra- and intermolecular regulation of biological processes. All physiological signaling pathways implying cascades of protein–protein binding events are initiated by an intramolecular allosteric transition occurring within a given receptor in response to a specific perturbation (*e.g.*, ligand binding). As a typical paradigm, the receptor tyrosine kinases (RTKs) act as sensors for extracellular ligands, the binding of which triggers receptor dimerization and activation of the kinase function, leading to the recruitment, phosphorylation and activation of multiple downstream signaling proteins, which ultimately govern the physiology of cells.[22] RTKs allosteric activity is tightly and finely regulated under normal physiological conditions. Deregulation of RTKs activity prompts perturbation in signaling pathways and causes a wide range of diseases, particularly cancers.[23]

At the molecular level, allostery is manifested as the regulation of a protein function by the binding of an effector (ligand/substrate/protein) at a site other than the active site of the protein. This mechanical perturbation triggers a specific conformational rearrangement at another spatially distant protein site. Discovered initially by the pioneering studies of the oxygen binding to hemoglobin by Monod,[24] allostery is now commonly recognized as a universal phenomenon in proteins, that is intimately related to their structure, dynamics and functions. The allosteric effects have different natures and control nearly all biological processes, *e.g.*, cellular network for signal transduction, metabolism activation, motor work and transcription control. Allostery regulates activation/deactivation in enzymes, opening/closing of the ion channels, oligomerization/monomerization (dimerization of RTKs), association/dissociation of macromolecular subunits (G proteins), loss/gain of affinity for a ligand (O_2 binding to hemoglobin) or a membrane, exposure/burial of the recognition site to another protein or nucleic acid.

It is believed that proteins have the remarkable property of transmitting intramolecular signals through specific fold-encoded pathways that form a network of covalently bonded and non-covalently bonded interactions between the protein residues.[24,25] The first theories considered allosteric communication as a domino effect where local structural distortions propagate step by step through a protein structure viewed as globally static.[26] The vision of protein structure evolved from *an ensemble of discrete conformations* to a *continuous landscape of conformations*, which led to reconsideration of the allosteric phenomenon as a consequence of the redistributions of protein conformational ensembles (shift in conformational preferences).[27,28] A population shift in the conformational space of a given protein can be induced by the binding of an effector, point mutations or environmental conditions (pH, voltage, ions concentration, *etc.*).

Allostery offers a highly specific way to modulate protein function. So far as allosteric events perturb the structure, and the energetic strain propagates and shifts the population, this can lead to changes in the shapes and properties of target binding sites. Consequently, combining equilibrium fluctuation concepts with genome-wide screens could considerably help drug discovery. However, allosteric signal transmission is difficult to detect experimentally and to model because it is often mediated by local structural changes propagating along multiple pathways. These concepts have emphasized the necessity for the development of computing tools able to account for both conformational global changes and local fluctuations to elaborate rational descriptions of allosteric coupling.

For such a purpose, we propose here an original approach based on the analysis of inter-residue dynamical correlations to localize the propagation of both structural and dynamical effects of a perturbation throughout a protein structure. The principle of this approach consists of building a modular network representation of a protein composed of clusters of residues representing *independent dynamic segments* (*IDSs*) and chains of residues acting as *communication pathways* (*CPs*) (Fig. 1). The representation is derived from the topology of the protein and the inter-residue dynamical correlations computed from molecular dynamics simulations. The utilities for a modular network representation of protein structures are implemented in an interactive package MONETA (MOdular

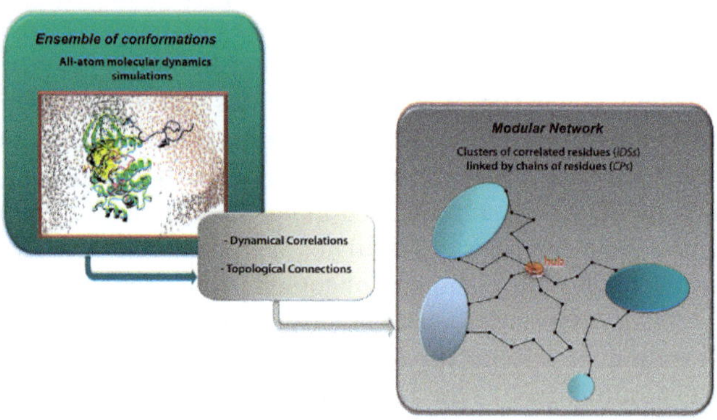

Fig. 1 Schematic representation of the Modular NETwork Analysis (MONETA) general principle. A modular network representation composed of clusters of residues and chains of residues is built from the dynamical correlations and topology calculated from a protein conformational ensemble. In MONETA, residue clusters or modules are delineated as *independent dynamic segments* (*IDSs*) as they represent the most striking features of the protein local dynamics. Chains of individual residues are designated as *communication pathways* (*CPs*) as they represent well-defined connectivity pathways along which interactions can be mediated at long distances in the protein. Information is propagated through *IDSs via* the modification of the local atomic fluctuations and through CPs *via* well-defined interactions. The highly connected residues, at the junction of many pathways, can be considered as "*hubs*" in the protein network.

NETwork Analysis) designed for generation of the communication landscape in a protein and its visualization/representation in 2D and 3D. These functionalities are very useful for exploring communications in allosterically regulated proteins.

2. Implementation and functionality

The first release of MONETA (MONETA 1.0) used for the analysis of communications between two distant regions, A-loop and juxtamembrane region (JMR), in the receptor tyrosine kinase KIT,[29] was significantly improved, optimized and enriched by additional functionalities. Optimization was oriented principally to (i) increasing performance, allowing significantly reduced computing time (by a factor of 30); (ii) reducing the number and volume of generated temporary files and (iii) introducing a 2D graphical representation of a communication network. The advanced version of MONETA (release MONETA 2.0)[30] contains novel options for data analysis, *i.e.*, display of intra-protein connections network in a dynamical graph for the rapid identification of biologically relevant *communication pathways* between residues or domains. The advanced version of MONETA was used for the analysis of allosteric communication in different proteins divergent by their size, form and molecular architecture.

2.1. MONETA composition and functionalities

MONETA analyses and extracts data from conformational ensembles computed by all-atom MD simulations of a protein to infer the topological connections (residues interactions) and the dynamical correlations between residues or domains (Fig. 1).

MONETA was implemented in the form of a package that automatically performs all computational steps and data analysis – from Principal Component Analysis (PCA) of the MD simulations to a viewing of communication landscape – through a python scripting interface to the software R,[31] PyMOL[32] and Gephi.[33] MONETA running requires some additional software – the *ptraj* module of AmberTools,[34] the module *g_mdmat* (or *g_mdmat_d*) of GROMACS,[35] the modules HBPLUS[36] and HBADD.[36] The package is currently composed of three parts: (i) the preparatory phase generates the data required for communication analysis from an MD trajectory and a 3D protein structure, (ii) the application phase performs the computation of intra-protein *communication pathways* and *independent dynamic segments*, and (iii) the visualization phase allows data analysis and graphical representation of the generated data (Fig. 2).

The user gets interactive and modular representations of intra-protein connections at the protein and residue scales and of *communication pathways* at the atomic scale. The analysis *per se* is a three-step procedure:

(i) Identification of protein regions displaying the most striking features of the protein's local dynamics. Such regions, or *independent dynamic segments*, constitute residue clusters in which atomic fluctuations are highly concerted within each cluster, although independent from the rest of the protein. These segments were previously described in the literature as conformationally sensitive to a disturbance of the global protein dynamics, capable of energy storage and transfer.[18,19,37] The *IDSs* are identified by a statistical technique, Local Feature

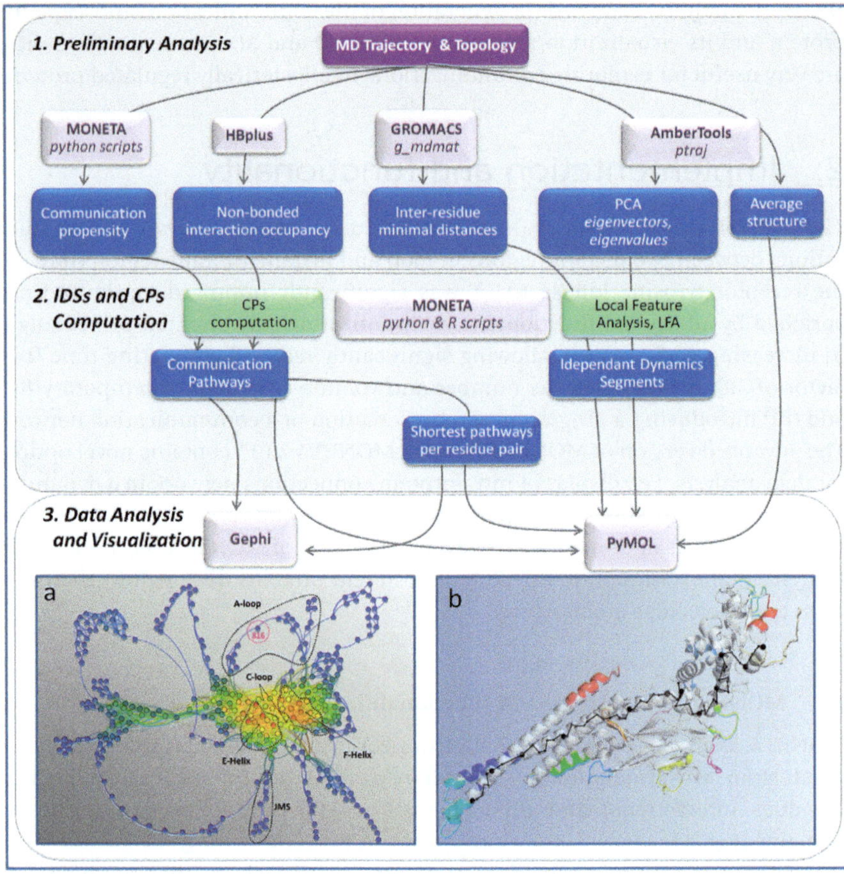

Fig. 2 Overview of the major analysis steps in the MONETA workflow. Each step of the MONETA procedure is delimited by an icon. The required inputs, parameters, outputs and scripts are identified by colors: initial mandatory inputs in purple, outputs in blue, MONETA computation steps in green, software and program in grey. Step 3 is illustrated by a 2D graph of the communication landscape in KIT (a) and by 3D representation of communication pathway in STAT5 (b). 2D and 3D graphs drawn with GEPHI and PyMOL modules incorporated in MONETA.

Analysis (LFA),[38] adapted for analysis of the atomic coordinate fluctuations from MD simulations.[39]

(ii) Detection of *communication pathways* linking the *IDSs* through non-bonded interactions between residues. The ability of the protein residues to communicate efficiently is evaluated by using the measure of communication propensity.[40] The communication between two residues is estimated by their commute time, expressed as the variance of their inter-residue distance over MD trajectories.[15,41] Chains of residues interacting by pair and displaying high communication propensities between them would represent pathways of well-defined interactions through which information would be transmitted efficiently. Such chains of residues are denoted as *communication pathways*.

(iii) Visualization of *IDSs* and *CPs* in a communication profile of the protein. This feature of MONETA is based on the automatic computing of all *communication pathways* from the non-bonded interactions between residues and the variance of inter-atomic distances (*i.e.*, dynamic correlation) in MD simulations. An implemented GEPHI[33] module translates these connectivity groups and pathways into a 2D network graph (Fig. 2). The 3D representations of *CPs* at the atomic level are performed by a PyMOL[32] module implemented in MONETA.

2.2. Theoretical background

Inter-residue contacts (topology). The topological description of a protein at the level of inter-residue contacts can be derived from distance matrices consisting of the average smallest distances between all residues pairs, computed using the *g_mdmat* program of GROMACS.[35] Two residues i and j are considered topologically close if the average smallest distance between them is lower than a given threshold $d_{cut} = 3.6$ Å. This value accounts for non-covalent contacts between protein residues such as hydrogen bonds, hydrophobic interactions and salt-bridges.

Local representation from PCA. The *IDSs*' identification is based on a statistical technique, Local Feature Analysis (LFA), originally developed for image analysis.[38] LFA processes data from Principal Component Analysis (PCA) and extracts local outputs of reduced dimensionality from the global PCA modes. The LFA formalism was adapted to study essential dynamics in proteins.[39] This adapted LFA was implemented in MONETA 2.0 for identification of the *independent dynamic segments*.

Identification IDSs. The PCA of a MD trajectory containing m conformations of a set of N atoms (*e.g.*, the Cα atoms of a protein composed of N residues) allows reconstruction of the displacement D_{α_i} of a coordinate α_i over the trajectory according to the projections A_r of the trajectory on each rth mode (eqn (1)). The dimensionality of the trajectory can be reduced by keeping only n first modes that describe the main part of the dynamics of the system ($n \ll 3N$) and reconstructing an approximated trajectory (eqn (2)).

$$D_{\alpha_i \text{recons}} = \sum_{r=1}^{3N} A_r \varphi_r(i), \text{ with } \varphi_r \text{ the eigenvectors describing the } r^{\text{th}} \text{ PCA mode} \tag{1}$$

$$D'_{\alpha_i \text{recons}} = \sum_{r=1}^{n} A_r \varphi_r(i) \tag{2}$$

$$A_r = \sum_{i=1}^{3N} \varphi_r(i) D_{\alpha_i} = \sum_{i=1}^{3N} K^{\text{PCA}}(r,i) D_{\alpha_i} \tag{3}$$

In eqn (3), K^{PCA} represents the kernel of the transformation of the trajectory from Cartesian global coordinates consisting in the PCA modes. The PCA offers a reduced dimensionality, however, it is nonlocal. The principle of LFA is to describe movements of reduced dimensionality, not global as the PCA modes are, but local. A new kernel K^{LFA} of the transformation is derived by analogy with the PCA (eqn (4)).

$$K^{\text{LFA}}(j,i) = \sum_{r=1}^{n} \varphi_r(j) \frac{1}{\sqrt{\lambda_r}} \varphi_r(i), \text{ where } \lambda_r \text{ is the eigenvalue corresponding to } \varphi_r \tag{4}$$

Similarly to PCA outputs A_r (eqn (3)), local outputs O_j are defined as:

$$O_j = \sum_{r=1}^{n} \frac{A_r}{\sqrt{\lambda_r}} \varphi_r(j) \tag{5}$$

As PCA outputs A_r are decorrelated by frequency (corresponding to the index r), LFA outputs O_j are decorrelated by space when n tends to $3N$. In the general case as $n \ll 3N$, LFA outputs O_j are minimally correlated. The residual correlations are expressed as eqn (6).[38]

$$< O_i O_j > \;= \sum_{r=1}^{n} \varphi_r(ji)\varphi_r(j) \equiv P(i,j) \tag{6}$$

The $3N$ local features O_j could be computed from n PCA modes. To reduce the dimensionality of the displacement D' described by these LFA modes, the set of n features O_j which best approximates D' was selected. We refer to them as $O_{jm(m \text{ in } M)}$, where M is the set of n *seeds* corresponding to the coordinates around which are centered the O_{im}. The *seeds* are obtained by applying the sparsification algorithm described in ref. 39. Briefly, at each step, among the $(3{*}n{-}3{*}m)$ coordinates of the atoms not in the current set M, the $(m + 1)^{\text{th}}$ *seed* is chosen as the one whose atomic fluctuations are the least described by the atomic fluctuations of the *seeds* already in the set M. An additional requirement is that the $(m + 1)^{\text{th}}$ *seed* coordinate is not a coordinate of an atom (i) with another coordinate already in M or (ii) adjacent to an atom with a coordinate in M. This condition ensures that the new *seed* corresponds to an atom dynamically decorrelated from the *seeds* in M.

As each *IDS* describes a very high proportion of the protein fluctuations, several *seeds* can be identified in the proximity of each other, although they belong to overlapping dynamic segments. To limit this phenomenon, a dimensionality reduction step was added to the LFA algorithm according to the following protocol: if two *seeds* correspond to two residues located within 6 positions from each other in the peptide sequence, only one of the two *seeds* was kept, chosen as the one describing the largest part of protein fluctuations. The LFA procedure thus identifies a number of *seeds* lower or equal to the number n of PCA modes included in the procedure.

Growth of IDSs. Each *IDS* or S_m ($m = 1, 2, 3, \ldots$) is iteratively constructed around a *seed* as a group of neighboring residues with correlated movements. A residue i is added to S_m if (a) it is topologically close to at least one residue in S_m and (b) its average correlation with the residues in S_m is greater than a threshold P_{cut} (eqn (7)). The correlation between two residues h and k is evaluated as in eqn (8).

$$\frac{\Sigma_{j \in S_m} P(i,j)}{|S_m|} \geq P_{\text{cut}} \tag{7}$$

$$P(h,k) = \sum_{r=1}^{n} \sum_{d=1}^{3} \varphi_r(h_d)\varphi_r(k_d) \tag{8}$$

where d is the (x,y,z) coordinate index and n is the number of retained PCA modes.

Communication pathways. *CPs* are chains of neighboring residues whose communication propensities between each other are high. We define them according to the concept of «*communication propensity*».[42] The communication propensity of a pair of residues is inversely related to their *commute time* CT(i,j), expressed as a function of the variance of the inter-residue distance (9):

$$\mathrm{CT}(i,j) = <\left(d_{ij} - \overline{d}_{ij}\right)^2> \tag{9}$$

where $d_{ij} = |ri - rj|$ is the distance between the Cα atoms of residue i and residue j.

Intuitively, if a residue moves, it transmits this movement to another residue which moves in its turn, and these residues communicate all the more effectively that the delay separating their respective movements is short. From a trajectory of molecular dynamics, MONETA calculates for every pair of residues[29] (i) their *commute time* and (ii) the percentage INT(a,b) of conformations in which a non-covalent interaction exists between these two residues. Two residues communicate efficiently if their *commute time* is lower than a chosen value $\mathrm{CT}_{\mathrm{threshold}}$. A non-covalent interaction is considered as stable for INT(a,b) greater than a chosen threshold value $\mathrm{INT}_{\mathrm{threshold}}$. MONETA generates *CPs* iteratively according to the following algorithm:

- Start from one residue
- Create as many pathways $\{r_1, r_2 \ldots r_n\}$ as residue i's neighbors
- Grow each path $\{r_1, r_2 \ldots r_n\}$ iteratively such that (i) r_i and r_{i+1} are not covalently bonded, (ii) there exists a stable non-bonded interaction between r_i and r_{i+1}, and (iii) r_{i+1} communicates efficiently with every residue in $\{r_1, r_2 \ldots r_i\}$.

The way *communication pathways* are grown ensures that any two adjacent residues are connected by non-covalent interactions, that all residues in the same path are directly or indirectly non-covalently connected and that every residue in the *CP* is reachable from any other point in a short *commute time*.

2.3. Graphical implementation

Two-dimensional graphs. GEPHI[33] is an analytical tool designed for the representation of interacting identities as a bi-dimensional graphic network. MONETA produces a protein network graph (Fig. 2) where all residues of the protein are denoted by as many nodes linked by edges representing connections of different natures: peptide links, direct connections (the two residues are consecutive in the same paths) and indirect connections (there exists at least one path linking the two residues). The global shape of the graph is constructed according to the inter-residue connections, so as to reflect the global communication propensity of the protein.

Three-dimensional visualisation. A PyMOL[32] module integrated in MONETA permits visualization of the *communication pathways* between the residues in the 3D structure of a protein (Fig. 2). This representation gives access to a detailed vision of the *communication pathways* at the atomic level.

2.4. How does MONETA work?

The package distribution includes a step-by-step tutorial with description of input files, output files and shell scripts. The scripts can easily be modified to process user-provided input data. In brief, the first step of MONETA running requires a trajectory file obtained by MD simulations and the associated topology file. It generates all the prior data necessary for computing the *IDSs* and *CPs*:

• The structure onto which the MONETA representation will be drawn is by default chosen as the average structure of the protein in the MD simulation.

• Non-bonded interactions, *i.e.*, hydrogen bonds and hydrophobic contacts, are monitored along the trajectory and recorded using the program HBPLUS. MONETA runs HBPLUS on each conformation of the ensemble and computes the occurrence of non-covalently bonded interactions for each residue pair.

• The mean commute time of each residue pair over the MD simulation is computed.

The *g_mdmat* or *g_mdmat_d* program of GROMACS[43] is run by MONETA to compute, for each pair of residues, the minimal inter-residue distance averaged on the conformational ensemble.

A PCA is performed by use of the *ptraj* module of AMBER Tools to compute the covariance matrix, the PCA eigenvalues and eigenvectors.

To compute the data listed above with MONETA, the values of the required parameters listed below may be chosen by the user or computed automatically:

• The threshold for non-covalently bonded interaction occupancy ($INT_{threshold}$) beyond which a bond will be considered is 0.5 by default (*i.e.*, the interaction is present 50% of the time) but tunable if desired.

• The threshold for communication propensity ($CP_{threshold}$) – the inter-residue commute time below which the communication between these residues is considered as efficient – can be either arbitrarily chosen or computed as a chosen percentile of all the values of the communication propensity matrix.

• MONETA computes communications resulting from the 3D structure of a protein, rather than from its sequence. Consequently, a parameter sets the minimal distance along the sequence between two residues adjacent in a path. A value of 1 (residue), which turns off the «communications» mainly sustained by peptidic links, is recommended.

• The number of PCA modes to be considered in LFA is determined by setting a minimal part of the total atomic fluctuations of Cα to be covered by these modes.

• The user can choose to remove some protein residues from the analysis (typically the strongly fluctuating extremities).

In the second step, MONETA computes the *IDSs* and all the *CPs* within the protein. Then, MONETA computes for each residues pair {*i*,*j*} the number of paths containing the two residues and identifies the smallest paths between the two residues. The output files may be used for display of communication in the protein within the GEPHI[33] and PyMOL[32] modules.

PyMOL is automatically launched after *IDSs* and *CPs* computation. The structure file is loaded and represented as a green illustration. Independent dynamic segments are loaded as PyMOL selections and highlighted in blue (tunable for each *IDS* in PyMOL selection column). Two functions, find_paths and find_path2, for visualization of *communication pathways* are loaded. Using these functions, the user can display all the *CPs* generated from a specified residue of the protein (find_paths), or the CPs between two specified residues (find_path2). A new PyMOL object is generated for each computed path and displayed as chains of small spheres connected together by lines (Fig. 2–5). Within a given path, each residue member is illustrated by up to three spheres corresponding to: the atom closest to the preceding residue in the path, the Cα, and the atom closest to the following residue in the path. The initial residue is highlighted by a bigger sphere centered on its Cα.

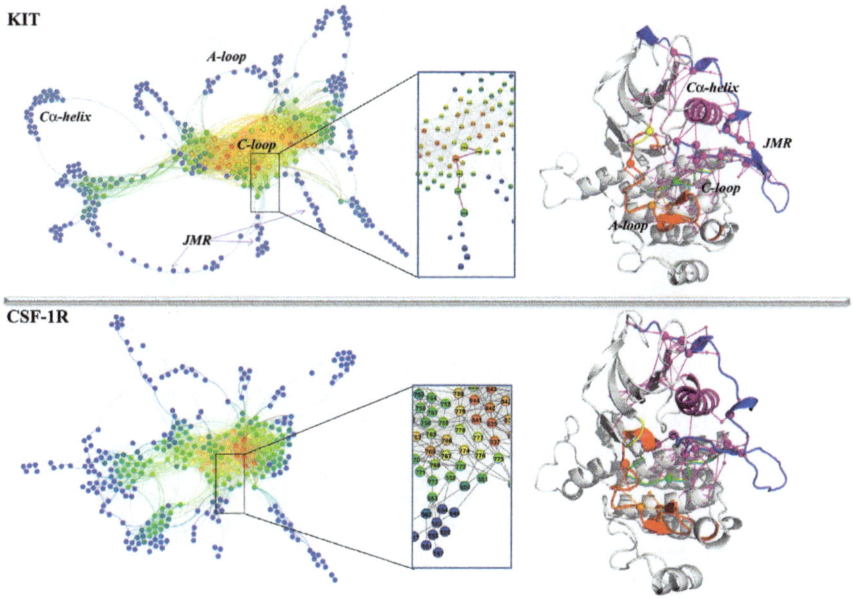

Fig. 3 *Communication pathways* of cytoplasmic region in the native receptors KIT (top) and CSF-1R (bottom). *Left panel*: 2D graphs of a global topology of the inter-residue communications. Residues are represented by points, *communication pathways* are depicted by bold lines and two connected residues by a thin line. Residues are coloured from blue through green and yellow to red according to their communication efficiency, estimated as the number of residues to which they are connected by at least one *CP*. *Middle panel*: large-scale view of the *CPs* in KIT and CSF-1R zoomed on *CPs* in the JMR, the A-loop and C-loop. *Right panel*: 3D structural mapping of the inter-residue communication in KIT and CSF-1R. For each protein, the average MD conformation is presented as an illustration. The key structural fragments of the receptors are highlighted in different colors: the A-loop is in red, the JMR is in blue, the Cα-helix is in magenta, the C- and P-loops are in green and yellow, respectively. Labels for the different regions are indicated on the top panel. *Communication pathways* between residue atoms (circles) are depicted by coloured lines: *CPs* formed by the A-loop residues in orange, and by the JMR residues in magenta. 2D and 3D graphs are drawn with GEPHI and PyMOL modules incorporated in MONETA.

GEPHI may be launched by the user to read the GEPHI outputs computed by MONETA and to create a 2D-representation of the communication network of the protein (Fig. 2, 3 and 5). GEPHI uses two types of elements: nodes and edges. Nodes are auto-closing beacons containing an identification number (ID) and a label. The links have necessarily one ID, one label, a source (node of departure) and a target (node of arrival). To generate the GEXF files, MONETA generates automatically (i) the list of nodes from the topological file of the protein and (ii) the list of inter-residue links from both the topological file (peptidic links) and the files containing the *communication pathways*. Residues (nodes) are represented as circles, and connections between residues (edges) are represented by more or less thin lines. By default, large edges represent direct connection of two residues adjacent in a path and thin edges represent either a peptide link or a connection between two residues that are in the same path but not adjacent. MONETA computes the communication propensity of each residue as the number of other residues with which it communicates, and reports these data on the 2D and/or 3D

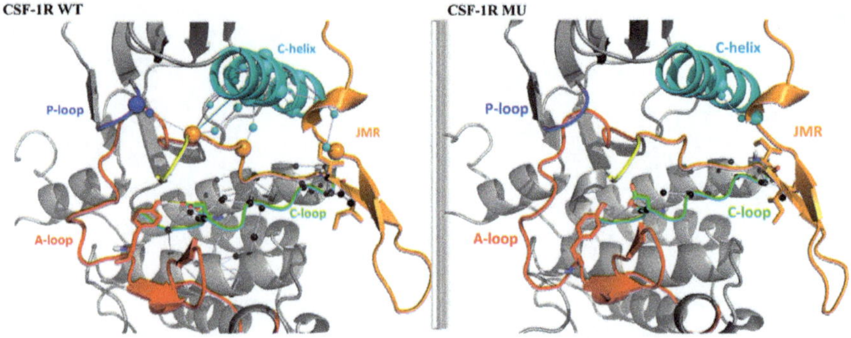

Fig. 4 Characteristic features of the *communication pathways* generated in the CSF-1R cytoplasmic region. 3D structural mapping of the inter-residues communication in the native (*left*) and mutated (*right*) receptor. For each protein, the average MD conformation is represented as an illustration. The key structural fragments of the receptors are highlighted in different colors: the A-loop is in red, the JMR is in orange, the Cα-helix is in cyan, the C- and P-loops are in green and blue, respectively. The *communication pathways* between residue atoms (circles) are depicted by coloured lines: *CPs* formed by the C-helix residues in cyan, and by the JMR residues in grey. 2D and 3D graphs are drawn with GEPHI and PyMOL modules incorporated in MONETA.

representation by scaled colors. These graphical parameters can be tuned at will. The Force Atlas layout provided by GEPHI allows drawing of the global shape of the protein network (Fig. 6).

3. Exploring and predicting allosteric communication in RTKs and in its signalling protein STAT5

Receptor tyrosine kinases (RTKs) are cell-surface transmembrane receptors that possess a tightly regulated tyrosine kinase activity within their cytoplasmic domain.[44] RTKs are crucial components in signal transduction pathways involved in cellular differentiation, growth, survival and metabolism (see ESI,† Scheme S1). The crystal structures of their cytoplasmic region in the activated forms are very similar,[45] however crystal structures of the inactive forms have revealed a remarkable variability in the kinase domain, which allows distinct activation mechanisms.[46] The description of the *communication pathways* mediating the allosteric regulation of the receptors' activity is a challenging task. Furthermore, the deregulation of RTKs' activity, mainly caused by mutations, is associated with various forms of cancers, neuronal disorders and inflammatory diseases. The motivation to understand disease processes and discover possible therapies have driven our study of the type III RTK subfamily that includes KIT, CSF-1R, PDGFR-α, PDGFR-β and FLT3[47,48] and of the protein STAT5 involved in cell signaling pathways, downstream of various oncogenic proteins, particularly, of KIT.[49]

Type III RTKs share a common architecture that includes an extracellular domain to which ligands bind, a transmembrane helix, an autoinhibitory juxtamembrane region (JMR) and a cytoplasmic region composed of the proximal and distal kinase domains (see ESI,† Scheme S1, A). Kinase activation, initiated through binding of its cognate ligand, stabilizes a dimeric form of the receptor, which

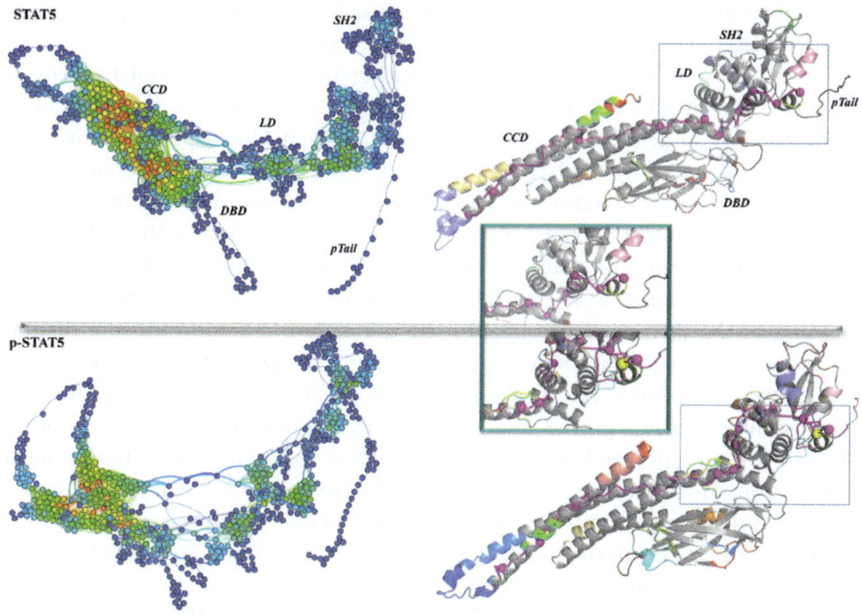

Fig. 5 Interaction network and modular network representation in the unphosphorylated (top) and the phosphorylated (bottom) STAT5. *Left panels*: 2D graphs of a global topology of the inter-residue communications. Residues are represented by points, *communication pathways* are depicted by bold lines and two connected residues by a thin line. Residues are coloured from blue through green and yellow to red according to their communication efficiency, estimated as the number of residues to which they are connected by at least one *CP*. *Right panels*: 3D structural mapping of the inter-residue communication in STAT5. For each protein (unphosphorylated and phosphorylated), the average MD conformation is represented as an illustration. The *independent dynamic fragments* are highlighted in different colors. *Communication pathways* between residues atoms (circles) are depicted in magenta. 2D and 3D graphs are drawn with GEPHI and PyMOL modules incorporated in MONETA. Large-scale view of the *CPs* in STAT5 zoomed on the pathway within the LD and SH2 domains indicating significant differences. Labels for the different regions of the proteins are indicated on the top panel.

Fig. 6 Example of specific representation of inter-residues connection. Residues 9, 84, 25 and 13 represented by circles belong to a unique communication pathway and are linked in green (connected by at least one path or indirect connection) or in blue (adjacent in a path or direct connection). Residues 9 and 10 are linked in orange (peptide bond), but residue 10 does not belong to the path.

induces extensive conformational rearrangements involving the principal regulating fragments, A-loop and JMR[45,50] and facilitates trans-autophosphorylation of tyrosine residues. These processes simulate tyrosine activity and the creation of binding sites for downstream signaling proteins (see ESI,† Scheme S1, B).

Several gain-of-function point mutations induce tyrosine kinase constitutive (ligand-independent) activation and/or resistance to the tyrosine kinase inhibitors. For example, the equivalent mutations, D816V in KIT and D802V in CSF-1R, induce a similar resistance effect, although only D816V confers a proliferative signal in KIT. The KIT mutations positioned in the JMR (V560G) or at the proximity of the Cα-helix (S628N) are highly oncogenic and provoke violent forms of cancers.[51,52] Cancer-related mutations frequently observed as a replacement by different amino acid residues (D816V/H/N/Y/E/I or V560G/D) have a divergent impact on the severity of disease or drug resistance. Another issue concerns the role of the allosteric communication between the JMR and A-loop on RTKs' signaling. We proved that disruption of this communication in the KIT mutant provokes a structural reorganization in the JMR, distant by more than 15 Å from the point mutation.[53] This important structural reorganization evidenced as a folding of the β-sheet of the JMR in the KIT mutant should induce a distinct adaptation of the phosphotyrosine-based sites in the JMR affecting downstream signaling, which might not be the case in the CSF-1R mutant. This hypothesis is consistent with the reported direct KIT–STAT5 interaction in the context of the KIT D816V mutant (see ESI,† Scheme S1, D) that does not occur with the native KIT.[49] Consequently, description of the mutation-induced effects at the atomic level and discrimination between impacts of different mutations will contribute significantly to the understanding of mechanisms leading to deregulation of functional activation and signaling of RTKs.

We used MONETA to describe the allosteric regulation in the native receptors KIT and CSF-1R and in their numerous clinically observed mutants.[29,54,55]

First, the general landscape of the *communication pathways* and the mapping of communication efficiency of residues, illustrated as 2D and 3D graphs (Fig. 3) show evidence of the similarities between the two proteins and their particular features. Second, a detailed inspection of the *communication pathways* allowed us to establish the allosteric communication between (i) the JMR and the A-loop and (ii) the JMR and the Cα-helix, the major regulatory fragments in RTKs. Third, we established that all studied mutations (D816V/H/N/Y, V560G/D, S628N in KIT and D802V in CSF-1R) provoke (a) a diminishing of communication network in the mutants respectively to the native proteins, (b) the disruption of allosteric communication between essential regulatory fragments of the receptors (Fig. 4), and (c) a differential impact of the equivalent mutations on the communication in two homologous RTKs. Particularly, the differences between communication in KIT D816V and in CSF-1R D802V mutants indicate a divergent role of the equivalent mutations, which could explain why D802V mutation is not naturally found in cancer.

In addition, MONETA-guided *in silico* mutagenesis as a way to restore the allosteric communication destroyed by oncogenic mutation.[29]

MONETA was also applied on STAT5 proteins (STAT5a and STAT5b) and allowed identification of the *communication pathways* connecting two spacially distant sites (>100 Å) of the protein in a monomeric state (Fig. 5). Phosphorylation of a specific tyrosine in the SH2 receptor binding domain of STAT5 controls the allosteric regulation across this signaling protein mainly through the long α-helices of a Coiled-Coil Domain (CCD) acting as communication fibers. The different *communication pathways* within the Src-Homology domain 2 (SH2) and the Linker Domain (LD) in the unphosphorylated and the phosphorylated STAT5 proteins revealed the peculiar allosteric regulation in the activated and non-activated protein. Moreover, the *communication pathways* in STAT5a and STAT5b evidenced a different role of phosphorylation of the specific tyrosine residue in these very similar proteins. It has been reported that limited structural dissimilarities between the STAT5a and STAT5b undoubtedly influence gene regulation,[56] and the sequence-dependent asymmetries in STAT5 communication revealed by MONETA can be an underlying factor. Differences in a phosphorylation effect on the STAT5a and STAT5b *communication pathways* are other potential sources of functional disparity and the signal amplitude, frequency or duration which can also be significant.[57]

4. Conclusions

MONETA is a freely available, easy-to-use and fully documented integrated package for analysis of the allosterically regulated proteins. The MONETA modules extract data from all-atom molecular simulations to detect, annotate and interpret local motions of functional relevance. It identifies dynamically coupled residues, *independent dynamic segments*, and builds a modular network representation composed of *IDSs* linked together by chains of residues, *communication pathways.* Such dual representation based on dynamics correlations and topological descriptors depicts signal propagation between allosterically coupled distant protein sites. This approach is an essential step toward the understanding of protein allosteric regulation and function.

The examples of a comprehensive analysis of communication in RTKs (KIT and CSF-1R) furnished here provide evidence that the well-established allosteric communication in the native proteins between the major regulating regions, A-loop, JMR and C-helix, was disrupted by the point mutation localized in the A-loop. A difference of communication in RTKs possessing equivalent mutations permits us to distinguish between oncogenic mutation in KIT and resistance mutation in CSF-1R. The influence of phosphorylation on the allosteric communication route established in STAT5 opens a door for study of KIT–STAT5 signaling. An accurate description of the effector-induced (effector is a point mutation or phosphorylation) effects in terms of communication provides a solid basis for further guiding the discovery of next-generation allosteric drugs, capable of controlling protein functions.

A "restoring" second mutation in the KIT mutant, predicted by MONETA, allowed modulation of the communication between two remote principally regulatory segments, the A-loop and the JMR. The strong correlation between such communication and the structural and dynamical features of the receptors was established, illustrating an accurate predictive ability of MONETA.

Graphical tools, PyMOL and GEPHI, integrated in MONETA offer interactive 2D and 3D representations of communication at different scales – from a global landscape describing the inter-domains communication to inter-residues and inter-atomic pathways – to fully characterize the allosteric effects' propagation in proteins. The next step for the MONETA application will be the description of downstream signaling proteins' activation.

MONETA addresses the need for automated functional analysis of allosteric communication emerging from molecular dynamics simulations. MONETA provides a solid basis for further guiding the discovery of next-generation allosteric drugs, with the aim to modulate or control protein function. The source code of MONETA is available by a simple request. Contacts: Luba.Tchertanov@lbpa.ens-cachan.fr.

Author contributions

Conceived and designed the experiments: LT. Performed the experiments: AA FL EL. Analyzed the data: AA ICB FL YG EL. Wrote paper: LT ICB

Acknowledgements

This work was supported in part by funding from Ecole Normale Supérieure de Cachan, Institut Farman (ENS Cachan), Centre National de la Recherche Scientifique (CNRS), Laboratory of Excellence in Research on Medication and Innovative Therapeutics (LERMIT) and the French State Agency for Innovation (OSEO).

References

1 R. Nussinov, The spatial structure of cell signaling systems, *Phys. Biol.*, 2013, **10**, 045004.

2 C. Lepoivre, A. Bergon, F. Lopez, N. B. Perumal, C. Nguyen, J. Imbert and D. Puthier, TranscriptomeBrowser 3.0: introducing a new compendium of molecular interactions and a new visualization tool for the study of gene regulatory networks, *BMC Bioinformatics*, 2012, **13**, 19.

3 J. R. Perkins, J. Lees, A. Antunes-Martins, I. Diboun, S. B. McMahon, D. L. Bennett and C. Orengo, PainNetworks: a web-based resource for the visualisation of pain-related genes in the context of their network associations, *Pain*, 2013, **154**, 2586.

4 A. Goel, S. S. Li and M. R. Wilkins, Four-dimensional visualisation and analysis of protein-protein interaction networks, *Proteomics*, 2011, **11**, 2672–2682.

5 A. L. Barabasi, N. Gulbahce and J. Loscalzo, Network medicine: a network-based approach to human disease, *Nat. Rev. Genet.*, 2011, **12**, 56–68.

6 N. T. Doncheva, Y. Assenov, F. S. Domingues and M. Albrecht, Topological analysis and interactive visualization of biological networks and protein structures, *Nat. Protoc.*, 2012, **7**, 670–685.

7 S. Tang, J. C. Liao, A. R. Dunn, R. B. Altman, J. A. Spudich and J. P. Schmidt, Predicting allosteric communication in myosin *via* a pathway of conserved residues, *J. Mol. Biol.*, 2007, **373**, 1361–1373.

8 A. R. Atilgan, D. Turgut and C. Atilgan, Screened nonbonded interactions in native proteins manipulate optimal paths for robust residue communication, *Biophys. J.*, 2007, **92**, 3052–3062.

9 A. R. Atilgan, P. Akan and C. Baysal, Small-world communication of residues and significance for protein dynamics, *Biophys. J.*, 2004, **86**, 85–91.

10 P. S. Gandhi, Z. Chen, F. S. Mathews and C. E. Di, Structural identification of the pathway of long-range communication in an allosteric enzyme, *Proc. Natl. Acad. Sci. U. S. A.*, 2008, **105**, 1832–1837.

11 A. Goncearenco, S. Mitternacht, T. Yong, B. Eisenhaber, F. Eisenhaber and I. N. Berezovsky, SPACER: Server for predicting allosteric communication and effects of regulation, *Nucleic Acids Res.*, 2013, **41**, W266–W272.

12 M. Pasi, M. Tiberti, A. Arrigoni and E. Papaleo, xPyder: a PyMOL plugin to analyze coupled residues and their networks in protein structures, *J. Chem. Inf. Model.*, 2012, **52**, 1865–1874.

13 A. Ghosh and S. Vishveshwara, A study of communication pathways in methionyl- tRNA synthetase by molecular dynamics simulations and

structure network analysis, *Proc. Natl. Acad. Sci. U. S. A.*, 2007, **104**, 15711–15716.

14 A. Ghosh, R. Sakaguchi, C. Liu, S. Vishveshwara and Y. M. Hou, Allosteric communication in cysteinyl tRNA synthetase: a network of direct and indirect readout, *J. Biol. Chem.*, 2011, **286**, 37721–37731.

15 A. Dixit and G. M. Verkhivker, Computational modeling of allosteric communication reveals organizing principles of mutation-induced signaling in ABL and EGFR kinases, *PLoS Comput. Biol.*, 2011, **7**, e1002179.

16 W. Zheng and S. Doniach, A comparative study of motor-protein motions by using a simple elastic-network model, *Proc. Natl. Acad. Sci. U. S. A.*, 2003, **100**, 13253–13258.

17 A. Panjkovich and X. Daura, PARS: a web server for the prediction of Protein Allosteric and Regulatory Sites, *Bioinformatics*, 2014.

18 B. Juanico, Y. H. Sanejouand, F. Piazza and R. P. De Los, Discrete breathers in nonlinear network models of proteins, *Phys. Rev. Lett.*, 2007, **99**, 238104.

19 S. Luccioli, A. Imparato, S. Lepri, F. Piazza and A. Torcini, Discrete breathers in a realistic coarse-grained model of proteins, *Phys. Biol.*, 2011, **8**, 046008.

20 F. Piazza and Y. H. Sanejouand, Discrete breathers in protein structures, *Phys. Biol.*, 2008, **5**, 026001.

21 S. W. Lockless and R. Ranganathan, Evolutionarily conserved pathways of energetic connectivity in protein families, *Science*, 1999, **286**, 295–299.

22 G. Manning, D. B. Whyte, R. Martinez, T. Hunter and S. Sudarsanam, The protein kinase complement of the human genome, *Science*, 2002, **298**, 1912–1934.

23 J. B. Casaletto and A. I. McClatchey, Spatial regulation of receptor tyrosine kinases in development and cancer, *Nat. Rev. Cancer*, 2012, **12**, 387–400.

24 J. Monod, J. Wyman and J. P. Changeux, On the nature of allosteric transitions: a plausible model, *J. Mol. Biol.*, 1965, **12**, 88–118.

25 D. E. Koshland, Jr., G. Nemethy and D. Filmer, Comparison of experimental binding data and theoretical models in proteins containing subunits, *Biochemistry*, 1966, **5**, 365–385.

26 M. F. Perutz, Stereochemistry of cooperative effects in haemoglobin, *Nature*, 1970, **228**, 726–739.

27 Q. Cui and M. Karplus, Allostery and cooperativity revisited, *Protein Sci.*, 2008, **17**, 1295–1307.

28 C. Tsai, J.; del, S. A.; Nussinov, R. Protein allostery, signal transmission and dynamics: a classification scheme of allosteric mechanisms, *Mol. BioSyst.*, 2009, **5**, 207–216.

29 E. Laine, C. Auclair and L. Tchertanov, Allosteric communication across the native and mutated KIT receptor tyrosine kinase, *PLoS Comput. Biol.*, 2012, **8**, e1002661.

30 L. Tchertanov, E. Laine, A. Allain and I. Chauvot de Beauchêne, *MONETA – Modular Network Analysis*, 2014, ENS Cachan.

31 R Development Core Team, *R: A Language and Environment for Statistical Computing*, 2013, R. Foundation for Statistical Computing, Vienna, Austria, ISBN 3-900051-07-0.

32 W. L. DeLano, *The PyMOL Molecular Grapics System*, 2002.

33 M. Bastian, S. Heymann and M. Jacomy, *Gephi: an open source software for exploring and manipulating networks*, 2009, International AAAI Conference on Weblogs and Social Media.

34 D. A. Case, T. E. Cheatham, T. Darden, H. Gohlke, R. Luo, K. M. Merz, A. Onufriev, C. Simmerling, B. Wang and R. J. Woods, The Amber biomolecular simulation programs, *J. Comput. Chem.*, 2005, **26**, 1668–1688.
35 S. Pronk, S. Pall, R. Schulz, P. Larsson, P. Bjelkmar, R. Apostolov, M. R. Shirts, J. C. Smith, P. M. Kasson, D. van der Spoel, B. Hess and E. Lindahl, GROMACS 4.5: A high-throughput and highly parallel open source molecular simulation toolkit, *Bioinformatics*, 2013, **29**(7), 845–854.
36 I. K. Mcdonald and J. M. Thornton, Satisfying hydrogen bonding potential in proteins, *J. Mol. Biol.*, 1994, **238**, 777–793.
37 F. Piazza and Y. H. Sanejouand, Long-range energy transfer in proteins, *Phys. Biol.*, 2009, **6**, 046014.
38 P. S. Penev and J. J. Atick, Local feture analysis: a general statistical theory for object representation, *Network: Comput. Neural Syst.*, 1996, **7**, 477–500.
39 Z. Y. Zhang and W. Wriggers, Local feature analysis: A statistical theory for reproducible essential dynamics of large macromolecules, *Proteins: Struct., Funct., Bioinf.*, 2006, **64**, 391–403.
40 C. Chennubhotla, Z. Yang and I. Bahar, Coupling between global dynamics and signal transduction pathways: a mechanism of allostery for chaperonin GroEL, *Mol. BioSyst.*, 2008, **4**, 287–292.
41 G. Morra, G. Verkhivker and G. Colombo, Modeling signal propagation mechanisms and ligand-based conformational dynamics of the Hsp90 molecular chaperone full-length dimer, *PLoS Comput. Biol.*, 2009, **5**, e1000323.
42 C. Chennubhotla and I. Bahar, Signal propagation in proteins and relation to equilibrium fluctuations, *PLoS Comput. Biol.*, 2007, **3**, 1716–1726.
43 M. J. Abraham and J. E. Gready, Optimization of parameters for molecular dynamics simulation using smooth particle-mesh Ewald in GROMACS 4.5, *J. Comput. Chem.*, 2011, **32**, 2031–2040.
44 P. Blume-Jensen and T. Hunter, Oncogenic kinase signalling, *Nature*, 2001, **411**, 355–365.
45 M. Huse and J. Kuriyan, The conformational plasticity of protein kinases, *Cell*, 2002, **109**, 275–282.
46 M. A. Lemmon and J. Schlessinger, Cell Signaling by Receptor Tyrosine Kinases, *Cell*, 2010, **141**, 1117–1134.
47 D. R. Robinson, Y. M. Wu and S. F. Lin, The protein tyrosine kinase family of the human genome, *Oncogene*, 2000, **19**, 5548–5557.
48 A. Ullrich and J. Schlessinger, Signal transduction by receptors with tyrosine kinase activity, *Cell*, 1990, **61**, 203–212.
49 A. Chaix, S. Lopez, E. Voisset, L. Gros, P. Dubreuil and S. P. De, Mechanisms of STAT protein activation by oncogenic KIT mutants in neoplastic mast cells, *J. Biol. Chem.*, 2011, **286**, 5956–5966.
50 B. Nolen, S. Taylor and G. Ghosh, Regulation of protein kinases; controlling activity through activation segment conformation, *Mol. Cell*, 2004, **15**, 661–675.
51 M. J. Frost, P. T. Ferrao, T. P. Hughes and L. K. Ashman, Juxtamembrane mutant V560GKit is more sensitive to Imatinib (STI571) compared with wild-type c-kit whereas the kinase domain mutant D816VKit is resistant, *Mol. Cancer Ther.*, 2002, **1**, 1115–1124.
52 M. Vita, J. Tisserand, I. Chauvot de Beauchêne, N. Panel, L. Tchertanov, J. Agopian, L. Mascam-Mancini, B. Fouet, B. Fournier, P. Dubreuil,

F. Bertucci and P. De Sepulveda, Characterization of S628N, a novel *KIT* mutation found in a metastatic melanoma, *JAMA Derm.*, 2014, accepted.

53 E. Laine, d. B. I. Chauvot, D. Perahia, C. Auclair and L. Tchertanov, Mutation D816V Alters the Internal Structure and Dynamics of c-KIT Receptor Cytoplasmic Region: Implications for Dimerization and Activation Mechanisms, *PLoS Comput. Biol.*, 2011, **7**, e1002068.

54 I. Chauvot de Beauchêne, A. Allain, E. Laine, P. Dubreuil and L. Tchertanov, Hotspot mutations in KIT receptor differentially modulate its allosteric regulation: impact on activation and drug sensitivity, *PLoS Comput. Biol.*, 2014, resubmitted after revision.

55 P. Da Silva Figueiredo Celestino Gomes, N. Panel, E. Laine, P. G. Pascutti, E. Solary and L. Tchertanov, Differential Effects of CSF-1R D802V and KIT D816V Homologous Mutations on Receptor Tertiary Structure and Allosteric Communication, *PLoS One*, 2014, **9**(5), e97519.

56 P. M. Grimley, F. Dong and H. Rui, Stat5a and Stat5b: fraternal twins of signal transduction and transcriptional activation, *Cytokine Growth Factor Rev.*, 1999, **10**, 131–157.

57 K. Paukku and O. Silvennoinen, STATs as critical mediators of signal transduction and transcription: lessons learned from STAT5, *Cytokine Growth Factor Rev.*, 2004, **15**, 435–455.

Faraday Discussions

PAPER

Haptic-driven, interactive drug design: implementing a GPU-based approach to evaluate the induced fit effect

Athanasios Anthopoulos,[ab] Gaia Pasqualetto,[a] Ian Grimstead[b] and Andrea Brancale*[a]

Received 14th December 2013, Accepted 13th March 2014
DOI: 10.1039/c3fd00139c

In this paper, we describe a hybrid meta-heuristics method of energy minimization and conformational sampling and its application into our haptic-driven molecular modelling simulator. The proposed method has been designed to suit real-time molecular docking simulations, where the time-lapse between two successive ligand poses is relatively small. In these situations, the energy minimization problem becomes increasingly complex and chaotic. The algorithm is tuned to take advantage of recent advances in GPU computing with asynchronous kernel execution, which has allowed us to include full protein flexibility in the real-time interactive, haptic-driven simulations. Finally, in this paper, we will also discuss the implementation of such high-performance computing approaches in our software, discussing the results of our initial validation studies, highlighting the advantages and limitations of such interactive methodology.

1. Introduction

The assessment of protein flexibility upon ligand binding remains one of the most challenging aspects of computer-aided drug design. Indeed, in the last decade, several methodologies that attempt to mimic the induced fit effect have been developed and implemented in several software packages, from the simple exploration of rotamers of predefined amino acid side chains to the prediction of backbone flexibility using short molecular dynamics runs.[1] In all cases, molecular conformations are updated using the appropriate algorithm, depending on the degree of flexibility that the application offers.[2] The objective of finding stable (or preferred) conformations for the system requires locating those conformations that lie on minimum points of the energy surface.[1] However, this is a complex multi-dimensional problem. For a system with N atoms, its potential energy is a function of $3N$ Cartesian co-ordinates. Therefore, the molecular energy surface is

[a]*School of Pharmacy and Pharmaceutical Sciences, Cardiff University, Cardiff, CF10 3NB, UK. E-mail: brancalea@cardiff.ac.uk*
[b]*School of Computer Science & Informatics, Cardiff University, Cardiff, UK, CF24 3AA*

rather chaotic and searching for the global minimum cannot always be feasible. Failing that, exploring the potential energy surface is the way to find a set of optimal solutions (local minimums) where the best solution can be chosen. The search for optimal solutions can be facilitated with the use of meta-heuristics.[3]

Meta-heuristic methods cover a wide spectrum of algorithmic techniques. Evolutionary algorithms (EA) and genetic algorithms (GA) are methods able to iteratively generate a set of solutions. Evolutionary algorithms are amongst the most popular meta-heuristics and attempt to solve the optimization problem by generating a population of possible solutions from a parent (mutation). From the generated population, the fittest is chosen to generate the new pool of solutions and so forth. Genetic algorithms adopt a similar approach, with the addition of the concept of crossover between two solutions to create the new generation.[4] These solution generations will be gradually improving with respect to the strategy adopted to move across the energy landscape.

The exploration of potential energy surfaces requires the adoption of uphill moves in order to recover from local minima. Such strategies for uphill move selection include simulated annealing (SA), Monte Carlo (MC), iterated local search (ILS), guided local search (GLS) and many more. Simulated annealing uses a temperature variable to generate a probability function for accepting solutions that go uphill on the energy surface.[5] Monte Carlo methods adopt a similar approach to escape local minima by accepting uphill moves based on a probability function such as the Metropolis criterion.[6] ILS reaches optimal solutions by iteratively performing a local search to reach a local minimum;[7] the way it escapes a local minimum is by perturbing the discovered solution, in hope that the perturbation mechanism will bring the system into a well with a lower minimum than the parent one. GLS escapes from local minima by gradually increasing the global minimum energy value. This way, the system will escape from a well if it is trapped, by taking an uphill move close to the minimum, which will result in a new landscape with the hope that a better minimum solution will be discovered.

Various meta-heuristics have been used by many recent molecular docking applications. AutoDock Vina used a hybrid ILS strategy using the Metropolis criterion to accept optimal solutions and predict the binding affinity of protein–ligand systems.[8] Variations of the genetic algorithm have been implemented in various published works to predict molecular structures.[9,10] Brasil used an evolutionary strategy to solve this problem[11] and Cai and Shao used a hybrid method of EA combined with SA for structure prediction.[12] ParadisEO approached the docking problem with a parallel GA in order to leverage additional processing power and improve its performance and efficiency.[13]

In our group, we are developing an interactive molecular docking program, which allows the user to place, in real time, a small molecule in a biological target, while feeling the molecular interactions it establishes through a haptic device.[14,15] In our original system, the target structure was kept rigid, while the ligand was kept flexible and the system underwent continuous energy minimisation *via* a steepest descent algorithm. Recently, we have explored the possibility of including a degree of protein flexibility in the interactive simulation, by taking advantage of the computational power of GPGPUs (General-Purpose Graphics Processing Unit).[16] The main challenge in our approach is that if we want to achieve a level of interactivity that provides the user with a natural experience, our algorithm should be able to obtain a stable, low energy, protein–ligand conformation within a restricted timeframe. In our software, the ligand receives a signal to move at fixed time intervals $t = 33$ ms, which matches the rendering rate on screen. When the signal is received, the ligand atoms are translated into their new position. On the completion of ligand translation, the aim is to provide the system (ligand and protein) with the best atomic structure when the time-lapse t expires and before the new haptic pointer signals to translate the ligand to its next position (minimization cycle).

In this paper, we present a hybrid meta-heuristic method of exploring molecular energy surfaces with a focus of solving this problem as a set of independent protein ligand poses generated in a restricted time frame.

2. Results and discussion

2.1 CUDA streams

The recent generation of Kepler cards (GK110) has introduced new features in GPU computing, including the ability to run up to 32 kernels asynchronously achieving execution overlaps very close to that of a parallel execution. This introduces an additional dimension of parallelism in CUDA and initiates the need for new optimization techniques in CUDA-parallel programming. To be more specific, considering this new feature, an algorithm designer should not treat each kernel individually any more, but as part of a wider context able to utilize the graphics card more efficiently. However, there are some rules and constraints on the way that GPUs parallelize kernel execution: at the present time, NVIDIA GPUs are able to run 16 scheduled blocks per streaming multiprocessor (SM). In the case of a K20 card, that gives a total of 208 blocks able to run concurrently at any time in its 13 SMs. From a resources perspective, the shared memory is limited at 48 kb, so asynchronously scheduled kernels in each stream must not exceed this limit. If it does, then the parallel execution is restricted to the number of streams whose total shared memory usage is below the 48 kb limit. Regarding register usage, similar restrictions apply and if the total number of registers used per thread exceeds the card's limit (255 for GK110s) then some register overspill might take place and the slower performing local memory might be used. Finally, overlapping streams can only write on independent memory addresses; this is not a major problem as arrays can be copied in multiple buffers for each stream to operate on independently.

2.2 Potential energy and force calculations

The energy (eqn (1)) and force gradients (eqn (2)) computations to facilitate the minima search and conformation updates for our proposed method were performed using the MMFF94s force field and implemented in CUDA. E_i is the energy of atom i.

$$E = \sum E_{\text{bonded}} + \sum E_{\text{non-bonded}} \tag{1}$$

$$\vec{F_i} = -\nabla E_i \tag{2}$$

Regarding the non-bonded interactions, atoms were geometrically mapped (binned) into a 3D load-balanced irregular grid.[16,17] Calculations then took place with each CUDA block responsible for processing the atomic interactions between each home cell and its neighbours.[16]

Once the force vector is available for each atom i, a scalar value (step-value) needs to be found in order to update the system into the new conformation. The conformational update is performed for each atom in the system in vector space as follows:

$$\overrightarrow{C_i'} = \overrightarrow{C_i} + \lambda \overrightarrow{F_i} \tag{3}$$

where $\overrightarrow{C_i'}$ is the update position of an atom i and $\overrightarrow{C_i}$ is its position before the update.

2.3 The greedy algorithm implementation

Given that each minimization cycle must be performed within a 33 ms time frame, the initial plan was to find a local minimum and terminate searching in the hope that the minimum would correspond to a stable molecular conformation. For this reason, we used a simple steepest descent (SD) step minimization approach. Its characteristic is that it takes only downhill moves (the greedy approach), which results in converging into the first local minimum it finds. The conformation sampling that we obtained from this method was slow and sometimes erroneous; it only helped us understand some additional aspects of the problem we were trying to solve.

An example conformation produced by the algorithm is presented in Fig. 1 and it is obvious that the atom geometry is wrong. The atoms in the histidine aromatic ring should be co-planar, as the urea group on the ligand. The test system was relatively small (<2500) atoms and there were no performance restrictions. The SD algorithm was converging before t expired and the system would quickly optimize its configuration, within a few minimization cycles, to reach often an erroneous geometry as demonstrated in Fig. 1. Further improvements on the system's configuration would be either really slow or not evident and the conformation in the pocket area would always be non-ideal. As a first thought, one could suspect

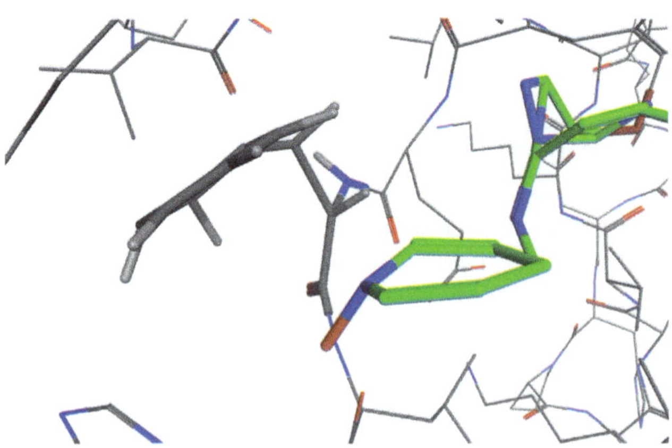

Fig. 1 A snapshot of a ligand approaching a protein using a greedy local search approach. The algorithm is continuously improving after the initial impact, but settling at erroneous configurations (evident geometric distortion on the protein and the ligand).

that this would be an accuracy problem on the force field calculations. However, probing the ligand into a different area, inducing further conformational updates, would quickly restore the previously distorted area and cause the same problems in the area of latest interaction.

From the above, it is obvious that for the type of protein–ligand simulations performed by our application, finding the first potential energy minimum on the molecular energy landscape would not guarantee a reasonable conformation around the area of interaction between the two molecules. That might be because the rest of the protein could be well minimized and no scalar value for eqn (3) would exist, able to move all atoms in vector space such that the system would reach a lower energy value. What is also worth noting is that keeping the ligand stalled in its position in order to allow the system to slowly recover would not always work either. That gave the impression that every new minimization cycle would pick up from almost where it left off and a new minimum could not be found. As a result, further conformational updates would either not be performed or they would be too few with a very small lambda value (*i.e.* $\lambda = 10^{-6}$) which would have a quite insignificant impact on the system's configuration.

In our attempts to design an efficient energy surface algorithm, we also considered Monte Carlo and simulated annealing methods. However, given the strict time constraints and the finite amount of steps as a result, perturbations relying on a probability would expose our solution to the risk of wasting moves inside a single valley. This means that during a 33 ms cycle these techniques would under-sample the potential energy surface, compared to faster sampling techniques like iterative local search. Both methods would be good candidates for the problem of finding the best conformation for a protein ligand pose on a non-interactive application without strict time constraints, but probably this is not the case in our scenario.

The first attempt to solve this problem was to perform a "shuffle" technique. This means that at the beginning of each minimization cycle and before searching for a local minimum, all atoms in the system would move along their force vector trajectories using a certain value. Experimenting and fine-tuning with different λ values showed that a good value could be found in the 10^{-3} magnitude. In order to avoid big atom moves, a constraint was placed on the maximum value of displacement. This technique improved the situation, but raised concerns for the overall quality of the system's conformation as a result of the frequent shuffling.

In a second attempt, we decided to introduce an uphill move approach, by accepting a random small uphill step as a transient state to recover the system from a local minimum. In this way, we could explore the potential energy landscapes for other valleys in the hope that a lower energy value can be found. This technique improved the local restoration defect evident in previous experiments. The potential energy surface (PES) exploration would not always result in a lower energy configuration, but the configurations visualized within the interaction area between the two molecules improved significantly. This is due to the fact that exploring the energy landscape in this way triggered a higher number of conformation updates. This then caused the atoms within the interaction area to move along their force vectors a number of times and eventually form a stable conformation, with correct bond lengths and angles on both molecules.

These experiments led us to reformulate our objectives to seek for a method able to search the potential energy landscape for a good minimum and also restore the area of interaction between the two molecules in a single minimization cycle. The new algorithm should be able to evaluate candidate solutions in efficient manner. It should also ensure that within each minimization cycle there would be enough conformational updates to restore the area of interaction.

The resulted implementation uses an evolutionary algorithm to generate a set of candidate solutions to the problem (system conformations). The candidates are evaluated by calculating their potential energy asynchronously using CUDA streams, which allows for a degree of performance gain. The potential energy value of each candidate is used as an indicator of fitness. The fittest candidate is the one whose potential energy value is lower than the latest record of potential energy minimum (old minimum). If no child solution has a value lower than the old minimum, then the fittest child is chosen with criteria set by our uphill move strategy. The fittest child becomes the parent of the new solution's generation and the system's configuration is updated adopting its co-ordinates. Finally after the conformational update, the force vectors for the newly updated system are calculated.

2.4 Fast surface exploration and induced local restoration

Initially our algorithm would find its first valley and aim downhill until it reaches a minimum. On reaching a minimum, the parent solution would be unable to generate a population with a candidate solution whose energy value is less than the old minimum. At this point, a strategy for escaping the local minima is needed, which would be to immediately try a different valley in hope of finding a better minimum and induce stable conformational updates at the same time. The perturbation strategy of Iterated Local Search (ILS) is ideal for quick jumping from one valley to another by accepting an uphill move when the system lies at a local minimum. This is achieved by choosing a scalar λ value that is able to perform a shuffle operation in a similar way as explained earlier.

ILS initially gave promising results, however, there were two main problems with it. The first is that it would occasionally show a visual pulsating effect, where atoms would seem to perform a fast oscillation. That was because the time-lapse of the algorithm would expire immediately after a perturbation and the system configuration on exiting the algorithm would not be optimal. This problem could be easily solved with a ghost particles buffer holding atom co-ordinates after a perturbation move. The post-perturbation generations would inherit and update co-ordinates on the ghost buffer up until a new minimum is reached that is lower than the one associated with the real co-ordinates array. This solved the pulsating visual effect, but initiated the second problem that we described, where ILS would often find a good minimum within the first few moves, and the background search following the last visited minimum would not be able to find a better solution. This would lead to the aforementioned problems of too few conformational updates and hence inefficient minimization cycles.

Guided local search is ideal for inducing updates on the principle that the old minimum is being raised by a small percentage at every solution evaluation (penalty function). Eventually, the algorithm can escape from the well, but will spend some time in it waiting for its penalty function to reach the appropriate levels. Experiments showed that this method underperformed compared to ILS for our simulations, as shown in the results section.

Our final solution came as an amalgamation of the two approaches. We introduced a penalty value p and an energy function E, which is updated at every step i: $E(i + 1) = E(i)(1 + p)$. The penalty value is usually in the magnitude range of 0.001–0.01, which means that on every iteration the old minimum bar is slightly raised so that it can allow for an update of the system's real co-ordinates. This modification aims to keep the fast track of valley exploration that ILS can offer and introduce enhanced conformational update abilities to our approach. We will refer to this algorithm as IGLS (Iterated Guided Local Search).

In essence, IGLS achieves the following: it accepts updates for energy values close to a recently visited lowest energy configuration, for subsequent solution populations. Failing this, when subsequent solution generations do not produce an update, it prevents the algorithm from wasting too many generations without performing an update, which in turn benefits the visual effect of our application as well as the interacting molecules' configuration, especially within their interaction zone.

The implementation of the above technique is performed by using two energy variables: current_lowest and the global_lowest. Every time a new minimum is reached, both the current_lowest and the global_lowest take its value. When a local minimum is reached and the new generation cannot provide a candidate with a lower energy value, perturbation takes place for a new well to be explored and the current_lowest takes the corresponding energy value. While in the new valley, the global_lowest value is raised at every energy evaluation by a small factor (such as 0.5%), while the current_lowest keeps reaching lower values, until we reach the new local minimum. Now at this point there are two possible outcomes: the first is that the new local minimum (current_lowest), is lower than the global_lowest and a conformation update is being performed. Otherwise, a new perturbation move is being performed. However, the probability of finding a new minimum this way is directly proportional to the number of energy evaluations performed. This way, a greater ratio of contributing moves compared to the total number of moves is achieved, which in turn assists in solving the local restoration problem.

This algorithm allows for quick exploration of molecular energy surfaces, by encouraging more conformational updates than a purely ILS strategy would perform. The visual result is significantly improved, as there is rapid local restoration on the impact area.

2.5 Asynchronous execution

The performance of the above algorithm is important for our application as the potential energy landscape exploration is constrained to run for a small timeframe t. The more moves we are able to perform in the landscape, the better the quality the resulting solution will be. In addition, the more parent solutions the algorithm generates, the higher the probability of discovering lower energy minimums. From the above, we understand that performance is vital for our simulations.

Initially, our force and energy kernels as well as a binning method to an irregular grid approach were designed to minimize the execution time on GK104

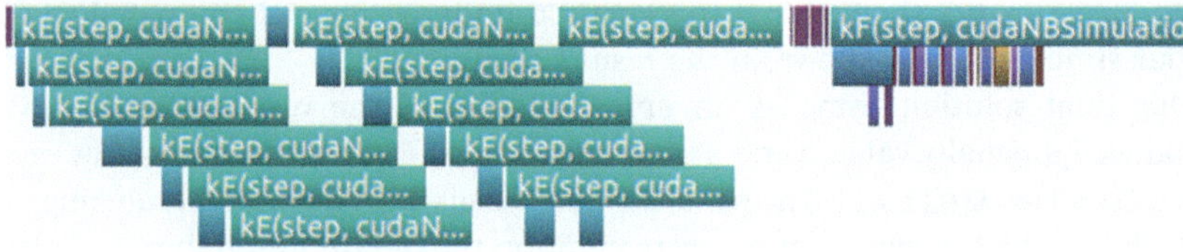

Fig. 2 Visual profiling data for a move. Asynchronous execution of *s* streams where the total block and shared memory conditions are not met. Overlapping is divided into stream groups that meet the hardware's block and shared memory restrictions. The kE blocks represent the non-bonded energy kernels, and the corresponding ones under the kF block are the bonded forces and the 3D grid binning kernels.

chipsets. After the introduction of GK110, concurrent streaming capabilities were enhanced, giving the option of running up to 32 kernels asynchronously, when the criteria listed in the background section were met.

For this reason, we designed an algorithm to take advantage of asynchronous kernel execution using CUDA-streams (Hyper-Q according to NVIDIA terminology) and improve the performance of the existing CUDA functions. In order to achieve peak performance, all the aforementioned kernels (energy, force, binning) were refactored in order to meet the criteria for concurrent execution and without sacrificing their individual performance. On the current generation of graphics cards (GK110), we can now evaluate the fit of a whole generation of solutions asynchronously and almost in parallel for smaller protein–ligand systems. This means that the first energy evaluation kernel hides the latency of the remaining s-1 evaluation kernels scheduled to run asynchronously in s streams within a single move (a population evaluation series, followed by a force calculation for the fittest child), where $s \leq 16$. However, if the conditions mentioned in the background section are not met, then the number of asynchronous streams running concurrently at any time drops as demonstrated in Fig. 2.

In addition, we can hide the latency of binning the atoms into a 3D irregular grid for the cell-list generation in the force gradients kernel. This means that binning overheads are no longer an issue and binning can be safely performed at every single move, with a small skin value δ' that caters for the biggest displacement that we anticipate during each move (cut-offlist = cut-off + δ').

2.6 Performance of the algorithm

The performance of our CUDA-streams implemented hybrid approach has been benchmarked against a synchronous execution version in CUDA of the above algorithm. In order to perform a fair comparison, benchmarks were scheduled such that both versions would perform the exact same amount of energy and force gradient kernel calls. The results are listed in Fig. 3 for 4–16 concurrent streams. From the figures, we can observe that there are performance gains fluctuating from 0.2×–4× depending solely on the system size and number of streams. Regarding shared memory usage, both our non-bonded energy and force gradient kernels are using 9 kB whereas the bonded energy and forces' kernels do not use shared memory. That means that our concurrent execution on energy evaluations at any time is restricted to five kernels as a sixth kernel would overtake the 48 kB shared memory per SM limit.

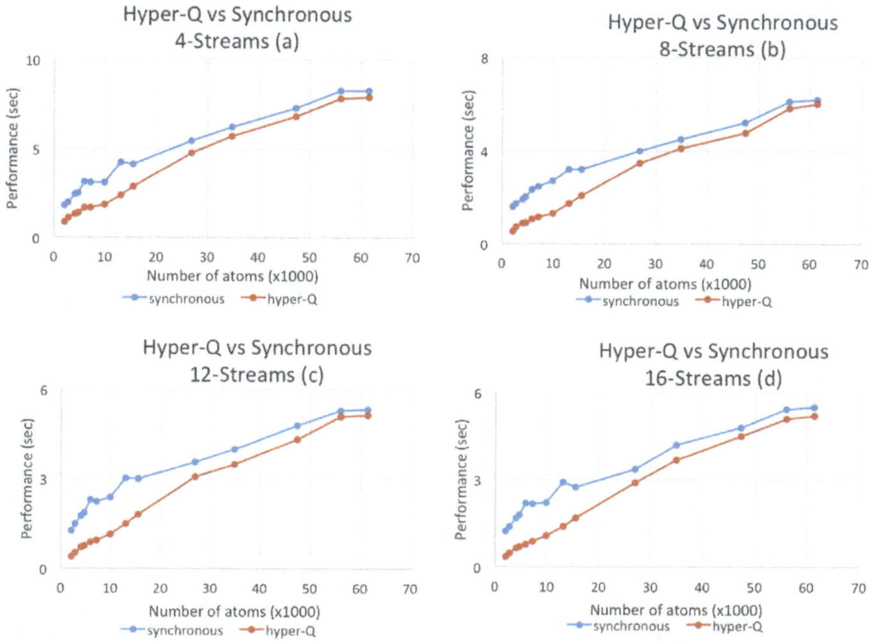

Fig. 3 (a–d). Performance benchmarks between CUDA-asynchronous and CUDA-synchronous stream versions. The same amount of energy kernels (1000) and force gradient kernels (1000/nstreams) are evaluated from both approaches in each graph.

Our algorithm achieves five concurrent energy evaluations for system sizes up to 10 000 atoms. This is because the energy kernels divide execution into blocks of 256 threads, where each thread is associated with one atom. This equates into a maximum of 40 blocks per stream, which is good enough to saturate the limit of five concurrent streams given from shared memory restrictions ($5 \times 40 < 208$ concurrent blocks limit). Beyond that limit, the number of blocks needed to process the energy kernels rises above 208 and that means that CUDA cannot handle the total number of blocks scheduled for asynchronous executions, hence concurrency efficiency starts deteriorating.

Finally, for systems above 50 000 atoms, concurrency almost vanishes on the energy kernels as the number of blocks needed to process one system alone is close to the architecture's limit. However, concurrency in the forces-binning stream pair still holds as our binning method has been designed using a series of fast executing kernels able to run using only a few blocks. In addition, bonded forces' kernels require fewer thread blocks and hence can contribute towards some asynchronous execution too. This is why even in the case of large systems (50 000–65 000 atoms) there is still a small performance gain.

Tables 1a–b demonstrate a comparison of the elapsed time of execution for the calculations demonstrated in Fig. 3 between CUDA-synchronous and CUDA-asynchronous executions. In addition, it appends execution times for the serial implementation on a workstation equipped with a Tesla K-20 GPGPU card and Intel Xeon E5-4620 CPUs running at 2.20 GHz. The serial execution uses the OpenBabel library for the MMFF94s calculations.

Table 1 (a–b). Results for CUDA-synchronous, CUDA-asynchronous and serial executions

4-Streams (a)

Proteins	K (×1000) atoms	Synchronous (sec)	Hyper-Q (sec)	Serial (sec)
1UGM	2.103	1.8	0.87	127.5
3RDD	2.786	1.976	1.115	235
3F9E	4.74	2.508	1.378	647.5
3FAU	5.917	3.126	1.677	1017.5
2O0O	7.165	3.074	1.662	1447.5
3O05	13.053	4.216	2.368	7430
2EAR	15.528	4.14	2.826	7450
1TBG	26.927	5.46	4.74	32277.5

12-Streams (b)

Protein	K (×1000) atoms	Synchronous (sec)	Hyper-Q (sec)	Serial (sec)
1UGM	2.103	1.272	0.393	89.16667
3RDD	2.786	1.502	0.511	165
3F9E	4.74	1.862	0.74	455.8333
3FAU	5.917	2.307	0.868	719.1667
2O0O	7.165	2.248	0.941	1015.833
3O05	13.053	3.022	1.486	5870
2EAR	15.528	2.999	1.8	5423.333
1TBG	26.927	3.559	3.062	25385.83

2.7 Heuristic ability of the algorithm

Fig. 4a–d provides a schematic overview of the four algorithms evaluated in this report. The graphs depict the steps taken by each algorithm in order to complete a 33 ms computational cycle. In order to demonstrate how IGLS solves our problem, the starting configuration was chosen after a run of 1000 steps of the greedy algorithm. This way, the starting conformation lies at or very near a local minimum and Fig. 4a–c demonstrate how each of the three proposed techniques escape it and explore the landscape. For these calculations, the parameters used are now described.

Eight equally spaced values are chosen in the range 0.1–0.000001 and allocated to each computational stream; perturbation for both GLS and ILGS is happening, choosing the child solution with index four. There has been some experimentation with different parameters for the perturbation step, but the results are similar. The essence of perturbation is to escape the current valley and explore the rest of the landscape and we have no knowledge where each child solution will bring us. The penalty value was set to 0.05 for both GLS and IGLS. In IGLS, it is applied every time a perturbation takes place and, in GLS, every time a new global minimum is not reached. A good example of the benefits of our approach is demonstrated in Fig. 4a and b where ILS and IGLS are compared head-to-head. In steps three and four, both approaches are perturbed and in step five the proposed

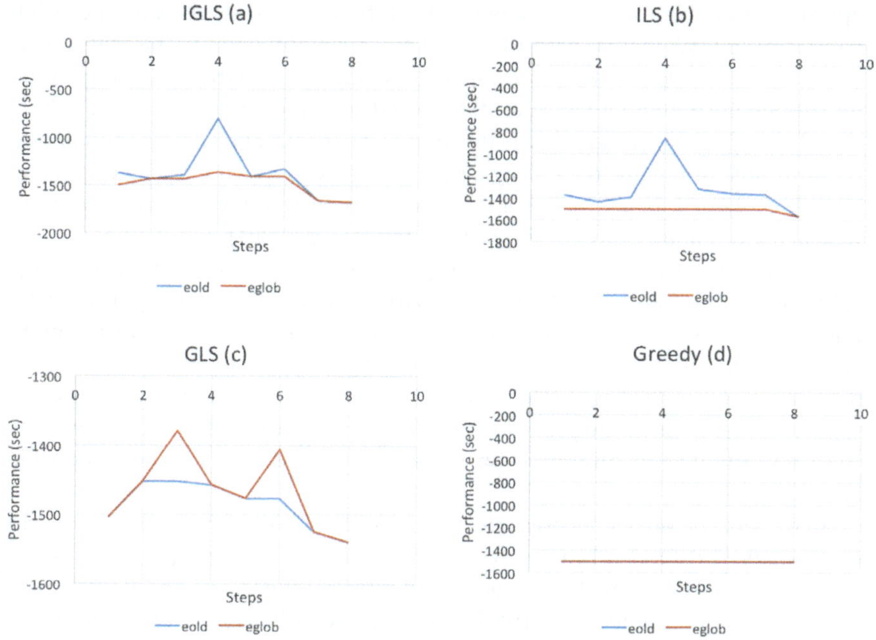

Fig. 4 a–d. A schematic overview figure of the four algorithms compared in this report, depicting how a calculation circle of $t = 33$ ms is subdivided into 8 steps of a group of energy kernel evaluation streams for the 1UGM.pdb protein file. The system is already minimized with a local search method prior to running the experiment.

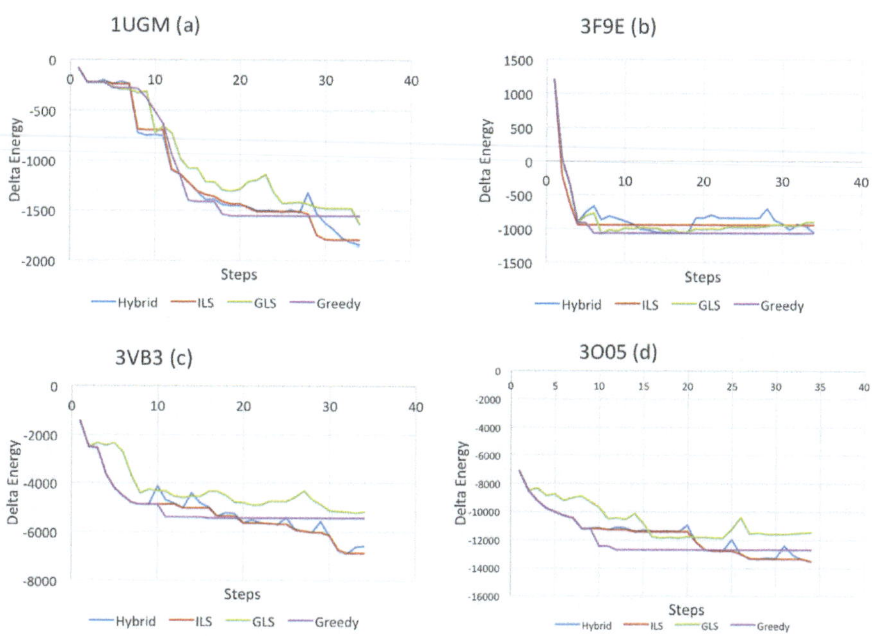

Fig. 5 (a–d). Comparison of the 4 different approaches with the ligand placed into a pocket.

approach was able to accept a very good solution by slightly relaxing the eGlob constraint (global_lowest) with the aid of the penalty function. The reason why GLS was discarded is because it can spend too much time in the same well waiting for the penalty function to rise over the latest discovered energy levels. We believe that the proposed method inherits the merits of both ILS and GLS for the solution of our problem. Finally, the greedy algorithm had converged before step 1 as we already explained, hence we see a straight line depicting the best child solution repeatedly found on each circle (probably for $\lambda = 0.000001$).

Another set of experiments was designed to show the fluctuation of energy values using the four algorithms in comparison. In order to perform this evaluation, we ran two sets of experiments. The first set listed in Fig. 5a–d demonstrates the heuristic ability of our approach for ligands positioned close enough to a protein in order to trigger interactions. The second set of experiments listed in Fig. 6a–d is similar to the previous set, with the only difference that it performs the same number of minimization steps for proteins in their initial pdb file structure.

In the simulation results presented in Fig. 5a and 5c, we can see that the hybrid method and the ILS clearly outperform the GLS and the greedy approach. It is also interesting to see that in all four figures the curves of the hybrid and ILS methods having a similar trend, with the hybrid method following a more turbulent trajectory due to its penalty function. Fig. 5b shows an interesting case as we can see that ILS is trapped at a minimum, which it cannot escape from. From Table 2, we can see that for the experiment with 3F9E.pdb structure, ILS could only reach four minima and hence perform four updates. This explains why this method cannot perform well on our simulations, as this is a common scenario. Four updates are usually not enough to bring the system back to a stable conformation.

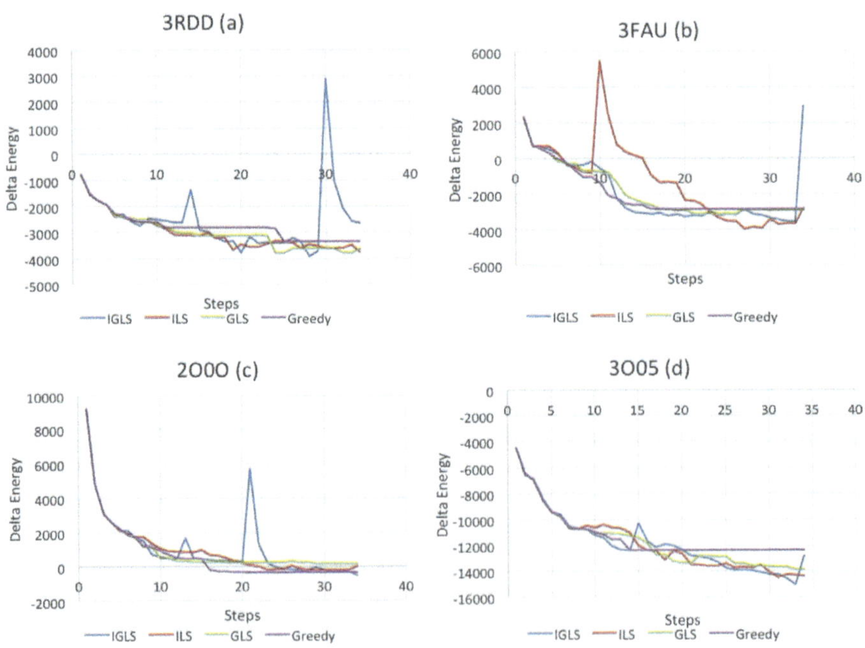

Fig. 6 (a–d). Comparison of the 4 different approaches with four un-minimized proteins.

Table 2 (a) Number of conformational updates for each protein for the four different algorithmic approaches in Fig. 5a–d. (b) Number of conformational updates for each protein for the four different algorithmic approaches in Fig. 6a–d

(a)

	Hybrid	ILS	GLS	Greedy
1UGM	27	24	20	15
3VB3	26	20	22	11
3F9E	18	4	17	5
3OOP	26	20	18	10

(b)

	Hybrid	ILS	GLS	Greedy
3RDD	23	20	25	13
3FAU	25	20	25	15
2O0O	29	22	19	18
3O0P	27	20	25	12

Our method on the other hand, using the penalty function, can escape such a minimum and carries on exploring different energy landscapes, finding new minimums and updating the system into better conformations. Regarding GLS on its own, we can see it always underperforms compared to the hybrid method and it only performs better than the ILS in the 3F9E experiment, where ILS gets stuck in a well.

The four graphs in Fig. 6 present similar behavior. Our hybrid method, together with ILS are the best performing ones as they always have the ability to discover deeper valleys in comparison to the other two methods.

From these results, it is evident that the hybrid approach is the most suitable for the type of simulations that our software performs for its ability to find a very good potential energy minimum by inducing more conformational updates than it would following a pure ILS strategy. Also, the GLS element in it does not really mean that the resulting global minimum would be lower than the corresponding ILS, nor is the opposite true. The hybrid method seems promising in solving other optimization problems too due to its ability to change over different energy landscapes and evaluate several different minima at a small temporary potential energy cost.

Overall, the improved hybrid algorithm was fast and responsive in our simulations. There were no evident distortions in any of the molecules' structure and interactions between ligand and protein were modelled accurately in real time. This algorithm is capable of simulating systems of up to 30 000 atoms, with the interactivity being inversely proportional to the system size. For larger systems of 16 000–30 000 atoms, the user has to stall the ligand inside a pocket for the minimization algorithm to take effect. The overall conformation quality can improve with the addition of a long range electrostatics routine. Finally, as the algorithm relies almost solely on floating point operations and future GPGPU generations promise even better FLOP capabilities, we could expect our approach to perform even better on forthcoming cards.

Table 3 Crystal structures used in the haptic-driven simulations

Protein	PDB	Ligand Name
CDK2	1HCL[18]	—
	1GZ8[19]	MBP
	1JVP[20]	LIG
	2R3F[21]	SC8
	2R3H[21]	SCE
	2R3I[21]	SCF
	2R3R[21]	6SC
	4EK4[22]	1CK
	4FKL[22]	CK2
HCV NS5B	1NB4[23]	—
	3UPH[24]	0C1
	3U4O[25]	08E
	3GNW[26]	XNC
	4J02[27]	1JE
	4J06[27]	1JG
	4JJU[28]	1MB
CDK2	1HCL[18]	—
	1GZ8[19]	MBP
	1JVP[20]	LIG
	2R3F[21]	SC8
HIV RT	1RTJ[29]	—
	1VRT[30]	NVP Fragment

2.8 Algorithm practical applications

To further validate and assess our system, we have also carried out a set of simulations that aimed to mimic the actual use of the software. To evaluate the induced fit effect, we have selected a series of known biological targets for which a series of crystal structures are available (Table 3) as the apo-form and co-crystallised with one or more ligands.

Our approach was to extract a ligand from a specific complex, then place it in the apo-form of the corresponding protein using the haptic-driven simulator. The results obtained from the docking were then compared with the structure of the complex from which the ligand was extracted, measuring both the root mean square deviation (RMSD) of the ligand and the protein residues of the binding pocket (superposition for the different proteins was performed on the whole structure using the backbone CA as reference).[31] Furthermore, the pocket residues were also compared with the corresponding amino acids in the apo-form and their RMSD calculated. In this manner, we could estimate how much of the protein had reacted to the ligand binding and if the induced fit effect was comparable to the actual crystallographic structure of the complex.

The protein we have chosen presents three scenarios: CDK2 presents a fairly rigid binding pocket and only a few minor side chain movements are visible between the apo-form and the ligand–protein complexes. The two chosen allosteric pockets of the HCV NS5B present a more evident induced fit effect and, in some cases, there is also a clear protein backbone movement; the HIV reverse transcriptase presents one of the most dramatic effects of induced fit, as the binding pocket of the non nucleoside reverse transcriptase inhibitors (NNRTIs) is not even visible in the apo-form. All the simulations were performed on a 2009

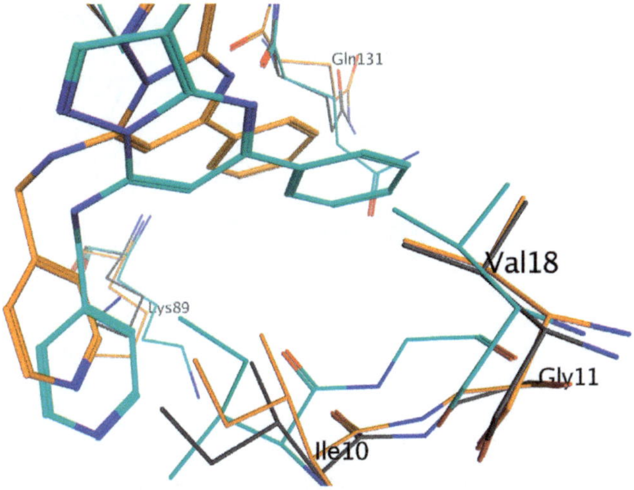

Fig. 7 Snapshot of the SFC ligand manually docked in the CDK2 as examples of residue movements. The apo-structure is represented in grey; the crystallised complex represented in cyan; the structure obtained from the simulation is indicated in orange.

quad-core MacPro, running Ubuntu 12.04, equipped with a NVIDIA Geforce 680 and a Phantom Omni haptic device. As we consider the idea of developing a natural and intuitive interface at the core of our software development efforts, we have asked an undergraduate Medicinal Chemistry student to perform the simulations.

2.8.1 CDK2 results. For CDK2 we have run 8 simulations, one for each ligand reported in Table 3. In terms of ligand placement, the haptic-driven simulation performed relatively well; the ligand RMSD values ranged between 0.89 Å and

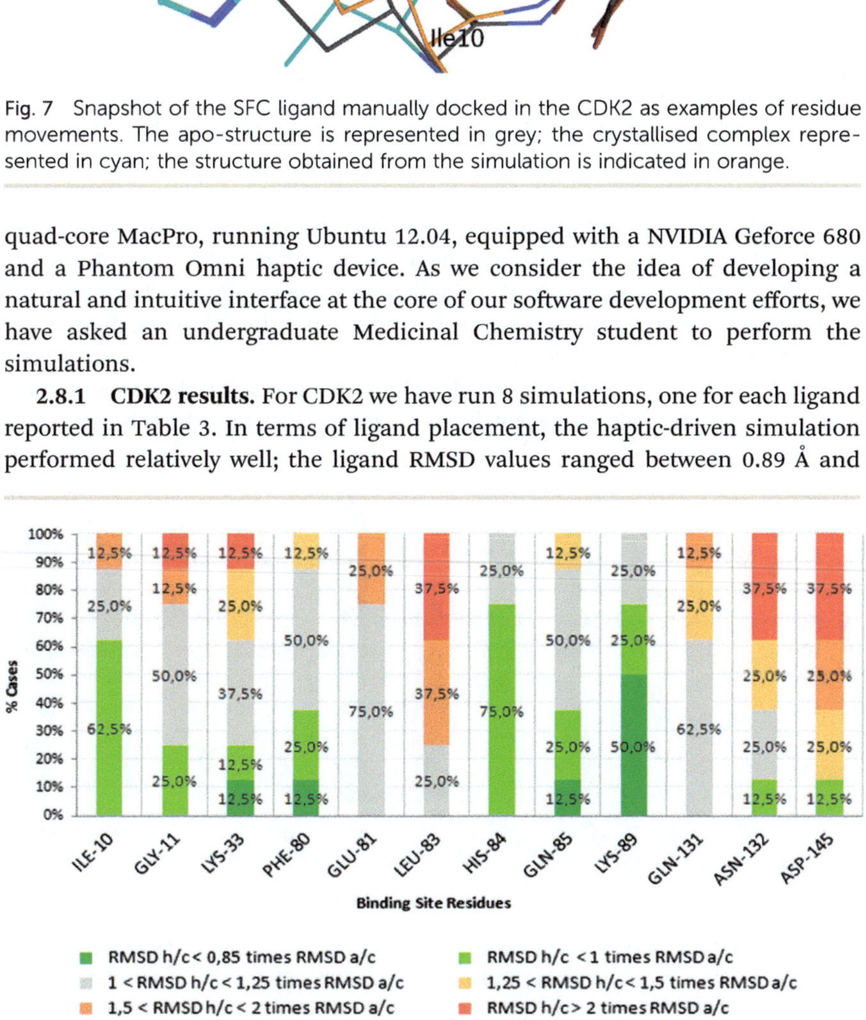

Fig. 8 Visual representation of how CDK2 residues had moved during the simulation. The percentage of structures (8 structures – 100%) whose residue RMSD-cryst/haptic (referred as RMSD c/h in the legend) is lower than correspondent residue RMSD-apo/cryst (referred as RMSD a/c) are shown in green, while higher ones are shown in red.

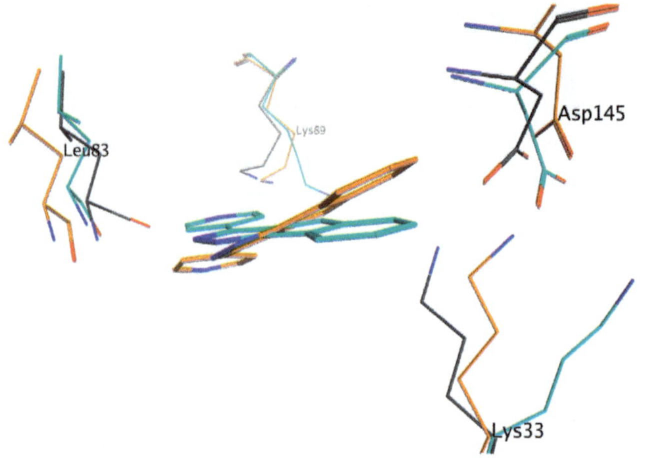

Fig. 9 Snapshot of ligand "LIG" in the CDK2 binding site. Lys33 moves toward the position present in the crystallized structure. The apo-structure is represented in grey; the crystallised complex is represented in cyan; the structure obtained from the simulation is indicated in orange.

2.23 Å. The binding site residues' RMSD also produced some very interesting results: Fig. 7 shows two examples of the results obtained and it is possible to see how the protein has moved in the simulation. Interestingly, the biggest movements are seen on the residues that are indeed the most flexible when the apo and the complex structures are compared.

Fig. 8 shows how the different residues have changed and if their movement is toward the conformation present in the complex structure. In this case, it is

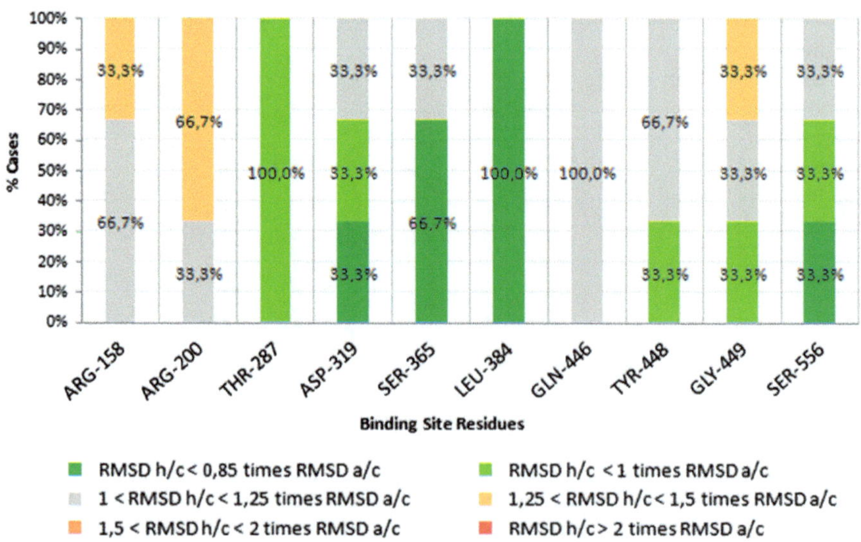

Fig. 10 Visual representation of how HCV NS5B residues had moved during the simulation. The percentage of structures whose residue RMSD-cryst/haptic (referred as RMSD c/h in the legend) is lower than correspondent residue RMSD-apo/cryst (referred as RMSD a/c) are shown in green while higher ones are shown in red.

possible to see a more complex scenario, where some residues move towards the corresponding position in the complex structure (indicated in green) and others that move away from that ideal position (indicated in red).

However, we should point out that the interval parameters represented in Fig. 8 are not absolute values but they represent how far away the haptic-generated structure is to the crystallised structure, relative to the RMSD of the crystallised structure *vs.* the apo-structure. As most of the residues virtually retain the same conformation in all the different structures, their actual absolute RMSD-apo/cryst values are very small. Hence, any minor, favourable or unfavourable, movement generated from the haptic-driven simulation will be flagged as significant. An exception is represented by Lys33, which moves considerably from the apo-structure to the 1JVP crystal structure (RMSD of 2.45 Å). In the haptic driven simulation, the residue moves toward the position present in the crystal giving a RMSD-haptic/cryst of 1.75 Å (Fig. 9). In this case, it is clearly evident the induced fit effect is generated by the haptic-driven ligand placement.

2.8.2 HCV NS5B results. The two pockets selected for this target (Palm and Thumb-2) present a more obvious induced fit effect, compared to what has been observed with CDK2. Indeed, the results obtained are more informative (Fig. 10) as significant conformational changes from the apo-structure to both the haptic generated structure and the crystal complex are evident for a good number of residues.

The data presented in Fig. 10 suggests that there is a general tendency of the binding site residues to move in the direction of their respective complex conformation. In one specific case, even a relevant backbone movement is present (Fig. 11). Interestingly, some residues move away from the ideal final position and this seems to be happening for two reasons: the placement of specific ligands is not accurate (the ligands' RMSD range between 1.10 Å and 3.10 Å); some specific residues should undergo a considerable conformational change and the algorithm is not yet able to explore the full rotamer space available to the residue.

2.8.3 HIV RT results. The HIV reverse transcriptase presents a very different scenario: the NNRTIs pocket is not structured or visible in the apo-form. The

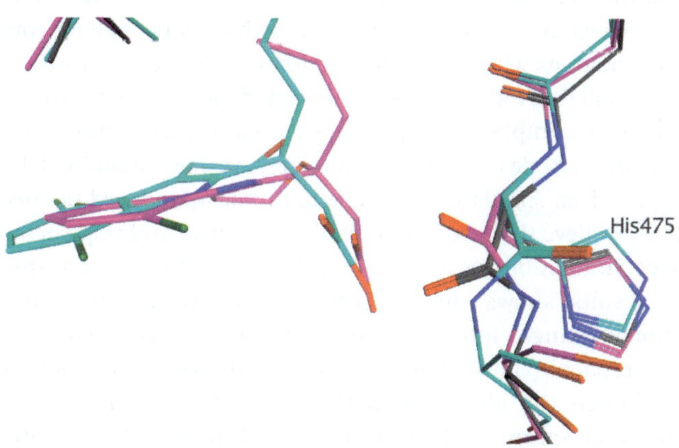

Fig. 11 Snapshot of 1JE in the HCV NS5B binding pocket. His475 backbone movement is highlighted. The apo-structure is represented in grey; the crystallised complex is represented in cyan; the structure obtained from the simulation is indicated in purple.

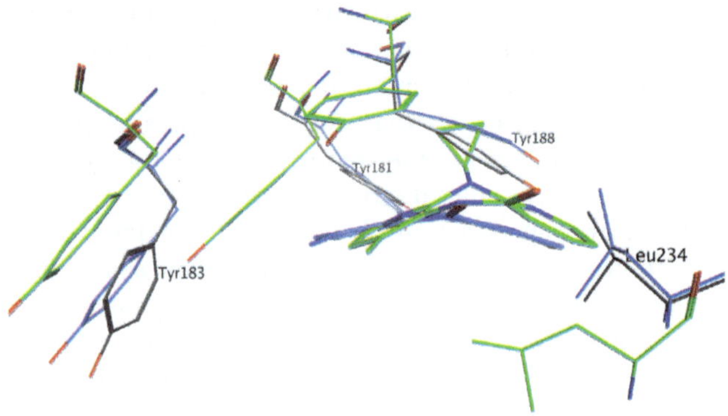

Fig. 12 Snapshot of ligand NVP in the HIV RT. The apo-structure is represented in grey; the crystallised complex is represented in green; the structure obtained from the simulation is indicated in blue.

extent of the induced fit effect, in this case, is really remarkable. The haptic-driven placement results are shown in Fig. 12 and they clearly demonstrate the complexity of these simulations. The user can push the ligand inside the pocket and all the RMSD values shows a positive move.

To a certain extent, this is not surprising, considering how different the apo-structure and the crystal complex are. Indeed, we can see that although the haptic-driven simulation is able to induce a substantial rearrangement in the protein to accommodate the ligand in a reasonable manner, the extent of the protein structural change upon ligand binding in reality is so extensive that, once again, the algorithm cannot explore efficiently such conformational space.

3. Conclusions

In this paper we have reported a hybrid evolutionary strategy for exploring molecular energy landscapes in a small elapsed time period. Our algorithm was tuned to take advantage of the extra parallelism that CUDA streams can provide as the force field used was also coded in CUDA. The proposed method works well for conformational sampling in situations where the exit criterion for the meta-heuristic is the expiration of a very small timeframe. We have also tested the algorithm in a possible actual usage scenario, by implementing it in our haptic-driven molecular modelling simulator. Using a series of crystal structures, we have examined how the fully flexible haptic-driven simulation was able to mimic an induced fit effect.

Overall, we believe the results obtained are positive and very encouraging. The haptic-driven simulations occur in a very natural and intuitive fashion. Furthermore, our results shows that we can induce meaningful and appropriate conformational changes into a protein by placing a ligand using an interactive, real time approach. The data obtained also shows some limitations of this method. Although the algorithm performs extremely well, it is possibly too conservative when sampling the conformational space of the protein residues (of course, a more powerful GPGPU card would explore more conformational states). This should not be considered always a negative aspect as, for example, this more cautious exploration would preserve some important structural information in a binding pocket in those cases when the user applies a strong force to the ligand through the haptic device. However, as in the case of HIV RT, sometime the changes are indeed dramatic and they would require an algorithm able to sample a considerably bigger conformational space to obtain an accurate prediction.

In comparison with a classical manual molecular docking methodology, which often is performed in three separate stages (placement, energy minimization and evaluation), the integrated and interactive nature of our methodology allows the researcher also to explore other aspects of the binding process. For example, the user could investigate and evaluate, to a certain extent, the results generated by approaching the desired active site from different directions, as the actual path taken by the ligand could affect the protein conformational changes. Clearly, an accurate estimate of this process would require a significantly higher computational power than the one available to our system, which currently can only provide a small insight in this part of the binding event, as the results obtained for the HIV RT suggest.

However, even if we believe our approach represents a step forward in addressing several important aspects of protein–ligand binding, there are still several issues to be addressed. In particular, efficiently exploring the ligand conformational space within a very short timeframe and in an interactive environment remains a challenge. In our system, the results for flexible ligands (>2 rotatable bonds) could be very different based on the starting conformation of the ligand. Solvation, as in many other docking algorithms, is also difficult to evaluate. It is true that an implicit model could be used, but often the information provided by explicit water molecules could be essential to understand the accurate binding of a ligand.

In conclusion, we believe the results presented here are very promising and encouraging. We are now taking the next step in the validation of our system and the interactive approach, by comparing the haptic-driven simulator to other commercial and non-commercial flexible docking packages.

References

1 M. A. Lill, *Biochemistry*, 2011, **50**, 6157.

2 W. Sinko, S. Lindert and J. A. McCammon, *Chem. Biol. Drug Des.*, 2013, **81**, 41.

3 C. Blum and A. Roli, *ACM Comput. Surv.*, 2003, **35**(3), 268.

4 D. E. Goldberg, *Genetic algorithms in search, optimization, and machine learning*, Addison-Wesley Pub. Co, Reading, Mass., 1989.

5 S. Kirkpatrick, C. D. Gelatt and M. P. Vecchi, *Science*, 1983, **220**, 671.

6 N. Metropolis, A. W. Rosenbluth, M. N. Rosenbluth, A. H. Teller and E. Teller, *J. Chem. Phys.*, 1953, **21**, 1087.

7 R. Battiti, *Reactive search and intelligent optimization*, Springer, New York, 1st edn, 2008.

8 O. Trott and A. J. Olson, *J. Comput. Chem.*, 2010, **31**, 455.

9 J. Fuhrmann, A. Rurainski, H.-P. Lenhof and D. Neumann, *J. Comput. Chem.*, 2010, **31**, 1911.

10 S. M. Long, T. T. Tran, P. Adams, P. Darwen and M. L. Smythe, *J. Comput. Chem.*, 2011, **32**, 1541.

11 C. R. S. Brasil, A. C. B. Delbem and F. L. B. da Silva, *J. Comput. Chem.*, 2013, **34**, 1719.

12 W. Cai and Xand Shao, *J. Comput. Chem.*, 2002, **23**, 427.
13 R. Murty and D. Okunbor, *Parallel Comput.*, 1999, **25**, 217.
14 N. Zonta, I. J. Grimstead, N. J. Avis and A. Brancale, *J. Mol. Model.*, 2009, **15**, 193. 2009.
15 A. Ricci, A. Anthopoulos, A. Massarotti, I. Grimstead and A. Brancale, *Future Med. Chem.*, 2012, **4**, 1219.
16 A. Anthopoulos, I. Grimstead and A. Brancale, *J. Comput. Chem.*, 2013, **34**, 2249.
17 S. Pall and B. Hess, *Comput. Phys. Commun.*, 2013, **184**, 2641.
18 U. Schulze-Gahmen, H. L. De Bondt and S. H. Kim, *J. Med. Chem.*, 1996, **39**, 4540–4546.
19 A. E. Gibson, C. E. Arris, J. Bentley, F. T. Boyle, N. J. Curtin, T. G. Davies, J. A. Endicott, B. T. Golding, S. Grant, R. J. Griffin, P. Jewsbury, L. N. Johnson, V. Mesguiche, D. R. Newell, M. E. Noble, J. A. Tucker and H. J. Whitfield, *J. Med. Chem.*, 2002, **45**, 3381.
20 P. Furet, T. Meyer, A. Strauss, S. Raccuglia and J. M. Rondeau, J.M., *Bioorg. Med. Chem. Lett.*, 2002, **12**, 221.
21 T. O. Fischmann, A. Hruza, J. S. Duca, L. Ramanathan, T. Mayhood, W. T. Windsor, H. V. Le, T. J. Guzi, M. P. Dwyer, K. Paruch, R. J. Doll, E. Lees, D. Parry, W. Seghezzi and V. Madison, *Biopolymers*, 2008, **89**, 372.
22 Y. N. Kang and J. A. Stuckey, *To be Published*.
23 D. O'Farrell, R. Trowbridge, D. Rowlands and J. Jager, *J. Mol. Biol.*, 2003, **326**, 1025.
24 F. Velazquez, S. Venkatraman, C. A. Lesburg, J. Duca, S. B. Rosenblum, J. A. Kozlowski and F. G. Njoroge, *Org. Lett.*, 2012, **14**, 556.
25 G. N. Anilkumar, O. Selyutin, S. B. Rosenblum, Q. Zeng, Y. Jiang, T. Y. Chan, H. Pu, L. Wang, F. Bennett, K. X. Chen, C. A. Lesburg, J. Duca, S. Gavalas, Y. Huang, P. Pinto, M. Sannigrahi, F. Velazquez, S. Venkatraman, B. Vibulbhan, S. Agrawal, E. Ferrari, C. K. Jiang, H. C. Huang, N. Y. Shih, F. G. Njoroge and J. A. Kozlowski, *Bioorg. Med. Chem. Lett.*, 2012, **22**, 713.
26 K. Vandyck, M. D. Cummings, O. Nyanguile, C. W. Boutton, S. Vendeville, D. McGowan, B. Devogelaere, K. Amssoms, S. Last, K. Rombauts, A. Tahri, P. Lory, L. Hu, D. A. Beauchamp, K. Simmen and P. Raboisson, *J. Med. Chem.*, 2009, **52**, 4099.
27 T. A. Stammers, R. Coulombe, J. Rancourt, B. Thavonekham, G. Fazal, S. Goulet, A. Jakalian, D. Wernic, Y. Tsantrizos, M. A. Poupart, M. Bos, G. McKercher, L. Thauvette, G. Kukolj and P. L. Beaulieu, *Bioorg. Med. Chem. Lett.*, 2013, **23**, 2585.
28 P. L. Beaulieu, R. Coulombe, J. Duan, G. Fazal, C. Godbout, O. Hucke, A. Jakalian, M. A. Joly, O. Lepage, M. Llinas-Brunet, J. Naud, M. Poirier, N. Rioux, B. Thavonekham, G. Kukolj and T. A. Stammers, *Bioorg. Med. Chem. Lett.*, 2013, **23**, 4132.
29 R. Esnouf, J. Ren, C. Ross, Y. Jones, D. Stammers and D. Stuart, *Nat. Struct. Biol.*, 1995, **2**, 303.
30 J. Ren, R. Esnouf, E. Garman, D. Somers, C. Ross, I. Kirby, J. Keeling, G. Darby, Y. Jones and D. Stuart, *Nat. Struct. Biol.*, 1995, **2**, 293.
31 Molecular Operating Environment 2012.10. Chemical Computing Group, Montreal, Canada. http://www.chemcomp.com.

Faraday Discussions

PAPER

Accelerating electrostatic pair methods on graphical processing units to study molecules in supercritical carbon dioxide†

John A. Baker and Jonathan. D. Hirst*

Received 14th February 2014, Accepted 5th March 2014
DOI: 10.1039/c4fd00012a

Traditionally, electrostatic interactions are modelled using Ewald techniques, which provide a good approximation, but are poorly suited to GPU architectures. We use the GPU versions of the LAMMPS MD package to implement and assess the Wolf summation method. We compute transport and structural properties of pure carbon dioxide and mixtures of carbon dioxide with either methane or difluoromethane. The diffusion of pure carbon dioxide is indistinguishable when using the Wolf summation method instead of PPPM on GPUs. The optimum value of the potential damping parameter, α, is 0.075. We observe a decrease in accuracy when the system polarity increases, yet the method is robust for mildly polar systems. We anticipate the method can be used for a number of techniques, and applied to a variety of systems. Substitution of PPPM can yield a two-fold decrease in the wall-clock time.

1. Introduction

GPU architectures and their use in molecular dynamics (MD) simulations have attracted much recent attention.[1–4] Maximizing computational throughput on graphical processing units (GPUs)/hybrid architectures is of great interest. Decreasing the wall-clock time for each MD time step allows longer timescales to be sampled, giving greater confidence in the convergence of ensemble averages. GPUs have been widely adopted in various computational disciplines, due to the highly parallel design, low power consumption (FLOPS/Watt) and commodity cost (£/FLOPs). The GPU architecture is optimised for processing massive amounts of parallel calculations, and so has many more cores than a CPU, but at the sacrifice of memory capabilities.

Electrostatic interactions can be divided into first-order effects, which comprise point charge interactions decaying reciprocally with respect to intermolecular distance and higher-order interactions that decay more rapidly.

School of Chemistry, University of Nottingham, University Park, Nottingham, NG7 2RD, UK. E-mail: Jonathan.Hirst@nottingham.ac.uk; Fax: +44 1159 513562; Tel: +44 1159 513478

† Electronic Supplementary Information (ESI) available: See DOI: 10.1039/c4fd00012a

Electrostatic interactions can be modelled explicitly by various methods, including: cut-off truncation,[5–7] switched/shifted cut-off truncation,[8] Ewald[9] and its mesh derivatives, particle mesh Ewald (PME),[10] particle–particle particle mesh (PPPM)[11] and smoothed particle mesh Ewald (SPME).[12] The Ewald summation provides a good approximation for the electrostatic energy, as the algorithm accounts for periodicity of the simulation domain, but it scales as $O(N^{3/2})$. An approximation that is frequently adopted is PPPM, due to its superior scaling, $O(N\log N)$, over the Ewald summation. SPME,[13] PPPM[14] and multi-level summation[15] have been implemented on the GPU, yet the requirement for fast Fourier transforms (FFTs) intrinsic to Ewald methods, like SPME and PPPM, reduces the parallelism[14] and leads to poor scaling on GPUs. Pair-wise algorithms derived from cut-off techniques show superior scaling $O(N)$ and greater effectiveness on GPUs.

Cut-off techniques can reproduce experimental quantities, for instance, the Madelung constant. Wolf *et al.*[16] reported that the electrostatic (Coulombic) potential for condensed systems was effectively short-ranged, and the energies are in agreement with Ewald methods when the cut-off sphere reached neutrality. This was observed whilst calculating the Madelung constant for rocksalt (NaCl), where the relationship between truncation distance and electrostatic energy was investigated. When the cut-off sphere was truncated at a distance to achieve charge neutrality, the electrostatic energy was significantly closer to the Madelung energy. The repeating lattice structure of NaCl is well-suited for treatment with the Wolf method, but the supercritical phase of polar solutes can be more difficult to model as the atomic partial charges within the cut-off sphere are dynamic. The resulting shifted Coulomb potential achieves charge neutrality by projecting each atom charge, q_j, from q_i onto the edge of the solvation sphere. Therefore, every jth atom in the solvation sphere of atom i has a charge of equal but opposite sign set at the cut-off (R_c). This results in an artificially neutral solvation sphere for every ith atom, which effectively makes the system charge neutral.

$$E_{\text{Wolf}} = \frac{1}{2}\sum_{i=1}^{N}\sum_{\substack{j(i)\\ r_{ij}<R_c}} \left[\left(\frac{q_i q_j erfc(\alpha r_{ij})}{r_{ij}} - \lim_{r_{ij}\to R_c}\left\{\frac{q_i q_j erfc(\alpha R_c)}{R_c}\right\}\right)\right] - \left(\frac{erfc(\alpha R_c)}{2R_c} + \frac{\alpha}{\sqrt{\pi}}\right)\sum_{i=1}^{N} q_i^2$$

where q_i and q_j are point charges, r_{ij} is the intermolecular distance between atoms i and j, α is the damping coefficient in Å^{-1} and R_c is the cut-off distance. To obtain accurate electrostatic contributions, a damping function is applied. The electrostatic energy decay oscillates around a rate of $1/R_c$. Introducing the damping function quickly flattens the oscillations as the cut-off increases, effectively determining how fast the complementary error function falls from unity at $r_{ij} = 0$, to zero at the cut-off.[16] The damping function adopted is the complementary error function, as it has a close connection to the Ewald sum.[17] The coefficient of the error function, α, denotes the rate at which convergence is achieved. A large α value will converge the energy rapidly using a short cut-off, but the errors can be larger. A smaller α leads to less contamination of the potential, but the sum will fluctuate more rapidly. Assigning a value of 0 to α, results in the truncated shifted-force (SF) potential. This SF calculation is faster than the Wolf method, but the selection of α can enhance the accuracy of electrostatic forces and energies by optimizing agreement to Ewald methods.[18]

Simple cut-off based methods are unreliable for computing the forces, as the potential truncates abruptly at the cut-off, which causes forces to be undefined if the molecule lies at the cut-off boundary. The reliance upon Ewald methods instead of efficient cut-off methods has been discussed by Fennel and Gezelter,[17] who compared the accuracy of damped shifted-force (DSF) and shifted potential cut-off methods for water, argon in water, and NaCl. Benchmarking and validation of the DSF potential was applied to polyelectrolyte brushes on GPUs; speedups were achieved of a factor between 1.1 and 3.9, depending on the system and size of the cut-off with respect to PPPM.[19] Group based cut-offs, where the cut-off is dynamically allocated to ensure entire molecules are within the solvation shell, were investigated,[17] but the energies deviated significantly from those obtained using PME.

In this work we assess the applicability of the Wolf method to model electrostatic interactions for various systems in the supercritical region of carbon dioxide. Supercritical carbon dioxide ($scCO_2$) is an attractive 'green solvent' used in many industrial processes, such as caffeine extraction,[20] polymer solvation[21] and synthesis,[22,23] enzyme catalysis[24,25] and stabilization[26,27] and transition metal catalysis.[28,29] Although carbon dioxide does not possess a dipole, it has a significant quadrupole[30] (13.4×10^{-40} Cm^2), which means that electrostatic interactions are an important component of interactions involving $scCO_2$. Su and Maroncelli[31] observed that neglecting electrostatic interactions, *i.e.*, considering only Lennard-Jones interactions, led to a systematic 14% error in the solvation free energies of polymer-$scCO_2$ systems. This observation is attributed to quadrupole–dipole and quadrupole–quadrupole interactions,[31] which are inherently modelled in all point charge models. A nonpolar fluid should be well suited for treatment using the Wolf method, as quadrupolar interactions decay more rapidly than dipolar ones. This bodes well for the efficient electrostatic treatment of systems solvated by carbon dioxide by utilizing the Wolf method on GPU architectures. We also investigate the computational cost of the DSF method, with respect to the Wolf method.

Fluorous polymers are well known to dissolve in $scCO_2$,[32,33] which has been partially explained by ^{19}F-NMR experiments, which suggested a number of specific interactions between carbon dioxide and the fluorous solute that increase the solubility.[34] Many biomolecular systems are stable in the presence of $scCO_2$, but this is dependent on the species, the water content, and experimental conditions.[24] Protein stability can be observed when $scCO_2$ solvates the hydrophobic residues, and water solvates the polar/hydrophilic residues.[26] Water possesses a strong dipole moment, which will make it less amenable to treatment with the Wolf method. We aim to follow this work with an investigation for protein systems solvated by water and carbon dioxide.

To assess the applicability of the Wolf method to systems containing carbon dioxide, important physical quantities, such as PVT relationships and diffusion coefficients, can be calculated and compared with the values calculated using PPPM. The applicability of the Wolf method to study methane plus carbon dioxide gas hydrates has been reported, and the results show good agreement between lattice sum and reaction field methods.[35] Analysis of the convergence behaviour of the Wolf method by Angoshtari and Yavari[36] shows the method to be robust, but convergence can be problematic if poor choices are made for the cut-off or α values. In our study, we investigate the effect of increasing the polarity of the fluid by incorporating difluoromethane molecules, which possess a strong dipole. For low polarity systems the Wolf method should be well suited, but as the polarity increases the effect of long-range dipole interactions will become important and we assess the point where Ewald techniques will become necessary.

2. Method

LAMMPS is a multi-purpose MD code,[37] which is widely used in the fields of atomistic, coarse-grained and mesoscopic simulations. The code can process massive numbers of particles per simulation, which it achieves using an optimized spatial decomposition technique. The GPU optimized version, written in CUDA C, can run over multiple GPUs, either in conjunction with the CPU or entirely on the GPU. We have implemented the Wolf method into LAMMPS-CUDA the GPU-exclusive version of LAMMPS and LAMMPS-GPU the CPU/GPU implementation, as a new potential incorporating Lennard-Jones and electrostatic interactions. LAMMPS-CUDA was written exclusively for use with CUDA, but LAMMPS-GPU can be compiled and run on AMD GPUs and other applicable accelerators using OpenCL. The new pair style is implemented as a double potential function in LAMMPS as lj/charmm/coul/wolf for CPU, lj/charm/coul/wolf/cuda for LAMMPS-CUDA and lj/charm/coul/wolf/gpu for LAMMPS-GPU. The LAMMPS-CUDA version is solely GPU based, with the exception of file I/O and pre/post simulation setup. The GPU version of LAMMPS utilizes the GPU for the force and/or neighbour list generation, whilst all other operations and I/O are performed on the CPU. The GPU version allows for n CPUs to be used per GPU, whilst the LAMMPS-CUDA version only allows for one CPU per GPU. An important difference between packages is that the GPU version on LAMMPS uses the CPU to calculate FFTs for PPPM, whilst the LAMMPS-CUDA version performs the calculation on the GPU. We have also implemented the DSF implementations, lj/charm/coul/dsf/cuda and lj/charm/coul/dsf/gpu, for benchmarking purposes. We used the CHARMM[38] Lennard-Jones potential in the AB form with a switching function in conjunction with the Wolf method. A cut-off of 2.5σ was used for the Lennard-Jones potential, and tapered to zero at 2.65σ, where σ was calculated

Table 1 EPM2 force field parameters for carbon dioxide[39] and Palmer and Anchell parameters for methane and difluoromethane.[40] ($A^{1/12}$) and ($B^{1/6}$) have units ((kcal mol)$^{1/12}$ Å) and ((kcal mol)$^{1/6}$ Å) respectively

Atom site	$A^{1/12}$	$B^{1/6}$	q (|e|)
C (CO_2)	2.448	2.173	−0.3256
O (CO_2)	2.922	2.815	+0.6512
C (CH_4)	3.200	3.200	−0.4160
H (CH_4)	1.910	1.390	+0.1040
C (CH_2F_2)	2.900	3.590	+0.0500
H (CH_2F_2)	1.712	0.000	+0.1550
F (CH_2F_2)	2.650	2.237	−0.1800

from $A^{1/12}$ in Table 1. We used the rigid EPM2 atomistic force field[39] to represent carbon dioxide and parameters derived by Palmer and Anchell[40] to represent Lennard-Jones and charge interactions for methane and difluoromethane.

We utilized the LAMMPS software to simulate $scCO_2$ at several densities, comparing the results obtained by using PPPM or the Wolf method. We selected a tolerance setting of 0.0001 for the calculations involving PPPM that enables the root mean square error to be within a factor of 10 000 of the reference force, which is calculated analytically for short-range interactions[41] and in *k*-space.[42] The optimum value of α was investigated for binary mixtures of carbon dioxide and methane or difluoromethane. The criterion for selecting α is the level of agreement between the Coulombic energies computed by PPPM and Wolf. To consider the effects of neglecting only the periodic long-range Coulombic effects, we treat the cut-off as being half the length of the simulation box. We also investigate the level of approximation that arises from use of a quarter box cut-off in conjunction with the Wolf method.

2.1 Benchmarking GPU electrostatics

Two boxes of 10 000 and 50 000 carbon dioxide molecules were constructed with a molar density of 0.01 mol cm^{-3} and minimized for 1000 iterations using the conjugate gradient method, to an energy tolerance of 1×10^{-6} kcal mol^{-1}. Both systems were heated from 0 K to 308.2 K using the Nosè–Hoover thermostat[43] with a damping time of 500 fs for 2 ns, followed by 5 ns of equilibration at 308.2 K. We evaluated the performance of the Wolf method on two GPU architectures, Tesla (Tesla C1060) and Kepler (Tesla K10). The K10, released in 2012, is designed for high throughput calculations performed in single precision. The number of giga floating-point operations per second (GFLOPs) is 4577 in single precision, compared to 933 for the C1060. This improvement was achieved in part by increasing the number of cores in the streaming multiprocessor from 8 (Tesla), 32 (Fermi) to 192 (Kepler). Memory bandwidth on the GPU to global memory is 103 GB s^{-1} for the C1060, but 320 GB s^{-1} for Kepler K10. We include an eight core Intel Xeon (E5-2609) CPU for benchmarking purposes that has an approximate 77 GFLOPs with 34 GB s^{-1} bandwidth to RAM. Calculations were performed using single-precision on two separate nodes, both comprise an eight core Xeon CPU and either a Kepler K10 or two Tesla C1060. Multiple GPU tests were only carried out using the C1060, which yielded an almost two-fold increase in performance, which corresponds well to the linear dependency. We use the FFTW 3.3.1 library in single-precision for all *k*-space calculations of PPPM. We considered two modes for calculations using LAMMPS-GPU, one where all force and neighbour calculations are performed on the GPU, and the other where the force and neighbour calculations are dynamically assigned between CPU and GPU. We used the CUDA 5.0 Tokito (GPU driver v. 304.54), which has produced a noticeable improvement in performance over CUDA 4.0. A new feature, CUDA dynamic parallelism, allows kernels running on the GPU to spawn more grids and to continue to generate work depending on the calculation.[44] This feature has not been incorporated in our study.

2.2 Pure carbon dioxide

The 10 000 carbon dioxide molecule system was used to generate densities (box lengths) of 0.001 (255.1 Å), 0.002 (202.5 Å), 0.004 (160.72 Å), 0.005 (149.2 Å),

0.01 (118.42 Å) and 0.02 (94.0 Å) mol cm^{-3}, all within the supercritical region at 308.2 K. Each box was minimized for 1000 iterations using the conjugate gradient method, with an energy tolerance of 1×10^{-6} kcal mol^{-1}. The system was heated in the NVT ensemble using the Nosè–Hoover thermostat[43] with cubic periodic boundary conditions from 0 K to 308.2 K for 1 ns using a 1 fs time step. This was followed by 1 ns of equilibration. Integration was performed using the time-reversible velocity Verlet algorithm.[45] The procedure was repeated nine times to generate different equilibrated configurations for the purposes of error analysis, and the resulting standard deviation is noted in the error bars. The mean square displacement (MSD) of the centre of mass of carbon dioxide was obtained by calculating the gradient of the linear portion of the relationship between MSD and lag time over 10 ns production dynamics in the NVT ensemble. The pressures were obtained every 50 fs, and averaged over 10 ns to calculate the PVT relationship.

2.3 Carbon dioxide and solute

To study the effects of polar and non-polar solutes, we considered different quantities of methane and difluoromethane molecules. Five boxes of 10 000 carbon dioxide molecules were constructed with mole fractions $\chi_{(\text{solute})} = 0.0001, 0.001, 0.01, 0.09$ and 0.5, where the solute was either methane or difluoromethane. These systems correspond to box lengths (Å) of 95.01, 95.99, 97.49, 99.27 and 125.31 for CH_4/CO_2 and 81.64, 81.78, 81.83, 84.44 and 105.16 for CH_2F_2/CO_2. Each system was minimized and heated to 308.2 K using the same procedure as above. The pressure was equilibrated to 80 atmospheres (corresponding to a molar density of ~0.02 mol cm^{-3}), using the Berendsen barostat[46] with a 1 ps damping time for 20 ns with a 1 fs time step. The procedure was repeated nine times to generate different equilibrated configurations for the purposes of error analysis, and the resulting standard deviation is noted in the error bars. The MSDs of carbon dioxide and solutes were obtained from the linear portion of the relationship over a production run of 10 ns. We compare the errors in Coulombic energy with respect to α for $\chi_{(\text{solute})} = 0.09$ with PPPM and Wolf, which we performed by decomposing the energy into group contributions. We investigate the relationship between system polarity and diffusion coefficients, and the total Coulombic energy.

One method of quantifying the transport properties of a system is the MSD, and thus the diffusion coefficient using the Einstein relationship.

$$D = \lim_{t \to t_{\text{lag}}} \frac{1}{6t} \left\langle [r_i(t) - r_i(0)]^2 \right\rangle$$

where D is the macroscopic diffusion coefficient, $r_i(t)$ is the position of the centre of mass at time t, $r_i(0)$ is the initial position of the centre of mass, t_{lag} is the lag time, and $\langle [r_i(t) - r_i(0)]^2 \rangle$ is the ensemble averaged MSD. The diffusion coefficient is calculated from the slope of the MSD against lag time,[47] which is a measure of the atomic displacements through time.

The pressure of a system can be calculated using the virial equation (below), and the PVT relationship has an influence on the diffusion coefficients.[48] The long-range part of PPPM has a different contribution to the pressure.[49]

$$PV = Nk_{\text{B}}T + \frac{1}{x} \left\langle \sum_{i=1}^{N} \boldsymbol{r}_i . \boldsymbol{F}_i \right\rangle$$

where P is the pressure of the system, V is the volume, N is the number of particles, x is the dimensionality, K_B is the Boltzmann constant and the term in brackets is the total intermolecular force multiplied by the interaction distances.

3. Results

3.1 Benchmarking

To compare the efficiency of the Wolf method, we measured the computational throughput of the electrostatic routines using the 10 000 and 50 000 molecule systems. The timings of the implementations are calculated with respect to the number of neighbours in the cut-off sphere. Fig. 1(a) and 1(b) compare the throughput for different architectures, and show the acceleration gained when substituting PPPM for Wolf using LAMMPS-CUDA. The computational throughput is almost twice as high with the Wolf treatment than with PPPM. The Wolf method is slower than PPPM on the CPU, which is unexpected, but the CPU is well suited to handle FFTs that feature in PPPM. This is due to faster memory accesses between cache and compute cores, and the less parallel nature of the algorithm. The DSF method is marginally slower than PPPM in LAMMPS-CUDA, but is marginally faster than the Wolf method for the LAMMPS-GPU

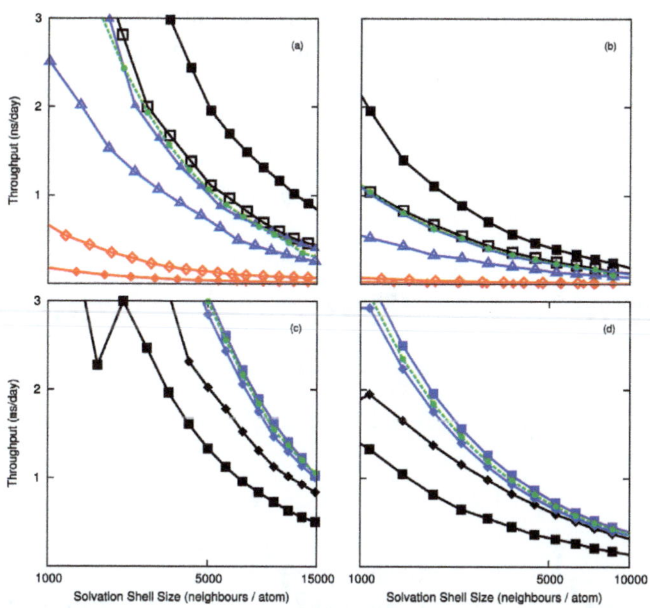

Fig. 1 Computational throughput as a function of number of neighbours for (a,c) 10 000 CO_2 molecules (B-D) 50 000 CO_2 molecules. Fig. (a) and (B) were calculated using the LAMMPS-CUDA package, and (c) and (d) were calculated using the LAMMPS-GPU package. In (a) and (B), three electrostatic treatments were considered, Wolf (filled shapes), PPPM (unfilled shapes) and DSF (green circles) for different architectures. (Triangles = Tesla C1060, squares = Kepler K10, diamonds = eight core Intel Xeon). In figures (c) and (d) the force/neighbour implementation is shown where the GPU is used exclusively (diamonds) and dynamically assigned between CPU/GPU (squares). PPPM is shown in black, the Wolf method is shown in blue, and the DSF force/neigh exclusively GPU is shown in green.

implementation. The difference between pair styles using LAMMPS-GPU is much less pronounced than LAMMPS-CUDA, where the acceleration gained by using the Wolf or DSF method is approximately 1.5 times faster. The difference between dynamic partitioning and exclusive GPU computation is small for the cut-off methods, and the dynamic mode is marginally quicker. For PPPM the opposite is observed, which could be due to the CPU already being used for the *k*-space calculations. The Kepler K10 GPU shows the greatest acceleration, which bodes well for future GPU releases.

3.2 Optimum alpha and cut-off parameters

We consider first the accuracy of the electrostatic energy and its convergence as a function of α and cut-off distance. We then turn to the properties computed from the simulation, considering structural, dynamic and thermodynamic aspects. The simulations have all been executed in the supercritical region of carbon dioxide, and therefore each molecule has a large number of neighbours in the solvation sphere. We compare the effects of selecting α values between zero and 0.3 in intervals of 0.025. We decompose the energy interactions into pairwise group contributions to the total Coulombic energy for CH_4 in carbon dioxide, and CH_2F_2 in carbon dioxide. The relationship between α and the accuracy of the Wolf summation method is shown in Fig. 2.

We observe for half box cut-offs the best agreement for Wolf is at low values of α, where for both the polar and non-polar systems the optimum value for α is

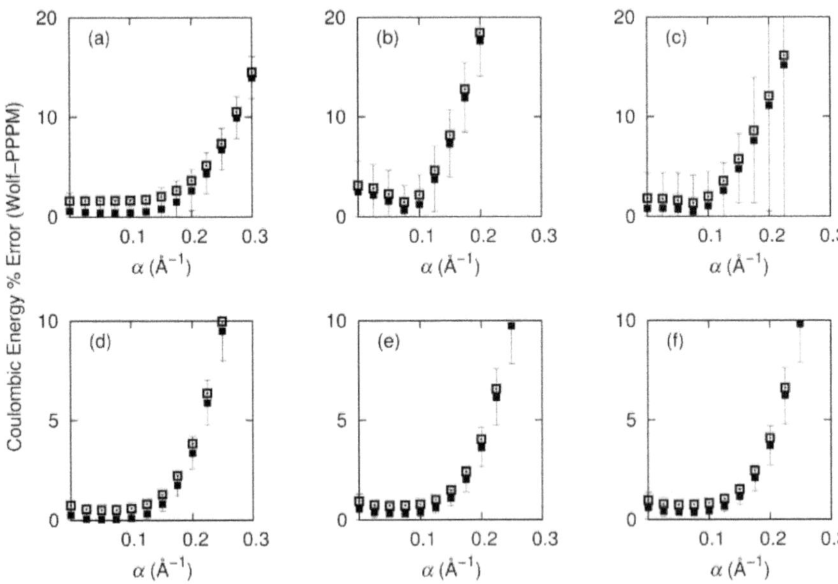

Fig. 2 Percentage error in Coulombic energy between PPPM and Wolf with respect to α for CH_4/CO_2 and CH_2F_2/CO_2 binary mixtures for χ(solute) = 0.1 at 308.2 K, 80 atmospheres. The results are decomposed into (a) (CH_2F_2) CO_2–CO_2 pair errors (B) (CH_2F_2) CH_2F_2–CH_2F_2 pair errors, (c) (CH_2F_2) CO_2–CH_2F_2 pair errors, (d) (CH_4) CO_2–CO_2 pair errors, (e) (CH_4) CH_4–CH_4 pair errors and (f) (CH_4) CO_2–CH_4 pair errors. The black squares and error bars indicate the average error and uncertainty for a half box cut-off, whilst the empty squares show the average error for a quarter box cut-off.

0.075. For non-polar systems increasing α beyond 0.2 results in an ~2–3% error (with respect to the PPPM Coulombic energy), whilst for the polar system the average error is ~20%. The non-polar CH_4/CO_2 system resulted in the lowest average errors of 0.05% for CO_2–CO_2 interactions, 0.39% for CO_2–CH_4 and 0.34% for CH_4–CH_4. We observe a similar trend for the polar system of CO_2/CH_2F_2 with increased average errors of 0.44% for CO_2–CO_2, 0.47% for CO_2–CH_2F_2 and 0.67% for CH_2F_2–CH_2F_2. At $\alpha = 0.075$ the greatest variance in the non-polar system is $\sigma^2 = 0.2$ for CH_4–CH_4, and for the polar system the maximum variance is $\sigma^2 = 2.94$ for CH_2F_2–CH_2F_2.

A quarter box cut-off gives greater errors in the Coulombic energies, but follows the same trends as the half box cut-off. For the non-polar system the optimum value for α is 0.05, which is seen for all pair interactions of methane and carbon dioxide.

The average error increases to 0.5% for CO_2–CO_2 and to ~0.75% for energies involving interactions with methane. The errors vary more when used with the polar system over a wide range of α, but at $\alpha = 0.075$ the error for all interactions is below 2%. The error increases sharply for the polar system with varying α. For interactions between CH_2F_2–CH_2F_2 the variance is higher than for methane, which can be attributed to the high polarity of both substituents.

3.3 Pure carbon dioxide

Diffusion coefficients of $scCO_2$ calculated from MD simulations using the Ewald sum[50] have been previously compared with experimental results.[51] Our simulations show that the Wolf method gives diffusion coefficients comparable to that of PPPM (Fig. 3b). Calculations were concurrently run on the CPU using the Wolf method, and the energies were within three decimal places. Simulations were also performed at 323.2 K, which show good agreement between PPPM and the Wolf method.† Low density boxes using the Wolf method show the best agreement, but all the densities investigated are within the bounds of error of PPPM. Pressures obtained over a range of densities coincide well with an equation of state,[52] for both the Wolf and PPPM implementations (Fig. 3a). Both electrostatic methods

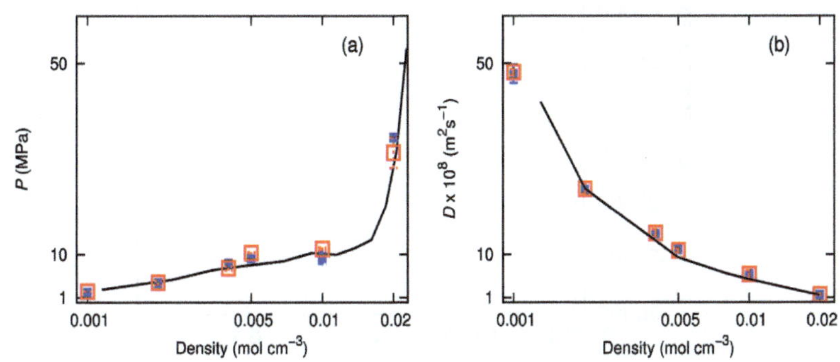

Fig. 3 (a) Diffusion coefficients for pure carbon dioxide obtained for PPPM (empty red squares) and Wolf (filled blue squares), compared with experimental (black line) at 308.2 K. b) PVT relationship of pure carbon dioxide simulated using PPPM (empty red squares) or the Wolf method (filled blue squares) compared with experimental (black line) at 308.2 K.

capture the PVT relationship properties at high and low densities, for both temperatures studied.

3.4 Binary mixture of carbon dioxide with difluoromethane or methane

The Coulombic energy for the polar systems is compared for different mole fractions of difluoromethane using the Wolf method and PPPM. As the system becomes more polar, the error of the Wolf method with respect to the PPPM value becomes greater. Fig. 4 shows the total PPPM Coulombic energy, and results for $\alpha = 0.2$ and $\alpha = 0.075$. For the system containing 100 difluoromethane molecules, the error is within 0.2% and 1.8% for $\alpha = 0.075$ and $\alpha = 0.2$ respectively. As the composition tends towards a 1 : 1 ratio, the errors increase, which indicates, as anticipated, that the Wolf approach is not suitable when the system becomes highly polar. The errors for $\alpha = 0.075$ are reasonable (0.9%) but for $\alpha = 0.2$ the errors are 3%. We observe better agreement for methane. For the 1 : 1 binary mixture, the error in $\alpha = 0.2$ is 1.8% and $\alpha = 0.075$ is 0.15%. The total Coulombic energy for difluoromethane is more negative than for methane, which can be explained by NPT dynamics resulting in a lower volume box and therefore closer contacts.

Experimental results from O'Hern and Martin[51] indicate pure carbon dioxide has a diffusion coefficient of 5 ($\times$ 10^8 m^2 s^{-1}) at 308.2 K, 80 atmospheres. The non-polar binary mixture of CH_4/CO_2 has a diffusion coefficient of ~5 ($\times$ 10^8 m^2 s^{-1}) for the inclusion of one methane molecule in 10 000 molecules of carbon dioxide

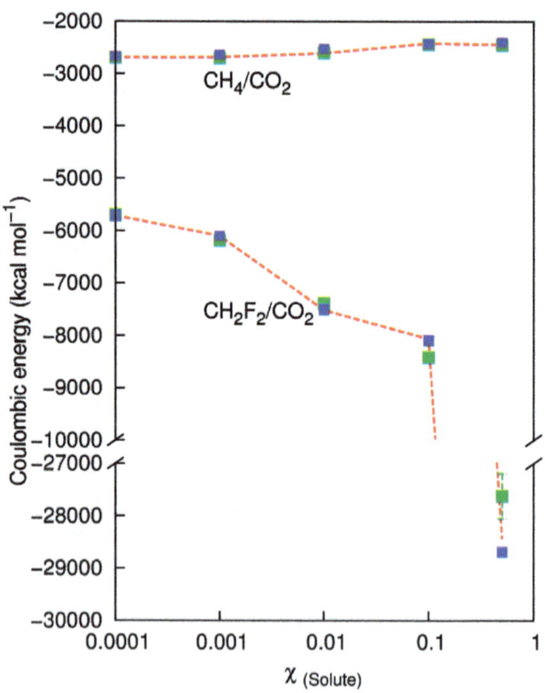

Fig. 4 Total Coulombic energy for methane and difluoromethane in carbon dioxide at 308.2 K, 80 atmospheres. The dashed red line indicates the total PPPM energy, whilst the green squares indicate $\alpha = 0.2$ and the blue squares indicate $\alpha = 0.075$.

(Fig. 5). As the number of methane molecules increases, the overall diffusion coefficients remain relatively constant. Diffusion coefficients for CH_4 and carbon dioxide are similar, with an average increase of ~0.6 ($\times$ 10^8 m^2 s^{-1}) between carbon dioxide and CH_4. As $\chi_{(solute)}$ increases, the accuracy of the solute diffusion coefficients increase, due to better averaging from more solute molecules. With the exception of $\chi_{(solute)} = 0.0001$ where averaging is poor, the agreement between Wolf and PPPM is good. We observe a reduction in diffusion coefficients for difluoromethane as the polarity increases. The values decay from 4.5 ($\times$ 10^8 m^2 s^{-1}), to 1.3 ($\times$ 10^8 m^2 s^{-1}), which can be attributed to favourable interactions between solute and solvent. Agreement between Wolf and PPPM is good, although there is an observable increase in the error in the diffusion coefficients for difluoromethane when the fraction of solute reaches 10%. This indicates the difluoromethane interacts strongly with carbon dioxide, thus limiting diffusion.

To characterise the interactions between difluoromethane and carbon dioxide, we calculate the radial distribution (RDF) between the centre of mass for CH_4/CO_2 and CH_2F_2/CO_2 (Fig. 6) and the associated residence times and coordination numbers are shown in Table 2.

The RDF shows a larger density of carbon dioxide molecules in the difluoromethane mixture in the first and second solvation shell compared to methane. The same trend was noticed in the RDF by Do *et al.*,[53] where the first solvation shell has a higher density for difluoromethane than for methane. The number of carbon dioxide molecules present in the first and second solvation shells is ~30% higher for difluoromethane, and carbon dioxide resides about four times longer compared to methane. This indicates that carbon dioxide has a higher affinity for difluoromethane.

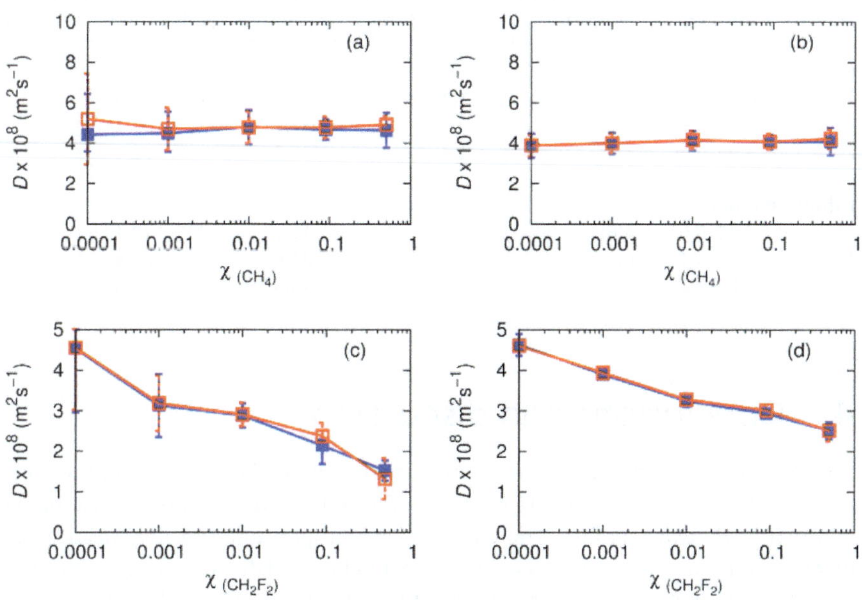

Fig. 5 Diffusion coefficients for (a) CH_4 in CO_2/CH_4 (B) CO_2 in CO_2/CH_4 (c) CH_2F_2 in CO_2/CH_2F_2 and (d) CO_2 in CO_2/CH_2F_2 at 308.2 K, 80 atmospheres. Filled blue squares and error bars indicate diffusion coefficients obtained using PPPM, and empty red squares and error bars indicate diffusion coefficients obtained using the Wolf summation method.

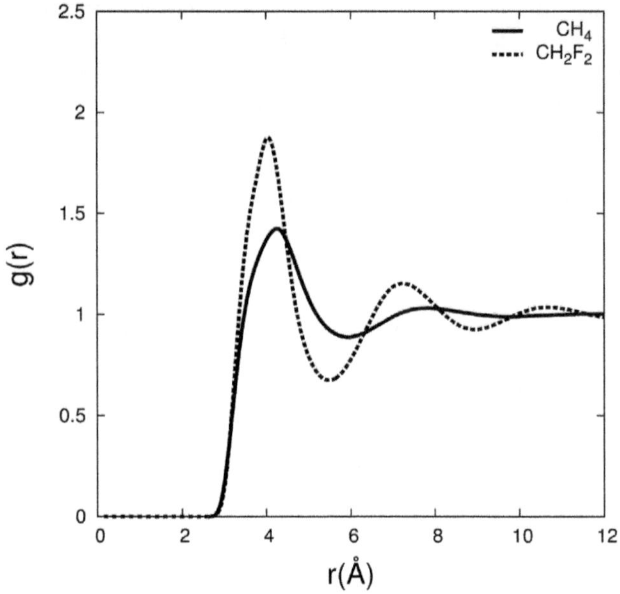

Fig. 6 Radial distribution function between the centre of mass of carbon dioxide and CH_4 (solid line), and CH_2F_2 (dashed line) for $\chi_{(solute)} = 0.1$ at 308.2 K, 80 atmospheres.

Table 2 Residence times and coordination for the centre of mass of carbon dioxide in CH_4 and CH_2F_2 for $\chi_{(solute)} = 0.1$ at 308.2 K, 80 atmospheres

Shell	Distance limit (Å)	Residence time - Wolf (ps)	Residence time - PPPM (ps)	Coordination number - Wolf	Coordination number - PPPM
Methane					
1st	0.00–5.95	1.2	1.2	8.6	8.6
1st + 2nd	0.00–9.60	2.6	2.5	37.2	37.1
2nd	5.95–9.60	1.3	1.3	28.6	28.5
Difluoromethane					
1st	0.00–5.65	4.6	4.6	12.1	12.1
1st + 2nd	0.00–9.24	10.5	10.3	55.6	55.6
2nd	5.65–9.24	5.9	5.7	45.5	45.5

4. Conclusions and discussion

The Wolf method shows good agreement with PPPM when modelling the electrostatic interactions of $scCO_2$ on GPUs for non-polar systems, whilst being approximately twice as fast. The choice of α is important, and may need to be investigated on a case by case basis to enable satisfactory agreement. For modelling carbon dioxide in the supercritical region it is advisable to use a half-box cut-off and a low value for α. In this investigation, all values of α less than 0.15 produced errors less than 2% for non-polar interactions, whilst polar interactions require α to be less than 0.1. Upon increasing the polarity of the system, the potential begins to degrade. Errors of the Wolf method with respect to PPPM are approximately 0.2% when considering a 100 : 1 mixture of carbon dioxide and difluoromethane with $\alpha = 0.075$. We observe a strong affinity of carbon dioxide to difluoromethane compared to methane, which can be seen by a decline in diffusion coefficients with increasing mole fraction of solute. Carbon dioxide resides about four times longer in the solvation sphere of difluoromethane compared to methane.

We can conclude that the significance of using the Wolf method on GPUs allows simulations to reach timescales twice as long as those run with PPPM, without significant loss in accuracy for a carefully chosen value of α for non-polar and mildly polar systems. We aim to follow up the investigation with further analysis of solvent–solute interactions and the study of fluorinated polymers, which have high solubilities in $scCO_2$. We will be investigating the free energy changes of fluorinated polymers, with an aim of further understanding the high affinity of fluorous polymers for carbon dioxide. Many free energy methods require long timescales in order to reach convergence; utilizing the Wolf method for this purpose will help achieve this goal.

Acknowledgements

This work was supported by the EPSRC, grants EP/I006559/1 and EP/K000128/1. We are grateful for access to the University of Nottingham High Performance Computing facility and MidPlus High Performance Computing facility. We appreciate useful discussions with Professor Paul Kelly and his co-workers (Imperial College) and Dr Chris Skylaris (University of Southampton). We thank Drs Richard Wheatley and Hainam Do for their critical reading of the manuscript.

References

1 J. E. Stone, J. C. Phillips, P. L. Freddolino, D. J. Hardy, L. G. Trabuco and K. Schulten, *J. Comput. Chem.*, 2007, **28**, 2618.

2 I. Buch, M. J. Harvey, T. Giorgino, D. P. Anderson and G. De Fabritiis, *J. Chem. Inf. Model.*, 2010, **50**, 397.

3 J. A. Baker and J. D. Hirst, *Mol. Inf.*, 2011, **30**, 498.

4 M. J. Harvey and G. De Fabritiis, *Wiley Interdiscip. Rev.: Comput. Mol. Sci.*, 2012, **2**, 734.

5 M. P. Allen and D. J. Tildesley, *Computer Simulation of Liquids*, Oxford University Press, 1989.

6 W. F. van Gunsteren and H. J. C. Berendsen, *Angew. Chem., Int. Ed. Engl.*, 1990, **29**, 992.

7 M. Bergdorf, C. Peter and P. H. Hünenberger, *J. Chem. Phys.*, 2003, **119**, 9129.

8 B. R. Brooks, R. E. Bruccoleri, B. D. Olafson, D. J. States, S. Swaminathan and M. Karplus, *J. Comput. Chem.*, 1983, **4**, 187.

9 P. P. Ewald, *Ann. Phys.*, 1921, **369**, 253–287.

10 R. Hockney and J. Eastwood, *Computer Simulation Using Particles*, CRC Press, 1981.

11 T. Darden, D. York and L. Pedersen, *J. Chem. Phys.*, 1993, **98**, 10089.

12 U. Essmann, L. Perera, M. L. Berkowitz, T. Darden, H. Lee and L. G. Pedersen, *J. Chem. Phys.*, 1995, **103**, 8577.

13 M. J. Harvey and G. De Fabritiis, *J. Chem. Theory Comput.*, 2009, **5**, 2371.
14 W. M. Brown, A. Kohlmeyer, S. J. Plimpton and A. N. Tharrington, *Comput. Phys. Commun.*, 2012, **183**, 449.
15 D. J. Hardy, J. E. Stone and K. Schulten, *Parallel Comput.*, 2009, **35**, 164.
16 D. Wolf, P. Keblinski, S. R. Phillpot and J. Eggebrecht, *J. Chem. Phys.*, 1999, **110**, 8254.
17 C. J. Fennell and J. D. Gezelter, *J. Chem. Phys.*, 2006, **124**, 234104.
18 J. S. Hansen, T. B. Schrøder and J. C. Dyre, *J. Phys. Chem. B*, 2012, **116**, 5738.
19 T. D. Nguyen, J.-M. Y. Carrillo, A. V. Dobrynin and W. M. Brown, *J. Chem. Theory Comput.*, 2013, **9**, 73.
20 H. Peker, M. P. Srinivasan, J. M. Smith and B. J. McCoy, *AIChE J.*, 1992, **38**, 761.
21 Z. Li and C. K. Hall, *Langmuir*, 2005, **21**, 7579.
22 J. Jennings, M. Beija, A. P. Richez, S. D. Cooper, P. E. Mignot, K. J. Thurecht, K. S. Jack and S. M. Howdle, *J. Am. Chem. Soc.*, 2012, **134**, 4772.
23 H. Lee, E. Terry, M. Zong, N. Arrowsmith, S. Perrier, K. J. Thurecht and S. M. Howdle, *J. Am. Chem. Soc.*, 2008, **130**, 12242.
24 Z. Wimmer and M. Zarevúcka, *Int. J. Mol. Sci.*, 2010, **11**, 233.
25 A. Gießauf and T. Gamse, *J. Mol. Catal. B: Enzym.*, 2000, **9**, 57.
26 R. L. Silveira, J. Martínez, M. S. Skaf and L. Martínez, *J. Phys. Chem. B*, 2012, **116**, 5671.
27 H.-L. Liu, W.-C. Hsieh and H.-S. Liu, *Biotechnol. Prog.*, 2004, **20**, 930.
28 W. Leitner, *Acc. Chem. Res.*, 2002, **35**, 746.
29 D. J. Heldebrant and P. G. Jessop, *J. Am. Chem. Soc.*, 2003, **125**, 5600.
30 D. M. D'Alessandro, B. Smit and J. R. Long, *Angew. Chem., Int. Ed.*, 2010, **49**, 6058.
31 Z. Su and M. Maroncelli, *J. Chem. Phys.*, 2006, **124**, 164506.
32 J. M. Desimone, Z. Guan and C. S. Elsbernd, *Science*, 1992, **257**, 945.
33 S. P. Nalawade, F. Picchioni and L. P. B. M. Janssen, *Prog. Polym. Sci.*, 2006, **31**, 19.
34 A. Dardin, J. M. DeSimone and E. T. Samulski, *J. Phys. Chem. B*, 1998, **102**, 1775.
35 A. Sadeghifar, M. Dadvar, S. Karimi and A. F. Ghobadi, *J. Mol. Graphics Modell.*, 2012, **38**, 455.
36 A. Angoshtari and A. Yavari, *Phys. Lett. A*, 2011, **375**, 1281.
37 S. Plimpton, *J. Comput. Phys.*, 1995, **117**, 1.
38 A. D. MacKerell, D. Bashford, R. L. Dunbrack, J. D. Evanseck, M. J. Field, S. Fischer, J. Gao, H. Guo, S. Ha, D. Joseph-McCarthy, L. Kuchnir, K. Kuczera, F. T. K. Lau, C. Mattos, S. Michnick, T. Ngo, D. T. Nguyen, B. Prodhom, W. E. Reiher, B. Roux, M. Schlenkrich, J. C. Smith, R. Stote, J. Straub, M. Watanabe, J. Wiórkiewicz-Kuczera, D. Yin and M. Karplus, *J. Phys. Chem. B*, 1998, **102**, 3586.
39 J. G. Harris and K. H. Yung, *J. Phys. Chem.*, 1995, **99**, 12021.
40 B. J. Palmer and J. L. Anchell, *J. Phys. Chem.*, 1995, **99**, 12239.
41 J. Kolafa and J. W. Perram, *Mol. Simul.*, 1992, **9**, 351.
42 M. Deserno and C. Holm, *J. Chem. Phys.*, 1998, **109**, 7694.
43 W. Hoover, *Phys. Rev. A: At., Mol., Opt. Phys.*, 1985, **31**, 1695.
44 NVIDIA, *CUDA Dynamic Parallelism Programming Guide*, 2012.
45 W. C. Swope, *J. Chem. Phys.*, 1982, **76**, 637.

46 H. J. C. Berendsen, J. P. M. Postma, W. F. van Gunsteren, A. DiNola and J. R. Haak, *J. Chem. Phys.*, 1984, **81**, 3684.
47 X. Michalet, *Phys. Rev. E: Stat., Nonlinear, Soft Matter Phys.*, 2010, **82**, 041914.
48 Y. Iwai, H. Higashi, H. Uchida and Y. Arai, *Fluid Phase Equilib.*, 1997, **127**, 251.
49 S. Nosé and M. L. Klein, *Mol. Phys.*, 1983, **50**, 1055.
50 H. Higashi and K. Tamura, *Mol. Simul.*, 2010, **36**, 772.
51 H. A. O'Hern and J. J. Martin, *Ind. Eng. Chem.*, 1955, **47**, 2081.
52 R. Span and W. Wagner, *J. Phys. Chem. Ref. Data*, 1996, **25**, 1509.
53 H. Do, R. J. Wheatley and J. D. Hirst, *J. Phys. Chem. B*, 2010, **114**, 3879.

Faraday Discussions

PAPER

A real-time proximity querying algorithm for haptic-based molecular docking†

Georgios Iakovou, Steven Hayward* and Stephen Laycock*

Received 6th December 2013, Accepted 31st January 2014
DOI: 10.1039/c3fd00123g

Intermolecular binding underlies every metabolic and regulatory processes of the cell, and the therapeutic and pharmacological properties of drugs. Molecular docking systems model and simulate these interactions *in silico* and allow us to study the binding process. Haptic-based docking provides an immersive virtual docking environment where the user can interact with and guide the molecules to their binding pose. Moreover, it allows human perception, intuition and knowledge to assist and accelerate the docking process, and reduces incorrect binding poses. Crucial for interactive docking is the real-time calculation of interaction forces. For smooth and accurate haptic exploration and manipulation, force-feedback cues have to be updated at a rate of 1 kHz. Hence, force calculations must be performed within 1ms. To achieve this, modern haptic-based docking approaches often utilize pre-computed force grids and linear interpolation. However, such grids are time-consuming to pre-compute (especially for large molecules), memory hungry, can induce rough force transitions at cell boundaries and cannot be applied to flexible docking. Here we propose an efficient proximity querying method for computing intermolecular forces in real time. Our motivation is the eventual development of a haptic-based docking solution that can model molecular flexibility. Uniquely in a haptics application we use octrees to decompose the 3D search space in order to identify the set of interacting atoms within a cut-off distance. Force calculations are then performed on this set in real time. The implementation constructs the trees dynamically, and computes the interaction forces of large molecular structures (*i.e.* consisting of thousands of atoms) within haptic refresh rates. We have implemented this method in an immersive, haptic-based, rigid-body, molecular docking application called Haptimol_RD. The user can use the haptic device to orientate the molecules in space, sense the interaction forces on the device, and guide the molecules to their binding pose. Haptimol_RD is designed to run on consumer level hardware, *i.e.* there is no need for specialized/proprietary hardware.

School of Computing Sciences, University of East Anglia, Norwich, Norfolk, NR4 7TJ, UK. E-mail: s.laycock@uea.ac.uk; sjh@cmp.uea.ac.uk; Fax: +44 (0) 1603 593345; Tel: +44(0) 1603 593795; +44(0) 1603 593542

† Electronic supplementary information (ESI) available: Algorithms 1 and 2. See DOI: 10.1039/c3fd00123g

1 Introduction

Intermolecular complex formation underlies all the metabolic and regulatory processes of the cell, as well as the therapeutic effects and pharmacological properties of drugs. For the past 40 years, scientists have been studying such bindings and relied on experimental (*in vitro*) work and computational methods (*in silico*) to study, model, and replicate them. *Molecular docking* refers to the computational methods devised and employed by researchers and field practitioners in order to simulate (as accurately as possible) this natural process. The ultimate goal of docking is to fit two molecules (often referred to as receptor and ligand molecules) together in a viable configuration based on their topographic and physiochemical properties.

There are two main approaches to docking, one automated, the other interactive. Automated methods[1–3] utilize sophisticated, pose selection algorithms and rely only on computer power to carry them through. Conversely, interactive methods[4,5] allow human intervention, and their performance depends closely on human intuition, knowledge and expertise. Automated docking solutions comprise the majority of the applications available in the field. Nonetheless, they are time consuming (solutions to docking problems can take several hours),[4,6] often predict incorrect docking conformations,[7,8] and by their very nature are not able to benefit from human knowledge and intuition. Haptic devices[9] and interactive molecular visualization systems[5,10–13] address these issues by transferring some of the complexity of the molecular binding process from the computer to the human. Haptic-based docking systems simulate the docking process in a 3D virtual environment, where the user interacts with the virtual molecules and performs a knowledge-guided search and selection of the final binding pose. They provide an immersive virtual docking environment where the user can sense (*via* a haptic device) the interaction forces and guide the molecules to their binding pose. They also establish a learning environment for the study of the docking process, and of the underlying interactions. It has been shown that such docking systems allow human perception, intuition and knowledge to assist and accelerate the docking process, and reduce incorrect binding poses.[14]

Other applications of haptics in biomolecular science assist users to investigate the importance of haptic technology in e-learning and education,[15,16] interact with molecules during molecular dynamics simulations,[17,18] explore interactively the solvent accessible surface (ISAS) of a protein,[19] deform an elastic network model of a biomolecule by applying forces to individual atoms,[20] and interact with properties related to molecular quantum dynamics (wave-packet dynamics) and potential energy surfaces.[21]

A fundamental part in a haptic-based docking solution is the calculation of the interaction forces. The forces and torques acting between molecules is a consequence of nonbonded (noncovalent) interactions (assuming no chemical reaction takes place between them) which are due to van der Waals (VDW) and electrostatic interactions, if we ignore solvent induced forces. If we are treating the two molecules as rigid, these are the only interactions required although when molecules are considered to be flexible bonded interactions must be included either explicitly or implicitly.

For smooth and accurate haptic exploration and manipulation of the interaction forces, haptic technology necessitates that all force-feedback cues must maintain a refresh rate of 1 kHz due to the sensitivity of the human haptic sense. Hence, all force calculations must be performed within 1 ms. A brute force approach to calculate the VDW and electrostatic forces requires a time complexity of $O(\mathrm{NM})$, where N and M are the number of atoms in the receptor and ligand respectively. Such an approach is impractical on modern CPUs even for small molecules (*i.e.* molecules comprising several hundred of atoms each) due to the lack of processing power. Modern, high-end GPUs offer an alternative execution platform for this approach, and, given their processing power, have the potential to perform force calculations for large molecules (*i.e.* consisting of thousands of atoms) within the 1 ms constraint. However, our preliminary results (discussed in Section 5) suggest that the performance gains achieved by a GPU-based implementation of this approach, although noticeable, do not suffice for molecular structures containing more than a thousand atoms each.

Existing haptic-based docking approaches compute these forces on the CPU and as such, they often utilize pre-computed 3D force grids[4,22] and linear interpolation to accelerate the respective computations. These force grids address the 1 kHz refresh rate requirement efficiently, however, they are time-consuming to pre-compute (especially for large molecules), memory hungry, can induce rough force transitions at cell boundaries[22] and cannot be applied to flexible docking problems (since the force grids must be recomputed in real time as the molecule deforms).

In this work, we propose an efficient octree-based, proximity querying method that enables us to compute (on the CPU) in real time and at haptic refresh rates the intermolecular forces of docking. Our approach addresses efficiently and successfully all issues related to the pre-calculated force grids and can facilitate a haptics-driven docking of large molecular structures (*i.e.* consisting of thousands of atoms). As such, it can be applied equally in the study of protein–protein and protein–drug docking problems. We have implemented this method in a haptic-based, immersive, rigid body docking system called Haptimol_RD, which is designed to run on consumer level hardware, (*i.e.* there is no need for specialized/proprietary hardware). Our motivation is the eventual development of a haptic-based docking solution that can model molecular flexibility.

2 Previous work

The potential benefits of integrating haptic technology in molecular docking solutions have been under investigation since the late 60s.[23] Nonetheless, the progress made in this field has been slow. The main reasons for this were the lack of product commercialization (early haptic technology was proprietary), and the lack of the necessary computing power which rendered the use of haptic technology in docking solutions either prohibitively expensive and/or computationally infeasible. The emergence of powerful desktop computers, affordable haptic devices and open-source rendering APIs (Application Programming Interface), at the beginning of the 21st Century, alleviated these obstacles, and enabled molecular-docking researchers to incorporate haptic technologies in their studies. Though the number of related studies still remains small, it is anticipated that this number will increase as haptic technology becomes easier and cheaper to use and integrate.

Brooks et al.,[23] from the University of North Carolina, pioneered the field of haptic-based docking with the GROPE III project. They addressed and solved the force calculation problem by adopting Pattabiraman et al.'s pre-computed, energy-potential, grid method.[24] Similarly to Pattabiraman *et al.*, Brooks *et al.* pre-computed and stored the inter-atomic force at predefined 3D-grid cells. Each grid cell stored the summation of the VDW and electrostatic contributions to the force. They then computed the total force of the intermolecular interactions with a tri-linear interpolation on the appropriate force vectors. Bayazit et al.[25] applied a similar force calculation approach in their study. They developed a hybrid docking solution by integrating haptic technology into their automated motion planning method (OBPRM). In their solution, the haptic device allowed the user to sample the conformational space, sense the interaction forces, identify sites with low energy potentials, and connect these findings into a roadmap. This roadmap was then given as an input to the road planner that calculated the final docking path. Their energy and force calculations modelled only VDW interactions, and were accelerated with the use of a force grid.

Lee and Lyons[26] improved upon the force grid method by calculating separate force grids for each component of the non-bonded interactions (*i.e.* VDW and electrostatic). Moreover, the forces stored in the grid cells accounted for all possible pairwise atom-type combinations between the receptor and ligand molecules. Their approach provided a better approximation of the interaction forces, and enabled the user to scale, and turn on/off in real time the forces attributed to the VDW and electrostatic interactions. Their method has been adopted by many current, haptic-based docking solutions and studies related to the teaching of structural biology to students,[27] to computer-aided drug design,[28] and to overcoming the *trapping* problem in Molecular Dynamics simulations,[14] Like Lee and Lyons, Wollacott and Merz Jr.[22] with HAMStER utilized two different 3D-grids for force calculations (one for VDW and one for electrostatic), but they pre-computed the grids only around the active site of the receptor.

Nagata *et al.*[29] attempted to compute the force of all non-bonded interactions (VDW, electrostatic, and hydrogen bonds) in real time, without utilizing pre-computed grids. They concluded that (even for small molecules) their brute-force approach could not compute the binding interactions at haptic refresh rates, without a 100-fold increase in processor power available to them at the time. Within project CoRSAIRe, Ferey *et al.*[30] calculated the docking forces using a force grid for the electrostatic contribution and a brute force approach for the VDW contribution. Their brute force calculations, however, did not account for all interatomic VDW interactions, but only for those involving the surface atoms of the molecules. Hou and Sourina[31] and Sourina *et al.*[32] used a brute-force approach to calculate the VDW forces in their haptic-based, helix–helix docking system called HMolDock. They were able to dock with their system a aIIb helix (with 154 atoms) and a anti-aIIb helix (with 266 atoms).

Other approaches to the computation of the intermolecular interaction forces include the works of Daunay *et al.*, and Zonta *et al.* Daunay *et al.*[7] developed a haptic-based flexible docking application and utilized a simulation engine for the respective energy computations. The forces and torques, rendered on the haptic device, were not calculated; they were converted from the total energy potential by a novel force-field reconstruction method. Their system could not support "as is" 1 kHz, haptic rendering rates, and thus they relied on wave theory (and on the appropriate wave transformations) to bridge the rate disparities between haptic rendering (1 kHz) and force calculations. Finally, Zonta *et al.*[33] integrated the OpenBabel library into their docking system ZODIAC and used it for all force calculations. Using ZODIAC, they were able to calculate the forces at haptic refresh rates while docking a ligand of 16 atoms to a receptor of 5.5K atoms.

3 Methods

3.1 Force equations

The primary interest of this work is the real-time calculation of the VDW and electrostatic forces exerted between the receptor and ligand molecules during docking. The VDW interaction is modelled by the Lennard-Jones (LJ) potential and the electrostatic potential by Coulomb's law.

The VDW force acting on atom j as exerted by atom i and modelled by the LJ potential is given by,

$$\vec{F}_{ij}^{\mathrm{VDW}} = \left[12\frac{A_{ij}}{r_{ij}^{13}} - 6\frac{B_{ij}}{r_{ij}^{7}}\right]\hat{\vec{r}}_{ij} \tag{1}$$

where A_{ij} and B_{ij} are constants that depend on the type of interacting atoms, r_{ij} is the distance between these atoms (measured from their centre), and $\hat{\vec{r}}_{ij}$ is the unit vector in the direction from atom i to j. In eqn (1), the first term inside the brackets defines the repulsive part of the force, whereas the second term defines the attractive part – the dispersion force.

The force acting on atom j due to the electrostatic interaction with atom i as modelled by Coulomb's law is given by:

$$\vec{F}_{ij}^{\mathrm{ES}} = \frac{q_i q_j}{4\pi\varepsilon\varepsilon_0 r_{ij}^2}\hat{\vec{r}}_{ij} \tag{2}$$

where q_i and q_j are the atomic charges of the two atoms, ε_0 is the permittivity of free space, and ε is the relative permittivity dependent on the dielectric properties of the solvent. Given eqn (1) and (2) we compute the total force of the intermolecular interactions by summing up all pairwise, inter-atomic interactions involved. Namely, if N atoms from the receptor interact with M atoms from the ligand then the total force is given by,

$$\vec{F}_{\mathrm{Tot}} = \Sigma_i^N \Sigma_j^M \left(\left[12\frac{A_{ij}}{r_{ij}^{13}} - 6\frac{B_{ij}}{r_{ij}^{7}} + \frac{q_i q_j}{4\pi\varepsilon\varepsilon_0 r_{ij}^2}\right]\hat{\vec{r}}_{ij}\right) \tag{3}$$

Eqn (3) is the total force acting on the ligand due to its interactions with the receptor. The total force acting on the receptor due to its interactions with the ligand is of the same magnitude but in the opposite direction. There are related expressions for the torque acting on the molecules but these are not modelled as most haptic devices are unable to exert torques on the user. In our work, eqn (3) is applied only for those inter-atomic interactions within a cut-off distance.

We use the non-bonded parameters of the Gromos54a7[34] force field (as specified and implemented in Gromacs version 4.6.2[35]) to provide values for the parameters A_{ij}, B_{ij}, q_i and q_j. Specifically, we calculate the values of A_{ij} and B_{ij} as

$A_{ij} = \sqrt{A_i \times A_j}$ and $B_{ij} = \sqrt{B_i \times B_j}$ respectively, where A_i, B_i and A_j, B_j are the LJ parameters of atoms i and j as given by the Gromos54a7 force field. The Coulomb constant $\frac{1}{4\pi\varepsilon_0}$ is set equal to 138.935485 kJ mol^{-1} nm e^{-2} and we set ε equal to 1.0 for the purposes of this benchmarking study, *i.e.* we assume interactions take place *in vacuo*. The total force is measured in kJ mol^{-1} nm^{-1}. To render it on the haptic device we convert it first to Newtons by dividing by 6.02329×10^{11} since 1 N is equivalent to 6.02329×10^{11} kJ mol^{-1} nm^{-1}. We then scale it appropriately to ensure that a good range of forces can be felt by the user through the haptic device. In addition to Gromos54a7, our method can utilize other force fields such as AMBER,[36] CHARMM[37] and OPLS-aa.[38]

3.2 Octree-based proximity query for force calculations

For a real-time calculation of the docking forces we devised a method that uses a cut-off distance to reduce the set of interatomic interactions considered in eqn (3), and octrees[39] to obtain this set at haptic refresh rates. Octrees are spatial partitioning hierarchies that recursively divide the geometry of an object into smaller subunits (called octants), until they reach a certain subdivision depth. Initially the object's bounding box (often modelled as a cube) is subdivided into eight child octants of the same size, then each child is subdivided into eight smaller child octants, and this process continues recursively L times, where L is the subdivision depth. This recursive subdivision results in a tree structure of degree 8 and depth L that represents the object's geometry, and in which each leaf octant contains a part of that geometry (Fig. 1). Such treelike structures simplify the implementation, and accelerate the execution of costly operations[40] such as object intersection discovery, neighbour finding, proximity querying *etc.* With octrees, these operations are most often implemented as simple, recursive tree traversals of the underlying structures. Our method constructs and stores within different octrees a spatial decomposition of the tertiary (3D) structure for both receptor and ligand molecules, and then uses these tree structures to efficiently query the 3D space and identify all atom pairs i and j whose r_{ij} distance is within this cut-off (Fig. 2). Force calculations are then performed on this set in real time.

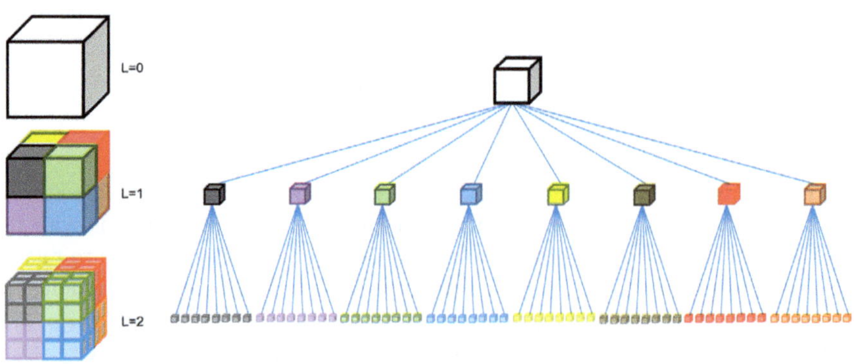

Fig. 1 An octree structure of depth 2 with its octant subdivisions. At depth 0 ($L = 0$) lies the root octant, at depth 1 are the eight child octants of the root, and at depth 2 are the eight leaf octants of those child octants. A complete octree of depth L contains 8^L leaf octants.

All tree traversals during octree construction and querying are performed recursively in a depth-first order starting from the root. Octrees have been applied in many different problems related to solid modelling, computer graphics, computer-aided design, computer vision and image processing, and robotics,[40,41] but there are no reports of their application in molecular-docking problems. Octrees perform very well in problems addressing deformable geometry,[42] and we want to exploit this property while studying and modelling molecular flexibility. In the following three subsections we describe our octree construction and querying algorithms, and outline how we compute the total force.

3.2.1 Octree construction. Our method constructs an octree, and populates it with atoms simultaneously. It requires as input a target subdivision level L, and the dimensions of the box that bounds the geometry to be partitioned. The method derives initially the dimensions of the molecule's minimum bounding cube, and sets this cube as the tree's root octant. It then utilizes Algorithm 1, ***AddAtomAndConstructChildOctants***,† to subdivide the tree into equally-sized child/leaf octants, and to populate them with atoms. Given an atom's coordinates, Algorithm 1† traverses the octree, and assigns the atom to the leaf octant that either intersects or contains this centre. If the path to this leaf octant (or the leaf itself) does not exist, the method creates the required octants (forming this path) dynamically, and subdivides the tree structure accordingly. This subdivision terminates when the target level L is reached, the respective leaf octant is created, and the atom is assigned successfully to it. When the atom is intersected by more

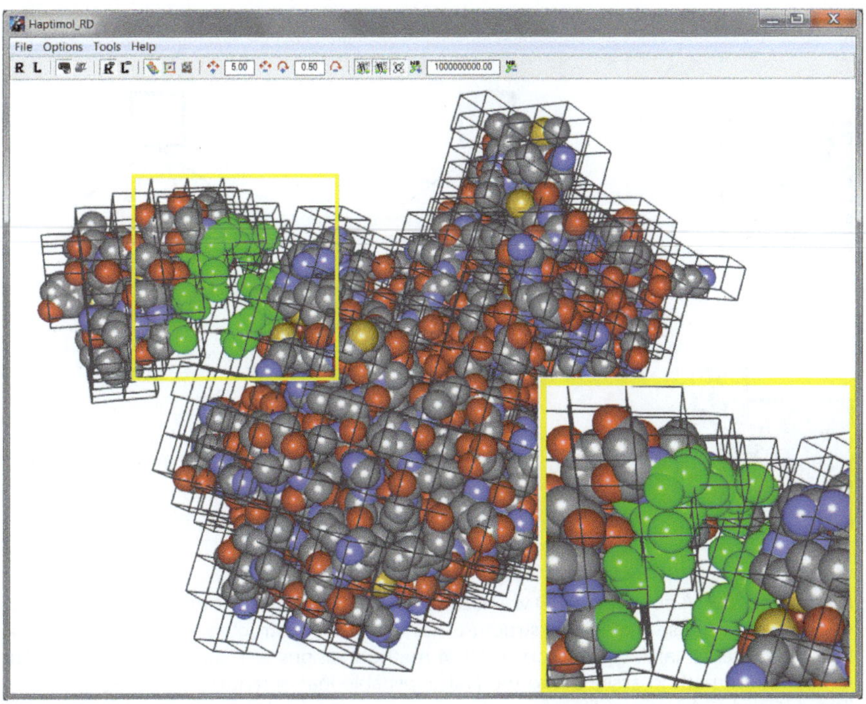

Fig. 2 Main image: Two proteins interact within Haptimol_RD. Green colours denote the atoms that interact within a given cut-off distance. The respective octree leaf octants are also depicted. Inset: Zoomed in view of the interacting atoms.

than one octant our construction algorithm assigns the atom to the first leaf octant traversed, and not to all of them, as is customary for octree-based collision detection algorithms (Fig. 3a and b). The query algorithm (see Section 3.2.2) ensures that the given atom will be considered for the final query set, irrespective of the child node it is assigned to. Furthermore, this one-to-one relationship between atoms and octants accelerates the performance of the query algorithm because each inter-atomic pair is considered only once, and thus the cost of handling duplicate pairs is avoided (Fig. 3c and d). Since an octant is created only when an atom intersects it, the resultant octree contains no empty octants (octants with no atoms), and thus it is compact and memory efficient, and helps the query algorithm avoid unnecessary octant traversals. Lastly, our experiments indicated (see Section 4.1) that when the octree depth was set to four, our approach attained a performance balance between its octree construction and querying times.

3.2.2 **Octree querying.** As the two molecules move in space, their geometry changes position and orientation. These changes are stored in the two viewing transformation matrices T_R and T_L, for the receptor and ligand molecules respectively. To query correctly the two octree structures, these matrices must be applied to both octant and atom coordinates stored within these trees. To save a substantial amount of matrix–vector multiplications, we combine T_R and T_L into one matrix using the relation $T_{New} = T_R^{-1}T_L$, and then apply T_{New} only to the ligand-related coordinates (we use $T_{New} = T_L^{-1}T_R$ and apply T_{New} to the receptor-related coordinates if the ligand is larger than the receptor). This approach maintains the relative orientation of both structures intact, and instead

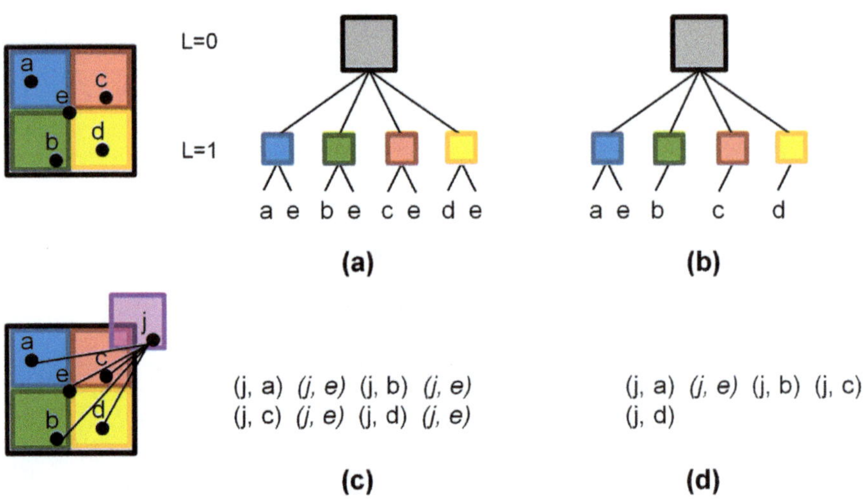

Fig. 3 Top row: A conceptual 2D visualization of an octree cube containing atoms a, b, c, d, e, and the corresponding tree structures (of depth 1) constructed. Atom e intersects all four leaf octants. (a) A typical construction method assigns e to all leaf octants. (b) Our construction method assigns e to the first applicable leaf octant traversed. Bottom row: The octree is intersected by an octree containing atom j. Atom j interacts with atoms a, b, c, d, and e. (c) Under the typical construction, the set of interacting atom pairs produced contains duplicates of (j,e). (d) Under our construction, the set contains only one instance of (j,e).

of transforming both molecules in space, it keeps the receptor geometry fixed and transforms only the ligand geometry with respect to it.

Given T_{New} and the two octree structures our method applies Algorithm 2, ***DeriveInteractingAtomPairsSet***,† to query pairwise and recursively the underlying octants, identify all interacting atom pairs within the cut-off distance, and return the set, S_{Pairs}, of these pairs. Using the atoms stored in S_{Pairs} (and their non-bonded parameters), the force calculation procedure derives the total force. The query algorithm starts from both octree roots and performs a pairwise traversal of their respective child octants. For each non-leaf octant pair examined, it updates the necessary octant coordinates using T_{New}, and then computes the net distance d_{Net} between the octant centres using:

$$d_{Net} = d_{Tot} - (r_R + r_L) \tag{4}$$

where d_{Tot} is the total distance between the octant centres, and r_R and r_L are the radii of their bounding spheres (Fig. 4a). If d_{Net} is less than or equal to the cut-off distance plus an error margin e_s (to safeguard the query from floating point arithmetic errors), the algorithm continues and examines recursively the children of this octant pair, and stops when it reaches the relevant leaf octants. At the leaf level, the algorithm computes all pairwise inter-atomic distances between the centres of the atoms stored in the respective leaf-octants, and saves in S_{Pairs} those atom pairs with a distance less than or equal to the cut-off.

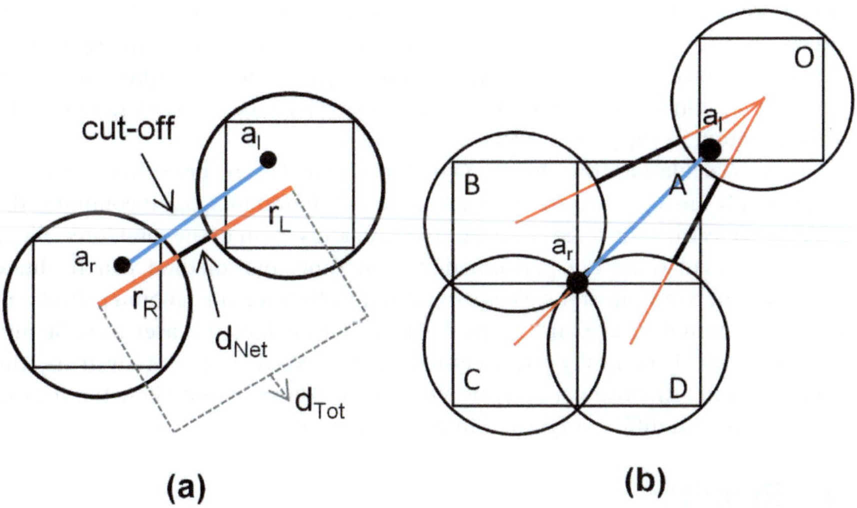

Fig. 4 (a) A 2D visualization of eqn (4). The two leaf octants belong to two different octrees and contain atoms a_r and a_l, respectively. These atoms are within the cut-off distance and thus interact. In such cases, d_{Net} will always be less than or equal to the cut-off, since it defines the closest distance between the two spheres (of radii r_R and r_L) bounding these atoms. (b) A diagram-based proof that our algorithm will handle the atom pair (a_r, a_l) correctly, irrespective of the leaf octant (A, B, C or D) atom a_r was initially assigned to. Even in the worst case scenario (a_r was assigned to C), the d_{Net} of octants C and O will be equal to the cut-off (blue line), whereas in the other three cases it will be less than the cut-off. In this example octant A is the best insertion case for a_r since it is in zero d_{Net} distance from octant O.

By utilizing eqn (4), the cut-off distance, and the octree hierarchy, our query strategy performs quick rejection tests on the underlying molecular geometry, and converges rapidly to those leaf-octants containing the interacting atom pairs. Our octant rejection test (*i.e.* d_{New} > (cut-off + e_s)) is a simple numerical test with no substantial computational cost. Moreover, it is invariant to octant orientation in space, since d_{Net} is computed based on octant bounding sphere radii and not on octant box dimensions (*i.e.* bounding cube dimensions). Since atoms are bounded by octant boxes and octant boxes are bounded by octant bounding spheres, two atoms will interact, if and only if the d_{Net} distance of their bounding octants is less or equal to the cut-off, regardless of octant orientation. Hence, if the d_{Net} distance between two octants is not within the cut-off it is safe to discard that part of molecular geometry from the solution set, and query it no further. The same reasoning applies to the special tree construction cases stated in Section 3.2.1 (*i.e.* when atoms are intersected by more than one octant). As previously mentioned, the construction algorithm assigns such atoms to the first leaf octant traversed. Let a_r be one such atom. If a_r is intersected by multiple octants, then its centre must lie on a common point shared by the bounding boxes/spheres of these octants. Let, now, a_l be an atom interacting with a_r. Clearly the d_{Net} distance between the octants containing a_l and a_r must be less than or equal to the cut-off. But since a_r lies on a shared point then the d_{Net} distance between the octant containing a_l and the remaining octants must also be less than or equal to the cut-off (Fig. 4b). As such, a_r will be queried and inserted in S_{Pairs} as expected, regardless of its placement within the octree during construction. Hence, our rejection test will prune correctly the octrees (and their underlying geometry), under all construction cases, special or trivial. Nonetheless, floating point precision errors might induce false fails in our test, and inadvertently lead to atom omissions. The use of e_s safeguards the rejection test and query algorithm from such erroneous outcomes. In our queries we let e_s be the VDW radius of the smallest atom in the given molecule.

3.2.3 Calculating the force. Our force calculation procedure traverses sequentially the S_{Pairs} set, and for each atom pair found in S_{Pairs}, it computes the VDW and electrostatic force contributions and adds them to the total force. Since all force calculations are performed in real time, our method can facilitate independent handling of the electrostatic and VDW forces in a manner similar to the one reported in Lee and Lyons.[26] Namely, it enables the user to scale and switch on/off dynamically the electrostatic and VDW forces, as well as, the repulsive and attractive parts of the VDW force. These option allow the user to experiment with different types of interactions easily.

4 Results

To test our approach we have implemented the methods discussed above in a haptic-based, rigid-docking application called Haptimol_RD. Haptimol_RD was developed using the OpenHaptics HD and OpenGL libraries, and the Visual C++ programming language. Using Haptimol_RD we conducted the following set of experiments:

1. benchmarking construction and querying performances
2. measuring querying performance during real-time rigid-docking simulations

To account for possible speed inconsistencies caused by background processes, we executed each benchmarking experiment ten times and reported the average. All of our tests were conducted on a 2.93GHz Intel Core i7 PC, running a 64bit version of Windows with 8GB RAM. The haptic device utilized in our simulations was the Geomagic Touch (formerly known as SensAble Technologies Phantom Omni).

4.1 Benchmarking performance

We report the octree construction and querying times on various molecular structures. The construction times must be considered if receptor flexibility is modelled, given that the trees would have to be constructed repeatedly and in real time as the molecule deforms. For rigid-body docking, the querying times are the only values of practical importance, since the trees need only be calculated once prior to the interactive session. We benchmarked our octree construction method using four proteins comprising different numbers of atoms. The protein structures were obtained from the Protein Data Bank (PDB).[43] The proteins used were *Crambin* (PDB code: 1CRN), *Lysozyme* (3HTB), *Alcohol Dehydrogenase* [containing only one of the two subunits, $1ADG^{1S}$], *Alcohol Dehydrogenase* [containing both subunits (dimer), $1ADG^{D}$]. For each test protein we constructed seven octrees of different depths (from 2 to 8), and recorded the number of atoms each protein comprises, the total number of octants created, and the total number of leaf octants (Table 1). For depths lower than four, our method constructed the respective octree structures within one millisecond, in all test cases. At depth four it supported millisecond/sub-millisecond construction times for proteins containing up to 3500 atoms. For depths larger than four, sub-millisecond construction times were attained only for small proteins (comprising several hundreds of atoms). Fig. 5 depicts these construction times per protein, for all seven depth values. We did not test our construction method for depth values larger than eight due to physical memory limitations.

Table 1 Octree construction times per molecule in milliseconds, for depth levels 3, 4, and 5. The table lists the number of atoms comprising each molecule, the total number of octants created, and the total number of octants at the leaf level. *D* stands for Dimer and 1*S* for one subunit

Molecule	# of heavy atoms	L	Tot. # of octants	Tot. # of leaf octants	Construction time (ms)
		3	132	97	0.1023
1CRN	327	4	372	240	0.1837
		5	699	327	0.2837
		3	165	124	0.2617
3HTB	1388	4	683	518	0.4879
		5	1810	1127	0.8329
		3	190	143	0.539
$1ADG^{1S}$	3445	4	888	698	1.0025
		5	3122	2234	1.8006
		3	137	108	1.0035
$1ADG^{D}$	7046	4	644	507	1.6403
		5	3135	2491	2.6987

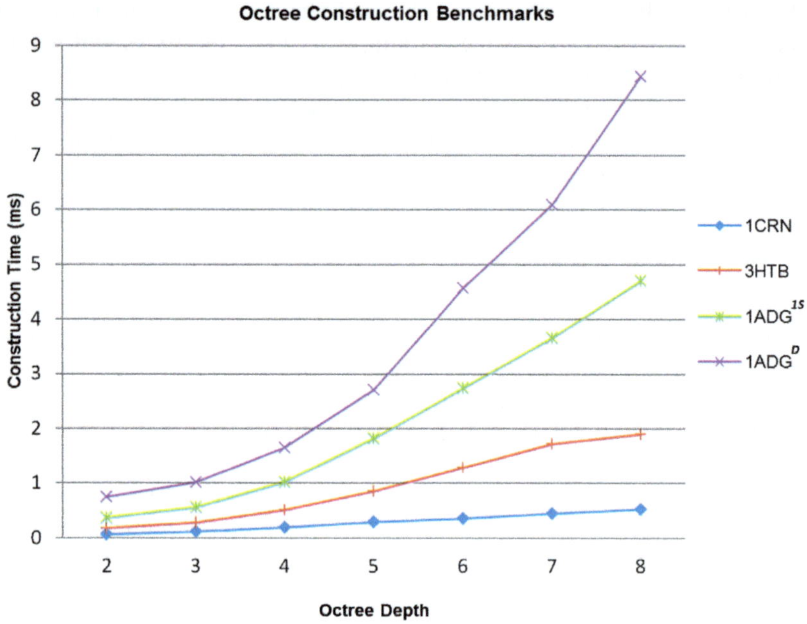

Fig. 5 Octree construction times per molecule in milliseconds, for octree depth levels 2–8.

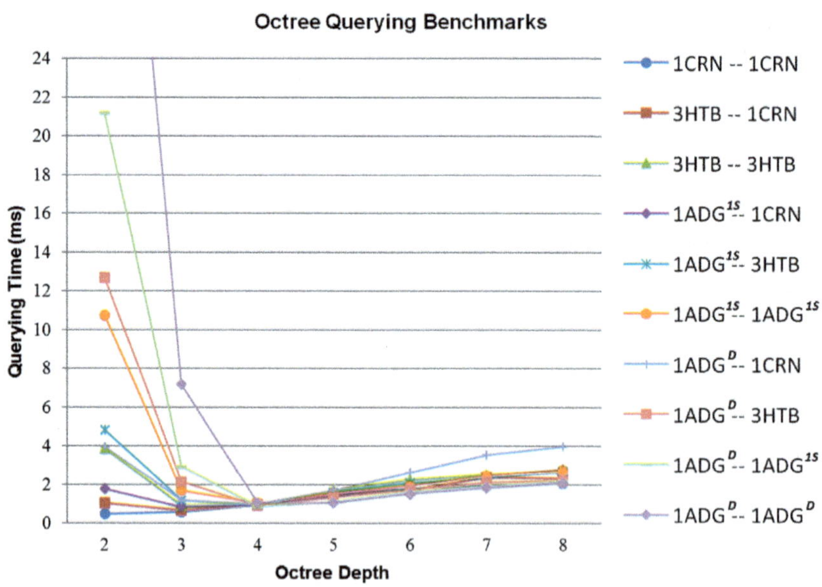

Fig. 6 Octree querying times of our ten interacting molecular pairs in milliseconds for the same seven octree depths (*i.e.* 2–8). The querying time for test case 1ADG(dimer)–1ADG(dimer) at depth 2 is not shown here (to avoid graph scaling and to improve graph readability). The time for this test case was 51.275 ms.

To benchmark our querying method we used the same four proteins and we devised ten test cases as shown in Fig. 6. Each test case consisted of two proteins (out of this set), the larger of which was assigned as the receptor. Based on their centres, the proteins were placed at the origin of the Cartesian space and the ligand molecule was then translated along the positive *x*-axis by a displacement distance δ_T. The δ_T distance varied per test case, and was chosen empirically. Namely, we displayed both proteins on the screen, moved the ligand along the positive *x*-axis while allowing it to intersect the receptor, and then visually selected the distance that generated a substantial amount of overlapping octants (and atoms), that allowed us to test adequately the performance of our querying method in relation to tree depth and protein size. Since such extensive atom overlapping would never occur during an actual docking simulation (because of the VDW repulsive forces), the querying response times recorded can also act as sufficient, upper-bound, performance indicators for our proximity query method. We used the value of 8 Å for the cut-off distance, and for each test case we recorded the δ_T distance used, the querying response time, the cardinality of the S_{Pairs} set, the total number of child and leaf octants traversed and the total number of inter-atomic distance calculations (Table 2). The querying response time is the time to determine the set, S_{Pairs}, the time to perform the force calculation being negligible in comparison. Fig. 6 depicts these querying times per test case, for all seven depth values. According to these results, our method achieved sub-millisecond querying response times for all test cases when the respective octree depths were set to four. Faster response times were attained for smaller octree depths in several cases, however our measurements indicated that at depths equal to four our approach maintained a performance balance between the construction and querying times. Evidently, the octree construction and querying costs are depth and geometry-size dependent, and impose a trade-off between construction speed and querying performance. This relationship is depicted in Tables 1 and 2 which list the construction and querying measurements pertinent to each test case, for depths 3, 4, and 5. Clearly as the size of the interacting proteins increases the construction cost also increases monotonically by almost twofold at every step, whereas the querying cost remains substantially lower at depths higher than four, than it is at depths three and lower. Measurements for depths 2, 6, 7 and 8 were not included for table clarity.

4.2 Real-time rigid-docking simulation

In addition to our benchmarking experiments we measured the performance of our proximity querying method in real-time, rigid-docking simulations, using three well known complexes, related to protein–protein and protein–drug docking. Although we did not model receptor flexibility in these simulations, we report the respective octree-construction times for the reader who wants to take into account these construction overheads (necessary if molecular flexibility is addressed). In our docking simulations, we used the complexes of *Epidermal Growth Factor* (EGF) with EGF receptor (EGFr), *Bovine Pancreatic Trypsin Inhibitor* (BPTI) with *Trypsin*, and anticancer drug *BAY43-9006* (sorafenib, Nexavar) with cancer target *B-raf* as defined in the 1NQL, 3OTJ, and 1UWH PDB files respectively. From these files, we extracted the 3D geometry of the six underlying molecules, in their binding conformations. Some molecules had gaps in their

Table 2 Octree querying times for ten interacting molecular pairs in milliseconds, for depth levels 3, 4, and 5. The table lists the displacement distance used in our experiments, the number of interacting atom pairs (S_{Pairs}) returned, the total number of child/leaf octants traversed, and the total number of atom pairs examined in order to generate the set S_{Pairs}. D stands for Dimer and 1S for one subunit

Interacting Molecules	δ_T(nm)	S_{Pairs}	L	Tot. # of octants travers.	Tot. # of leaf octants travers.	Tot. # of atom pairs exam.	Querying time (ms)
			3	2841	2450	30427	0.5666
1CRN–1CRN	1.55	4146	4	10405	7564	13792	0.9428
			5	18542	8137	8137	1.5023
			3	2076	1729	70700	0.6236
3HTB–1CRN	2	4040	4	8117	6041	22046	0.9337
			5	16690	8573	10234	1.4212
			3	1514	1114	123272	0.8613
3HTB–3HTB	2.9	3224	4	6707	5193	34134	0.9497
			5	14623	7916	11761	1.6833
			3	1474	1183	109636	0.8126
$1ADG^{1S}$–1CRN	2.8	4792	4	6885	5411	33116	0.9073
			5	16536	9651	14524	1.3562
			3	1140	848	211686	1.1711
$1ADG^{1S}$–3HTB	3.88	4338	4	5128	3988	48060	0.9148
			5	14077	8949	16695	1.587
			3	1008	670	315964	1.6513
$1ADG^{1S}$–$1ADG^{1S}$	4.29	3168	4	4254	3246	62847	1.0028
			5	11787	7533	17576	1.5472
			3	1040	838	218443	1.1536
$1ADG^{D}$–1CRN	3.27	8853	4	4824	3784	72293	0.9758
			5	16112	11288	31079	1.6743
			3	918	686	437063	2.096
$1ADG^{D}$–3HTB	4.2	2952	4	3670	2752	83642	0.8688
			5	10375	6705	22332	1.3097
			3	787	562	715797	2.902
$1ADG^{D}$–$1ADG^{1S}$	5.21	2569	4	2642	1855	99087	0.837
			5	8108	5466	23060	1.0944
			3	730	565	1719242	7.1563
$1ADG^{D}$–$1ADG^{D}$	5.58	2452	4	1830	1100	122310	0.9926
			5	5703	3873	25108	1.0454

geometry such as incomplete residues with missing atoms. These geometry gaps, however, were not significant enough to influence (positively or negatively) our performance measurements. We utilized Gromacs 4.6.2 (with the – *missing* flag when required) to obtain the non-bonded force parameters for all molecules except the drug sorafenib. For sorafenib we obtained the parameters through PRODRG server (http://davapc1.bioch.dundee.ac.uk/programs/prodrg/).[44] We conducted three docking simulations using Haptimol_RD, and the respective geometry and force parameter files. During the simulations, the user performed a haptic exploration of the receptor with the ligand, guided the ligand to its docking position and orientation (as defined in the original PDB file), and sensed the underlying intermolecular interactions on the haptic device. The simulation

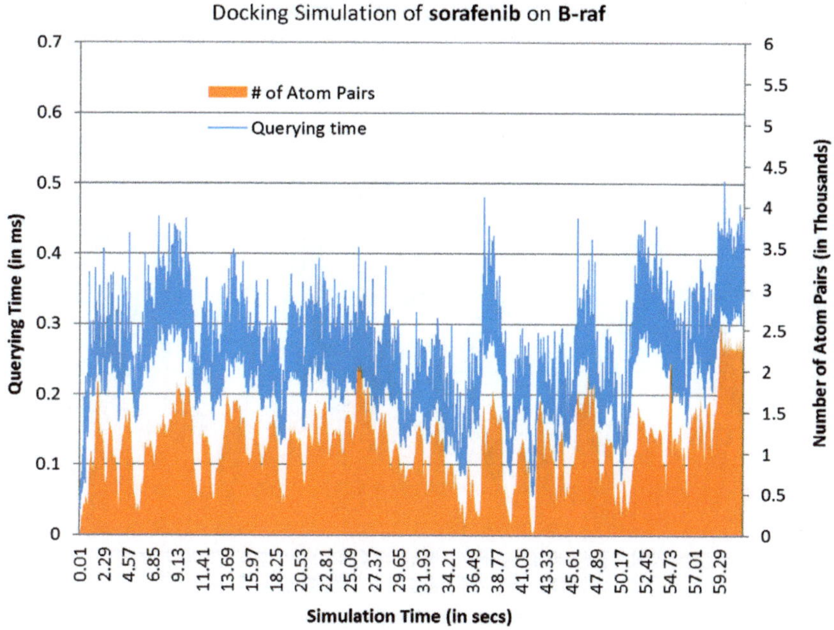

Fig. 7 A graph of the querying times and the number of atom pairs generated during a real-time, haptic-based docking simulation between the drug molecule sorafenib and the receptor protein B-raf.

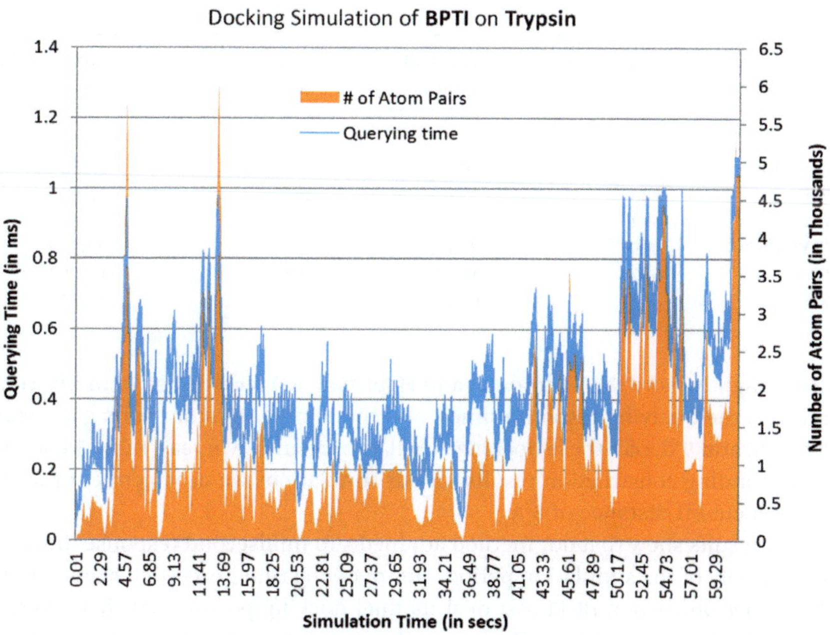

Fig. 8 A graph of the querying times and the number of atom pairs generated during a real-time, haptic-based docking simulation between the inhibitor BPTI and the receptor protein Trypsin.

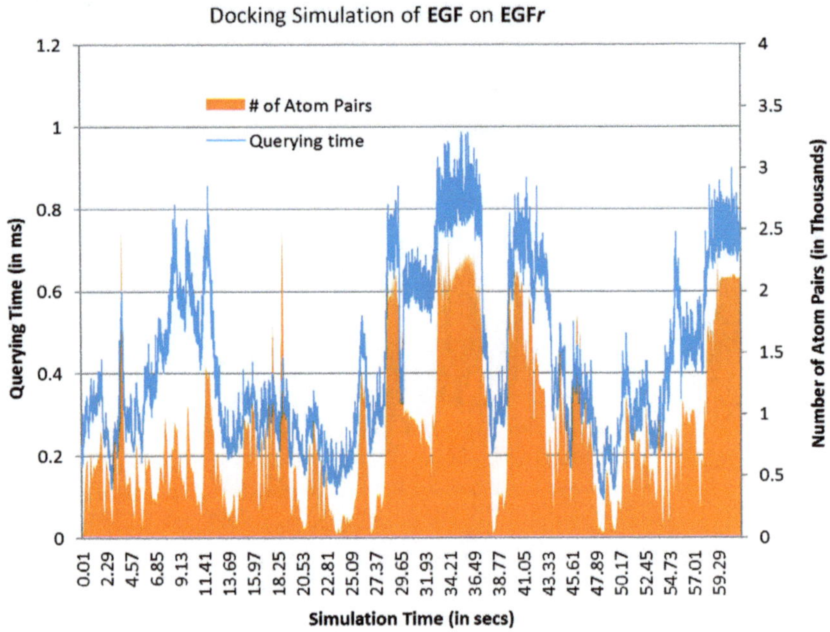

Fig. 9 A graph of the querying times and the number of atom pairs generated during a real-time, haptic-based docking simulation between EGF and the receptor protein EGF*r*.

Table 3 Octree construction times for the six molecules used in our real-time docking simulations

Molecule	# of heavy atoms	Tot. # of octants	Tot. # of leaf octants	Construction time (ms)
Sorafenib	48	100	45	0.0384
EGF	483	391	268	0.1929
BPTI	604	399	285	0.237
TRYPSIN	2094	843	656	0.6758
B-raf	5376	1012	795	1.3382
EGF*r*	5836	615	468	1.2572

lasted slightly more than a minute, and Haptimol_RD recorded at 10 millisecond intervals the querying response times, and the number of atom pairs generated. Fig. 7, 8, and 9 depict these docking simulation results, whereas Table 3 lists the corresponding construction results. In our simulations we used octrees of depth 4, and a cut-off distance of 8 Å.

Our results show that our method attained sub-millisecond response times for the majority of the simulation period. Our querying times exceeded slightly the 1 ms barrier only when BPTI assumed its final docking position. At that position our method performed a significant number of octant comparisons, induced by a substantial octree overlapping, because BPTI was docked deep into Trypsin's binding pocket. The response times in that case fluctuated between 1.015 and 1.092 ms.

5 Discussion

We have developed an octree-assisted, force calculation method applicable to haptic-based, rigid/flexible docking. When applied to rigid docking, our approach can facilitate larger molecular structures than the ones reported in previous studies, and it is therefore not constrained to protein–drug docking but can also be applied to protein–protein docking. Since tree construction times are irrelevant in rigid docking, our approach can maintain 1 ms force refresh rates for molecular pairs comprising 7K atoms each (*e.g.* $1ADG^{D}$-$1ADG^{D}$). On the other hand, flexible docking necessitates the real-time construction of the trees, and as such, 1 ms refresh rates would be achieved only for molecular pairs of up to 1.4k atoms each (*e.g.* 3HTB-3HTB). These sizes can be increased by more than twofold if we relax the 1 kHz refresh rate requirement down to 500 Hz (as noted in Otaduy and Lin[45]).

We also anticipate substantial performance gains if we adapt our method for a high-end GPU. Our belief stems from the observation that our GPU-based implementation from the brute force approach outperforming the CPU-based implementation by almost eightfold. Our GPU implementation (on an NVidia GeForce GTX 580) enabled us to compute the interaction force between eight hundred thousand atom pairs (simulating a receptor with one thousand atoms and a ligand with eight hundred atoms), whereas the CPU implementation (on a 2.93GHz Intel Core i7) constrained this set only to twenty thousand atom pairs (simulating a receptor with two hundred atoms, and a ligand with one hundred atoms). We expect to obtain analogous performance gains (if not better) when we transfer our octree proximity querying approach to the GPU. Therefore, a GPU-based implementation of our approach is the next logical step.

6 Conclusion

We have developed a method based on octrees for the real-time computation of the electrostatic and VDW force contributions in molecular docking. Our implementation calculates the total interaction force within haptic refresh rates using a cut-off distance, and overcomes the computational limitations of pre-computed force grids.

We have presented our octree construction and querying algorithms, and implemented and tested the performance of our method using different molecular structures. Our results show that we can achieve sub-millisecond force calculation responses for examples of interest from protein–protein and protein–drug docking.

Haptic-based docking enables intuitive and interactive exploration of binding poses which may give an advantage over automated approaches. It is expected that future research will attempt to improve the docking accuracy of such systems by incorporating more realistic force calculations (*e.g.* that model solvent effects implicitly), and addressing receptor–ligand flexibility.

References

1 N. Moitessier, P. Englebienne, D. Lee, J. Lawandi and C. R. Corbeil, *Br. J. Pharmacol.*, 2008, **153**, S7–S26.

2 M. Eisenstein and E. Katchalski-Katzir, *C. R. Biol.*, 2004, **327**, 409–420.
3 E. Yuriev, M. Agostino and P. A. Ramsland, *J. Mol. Recognit.*, 2011, **24**, 149–164.
4 M. Ouhyoung, *Ph.D. thesis*, University of North Carolina at Chapel Hill, Chapel Hill, NC, USA, 1990.
5 W. Humphrey, A. Dalke and K. Schulten, *J. Mol. Graphics*, 1996, **14**, 33–38.
6 D. Schneidman-Duhovny, Y. Inbar, R. Nussinov and H. J. Wolfson, *Nucleic Acids Res.*, 2005, **33**, W363–W367.
7 B. Daunay, A. Micaelli and S. Régnier, *Actes de ICRA'07 IEEE International Conference on Robotics and Automation*, Rome, Italie, 2007, pp. 840–845.
8 A. R. Leach, B. K. Shoichet and C. E. Peishoff, *J. Med. Chem.*, 2006, **49**, 5851–5855.
9 K. Salisbury, F. Conti and F. Barbagli, *IEEE Comput. Graphics Appl.*, 2004, **24**, 24–32.
10 R. Sayle and J. Milner-White, *Trends Biochem. Sci.*, 1995, **20**, 374–376.
11 E. F. Pettersen, T. D. Goddard, C. C. Huang, G. S. Couch, D. M. Greenblatt, E. C. Meng and T. E. Ferrin, *J. Comput. Chem.*, 2004, **25**, 1605–1612.
12 W. L. Delano, *The PyMOL Molecular Graphics System*, 2002, http://www.pymol.org.
13 T. E. Ferrin, C. C. Huang, L. E. Jarvis and R. Langridge, *J. Mol. Graphics*, 1988, **6**, 13–27.
14 E. Subasi and C. Basdogan, *Proceedings of EuroHaptics Conference*, 2006, 141–145.
15 A. Krenek, *Proceedings of Eurohaptics*, 2001, 142–145.
16 C. M. Sauer, W. A. Hastings and A. M. Okamura, *Proceedings of Eurohaptics*, 2004, 5–7.
17 N. Férey, O. Delalande, G. Grasseau and M. Baaden, *Proceedings of the 2008 ACM symposium on Virtual reality software and technology - VRST '08*, 2008, 91–94.
18 J. E. Stone, J. Gullingsrud and K. Schulten, *Proceedings of the 2001 symposium on Interactive 3D graphics - SI3D '01*, 2001, 191–194.
19 M. Stocks, S. Hayward and S. Laycock, *BMC Struct. Biol.*, 2009, **9**, 69–75.
20 M. B. Stocks, S. D. Laycock and S. Hayward, *J. Comput.-Aided Mol. Des.*, 2011, **25**, 203–211.
21 R. A. Davies, N. W. John, J. N. MacDonald and K. H. Hughes, *Proceedings of the tenth international conference on 3D Web technology - Web3D '05*, 2005, 143–150.
22 A. M. Wollacott and K. M. Merz Jr, *J. Mol. Graphics Modell.*, 2007, **25**, 801–805.
23 F. P. Brooks Jr, M. Ouh-Young, J. J. Batter and P. Jerome Kilpatrick, *ACM SIGGRAPH Comput. Graphics*, 1990, **24**, 177–185.
24 N. Pattabiraman, M. Levitt, T. Ferrin and R. Langridge, *J. Comput. Chem.*, 1985, **6**, 432–436.
25 O. B. Bayazit, G. Song and N. M. Amato, *Robotics and Automation*, 2001. Proceedings 2001 ICRA. IEEE International Conference on, 2001, pp. 954–959.
26 Y.-G. Lee and K. W. Lyons, *Comput.-Aided Des.*, 2004, **36**, 75–90.
27 P. Bivall, *Ph.D. thesis*, Linköping University, Linköping, Sweden, 2010.
28 S. K. Lai-Yuen and Y.-S. Lee, Computer Aided Design and Computer Graphics, *Ninth International Conference on*, 2005, **2005**, 199.
29 H. Nagata, H. Mizushima and H. Tanaka, *Bioinformatics*, 2002, **18**, 140–146.
30 N. Férey, J. Nelson, C. Martin, L. Picinali, G. Bouyer, A. Tek, P. Bourdot, J.-M. Burkhardt, B. F. Katz, M. Ammi, et al., *Virtual Reality*, 2009, **13**, 273–293.

31 X. Hou and O. Sourina, *Cyberworlds (CW)*, 2010 International Conference on, 2010, pp. 25–31.
32 O. Sourina, J. Torres and J. Wang, in *Transactions on Edutainment II*, Springer, 2009, pp. 105–118.
33 N. Zonta, I. J. Grimstead, N. J. Avis and A. Brancale, *J. Mol. Model.*, 2009, **15**, 193–196.
34 N. Schmid, A. P. Eichenberger, A. Choutko, S. Riniker, M. Winger, A. E. Mark and W. F. van Gunsteren, *Eur. Biophys. J.*, 2011, **40**, 843–856.
35 D. van der Spoel, E. Lindahl, B. Hess and the GROMACS development team, *GROMACS User Manual Version 4.6.2.*, University of Groningen, Royal Institute of Technology and Uppsala University, www.gromacs.org, 2013.
36 S. J. Weiner, P. A. Kollman, D. A. Case, U. C. Singh, C. Ghio, G. Alagona, S. Profeta and P. Weiner, *J. Am. Chem. Soc.*, 1984, **106**, 765–784.
37 B. R. Brooks, R. E. Bruccoleri, B. D. Olafson, S. Swaminathan, M. Karplus, et al., *J. Comput. Chem.*, 1983, **4**, 187–217.
38 R. C. Rizzo and W. L. Jorgensen, *J. Am. Chem. Soc.*, 1999, **121**, 4827–4836.
39 C. L. Jackins and S. L. Tanimoto, *Computer Graphics and Image Processing*, 1980, **14**, 249–270.
40 H. H. Chen and T. S. Huang, *Comput. Vision Graphics Image Process.*, 1988, **43**, 409–431.
41 M. C. Lin and D. Manocha, 2003.
42 M. Teschner, S. Kimmerle, B. Heidelberger, G. Zachmann, L. Raghupathi, A. Fuhrmann, M.-P. Cani, F. Faure, N. Magnenat-Thalmann, W. Strasser et al., *Computer Graphics Forum*, 2005, pp. 61–81.
43 H. M. Berman, J. Westbrook, Z. Feng, G. Gilliland, T. Bhat, H. Weissig, I. N. Shindyalov and P. E. Bourne, *Nucleic Acids Res.*, 2000, **28**, 235–242.
44 A. W. Schuttelkopf and D. M. Van Aalten, *Acta Crystallogr., Sect. D: Biol. Crystallogr.*, 2004, **60**, 1355–1363.
45 M. A. Otaduy and M. C. Lin, *ACM SIGGRAPH 2005 Courses*, 2005, p. 3.

Faraday Discussions

Computing power revolution and new algorithms: GP-GPUs, clouds and more: general discussion

DOI: 10.1039/c4fd90021a

Mr McIntosh-Smith opened the discussion of the papers by Mr Stone, Dr Nerukh and Dr Tchertanov by asking: With Petascale machines now becoming more common, and Exascale machines on the horizon, will simulation and visualisation applications be able to scale to take advantage of these new machines? What are the specific challenges? What will work well and what won't work so well?

Dr Tchertanov responded: The computer systems capable of reaching such performance will be the decisive components to perform the "high throughput" molecular dynamics simulations of very complex molecular systems at large scale (cells, tissues, small organisms) and even in real time enriched by the very large-scale performing graphical utilities, for example, the CAVEs (http://www.indiana.edu/~rcapub/v21n2/p28.html). Probably, the principal problems related with parallelization actually is the error propagation and it will be more pronounced with the increasing number of processors. A hybrid-reality visualisation (2D–3D approach) will be very challenging in the context of real-world scientific investigations.

Dr Nerukh answered: It is our firm belief, it will not be possible in the near future, or ever, to construct a machine that can simulate biological objects at biologically relevant times using all-atom MD (say, cells over minutes). Therefore, conceptually new methods are needed. Plus, MD provides "numerically experimental" data, and to extract knowledge from it, new concepts are needed. This is why "multiscaling" and more generally "multiphysics" is one of the ways forward.

Mr Stone replied: A number of molecular simulation tools, including NAMD in particular, provide good scalability for simulations of virus capsids and other large biomolecular complexes.[1]

The scalability of existing tools for single-copy simulations of much smaller molecular systems is more limited, due to the message passing latencies of the underlying communication interconnect hardware.

The performance limitations of current hardware can, in part, be overcome by so-called multiple copy algorithms, which are well suited to the limitations of available petascale computing hardware, making them usable for simulation of small biomolecular systems.[2]

Beyond the development of new methodological and algorithmic approaches, it is possible that future exascale machines will veer away from complete generality, toward architectures that are tailored specifically for particular application domains, such as the Anton machine by D. E. Shaw Research.[3]

At present, most visualization and analysis of molecular dynamics trajectories are performed on desktop workstations. In the case of petascale simulations, the size of the associated data is so large that is impractical to transfer, and it must be analyzed in-place on the supercomputer where the simulations are run. We have begun to adapt VMD for large scale parallel execution on petascale systems, thereby enabling analysis and visualization of petabytes of trajectory data.[4–6] During these efforts we have seen that it is possible to adapt many important analysis algorithms for massively parallel platforms, owing to the extensive data parallelism that is common to many molecular dynamics analysis and visualization techniques. Some of the challenges we have encountered to date have involved efficient use of heterogeneous computing architectures that combine multi-core CPUs with GPUs, finding ways to perform key data analysis and visualization tasks within the limited memory capacity of contemporary GPUs and supercomputer compute nodes, and hiding the complexities of distributed memory message passing and parallel programming from the end user of the tools, who ultimately wants to focus on science rather than on computing.

1 Gongpu Zhao, Juan R. Perilla, Ernest L. Yufenyuy, Xin Meng, Bo Chen, Jiying Ning, Jinwoo Ahn, Angela M. Gronenborn, Klaus Schulten, Christopher Aiken, and Peijun Zhang. Mature HIV-1 capsid structure by cryo-electron microscopy and all-atom molecular dynamics, *Nature*, 2013, **497**, 643–646.
2 Wei Jiang, James Phillips, Lei Huang, Mikolai Fajer, Yilin Meng, James Gumbart, Yun Luo, Klaus Schulten, and Benoit Roux, Generalized scalable multiple copy algorithms for molecular dynamics simulations in NAMD, *Comput. Phys. Commun.*, 2014, **185**, 908–916.
3 David E. Shaw, Martin M. Deneroff, Ron O. Dror, Jeffrey S. Kuskin, Richard H. Larson, John K. Salmon, Cliff Young, Brannon Batson, Kevin J. Bowers, Jack C. Chao, Michael P. Eastwood, Joseph Gagliardo, J.P. Grossman, C. Richard Ho, Douglas J. Ierardi, István Kolossváry, John L. Klepeis, Timothy Layman, Christine McLeavey, Mark A. Moraes, Rolf Mueller, Edward C. Priest, Yibing Shan, Jochen Spengler, Michael Theobald, Brian Towles, and Stanley C. Wang, Anton: A Special-Purpose Machine for Molecular Dynamics Simulation, Proceedings of the 34th Annual International Symposium on Computer Architecture (ISCA '07), New York, NY: ACM, 2007.
4 John E. Stone, Barry Isralewitz, and Klaus Schulten, Early Experiences Scaling VMD Molecular Visualization and Analysis Jobs on Blue Waters, Extreme Scaling Workshop (XSW), 2013, pp. 43–50.
5 John E. Stone, Kirby L. Vandivort, and Klaus Schulten, GPU-Accelerated Molecular Visualization on Petascale Supercomputing Platforms, UltraVis'13: Proceedings of the 8th International Workshop on Ultrascale Visualization, 2013, pp. 6:1–6:8.
6 John E. Stone, Ryan McGreevy, Barry Isralewitz, and Klaus Schulten, GPU-Accelerated Analysis and Visualization of Large Structures Solved by Molecular Dynamics Flexible Fitting, *Faraday Discuss.*, 2014, **169**, DOI: 10.1039/C4FD00005F

Dr Baaden commented: Cryo-EM determinations definitely provide stunning biological insights and represent an extremely exciting biophysical technique. Yet, there are limitations as envelopes derived from cryo-EM data have already undergone some processing and curation, are of relatively low resolution – compared for example to X-ray crystallography, even if this difference is closing in – and may incorporate ambiguities. In this paper, the evaluation of quality-of-fit of atomistic molecular dynamics simulation data to such envelopes is described. By design, the simulation provides very high resolution data. I wonder whether the simulation data that is fitted to the experimental envelopes is significantly more accurate in terms of resolution than the data it is fitted to, which may lead to overinterpretation. Hence, the apparent accuracy of the proposed algorithm may be significantly superior to the intrinsic degree of uncertainty due to the characteristics of the experimental method.

Mr Stone replied: The cross correlation algorithms presented in our paper provide multiple mechanisms by which uncertainty in the reference cryo-EM density map is accounted for when comparing with all-atom structures produced by MDFF. Regions of the cryo-EM density map with high noise or variance can be masked out by setting voxel values with the IEEE "not a number" or NaN floating point special value. Regions of the map with low density, far from the structure can be masked out according to a user-specified density threshold. Our cross correlation algorithms also test that a sufficient number of voxels have contributed to each cross correlation to be statistically meaningful. Spatial cross correlation maps are computed at only one eighth (12.5%) of the resolution and with 0.1% of the number of voxels relative to the original experimental cryo-EM reference density map, making it somewhat unlikely that visualizations of the spatial cross correlation maps would encourage over-interpretation. These algorithm features and design decisions provide the user with detailed control over the behaviour of the software, and with flexibility for handling density maps that are known to contain regions with high variance or noise, or structural ambiguities.

The MDFF method used for solving the structures that we are subsequently analyzing and visualizing attempts to improve upon the accuracy of the experimental cryo-EM density map by incorporating additional information not available in the density map itself.

In crystallography the term "super-resolution" refers to the ability to achieve higher coordinate accuracy than the resolution limit of the experimental X-ray diffraction data, which is similarly applicable here.[1] MDFF incorporates both generic structural information such as bond lengths and angles obtained from force fields, and structure-specific information from homology models, or any other type of initial models.

This combination of structural information allows MDFF to obtain structures that are indeed more accurate than would be provided by the EM data itself. Over-interpretation (or over-fitting) can be reduced by imposing or relaxing restraints (such as secondary structure) on part or all of the model during the fitting process. Furthermore, the extent of over-fitting can be quantified using cross-validation densities.[2]

The increased accuracy of the proposed models in turn affects the accuracy of the proposed cross correlation algorithm, in that the simulated density maps created for comparison are derived from this model. There are, however, several sources of error that influence the accuracy of the algorithm, how the simulated density is calculated, and how well it recreates the experimental data. The effects of these differences in density simulation are discussed briefly in the paper, and can affect subsequent analysis and visualization. The nature of the image processing performed the raw experimental cryo-EM images in addition to various other sources of ambiguity throughout all levels of any fitting process makes any definitive analysis difficult. Cross correlation values are very useful as relative measures of quality-of-fit, but all structures obtained through fitting methods should undergo further scrutiny in light of any additional experimental knowledge of the system in order to fully judge the accuracy of the model.

1 Gunnar F. Schröder, Michael Levitt, and Axel T. Brunger, Super-resolution biomolecular crystallography with low-resolution data, *Nature*, 2010, **464**, 1218–1222.
2 Benjamin Falkner and Gunnar F. Schröder, Cross-validation in cryo-EM-based structural modeling, *PNAS*, 2013, **110**, 8930–8935.

Dr Hayward asked Mr Stone: In the presentation a potential energy term, V_{ss}, was introduced but I could not find this in the paper. I asked the question whether this was something to do with secondary structure and this was confirmed. However, the reason for the addition of this term and its form was not elaborated upon. Although I now see in the introduction that mention was made of restraints being applied to secondary structures to preserve them during the fitting process, perhaps more details can be given.

Mr Stone replied: This is a detail of the MDFF fitting method that is not specifically related to evaluation of quality-of-fit so our paper only mentions it in passing. The use of restraints is described more fully in the MDFF method papers below. In the MDFF method, secondary structure restraints are used to maintain the integrity of secondary structure elements, to ensure stereochemical correctness, and to prevent over-fitting.[1–3] Restraints to ensure stereochemical correctness are also widely used in crystallography.

1 Leonardo G. Trabuco, Elizabeth Villa, Kakoli Mitra, Joachim Frank, and Klaus Schulten. Flexible fitting of atomic structures into electron microscopy maps using molecular dynamics, *Structure*, 2008,**16**, 673–683.
2 Leonardo G. Trabuco, Elizabeth Villa, Eduard Schreiner, Christopher B. Harrison, and Klaus Schulten, Molecular Dynamics Flexible Fitting: A practical guide to combine cryo-electron microscopy and X-ray crystallography, *Methods*, 2009, **49**, 174–180.
3 Kwok-Yan Chan, Leonardo G. Trabuco, Eduard Schreiner, and Klaus Schulten. Cryo-electron microscopy modeling by the molecular dynamics flexible fitting method, *Biopolymers*, 2012, **97**, 678–686.

Dr Baaden asked Mr Stone: Comparing simulation data to raw experimental data whenever possible has been suggested. This may represent a particular challenge with respect to cryo-EM experiments. In paper 10.1039/C4FD00005F the quality-of-fit of molecular dynamics trajectory data to experimentally-determined density maps is described. The cryo-EM envelopes are to some extent an interpretation of the raw experimental data, having gone through several image processing steps and some sorting and curation by the experimental teams. Deriving "images" similar to those acquired in cryo-EM experiments from the simulations might provide data for which at least some of these intermediate steps are eliminated and that may be closer and more legitimate to compare to the raw cryo-EM image data.

Mr Stone responded: The approach we describe in our paper is based in large part on the existing practices of experimental cryo-EM labs, where cross correlation metrics are used to compare experimental cryo-EM density maps with each other. The initial images captured during cryo-EM have a very, very low signal-to-noise ratio. The signal-to-noise ratios are often so low that no significant structural detail can be seen in a single CCD exposure. The raw cryo-EM images often look like white noise with only a slight "shadow" of the particles being imaged. It would appear to be very hard to make a useful comparison among individual CCD exposures and simulated images. I would assume that to use such an approach such as you are proposing, one would have to create a large number of synthetic cryo-EM images, and then go through the same image classification, registration, and co-adding process used for real cryo-EM images. At present, such an approach would be far too computationally costly to be practical for everyday use by researchers, but that could change with the passage of time. If the computational cost could be overcome, that would still leave problems related to reproducing the same imaging conditions, instrumentation-induced error, and image processing decisions that were made during the reconstruction of the experimental cryo-EM map that it is being compared with.

Dr Glowacki remarked: I wanted to reaffirm what John Stone said about the role that theory can play for complex systems nowadays, especially in complex systems. We are at a unique point in computational chemistry, where we have models that are capable of accuracy that, in some cases, can be even better than those provided by experiment – particularly in terms of time resolution and spatial resolution. Over the last few years, I have been involved in a series of studies on ultrafast solution phase dynamics.[1,2] In some cases, the theory has led to revision of the initial experimental interpretation, and new experimental insight. In many cases, experiments can provide phenomenological observables, but the underlying microscopic physics will not be obvious because the time and/or spatial resolution is limited. Of course experiments are extremely important to scientific progress, and this will never change. However, as computational approaches become better and better, it's important for the computational and modelling community to tell stories of how theory can lead experiment.

1 D. R. Glowacki, R. A. Rose, S. J. Greaves, A. J. Orr-Ewing and J. N. Harvey, *Nat. Chem.*, 2011, **3**, 850–855.
2 S. J. Greaves, R. A. Rose, T. A. A. Oliver, D. R. Glowacki, M. N. R. Ashfold, J. N. Harvey, I. P. Clark, G. M. Greetham, A. W. Parker, M. Towrie and A. J. Orr-Ewing, *Science*, 2011, **331**, 1423–1426.

Professor Olson commented: The amount of computer power for physics based simulations continues to grow, and the size of systems that can be studied with dynamics grows as well. However, there will always be biological systems that are beyond the reach of physical simulation. Meanwhile experimental information at all lengths and time-scales are providing immense amounts of data on complex biological systems. I believe that we need other, complementary methods of simulation to predict multi-scale biological phenomena that incorporates the knowledge contained in all of this data. I would suggest that knowledge-based and data informed agent-based simulations will be needed in addition to physics based simulations to address the structural dynamics of many biological systems.

Dr Zoppè continued the discussion of the paper by Mr Stone: EM data are obtained from averaging over thousands of sample molecules. But it is not certain that they are all in the same conformation, especially for large multiprotein objects. Do you think that it is possible to calculate a structure, select the samples that fit well, and separate them from others, that might really be different? In this way, going back and forth on the same set of acquisitions, it could be possible to refine the structure, and possibly obtain more than one conformation of the same complex.

Mr Stone replied to Dr Zoppè: In principle it is possible to obtain different structure conformations from the same density map, which can be distinguished using MDFF. In one such example, we solved an all-atom atomic structure using MDFF which, through careful analysis, showed that a conformation that was initially believed to have been captured in a particular cryo-EM density was actually an artifact created during processing of the raw EM images, where images associated with two different conformations had unknowingly been combined into one cryo-EM map. In cases like this, information from the MDFF-derived structure can then be used to guide experimentalists to re-process the raw EM data to create cleaner density maps, separating the two conformations into their own reconstructed density maps, and allowing MDFF simulations to be performed for the two conformations using the newly generated maps.

Dr Glowacki continued the discussion of the paper by Dr Nerukh: The hybrid hydrodynamics/molecular dynamics method that you describe seems very interesting. I'm curious what the conserved quantity is in the equation of motion that you're using. Within the atomically resolved region, is it appropriate to simply think of the interior as undergoing a standard NVE simulation? Are there any guidelines for how to choose the minimum/maximum size of the MD region, either in terms of number or atoms, or size of the cavity? I also wonder whether there are any guidelines on the maximum gradient for the s parameter in your eqn (17). I assume that too sharp a change in s is something to avoid. Is this correct?

Dr Nerukh answered: The conserved quantities are local 'macroscopic' densities, momenta, and energies (the last has not been implemented yet in our framework). But because the 'macroscopic' scales are at the border of 'microscopic', fluctuations are included explicitly in the hydrodynamic part. Therefore, it is not possible to consider the atomistic core as a standard NVE simulation – it exchanges matter and energy with the surroundings through fluctuations.

Dr Petrov asked: Does the framework of your hybrid model support a proper description of the electrostatic interactions? Is it possible to extend it to polar solvents and to account for dielectric coefficients of the continuum (hydrodynamic) region?

Dr Nerukh replied: Not explicitly at the moment, because we consider a non-charged hydrodynamic phase. However, this is also in our immediate plans to include the charges, which means that the hydrodynamics equations will be different.

The size of the atomistic interior should be such that the fluctuating hydrodynamics assumptions on the scales are fulfilled. In practice, that means that the elementary volume of the hydrodynamics grid should contain about 100 atoms.

Therefore, the purely atomistic core can not be smaller than this. There is no other restrictions on the size and shape of the atomistic interior – this is one of the advantages of the method.

Yes, you are right about the gradient. At the moment we assume that the dependence of s on the spacial coordinate is slow, and large gradients lead to artifacts. However, it is possible to consider s as a function of space and time explicitly, which will change the equations. This work is in our plans.

Professor Olson continued the discussion of the paper by Dr Nerukh: I am happy to see the combination of atomic mechanics and hydrodynamics in simulation. In the simulations that you have shown, all the particles are spherical, and thus there are no torques on them in the hydrodynamic simulation. How would the computations scale if non-spherical particles were considered? Also in mixing atomic and hydrodynamical simulation, how do you handle the time-scales of the mixed simulation?

Dr Nerukh replied: The non-spherical particles are treated as they are in standard MD simulations. At the hydrodynamic scale a non-spherical nature of the particles is irrelevant, only the averaged macroscopic properties of the material are considered. The time scales is an important issue. We have not discussed this in our paper in detail, however, this is conceptually important. We are working on it now. In a few words, the time scale (as well as the space scale) grows gradually when transforming from MD to hydrodynamics representation. This leads to the changes of concentration of the particles and their velocities. All this will be reported in our future publications soon.

Dr Glowacki asked Dr Nerukh: What do you see as the ideal target systems for your hydrodynamics/MD equations of motion? Over the past few years, I've been involved in carrying out NVE simulations of non-equilibrium reaction dynamics in liquids, where a chemical reaction creates nascent vibrational 'hotspots' that then cool through diffusion of the energy into the bulk solvent's degrees of freedom.[1–3] I wonder whether some of these systems might be good candidates for your dynamics method. Do you anticipate that the hydrodynamic/MD separation would in any way lead to complications in accurately modelling diffusion of energy from localized hotspots into the bulk? How is energy transport treated across the hydrodynamic/MD interfacial region?

1 D. R. Glowacki, A. J. Orr-Ewing and J. N. Harvey, *J. Chem. Phys.*, 2011, **134**, 214508.
2 D. R. Glowacki, R. A. Rose, S. J. Greaves, A. J. Orr-Ewing and J. N. Harvey, *Nat. Chem.*, 2011, **3**, 850–855
3 S. J. Greaves, R. A. Rose, T. A. A. Oliver, D. R. Glowacki, M. N. R. Ashfold, J. N. Harvey, I. P. Clark, G. M. Greetham, A. W. Parker, M. Towrie and A. J. Orr-Ewing, *Science*, 2011, **331**, 1423–1426.

Dr Nerukh responded: Our initial motivation for developing this method was to simulate proteins in water. So this is our ideal system. However, the framework is general enough to include non-equilibrium systems as well. We do not make any assumptions on equilibrium properties. The only assumption is the local conservation of mass, momentum, and energy. Globally the energy can diffuse. The energy diffuses between the hydrodynamics/MD subsystems as it would in these subsystems separately. There is no restriction or assumptions on the energy flow. We will look into the papers and consider the possibilities of applying our method to your systems. Thank you.

Dr Woods asked Dr Nerukh: First, a quick correction: In the conclusions I think that you mean 100ps per second, rather than 100ps per day.

I have two questions;

(1) Is it possible to use the shake algorithm for the atomistic part of the calculation, and what would the affect of shake be on the conservation of momentum at the boundary between the atomistic and hydrodynamic regions?

(2) How do you handle conservation of angular momentum? In the atomistic region, conservation of angular momentum occurs implicitly due to bonded forces. As these are removed or scaled down when a particle moves between the atomistic and hydrodynamic regions, how does the angular momentum of, *e.g.* a water molecule, change as its atoms sequentially move between the two regions?

Dr Nerukh answered:

(1) Yes, any algorithm of MD can be used in the framework. What is used is the 'macroscopic' properties (the density, momentum, and energy) calculated from the MD part as averages, therefore as long as these properties are calculated correctly the framework behaves properly.

(2) The angular momentum is conserved for the same reason – if the averages have a non-zero angular momentum, they are coupled to the corresponding hydrodynamics angular momentum.

Professor Olson continued the discussion of the paper by Dr Tchertanov: In your decomposition of protein structure into rigid and flexible regions, what is the range of sizes that you see in the rigid regions. Do they correspond to protein domains or subsets of secondary structures? How dependent is this decomposition on the length of the molecular dynamics simulations used?

Dr Tchertanov responded: Since MONETA has not been tested over a great variety of proteins, we could not predict exactly the range of sizes for the rigid and flexible parts of the protein, but for the studied systems, the IDSs (flexible regions) correspond to 30% of the protein length.

Mostly, the flexible regions (IDSs) are found among functional portions of the proteins, as in loop regions or extremities. The rigid parts correspond to highly ordered structured regions, as in helices or strands. The decomposition of the system into rigid and flexible regions is intrinsically dependent on the dynamic behaviour of the system during the simulations, not being directly correlated to its length.

Dr Glowacki asked Dr Tchertanov: The stochastic approach which you have utilized to build your allosteric network maps relies on a commute time parameter. And it seems to me that the allosteric networks that you would build will depend critically on the value of this parameter. Can you comment on guidelines for choosing the commute time parameter, how sensitive the networks are to the selected value, and how this impacts interpretation of the network for the purpose of comparing with experiment?

Dr Tchertanov replied: To build allosteric network maps we used a commute time (CT) parameter expressed as a function of the variance of the inter-residues distances over the MD simulations. Two residues communicate efficiently if their CT is lower than a chosen value CTthreshold. This threshold was chosen so that highly connected residues communicate efficiently with about 20% of the total number of protein residues, consistently with ref. 1.

The principle problem relates to an estimation of the non-covalent interaction (NCI) as "stable", depending on the percentage of conformations (occupancy) in which a given interaction exists. The threshold for NCI occupancy beyond which a bond is "stable" is 0.5 but tuneable if desired (between 0.09 and 0.1). The networks are sensitive to the CT threshold in a direct manner, in the sense that if we increase it, we'll increase the number of communication pathways.

We are now working to induce a better formalism allowing more unconstrained representation communication networks based on statistical correlations and graph theory.

1 G. Morra, G. Verkhivker, G. Colombo, Modeling signal propagation mechanisms and ligand-based conformational dynamics of the Hsp90 molecular chaperone full-length dimer. *PLoS: Comput. Biol.*, 2009, e1000323, DOI: 10.1371/journal.pcbi.1000323

Dr Zoppè asked Dr Tchertanov: The analysis made on kinase proteins show that your program can identify several independent dynamic segments. Many of them are known domains, with established function, which is a sign of the validity of the program. When used in proteins that are not as well studied as the RTK, the identification of such domains might help to gain new biological insights in the functional domains and activity of the protein.

Dr Tchertanov answered: Yes, you are right. A concept of relationships between the independent dynamic segments and their functions is central to our research. First, based on our studies of very similar receptors (KIT and CSF-1R) or signalling proteins (STAT5a and STAT5b), we evidenced that such a type of representation better characterizes the particular features of the protein, not manifested structurally (and dynamically) by using classical methods of molecular dynamics analysis. Second, in each analysed protein, these independent dynamic segments are coherent with functionally important regions of a given protein.

Third, the communication between these segments is very sensitive to the functionally significant perturbations (*i.e.*, mutation, phosphorylation...). We believe that (i) these statements can be extrapolated to the majority of allosterically regulated proteins and (ii) allosteric regulation and catalysis (or activation) have a common route. Consequently, identification of the independent dynamic segments and communication between them in non-well characterized or non-characterized protein will be a means to identify the functionally significant regions/fragments.

Mr Hall asked Dr Tchertanov: Some of the node-link diagrams in your paper, *e.g.*, those on the left hand side of Fig. 3, are difficult to read due to occlusion and the proximity of the nodes. Did you try alternative graph representations besides node-link diagrams, *e.g.*, adjacency matrices? Alternative representations may make the links in the graphs clearer or help to elucidate other aspects of your system.

Dr Tchertanov responded: The use of the GEPHI module was our first attempt (and not perfect) of data representation. We are now searching for alternative representations for graphs. Probably, time-dependent networks will be introduced.

Dr Glowacki asked Dr Tchertanov: Your analysis method allows you to generate a network of discrete states and connectivities within an allosteric network. Is it possible to derive rate coefficients for these networks – *i.e.*, characteristic transition times which connect each node along a vertex? If so, then it would be possible to formulate a kinetic master equation and solve a set of coupled differential equations to get the global population flux within the discretised states. This is an area of active interest for several groups across the world, including some of my own recent work.[1–3] Would it be possible to carry out such analyses of the allosteric networks which your analyses provide?

1 D. V. Shalashilin, G. S. Beddard, E. Paci and D. R. Glowacki, *J. Chem. Phys.*, 2012, **137**, 9.
2 D. R. Glowacki, E. Paci and D. V. Shalashilin, *J. Phys. Chem. B*, 2009, **113**, 16603–16611.
3 D. R. Glowacki, C. H. Liang, C. Morley, M. J. Pilling and S. H. Robertson, *J. Phys. Chem. A*, 2012, **116**, 9545–9560.

Dr Tchertanov replied: In the actual version of MONETA, it is not possible to derive rate coefficients for the networks, but we are currently working on this issue in collaboration with a mathematician group specialized in data processing. It would be very interesting, indeed, to improve MONETA to include this kind of analyses.

Dr Reiher opened the discussion of the paper by Dr Brancale: Considering the fact that you deposit energy into the system in a docking process when pushing a substrate onto its target, you may end up with artificial processes, *e.g.*, the induced conformational changes might be kinetically inaccessible but can be induced by the external force applied by the operator of the device. In our paper,[1] we have devoted a section to this issue because one does not want to explore molecular reactions that are at some given temperature thermally inaccessible. Is this also a problem for docking and, if yes, how do you deal with it in your algorithm? Also, entropic effects will play a role. Are there already concepts available to capture such effects in haptic docking studies?

1 M. P. Haag, M. Reiher, Studying Chemical Reactivity in a Virtual Environment, *Faraday Discuss.*, 2014, **169**, DOI: 10.1039/C4FD00021H

Dr Brancale replied: These are both important points, for which we do not have an elegant solution, yet. The introduction of artificial and non realistic conformations in the protein because of the forces applied through the haptic device is something that we have tried to address initially by allowing the target to restore its initial position once the ligand moved away from the point of interaction. Surprisingly, as we mention in the paper, the algorithm we described seems to be very resilient towards the forces introduced externally, probably because it is not a simple energy minimisation protocol, but it will always attempt to generate a reasonable conformation. The treatment of the entropic effects remains an unresolved challenge, at this point.

Mr Haag asked Dr Brancale: I would be interested in the experience you have concerning the question of how to incorporate the "other" hand, when one hand is manipulating the molecular system with a haptic device. You mentioned the possibility of the Leap Motion device. Do you have any experience with this special device and how would you recommend to use it? Is it precise enough to fix single atoms or molecular fragments in space? We have reported that in our experience (see our paper[1], section 6.2) that the limiting factor might be the subtle shaking of the user's hand. Are there possibilities to circumvent this problem?

1 M. P. Haag and M. Reiher, *Faraday Discuss.*, 2014, **169**, DOI:10.1039/C4FD00021H

Dr Brancale answered: We are currently experimenting with the Leap Motion Controller and, overall, we believe it has a very good potential to be used in combination with the haptic device. However, we are seeing the same effect you are reporting and at the moment, the extreme sensitivity of the sensor does not allow us to do any precision task in a natural manner. Our experiments are now focused on performing less demanding tasks, like basic rotation and translation of the system. Unfortunately, I do not yet have a full answer or suggestion regarding what would be the best way to deal with high precision tasks.

Dr O'Donoghue asked Professor Olson: I believe that depth-sensing cameras such as the Leap Motion show promise for improving 3D control of molecular graphics. Such cameras are now highly responsive, affordable, and compact, and at least one laptop on the market has a Leap Motion built in, with more on the way. My group has created the Molecular Control Toolkit (see attached file and video at https://www.youtube.com/watch?v=XU_6TWM9ke8), an open source framework designed to allow any depth sensor camera to control any molecular graphics system. We did a pilot user study where a group of molecular biologists used these devices for basic 3D manipulation in molecular graphics; the results were encouraging, suggesting that the responsiveness of the Leap Motion especially makes it already a viable alternative to traditional mouse and keyboard commands for 3D controls. In short, these technologies allow users a more intuitive way to achieve the full six degrees of freedom in 3D, by performing hand actions that directly rotate and position an imaginary model in their hands.

Your group has investigated similar scenarios featuring the combined use of 3D printed molecular models and augmented reality with traditional cameras. I'm curious to hear your opinion about the perspectives on the use of depth-sensing camera for molecular graphics. Also, what do you think are the perspectives on the combined use of 3D printed models with depth-sensors and with the kind of gyroscope sensors and accelerometers now widely available in smart phones?

Professor Olson commented: Sean O'Donoghue mentioned my work with tangible 3D printed molecular models as an alternative to haptic feedback. I would say that in general the ability to have a physical model in your hand that reflects the structure that you are manipulating has a number of perceptual and cognitive advantages. Perceptually you have the senses of touch and proprioception to augment your visual sense and cognitively you have the natural ability to manipulate a physical object without any intermediary device, such as the complex jointed arm of a haptic device. We have used augmented reality to add information such as the electrostatic field surrounding a protein surface to complement the physical shape of the molecular model. For example, see ref. 1 (and Fig. 1):

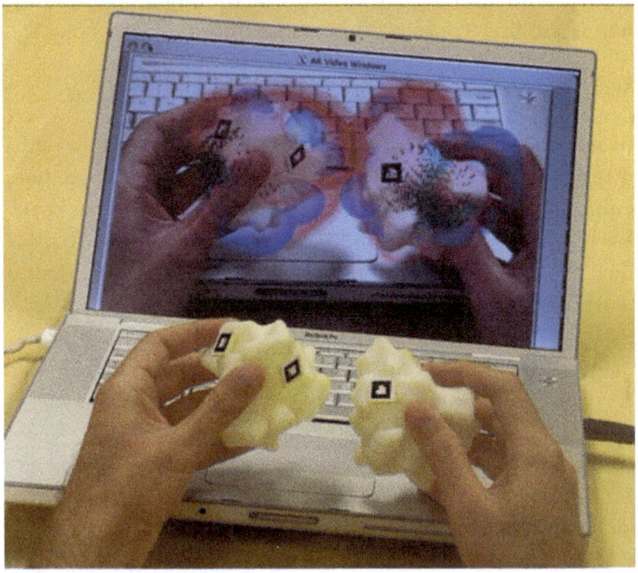

Fig. 1 Demonstration of combining and manipulating 3D printed molecular models with augmented reality display. Shown are two physical models of superoxide dismutase (SOD) monomers augmented with computer graphics generated electrostatic potential volumes (red–negative, blue–positive), as well as electrostatic field animations at the active site. Work from the Olson Lab.

1 A. Gillet, M. Sanner, D. Stoffler, A. J. Olson, Tangible interfaces for structural molecular biology, *Structure*, 2005, **13**, 483–491.

Dr O'Donoghue replied to Professor Olson: I agree there are advantages in using a tangible 3D printed model of a molecule to manipulate a representation of the same molecule on a computer screen. However, the right printed model won't always be easily available, so I'm curious to hear Art Olson's (and others) opinion about the use of surrogate objects, such as an iPhone or other devices with accelerometers and gyroscope sensors. Also, the advantage of using such 3D manipulation methods was somewhat undermined in the comment from Fred Brooks, when he stated that minimising human effort is always the winning strategy for user interaction. This would seem to argue for staying with mouse and trackpad. I am curious what others have to say on these two contrasting views of the future of user interaction for molecular graphics.

Professor Brooks remarked: We found the most powerful aid to 3-D perception was the kinetic depth effect.

We started out rotating our protein models, but it was too hard to keep track of where a particular residue was. So we went to rocking, and that worked great.

In the GRIP system, we provided a 2-D joystick horizontally mounted poking out at the user, with which the user could rock the molecule in any direction. We just doubled the angles specified in the 2pi-steradian working volume to enable any viewpoint in 4pi.

We found users universally kept their left hands on this joystick, using the righthand for keyboards, mouse, joysticks, menu selection. They twitched it all the time.

So we supplemented that with an automatic rocking mode, about the vertical axis, controlled by another 2-D joystick that specified rate and amplitude. Also popular.

Mr Stone said: I wanted to amplify what Professor Brooks mentioned about the benefits of using two hands for improved ease of use, combining a six-degree-of-freedom input device such as a SpaceNavigator or a Spaceball, both with haptic devices, as well as the more general case with a mouse. The use of a six-degree-of-freedom motion control input device such as a SpaceNavigator or a Spaceball in the non-dominant hand makes it easy to re-orient or position the molecular structure in the view or within a haptic workspace, while the dominant hand can perform detailed picking, pulling, or other operations using a haptic device, or in the more conventional case, a mouse. We have found this arrangement to be a much more efficient way to work, both with haptics and with mouse-based interfaces.

Dr Hayward replied: For novice users of haptic feedback devices we have found that adding a separate device, such as a mouse, for navigation makes the system difficult to control. In our experience with HaptiMol ISAS, where we use the haptic device for both navigation and exploration, users were able to explore and interact with very little instruction. We could imagine that for more experienced users a separate SpaceNavigator would be beneficial for some manipulations. However, care must be taken as manipulations through one hand (without force feedback) can induce forces on the other hand holding the haptic device. This feedback might be confusing for the user. Ultimately using something like two CyberForce devices, one for each hand, might be the best (but expensive) solution for molecular docking applications.

Mr Stone commented: I agree with the statements regarding the difficulty of interacting with input devices that are not calibrated to exactly match the coordinate system of the on-screen display. I personally found it much harder to perform 3-D picking operations using a haptic device on a desktop workstation, even when using a stereoscopic display, than doing the same thing in environments such as the CAVE, where the "wand" input device has been carefully calibrated to yield a 1-to-1 correspondence to what the user sees. In the case of VMD, the difficulty that I and others experienced with 3-D picking led us to develop a user interface that allowed users to avoid the need for 3-D picking by specifying a region of the structure to pull on through a text-based atom selection, in the same way that regions of structure to be associated with specific graphical representations are specified elsewhere in the program.

By allowing users to specify the part of the structure they wanted to pull on, it became unnecessary to perform 3-D picking in most cases, making the scientists more productive without actually solving the problem with lack of direct correspondence of the display and the input device coordinates.

Dr Hayward responded: From our experience with users of HaptiMol ISAS and HaptiMol ENM we have not found the issue of non-collocation of the graphics and haptics coordinate systems to be a major problem. To aid navigation we have found lighting and shadows to be particularly important.

Dr Raffin replied: Yes we agree that visual selection can be cumbersome. We did not mention it clearly in the paper, but Exaviz also supports a text-based atom selection, relying on the MDAnalysis selection engine.

Mr Stone addressed Professor Brooks and Dr Raffin: Continuing the topic of the difficulty of using interfaces that present different coordinate systems or scales for display and input, I wanted to ask Professor Brooks if he had any observations about the ease-of-use of 1-to-1 scale haptic interfaces that precisely match the coordinate system of the displayed scene. I know that University of North Carolina, Chapel Hill has done a lot of work with a variety of large format haptic input interfaces and immersive displays, and I was curious what experiences he might share on this topic.

Professor Brooks replied: We built a Protein Interactive Theater that incorporates separate 3-D screens for two users seated at right angles to each other, with the molecule stick figure suspended in the working space shared by both of them. Each had access to the single shared 3-D haptic device. The illusion was very strong; one user complained of being pricked by a sharp angle in the visual model.

In all our haptic work, including that of Ouh Young, we have made visual and haptic scales to correspond, so I can't give an experience on the difference it makes to have matched scales.

Dr Zoppè commented: When we consider the easy of use for new devices, we should also consider that things change: among us, some grew up using playstations and other devices, that for those that never used them (the older among us) are difficult and awkward to manipulate. I think that as new generations of young people come to the business of molecular simulations, they might find easier to use haptics then we do, simply because they have an instinctive ability, learnt as kids. For them, using a manual device, might become 'second nature' just like it is for us the act of writing

Dr Glowacki commented: I would like to remark how fantastic it is to witness the energy and enthusiasm driving this conversation. Certainly I've never been in a single room with so many folks who are actively pursuing the interface between high performance computing, new HCI technologies, and molecular simulation. It is very exciting. Professor Fred Brooks can probably comment on whether he has experienced an exchange of this sort during his many years working in this area. I would also like to address the difficulty of a 3rd person interaction which was just raised in relation to the widespread use of haptics at this meeting. I, like John Stone, found the haptic pointing and grabbing of the sort implemented in VMD rather difficult. I struggled with the interface because all of my human body and my movement intention was distilled into a single, localised virtual pointing device. In fact, this difficulty is very much what led me to the idea of the human Hamiltonian, which is the strategy utilized in dS. It offers a sort of compromise between fully immersive virtual reality, and third person interaction: you see a representation of your human form, but the representation is in the form of an energy avatar, with a net effect which is that of an extra term in the molecular Hamiltonian. The result is that some level of embodied human intuition is retained during interaction with the virtual space (a video of two users using their energy avatars to manipulate a simple 10-Alanine peptide is available at http://www.vimeo.com/81531449). The idea of the Human Hamiltonian may prove useful for people who are working in the area of interactive MD, particularly in association with something like a LEAP sensor. In Fig. 11 of our paper, we show a representation of human hands manipulating a 10-Alanine simulation: the hands look like hands, but behave like Hamiltonians. So again, some degree of embodiment is retained.

Mr Iakovou remarked: As all tools, the importance and applicability of haptic technology is closely affected by the simulation problem at hand. If the goal of the simulation is to educate the user about molecular structure and dynamics in a visual and user-intuitive way (like in dS) then haptics might be overkill. However, if the success of the simulation depends on the force feedback cues given to the user (*e.g.*, drilling a tooth, performing virtual surgery, path-planning and assembly-sequencing of industrial components, sensing the intermolecular interaction forces during docking) then haptic technology is by definition a crucial part of the solution.

Dr Raffin commented: Let me bring two citations regarding the benefits of first-person interaction (also called co-location) and haptics.

The first, co-authored by Professor Brooks, evaluates the benefits of co-location but also mentions that being able to rest on some solid support (a table for instance) helps for interaction.[1] The second compares having (or not) co-location with or without haptics for some simple 3D manipulation tasks. It shows that co-location is the most significant factor for improving performance, while haptics brings much fewer benefits.[2] Co-location was not always easy to obtain (requiring a Cave or a mirror and CRT screen), but today it can be achieved at low cost and with a limited effort using either a thin stereoscopic display or a head mounted display like the Occulus Rift. The only difficult point is to properly calibrate the device so that the real reference frame aligns with the virtual one.

1 Mark R. Mine, Frederick P. Brooks, Jr., and Carlo H, Sequin, Moving objects in space: exploiting proprioception in virtual-environment interaction, in Proceedings of the 24th annual conference on Computer graphics and interactive techniques (SIGGRAPH '97), 1997.

2 David Swapp, Vijay Pawar, C-line Locos, Interaction with co-located haptic feedback in virtual reality, *Virtual Reality*, 2006, **10**(1), 24–30.

Dr Chavent addressed Dr Brancale and Mr Iakovou: The discussion has focused a lot on the hardware but can the representation of the molecule improve the participants' results? Instead of only visualizing atoms as spheres could it be possible to render more complicated depictions such as a molecular surface?

Dr Brancale replied: Our original simulator had several rendering options available and, indeed, the clever use of these made the interaction more effective. These options are still available in the current, fully-flexible system, but the main issue now is that, as the graphic card is used to run the simulations, it cannot really be used for a more detailed rendering. For this reason, a second graphic card would be beneficial to exploit all the graphical features available in our simulator.

Mr Iakovou commented: We aren't aware of a user study investigating the benefits of different representations using interactive molecular docking tools but it would be interesting to investigate. One might expect that visualising the molecular surface to be advantageous as shape complementarity plays a crucial role in molecular recognition.

Dr Baaden continued the discussion: Papers DOI:10.1039/C3FD00139C and DOI:10.1039/C3FD00123G describe the use of haptics as a tool to help with docking and drug design. The haptic sensory channel is used to guide the process and represents a particular focus. I think it is important to mention that other multi-modal feedback may be required to render this process efficient. One may explore additional sensory channels such as vision or sound to provide complementary information that cannot easily be transmitted through the haptic channel, for example where one is located in space by using a visual feedback such as a light source or how good one is doing *e.g.*, in terms of interaction energy by using sound. Furthermore it is important to make sure that the feedback channels are not overloaded and do not generate more confusion for the user than actual benefit.

Dr Glowacki commented: Many of the interaction strategies for manipulating molecular simulations which have been presented at this meeting (and which are utilized more generally within the molecular simulation community) have focused on haptic devices. Following on from Dr. Baaden's remark, I think it's worth asking whether such devices afford maximum benefit. Is there a possibility that other interaction strategies are as useful as haptic devices in molecular simulation? I believe that further user studies are required to determine this. One of the few studies which has examined these issues was published by Professor Brooks in 1989. In testing haptic docking, he carried out user studies where the haptic motors were turned off for the control group. With the haptic motors on, users were able to accomplish simulation goals faster by approximately a factor of two.[1] This is certainly a benefit, but it's not an enormous benefit, and there is some inevitable noise in the statistics owing to the limited size of the user test group. It may be the case that other strategies *e.g.*, using camera based interfaces augmented with visual and auditory clues would have a comparable level of effectiveness.

1 M. Ouh-young, D. V. Beard, F. P. Brooks, Jr., Force Display Performs Better Than Visual Display in a Simple 3-D Docking Task, Proceedings of IEEE Robotics and Automation Conference, Scottsdale, AZ, 14–18 May 1989, pp. 1462–1466.

Dr Brancale replied: I think this is an important point and we are exploring this in our software (*e.g.* a traffic light system on ligand atoms based on binding scores) and our impression is that the overall user experience is greatly increased

Dr Montes asked Dr Hayward: What about using a dynamic controllable camera to set your point of view exactly where you like in order to have a better control and enhance the haptics experience?

Dr Hayward replied: A dynamic camera can enable the user to navigate precisely to areas of interest but adding an additional input device to control the camera can make the controls harder to manipulate particularly when using a haptic feedback device. The approach we have striven for enables the user to perform all interactions, including camera manipulations with the haptic device. In HaptiMol ISAS we have incorporated the probe camera which allows the user to obtain a view from the probe sphere in the direction of the haptic stylus, giving a "cockpit" style view.

Mr McIntosh-Smith asked: Haptic devices have had a clear impact on molecular simulation and visualisation. What other input devices do you see that have also had a significant impact in this area? For example, modern smartphones are now ubiquitous, and have high quality gyroscopes, compasses, *etc.* Modern tablets also have touch interfaces and high resolution screens that are now very familiar input devices for most people.

For any devices that you see having relevance to molecular simulation and visualisation, do you see this as complementing haptics or competing with them?

Mr Iakovou answered: Haptic technology enables the user to interact with the virtual world through their sense of touch. Namely, it enables the user to receive tactile (*e.g.*, feel an object's texture) and kinaesthetic feedback (*e.g.*, apply and receive forces) during his/her interaction with the virtual world. In the case of molecular docking simulations this translates to sensing the intermolecular interaction forces and torques in real time, while moving the ligand around the receptor. Our group believes that this haptic information along with the 3D visualization of the two molecules allows the user to perform a knowledge-guided search for the binding conformation, identify correctly the docking site on the receptor, and eventually bind the molecules into a new ensemble. On the other hand, modern input devices provide an alternative means to interact and explore the visual world (*e.g.*, touch screens, Kinect, *etc.*) but they lack this force feedback capability which we believe is an important component of an interactive molecular docking system. These other input devices could certainly be useful for visualisation tools in molecular graphics.

Dr Brancale replied: Based on our experience, I think small, easy to operate, devices, like mobile phones or the Leap Motion controller, can be used very efficiently in combination with haptic devices. In other cases, they could certainly provide a relatively accessible alternative to haptic controllers, but, at the moment, I believe the force feedback provided by devices like the Phantom still gives the user a more natural and interactive experience.

Professor Olson addressed Mr Iakovou: I would say that in general ligand docking is best done in an automated computer driven manner, since computers can more efficiently evaluate the typically high dimensional search space required to find the best binding modes of a ligand to a receptor. That having been said, interactive docking is a good way to explore ideas and hypotheses, where a quick test of an idea can lead to a more extensive study.

Mr Iakovou replied: In the context of drug design it is true that an interactive method would not be suitable for screening a large number of compounds for a particular target. These interactive methods can be used subsequent to automated methods when a small number of lead compounds have been identified. Namely, the user can utilize a haptic-based docking system to visualize these lead compounds in their docking conformations, feel the underlying interaction forces and improve upon or reject these conformations based on user knowledge, experience and expertise. Moreover, biologists and biochemists are often interested in a particular protein–ligand or protein–protein interaction and, as you say, these types of tools can provide an environment within which an expert can use their knowledge to test ideas and hypotheses. One thing that most automated methods are very poor at is teaching us about the process of docking itself, *e.g.* whether electrostatic steering is involved in the binding process. Haptic-based docking tools would be expected to be very useful in this respect.

Dr Reiher addressed Dr Brancale and Mr Iakovou: I would like to pass on the question that Professor Brooks asked me. That is, how can you show that the force feedback of a haptic device is an essential ingredient for the docking exploration process? Or is the haptic device simply a sophisticated input device whose force–feedback capability is not really needed in a docking study? In our implementation[1] of haptic quantum chemistry,[2] we use the force as THE descriptor to explore the shape of a potential energy surface (in particular, for the discrimination of different exploration paths that guide the operator energetically uphill to a transition state) and thus there is no way of doing reactivity exploration without force feedback. In docking, I imagine that the most important repulsive responses are repulsive regions of a Lennard-Jones potential or steric hindrance that are important to experience by force feedback. Is this true? And how important are attractive responses that drag the substrate to its host? What is the advantage of haptic devices over standard techniques in docking studies?

1 Moritz P. Haag, Alain C. Vaucher, Mael Bosson, Stephane Redon, Markus Reiher, Interactive Chemical Reactivity Exploration, *ChemPhysChem*, 2014, submitted, arXiv:1405.4036.
2 K. H. Marti, M. Reiher, Haptic Quantum Chemistry, *J. Comput. Chem.*, 2009, **30**, 2010

Mr Iakovou replied: A user study can quantify the importance (or otherwise) of haptic feedback in the docking exploration process. Currently, there are not many such studies in the field, but the ones that do exist, indicate clearly that haptic feedback can assist equally experts or students to improve their situational awareness in the docking process and achieve better/correct docking results[1]; select effectively the true binding site (among multiple candidates) and estimate the docking conformation[2]; and understand better the process of molecular recognition/binding.[3,4] Our group plans to conduct one such study with colleagues from the Biology department at the University of East Anglia, prior to the release of our haptic-based, rigid docking system called "Haptimol_RD". The modelling of van der Waals is important in order to test/identify geometric/surface complementarity, whereas the modelling of electrostatics allows you to gain a better awareness of the long range steering effect. We believe that both interactions are important in a haptic-based docking system, and we modelled them in our system for this reason. Finally, and in addition to the benefits mentioned earlier,[1–4] haptic-based docking systems can also enable expert users to utilize their knowledge to test new ideas and hypotheses.

1 F. P. Brooks, M. Ouh-Young, J. J. Batter, and P. J. Kilpatrick, Project GROPE Haptic displays for scientific visualization, SIGGRAPH '90 Proceedings of the 17th annual conference on Computer graphics and interactive techniques, 1990, ACM, New York, USA, pp. 177–185.
2 E. Subasi and C. Basdogan, A New Haptic Interaction and Visualization Approach for Rigid Molecular Docking in Virtual Environments, *Teleoperators Virtual Environ.*, 2008, **17**(1), 73–90.
3 Petter Bivall Persson, Shaaron Ainsworth and Lena A. E. Tibell, Do Haptic Representations Help Complex Molecular Learning? *Sci. Educ.*, 2011, **95**(4), 700–719.
4 Petter Bivall Persson, Matthew D. Cooper, Lena A. E. Tibell, Shaaron Ainsworth, Anders Ynnerman, and Bengt-Harald Jonsson, Designing and Evaluating a Haptic System for Biomolecular Education, 2007 IEEE Virtual Reality Conference, 10–14 March 2007, Charlotte, NC, USA.

Dr Brancale replied: I would start answering this by echoing the comment from Prof. Olson: automated docking is better than haptic driven docking of full drug-like, flexible, ligands in the vast majority of cases. However, in the placement of small fragments, or in the quick evaluation of a structural modification to test a new idea, the interactive approach, in my opinion, has an advantage (and consider that we are trying to sample interactively the target flexibility as well). The haptic feedback certainly provides the user with enriched information, but it is hard to quantify the impact of this on the quality of the results obtained. Regarding the forces experienced during the simulations, in our experience, the repulsive interactions are clearly very important and are immediately noticed by the user, who is often guided by them. However, especially after the user has acquired some familiarity with the device, attractive interactions also start to be recognised and appropriately used during the simulation.

Mr Stone commented: In regards to the discussion of tablets and mobile devices, we developed a "remote control" application to allow multiple phone or tablet devices to collaboratively control a VMD session running on a desktop workstation. Our original intent was to use the on-board accelerometers and gyros on the mobile device to allow detailed pointing and six-degree-of-freedom motion control for molecular models displayed by VMD, but we quickly discovered that there is a tremendous variation in the accuracy and usability of these on-board sensors from one mobile device to another. These experiences led us to focus on developing interfaces for the multi-touch displays and to leave the use of six-degree-of-freedom input as a goal for future devices with better on-board sensors.

Mr Stone continued the discussion of the paper by Dr Brancale by asking: I appreciated the approach taken with concurrent overlapping execution of multiple GPU force evaluation kernels, this is a nice effort. In the preprint of the paper, the units associated with the graphs in Fig. 3 and 4, and Table 1 are missing in several places. Are all of these units seconds?

Dr Brancale answered: Sorry. Yes the unit is seconds and the figures have been corrected at the proof stage.

Mr Stone opened the discussion of the paper by Mr Baker by remarking: The paper contains a statement that smoothed particle mesh Ewald (SPME), particle–particle particle mesh Ewald (PPME), and multilevel summation method (MSM) have been implemented on the GPU, but that the use of fast Fourier transform (FFT) intrinsic to Ewald methods reduces parallelism and leads to poor scaling on GPUs. I would like to point out that the multilevel summation method (MSM) is a real-space method that does not require the use of FFT or associated transpose operations, and it also has the benefit of allowing simulation of non-periodic, partially periodic (in only some dimensions), and periodic systems.[1–3]

1 David J. Hardy, Multilevel Summation for the Fast Evaluation of Forces for the Simulation of Biomolecules, *Ph.D. thesis*, University of Illinois at Urbana-Champaign, 2006.
2 John E. Stone, James C. Phillips, Peter L. Freddolino, David J. Hardy, Leonardo G. Trabuco, and Klaus Schulten, Accelerating molecular modeling applications with graphics processors, *J. Comput. Chem.*, 2007, **28**, 2618–2640.
3 David J. Hardy, John E. Stone, and Klaus Schulten, Multilevel summation of electrostatic potentials using graphics processing units, *J. Parallel Comput.*, 2009, **35**, 164–177.

Mr Baker replied: The inclusion of MSM in the list of methods that require FFTs was an oversight in editing, and has been removed.

Mr Thomas continued the discussion of the paper by Dr Brancale by stating: I understand that your protein models are not fully solvated, however in reality the binding pocket is likely to be solvated and will contain water molecules that will need to be forced out of the way in order to bind the ligand. Is any attempt made to model the forces that would be experienced due to the interactions with these water molecules as a user pushes a ligand into the binding pocket?

Dr Brancale responded: Explicit solvation of the model used in our simulator is, indeed, a real challenge and the lack of it a big approximation. Of course, automated docking systems do suffer the same problem. The problem is that the solvation and desolvation effects are extremely dynamic and extremely complex to address in the short time frame required to operate the haptic device. Unfortunately, I do not think we will be able to take them into account in our simulations any time soon.

Dr Woods asked Mr Baker: Following the title of one of the papers you cite, do you think that the Ewald sum is still necessary?

Mr Baker replied: Ewald methods are primarily used for accuracy and reliability purposes. We have shown that for solutes in supercritical carbon dioxide, it's possible to substitute the Ewald sum and suffering only a very minor loss in accuracy. In most cases the Ewald sum isn't necessary for production runs, but it can be used to ensure that a pair-wise method is optimised beforehand.

Dr Woods asked: Imagining a future exascale computer, do you think that a molecular dynamics code that will run on that computer will use the Ewald sum?

Mr Baker answered: As computational power increases, so does the size of the systems being studied. Whilst the Ewald derivatives remain the bottleneck in most molecular dynamics simulations, I would think that pair-wise electrostatic methods will be utilised more in future.

Dr Woods asked: In your paper you only compare energies. Have you tried comparing the magnitude and directions of forces?

Mr Baker responded: The force vectors have been investigated for these methods in the paper by Fennell and Gezelter.[1] We wanted to investigate the ensemble properties, such as, the diffusion coefficients and PVT relationships.

1 C. J. Fennell and D. J. Gezelter, Is the Ewald summation still necessary? Pairwise alternatives to the accepted standard for long-range electrostatics, *J. Chem. Phys.*, 2006, **124**(23), 234104.

Dr Woods addressed Mr Baker and Professor Hirst: Your work helps show that "long range" electrostatic interactions can actually be truncated at the nanoscale (10s of nanometers) and that this doesn't introduce large errors in the dynamics of the system. Given this observation, what do you think is the length range of interactions between the various bits of molecular machinery in the cell? Philosophically, is the speed of light really a restriction for modelling the whole cell, and do you think that the speed of light has any impact on the real physical behaviour of the cell?

Professor Hirst responded: There may be some communication mechanisms (perhaps effected through bond interactions or through significant potential gradients) which are surprisingly long-range and may extend beyond the cut-off distances we have considered. The speed of light and other physical constants, such as Planck's constant, are important for the photophysical behaviour of some constituents of cells.

Mr Baker commented: The speed of light is probably not a restriction on cell behaviour, certainly in terms of the direct influence on the conformational dynamics of the molecules within a cell.

Dr Glowacki stated: I would like to comment on how showing those outside the scientific establishment representations of the invisible atomic molecular world could lead to profound changes in how science operates in society. My own experience developing and prototyping dS has occurred in rather unconventional environments outside the academy. Probably the thing that I have been most surprised by is the fact that most people have no concept that the atomic world is dynamic. Professor Brooks came and asked me yesterday whether people learn about chemistry from dS, and my response to him was something like the following: "Certainly they walk away with the message that the invisible atomic world is dynamic." For most people, this is counter-intuitive. They are accustomed to seeing static representations of molecules in textbooks, and that's how they imagine them. They don't understand that the invisible world is driven by dynamics. But as scientists, we know that zero-point energy (one of the fundamental consequences of quantum mechanics) tells us that the universe is driven by perpetual dynamism. This is a very deep point for understanding how nature works, informing how we think about matters of science, metaphysics, and even philosophy/spirituality. I think that many of the people in this room are developing powerful mechanisms for revealing the dynamism of the atomic world, and I encourage them to do so.

Mr Stone commented: I wanted to briefly add to the discussion of educating the general public about the dynamic nature of molecules. We have held molecular visualization demonstrations for a variety of open house events at the University of Illinois, and in helping with these events I have repeatedly observed that members of the general public are surprised when they see molecular dynamics visualizations for the first time. Many times the attendees at these open houses indicate that they had always imagined molecules as static, unmoving, and crystalline in most respects, and they are dismayed to finally see how lively and dynamic biomolecules really are. I would just point out that these kinds of events for the general public are an excellent opportunity to expose a lot of people to a more realistic view of the nature of molecules, and of biomolecular structures in particular.

Dr Zoppè replied: I can add that we also have the same experience: when we can get people to watch proteins in their dynamical activities, we do make a strong impact on the concept of proteins. We performed in Milan, in April 2013, a 'Molecular Concert' in which we had some musicians playing with a background of two movies showing some MD simulations (PBTI, and the Glucose transporter) https://www.facebook.com/events/255944884542665/?ref=22. All were fixing the object, not knowing how to interpret it, and when I explained that they were actual proteins, of the kind that make our body work, they seemed to look with more interest. Here we should consider that for most people, the word 'protein' does not associate with an image of a single, still protein as seen in textbooks, but rather with the steak, or with muscle (their own) for some people.

So getting the public to watch proteins, with or without explanations, is certainly a way to change the primordial idea, making a cultural impact.

However, the impact of these representations is still quite limited to a few hundred people. The way to make a real strong impact, would be, in my opinion, something like a blockbuster movie, one that might have an impact for biology and proteins similar to what happened with palaeontology after *Jurassic Park*. After that movie, already 20 years ago, kids have been playing with many dinosaurs, knowing them by specific names, having them behave like hunters, grass-eaters and other specific features. If we could make a movie of that kind (and there is material abundant for a thrilling and entertaining plot), then we would have made a real impact. Of course, it takes Hollywood money, and a director of the calibre of Steven Spielberg, but it is not beyond possibility.

We have made a small step in this direction, by creating an animation of some events happening at the synaptic space to be seen in the Fall 2014 in a documentary movie titled 'The Dark Gene', and produced by German production company FilmTank (http://www.filmtank.de/p/the-dark-gene/).

Faraday Discussions

PAPER

Immune Attack players perform better on a test of cellular immunology and self confidence than their classmates who play a control video game†

Melanie Stegman*

Received 17th February 2014, Accepted 5th March 2014
DOI: 10.1039/c4fd00014e

Introduction

As molecular scientists, we understand the fundamentals of molecular behavior that underlie evolution, infection, physiology and environmental contamination. However, the basic fundamentals of molecular behavior are considered too abstract to teach to grade school or high school students in the context of cellular biology. Analogies are used to describe the functions of cells and their organelles. Concepts such as protein production and respiration are taught from a systemic, large scale perspective, typically without presenting the role of any individual proteins in these cellular processes. The molecular behavior of, say, ATP and its random diffusion is not taught in context with mitochondrial function. The next generation science teaching standards do not emphasize molecular behavior, and explicitly state that assessment of students should not include any information about individual proteins or a biochemical level of understanding of cellular processes.[7] Only students who choose to take an advanced level high school course or who opt into college level biology courses are exposed to the fundamentals of molecular biology.

This lack of detail and lack of exposure through high school to the fundamentals of molecular behavior leave the general public without the basic understanding required to grasp cellular biology or to understand new data. Because this new data often pertains to personal and public health decisions, these concepts are important for non-scientists to understand. Memorizing every step of glycolysis is not necessary. However, a fundamental understanding of how cellular processes are driven by concentrations of substrates, products, allosteric inhibitors; how enzymes, products and substrates diffuse in random directions, that through random interactions specific binding occurs due to very particular

FAS Learning Technology Program, 500 Wall Street #409, Seattle, Washington, 98121, United States. E-mail: mstegman@fas.org; Tel: +1-917-886-6079

† Electronic supplementary information (ESI) available: See DOI: 10.1039/c4fd00014e

aspects of shape, charge and other molecular forces, and that processes require a particular enzyme for each step, would serve students as a scaffold for a lifetime of learning.

Brunner's work on a game (The Ruby Realm) to teach photosynthesis to middle school students demonstrates that trying to teach cellular processes like photosynthesis without presenting the molecular interactions leads to misconceptions. She found that students consistently had trouble understanding that the mass of a plant comes from the sun and air, rather than the dirt, until she introduced the molecules and their interactions in the game. Using analogies and a large scale systematic approach (as recommended by most science teaching standards) did not help students comprehend photosynthesis because the large scale characteristics of dirt, air, and plant material are so different. By allowing students to "shoot" carbon dioxide molecules with sunlight and form glucose molecules with them, Brunner was able to teach a deeper understanding of photosynthesis to younger students.[12]

Klymkowsky and colleagues have demonstrated that misconceptions are commonly held by biology undergraduates as they enter college. Additionally, these misconceptions concern fundamental molecular behaviors, such as diffusion and the random motion of molecules and proteins.[5] Klymkowsky reported on a multi-year effort to address misconceptions in undergraduate biology classes and concluded that these misconceptions are difficult to erase and re-teach.[5] Teaching the basis of molecular behavior to younger people can help address this problem.

Jenkinson and McGill have shown that showing greater detail in biochemistry videos achieves a higher level of understanding.[2] Namely, they created four versions of a receptor and its ligand interacting in space. Where typical videos show a ligand and receptor only and place the ligand on a clear path to the receptor, Jenkinson and McGill increased the randomness of the ligands approach and also added water molecules that collide with the ligand. They created four variations of the receptor and ligand video. They found that the more complicated videos were best for evoking a deeper understanding of the process of a ligand randomly approaching and binding to its receptor. Jenkinson and McGill found that this deeper understanding wore off after two weeks, as the groups that saw the more complex videos failed to answer the deeper understanding questions correctly two weeks later.[3] Jenkinson and McGill state that complexity may be required to properly teach molecular interactions and that videos with audio and interactive elements may be the most effective.

What is missing is a tool to present the complex, abstract fundamentals of molecular behavior to younger students so that by the time we teach them the specifics of respiration or photosynthesis they can understand these processes deeply. The tool would 1) present a great deal of detail, presenting many types of objects that interact and have numerous traits that dictate their activities; 2) allow for interaction and puzzle solving: allowing students to manipulate, explore and use the objects they find to affect the state of the molecular world; 3) be engaging, fun and interesting to non-scientists so that the general public and younger students will take advantage of the tool and 4) attract students back to play again and again over a long period of time so that the deeper understanding does not wear off and so fundamentals used in the tool have an affect on what the students are currently learning in school.

The tool we need has already been created: the video game.[1,4,6,8,9] A video game with a story, set in a complex molecular world, in which the players not only watch, but also interact with and listen to in-game characters, may well be the best way to impart understanding of complex interactions among molecular entities. Complex games with multiple ways to achieve success often bring players back to replay, often over many years time. A game set in a real, complex world of molecules and cells may be an ideal way to introduce molecular behavior.

Successful commercial video games, that keep players coming back to replay, have a learning curve; a period of time and practice is required of the player to learn the game mechanism and the controls. A game complex enough to engage players in a story, with the ability to manipulate and experiment with objects will have a lengthy learning curve, as opposed to a casual vocabulary guessing game. Therefore, we are interested in knowing whether a complicated, third person shooter that involves controlling a ship in 3D space as well as shooting targets will have a learning curve that inhibits learning of the terms and concepts presented. Additionally, we are interested in knowing whether this learning curve inhibits players' confidence with the material.

Commercially successful video games do not have a list of multiple choice questions to answer or a vocabulary test. However, each level requires learning. Often some vocabulary is required to find objects and places on a map, or to choose weapons or to read tips about how to kill aliens. There are many potential ways that learning may occur in a video game. Therefore, we are interested in knowing whether we can demonstrate any retention of terms and concepts when the game itself does not ask any explicit questions of the player. Will players learn vocabulary that they are not forced to recall in the game? Will players learn about processes? Will they remember attributes of objects that they are not required to interact with? Will they be able to extrapolate what they learn to the real world? Will the game graphics, vocabulary, and situations prepare them for working with real materials on the subject? These questions are addressed by our evaluation of Immune Attack, and our data is presented in the results and discussion.

There is a precedent that concepts can be learned in a story-oriented, commercially successful game that is not designed to teach explicitly. Civilization is designed to be an engaging experience, not to teach facts about history.[14] However, Kurt Squire demonstrated that middle school students gained a valuable perspective on the situations, geographies, issues and technology innovations that lie at the heart of many events we learn about in history. In summary, Civilization players who have defended a large empire on a mountain from many smaller invading groups can change their strategy when they start their city on a low plains area.[9] In other words, the players have learned about how armies and geography interact in general. This general understanding of the complex interactions should provide the player with a scaffold for a deeper understanding of history. This type of general understanding of the interactions between molecules and cells is what we propose that a video game can provide. This understanding of molecular behavior is what we think will foster a deeper understanding of a health and biology related curriculum.

Immune Attack is a video game that requires players to activate specific proteins to cause specific behaviors of various white blood cells to win seven game levels. Immune Attack is a third person shooter-styled video game. Players fly a microbot and a nanobot through veins, through connective tissue and over the surface of white blood cells, while receiving instructions to "shoot" various proteins at appropriate times to solve various failings of a patient's immune system. Shooting proteins with an "EM emitter" causes the proteins to become active and perform their task.

Immune Attack is designed so that the action of the level is blended well with the science. Through a simple game mechanism (point and shoot), Immune Attack is able to present a wide range of molecular situations to players without introducing a new game mechanism. And the mechanism is reasonably well matched with the meaning of the game.

Proteins and cells can be manipulated in Immune Attack; they have roughly accurate sizes, shapes and clear roles to play in the action and drama of each level. In Immune Attack, molecules and cells are not just words we need to memorize, they are tools we need to use. Therefore, objects and concepts have scaffolding upon which to learn them. This should allow Immune Attack to present more detail and more abstract concepts to younger audiences. We also expect that the use of these objects in the game will foster students' abilities to recall their names and functions after playing.

Immune Attack is unique among learning games because 1) it presents advanced, abstract and fundamental concepts in cellular and molecular biology and 2) does not explicitly teach, but rather treats the material as if it were actually part of the story of a typical third person shooter video game. Immune Attack is a video game. It does not break its cover; there are no quizzes to pass, no mini games, no guarantee of formal learning. Learning happens within the context of the game: players only need to learn how to win the level in order to move on. Like a regular, commercial game, the player learns what they need to know and uses it immediately.

There is no explanation of what a protein is, instead we just get instructions to activate these proteins and what results. We see a response by the cell to the activation of proteins and we learn a concept. Therefore, Immune Attack presents a unique opportunity to begin to uncover what kind of learning and what kind of attitudes are imparted about biology through a commercial type video game.

Additionally, it should be noted that Immune Attack is designed for gamers. The tutorial is not intuitive for anyone who has not played this kind of complex video game already. Created in 2007–2008, Immune Attack is similar to what is referred to as a "hardcore" video game. It expects the player to devote a few minutes to listening, learning and practicing the controls. In the 6 years since Immune Attack was released, advances in tutorial design and game design have resulted in new ways to involve the player in the story while giving them time to practice the controls. So the tutorial is more integrated in the game and therefore more engaging for the player. Immune Attack, therefore, provides an excellent example of a game designed for a game playing audience. Any results we find for non-gamers may very well be improved upon by a more intuitive game design and more engaging tutorial.

Originally intended to teach immunology concepts to advanced high school students and beginning college students, Immune Attack teaches much more general content that immunology, rather it teaches the basics of molecular behavior. For example, Immune Attack shows players that individual proteins perform their own tasks and that proteins are not interchangeable. Additionally, Immune Attack shows that cells require a certain set of proteins in order to function and that missing a protein's function can cause disease. Additionally, 6000 grade school and high school teachers had registered to evaluate the game in their classrooms. Clearly, there was demand for the game for use with younger students. Therefore, we decided to evaluate the effectiveness of Immune Attack as a teaching tool for basic cellular and molecular biology with students in 7th–12th grades. (We present here our data from 10–12th grade students. We will present our data on 7th–9th grade students elsewhere.)

We address the following questions:

1. Can a video game (Immune Attack) teach concepts and terms of molecular biology, cellular biology and immunology?

2. Can a video game benefit students who do not play video games or who perform poorly in the test game?

3. Can a video game impart confidence, or in other words, a feeling of self-efficacy as regards the diagrams and graphics used to represent molecular and cellular biology data?

We tested Immune Attack players using their randomly assigned classmates as a control group. We developed an assessment of terms and concepts as well as an innovative test of confidence with molecular science. We found that Immune Attack players perform significantly better than control students on our test of terms and concepts. Additionally, players who reported playing 1 or more hours of video games per week, that is two-thirds of the students we tested, gained a significant amount of confidence with diagrams related to molecular science. On the other hand, students who were not gamers, who performed poorly in the game and who claimed the game was hard to play, showed no significant difference in confidence from the control students. However, while each of these groups (non-gamers, poor Immune Attack players, and those who reported that Immune Attack is hard to play) still had an average score that was significantly greater than the control on the terms and concepts test. That is, all students responded well to our test of terms and concepts, while most gained confidence with the material. And most noticeably, this complex game did not cause any players to lose confidence in their molecular biology abilities. Therefore, a complex game like Immune Attack may well be a useful medium for molecular science education.

Experimental methods

We built an assessment tool to measure 1) knowledge gains in the area of cellular and molecular immunology, 2) understanding of the game mechanism, 3) change in self confidence regarding the subject matter and 4) voluntary self reporting of demographic information. A complete list of these three sets of questions is available as ESI.† The assessment tool is taken – all three parts – by the students online. The online test takes about 30–40 min to complete. We created our own tests because it was completely unknown what we might discover. We tested the game in junior high and senior high schools across the US. We tested the game in classrooms, not because the game is designed for use in a classroom, but because we needed to know that the knowledge level of our test players and control players was as similar as possible.

Terms and concepts

We developed a 27 item multiple choice test of terms and concepts to test for gains in knowledge of molecular and cellular biology. We started with basic questions and received such encouraging results we spent the two years creating harder questions. We tested hundreds of students in grades 7–12 in American schools. Our assessment questions were tested numerous times over the course of two years and, in collaboration with statisticians Caroline Pinkham and David Silvernail at the University of Southern Maine, we had a multiple choice test with 27 items. The statistical report of details about the test of terms and concepts is included in the ESI.† The statistical report on the final iterations of this test are available as ESI.† The final Cronbach alpha score for these 27 knowledge questions is 0.79.

Choosing the concepts to focus on

During the test creation phase, 2009–2010, we asked students to answer open-ended questions, similarly to the method of Mike Klymkowsky.[5] The method of open-ended question asking is designed to bring out students' misconceptions. We used the students' misconceptions to focus our assessment questions, since there are so many possible topics to focus on in the game. For example, when we asked students "what is protein?" they often answered that "protein is found in chicken and egg whites" and that protein is an "important building block of the body." In other words, it appeared that students knew about proteins generally, but did not realize that each protein is unique and has a particular job to do. The game Immune Attack requires players to locate and shoot particular proteins, "Selectin" and "ICAM" in the very first level of the game. So we used this to design our test questions. There are basic knowledge questions to determine whether players remember the names of two proteins required to complete level 1. Knowing that proteins have individual names and activities is the first step to understanding the bigger concept that mutations in a single gene can lead to disease. Questions in this group we refer to as one protein/one job. Other concepts are the process of transmigration, the process of cytokinesis and the meaning of cell differentiation. The survey questions are submitted as ESI.†

Game mechanism

An understanding of the game mechanism is expected to be a prerequisite to learning anything from the game, and lack of game mechanism understanding may explain a lack of learning or confidence gains. Therefore, we also created a 10-item multiple choice test that is designed to test whether players know what the goals of the first three levels are. These questions simply ask the players what is the name of this object and what is in this image? We included questions about objects required for game success and objects that were not required for success. "What color are the monocytes?" Additionally, we showed images from the game such as the one in Fig. 1C and asked "What is this arrow pointing to?"

Confidence with cell biology/biochemical diagrams

Our test of self confidence is a series of images, diagrams and photos that are typical of the images scientists use to communicate their research results as well as college level immunology textbooks. Rather than ask students the typical questions about whether they see themselves as scientists, we wanted to ask "Does this look like something you would read?" We reasoned that students do not know what the words "molecular biology" mean, so they can't say whether they like it or would want to practice it or learn about it. Additionally, the game Immune Attack does not introduce a realistic science scenario and is not intended to present the career or activities of a scientist, rather all of the game takes place at the cellular and nanoscale levels. Immune Attack is, however, designed to present

A.

B.

C.

Fig. 1 Immune Attack screen shots. (A) Level one is about the process of transmigration. (B) The Monocyte "the big blue ball" will roll over the selectin proteins and slow down if the player shoots the selectin proteins. Selectin is shown in panel (C) with the white arrow drawn in to indicate the yellow Selectin protein.

cell and proteins in a manner typical of scientific diagrams. Cells are drawn to mimic schematic drawings used in journal articles and textbooks. Therefore, we reasoned that players would become more familiar with such diagrams and perhaps gain a feeling of self efficacy when presented with such diagrams in the future. So we designed questions that show the player cellular immunology-related diagrams and see if playing a video game about the topic would affect their feelings of self confidence with the material. The five confidence questions ask students to agree or disagree with the statement "I would be able to understand this if I read it and thought about it." See Fig. 5 for the images used. See Results and discussion for more detail.

Demographic information

As with every portion of this test, reporting of demographic information was voluntary. Students self-reported their grades in science, math and English, hours per week spent playing video games and which kinds. Students who played Immune Attack were also asked whether they enjoyed the game, found it "easy to play", would like to play such games in their science classes and whether they would recommend Immune Attack to their friend.

Protocol

Teachers reach us though our website, ImmuneAttack.org. We sent e-mails to 6000 people who registered as interested teachers prior to the game's release in May 2008. We received responses from about 100 teachers. Upon informing those 100 teachers of our three day controlled protocol, ten became our first in-class evaluators. Over the next 3 years, 2009 through 2012, several teachers participated each year.

Controlling for students' past experiences

Teachers were instructed to divide each of their classes randomly into two halves. Students were to play either Immune Attack or a control game for 40 min. The next week, the same halves played the games again. Twenty-four hours after the second play period, all students took our online test of cellular immunology, game mechanisms and self confidence (Fig. 2). Finally, teachers filled out another online survey in which they described in their own words how they divided their classes, how long the students played, and whether any deviations from the protocol or technical difficulties, *etc.* occurred. We were able to exclude data from teachers who did not describe a random method of dividing their class or who had technical difficulties playing the games. Teachers also distributed and collected consent forms. Our consent form explained that the games would be used in class and that all students can participate, but no data would be used from students who did not hand in their signed parental consent/student assent form.

The control games were related generally, but did not address any cellular or molecular biology. The games used were CSI: The Experience[15] (10–12th grades) or N-Squad[16] (7–9th grades). By dividing each classroom in half, we controlled for the background and educational experiences of the students as much as possible.

Week one				
40-minute game play in first week.				
Week two				
40-minute game play in second week	Online survey 24 hours after second game play			

Fig. 2 Schedule for classroom evaluations. Our evaluation protocol consisted of two 40-minute play periods and an online post test. Teachers reported to us through our online survey, filling in their own words how they accomplished the random assignment of their students to the two groups, how long the students actually spent with the games each day and if there were any technical problems.

Students self-reported their grades in math, science and English, how many hours per week they play video games and which kinds.

To ensure that the results were due to the game only, teachers were asked not to introduce the game or its topics. We also did not include any lecture or notes for teachers to read. We did, however, provide teachers with a written step by step description of what players do in the first three levels. This walkthrough is useful for teachers who wish to follow up on what students learn in the game. The walkthrough, FAQ and curriculum guides are available at ImmuneAttack.org.

Results and discussion

A statistical report characterizing the data collected is available as ESI.† Additionally, a complete list of all the questions and their answers are available in the ESI.†

Immune Attack players were statistically significantly more likely to score higher on our test of terms and concepts than their classmates who played control games. Fig. 3A shows that the distribution of scores on our test of terms and concepts is shifted to the right. Fig. 3B. shows that boys and girls benefited from playing Immune Attack to the same degree. Girls scored on average 1 point higher than boys, but this difference is not statistically significant. So gender and prior video game experience did not have an effect on student scores on our test of cellular and molecular biology terms and concepts.

We were interested in knowing whether a third-person shooter video game that requires two hands to control with a keyboard and mouse would perhaps intimidate or confuse non-gamers. We wanted to see if non-gamers would perform as well as gamers on our test of terms and concepts. Fig. 3C. shows that students who report playing zero hours of video games per week scored as well as students who report playing 1–5 or 6+ hours of video games. Consistent with this figure, the Pearson correlation and one-way ANOVA calculations in the statistical report also indicate no statistically significant correlation among the high scoring group and the hours of games per week students reported playing (see ESI†).

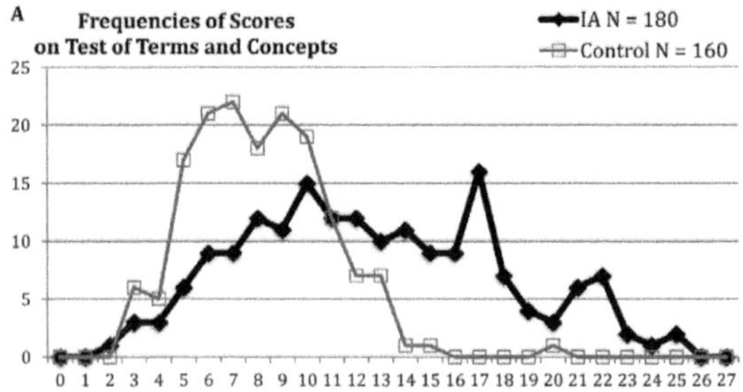

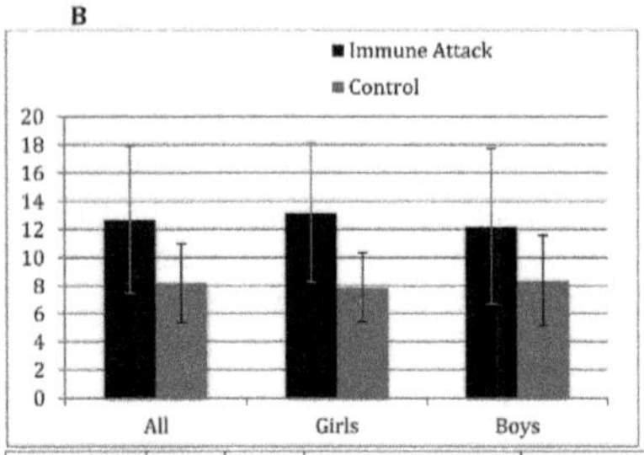

Terms and Concepts assessment	Average Score	n		T Test two sided, unknown variance	Effect Size
Immune Attack				P =	Cohen's d
All	12.7	180	All IA vs All Ctrl	7.77×10^{-21}	1.08
Girls	13.2	89	Girls IA vs Girls Ctrl	1.97×10^{-15}	1.36
Boys	12.2	88	Boys IA vs Boys Ctrl	1.74×10^{-07}	0.84
Control					
All	8.2	89			
Girls	7.9	69			
Boys	8.4	161			

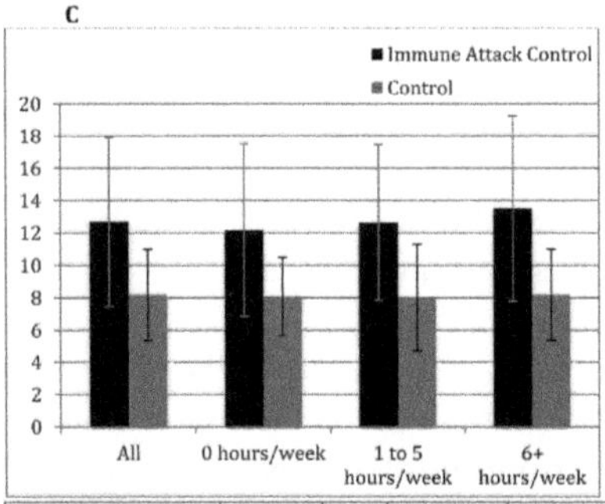

Terms and Concepts assessment	Average Score	n	Standard Deviation		T test 2 sided, unknown variance	Effect size
Immune Attack					P =	Cohen's d
All IA	12.7	180	5.2	All IA vs All Ctrl	$7.77x^{-21}$	0.81
0 hours/week	12.2	63	5.3	0 hr, IA vs Ctrl	$3.08x^{-07}$	0.73
1 to 5 hours/week	12.7	66	4.8	1-5 hr, IA vs Ctrl	$8.33x^{-09}$	0.87
6+ hours/week	13.5	50	5.7	6+ hr, IA vs Ctrl	$1.85x^{-07}$	0.88
Control						
All Ctrl	8.2	161	2.8			
0 hours/week	8.1	65	2.4			
1 to 5 hours/week	8.0	51	3.3			
6+ hours/week	8.2	42	2.8			

Fig. 3 Student scores on our test of terms and concepts presented in Immune Attack.

There is no correlation with gender or with the number of hours students report playing video games each week. However, we still want to determine whether a video game would work for all students. So we looked at other criteria we thought might affect students' scores on the terms and concepts test. These criteria are how the players reacted to the game itself.

We assumed that students who did not beat the first level of Immune Attack would be the group least likely to perform well on our test. We designed the test of terms and concepts to draw from the first three levels of Immune Attack only. Fig. 4A. shows the scores of Immune Attack players according to the level of the game they reported reaching. Our results show that all Immune Attack players, regardless of how many levels they reached, scored significantly better than the control game players.

However, Immune Attack players who reached the second half of the game scored even better on the terms and concepts than the players who only reached the first half of the game. Because the test of terms and concepts only covered the first three levels of Immune Attack, this difference is not due to the fact that level 5–7 players were exposed to more of the tested material. However, students who played through the second half of the game are exposed to more of the same concepts with new cells and new protein types (Fig. 4A).

Because learning from a video game is expected to require first an understanding of the mechanism of the game, we created a 10 question test. These questions about the game were straightforward: for example, students were shown Fig. 1C and asked "What is this arrow pointing to?" Control students were not asked these questions. Scores on the 27 terms and concepts test and the 10 game mechanisms test were very well correlated. There is a strong correlation between students' scores on the ten game mechanism questions and their scores on the test of concepts and terms (data not shown).

Additionally, responses to post-play questions, such as "do you agree that Immune Attack is easy to play?" were correlated with higher scores on the terms and concepts questions. So understanding the game mechanism and understanding how to play is important for remembering terms and concepts after the game.

Next, we address whether the game imparts a feeling of confidence with the material presented. We reasoned that seeing the images of cells and proteins in the game would make players feel more familiar with these types of images. The cells are drawn similar to typical schematics scientists use to communicate their data, so we were able to use images used by scientists to convey their data with the game players in our study.

Fig. 5 shows the five images used to test for student's self confidence with images related to cellular and molecular biology. Students were asked to agree or disagree with statement "I would be able to understand this is I read it and thought about it." The following answer choices were available for each question:

1. I disagree definitely.
2. I disagree somewhat.
3. I am neutral.
4. I agree somewhat.
5. I agree definitely.

We found that Immune Attack players do gain confidence as regards the transmigration and data images. In the transmigration image (5A), there is a high degree of complexity.[10] This image, from *Nature Immunology Reviews*, is designed for an expert audience. The second image, "Yellow Macrophages" (5B) from Janeway's *Immunobiology* textbook, is designed for undergraduate and graduate

A

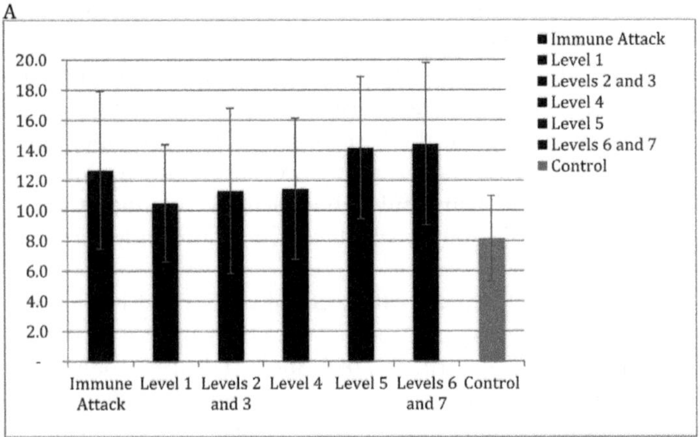

Score per level of game reached	Average Score	n	Standard Deviation		T Tests	Effect Size
All levels	12.7	180	5.2	IA vs CTRL	**7.77x10-21**	**0.81**
Level 1	10.5	25	3.9	1 vs CTRL	**6.89x10-03**	**0.54**
Levels 2 and 3	11.3	36	5.5	2 & 3 vs CTRL	**1.63x10-03**	**0.55**
Level 4	11.5	32	4.7	4 vs Control	**4.47x10-04**	**0.65**
Level 5	14.2	20	4.7	5 vs CTRL	**6.74x10-06**	**1.18**
Levels 6 and 7	14.4	67	5.4	6 & 7 vs CTRL	**9.71x10-14**	**1.09**
Control	8.2	161	2.8			

B

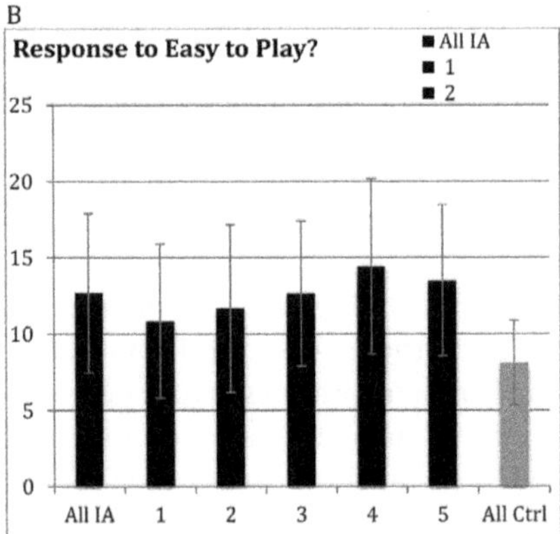

Terms and Concepts vs "IA is easy to play"	Average Score	n	Standard Deviation		T Tests	Effect Size
All IA	12.7	180	5.2	IA vs Ctrl	**$3.17x10^{-21}$**	**0.824**
1	10.9	27	5.0	1 vs Ctrl	**$1.01x10^{-02}$**	**0.507**
2	11.7	27	5.5	2 vs Ctrl	**$2.42x10^{-03}$**	**0.616**
3	12.7	55	4.8	3 vs Ctrl	**$4.62x10^{-09}$**	**0.887**
4	14.4	41	5.7	4 vs Ctrl	**$1.71x10^{-08}$**	**1.043**
5	13.5	23	4.9	5 vs Ctrl	**$2.97x10^{-05}$**	**1.017**
All Ctrl	8.1	161	2.8	6 vs Ctrl		

Fig. 4 Scores on test of terms and concepts by (A) level of Immune Attack reached and by (B) response to the question "Is Immune Attack easy to play?" Students were given images to select from in order to indicate which level of Immune Attack they reached—without needing to recall the name of the level. (See online survey, in the ESI†).

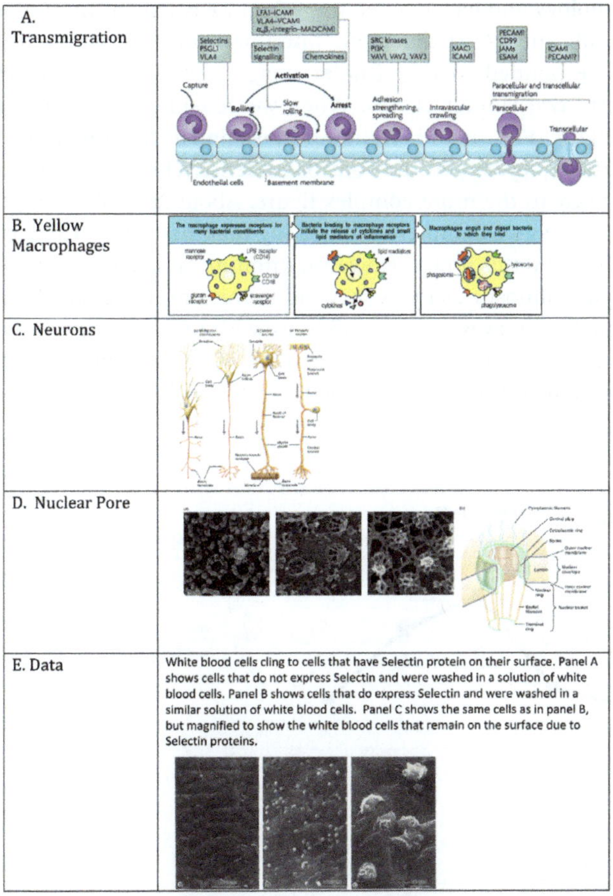

			T Test	Stdev	Effect size
A. Transmigration					
	IA	3.16	2.30x^{-07}	1.13	0.47
	Ctrl	2.50		1.21	
B. Yellow Macrophages					
	IA	3.23	1.19x^{-01}	1.10	0.14
	Ctrl	3.04		1.15	
C. Neurons					
	IA	3.10	1.09x^{-01}	1.16	0.38
	Ctrl	2.56		1.20	
D. Nuclear Pore					
	IA	2.84	3.14x^{-01}	1.23	0.088
	Ctrl	2.71		1.17	
E. Data					
	IA	3.16	3.69x10^{-02}	1.12	0.030
	Ctrl	3.12		1.25	

Fig. 5 Confidence questions. Five images were presented to Immune Attack and control game players. Under each picture was the statement "I would be able to understand this is I read and thought about it." The five answer choices were 1) I agree completely, 2) I agree somewhat, 3) I am neutral, 4) I disagree somewhat and 5) I disagree completely.

level immunology and has less detail.[11] When we consider the simpler image, we see in Fig. 6B and 7C–D that both Immune Attack players and their control classmates responded to the simpler image in the same way. The students' answers form two similar curves. When we consider the more complex transmigration figure (Fig. 7B and 7A) however, we find a clear difference between the Immune Attack players and the control students. The control students respond more negatively to the more complex figure. About 25% of the control students responded "I disagree definitely," indicating that they do not think they would be able to understand this image. However, Immune Attack players responded to the complex figure almost exactly as they did to the simpler textbook image.

We wanted to know whether confidence with related images would carry over to other image types. Images of neurons and of a three dimensional representation of a protein complex did not elicit a more positive response from Immune

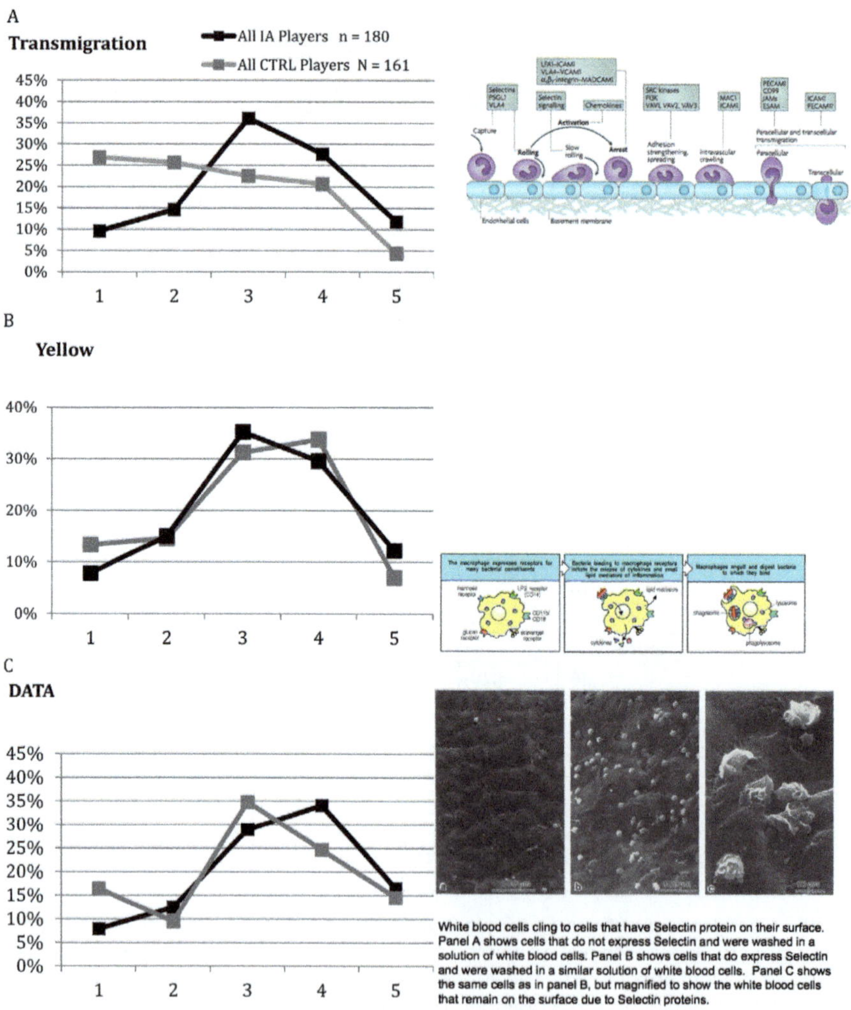

Fig. 6 Confidence data for (A) the complex transmigration diagram, (B) the simple textbook yellow macrophages diagram and (C) the data image.

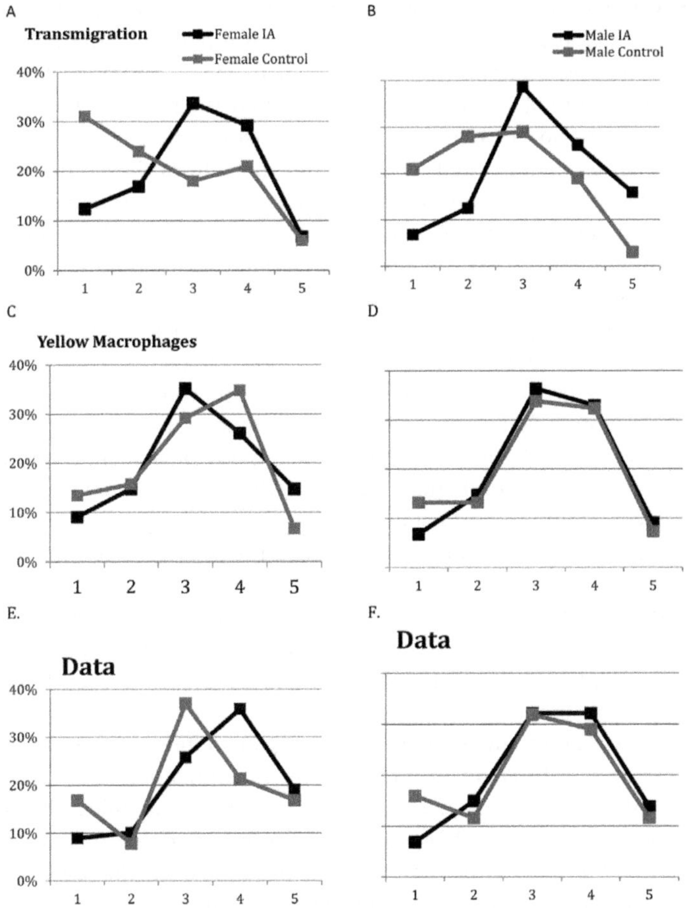

	Transmigration	Yellow Macrophages	Neurons	Nuclear Pore	Data
Immune Attack all	3.16	3.23	3.10	2.84	3.39
Girls	3.01	3.23	2.98	2.74	3.46
Boys	3.32	3.23	3.23	2.92	3.31
Control All	2.50	3.04	2.90	2.71	3.12
Girls	2.46	3.06	2.91	2.64	3.13
Boys	2.56	3.07	2.87	2.81	3.09
T TESTS					
IA all vs Ctrl All	**0.000**	1.123	1.413	0.469	**0.037**
IA Girls vs Ctrl Girls	**0.003**	0.324	0.704	0.573	0.079
IA Boys vs Ctrl Boys	**0.000**	0.386	**0.043**	0.576	0.241
Effect size					
IA all vs Ctrl All	**0.47**	0.14	0.14	0.09	0.19
IA Girls vs Ctrl Girls	0.38	0.12	0.05	0.07	0.22
IA Boys vs Ctrl Boys	**0.56**	0.12	0.27	0.07	0.16
IA Girls vs all control	0.36	0.13	0.05	0.02	0.23
IA Boys vs All Ctrl	**0.59**	0.14	0.24	0.14	0.14

Fig. 7 Confidence data for (A and B) the complex transmigration diagram, (C and D) the simple text book yellow macrophages diagram and (E and F) the data images broken down into girls' data and boys' data.

Attack players compared with the control. It may be that the positive effect on their responses is very specific to the type of image and the drawing style. Immune Attack cells look very similar to the diagram in Fig. 1, in which H&E stained features of the white blood cells are drawn, namely the nucleus shape and color. However, it may also be that the complex transmigration image was the only image complex enough to elicit a negative response from the control students.

We wanted to see if Immune Attack players could extrapolate their new understanding and confidence to actual data images. Fig. 5E is from William Muller's groundbreaking work, which showed that transmigration—the objective of level 1 of Immune Attack—is dependent on selectin—the protein players need to shoot to successfully transmigrate their monocyte.[13] The images in the panels will naturally look foreign, since the scanning electron microscope images are 3D, grey and do not show any of the schematic details shown in the game, like nuclei. However, we wrote a figure legend in simple language that uses terms from the game and also clearly describes the experiment being presented. Immune Attack players consistently respond more positively than control game players about their perceived ability to understand the data images (Fig. 7E–F). For the data figure, the effect size is small (0.19), because the average score does not shift greatly, while for the transmigration figure, the effect size is medium (0.47).

So a difference between Immune Attack players and controls is found in their reaction to the transmigration diagram and the data with a figure legend. Immune Attack players respond more positively on the whole to both images. To investigate which groups of students gained confidence, we looked at boys *vs.* girls in Fig. 7A–F.and then at gamers *vs.* non-gamers in Fig. 8A–E.

Considering first the effect of gender for the transmigration image, both female and male Immune Attack players responded statistically significantly more positively to the transmigration image compared to matched controls, although the difference has a smaller effect size for girls than boys. For the data with figure legend, the increase in confidence is not significant unless all 180 Immune Attack students are compared to the control. This is primarily because the averages between the test and control means are so small. However, in Fig. 6C and Fig. 7E we can see that the curve of student responses is shifted to the more positive after exposure to Immune Attack. This shift effect is stronger for girls than for boys.

Considering the effect of hours of game play reported per week, we find in Fig. 8A that non-gamers (0 h per week reported) do not differ from non-gamer control students. Students who report playing 1 or more hours per week, which is two thirds of the students tested, show a statistically significant shift in their responses to the transmigration image. For the data image, results are only significant when all hours of games per week are considered together.

It is interesting to note, gamers (students who report playing 6 or more hours per week) in the control group were consistently more negative about their ability to understand these images. Perhaps video gamer players can more accurately assess their ability to understand a new diagram (Fig. 8A–E).

Finally, we looked at whether students' scores on the terms and concepts test are correlated with their confidence scores (Fig. 9). We found that the students who scored 9–16 and 17–27 questions correctly also responded more favorably to the transmigration diagram and the data images. The confidence measurement for students who scored 8 or fewer correct answers on the terms and concepts test could not be distinguished form the control response.

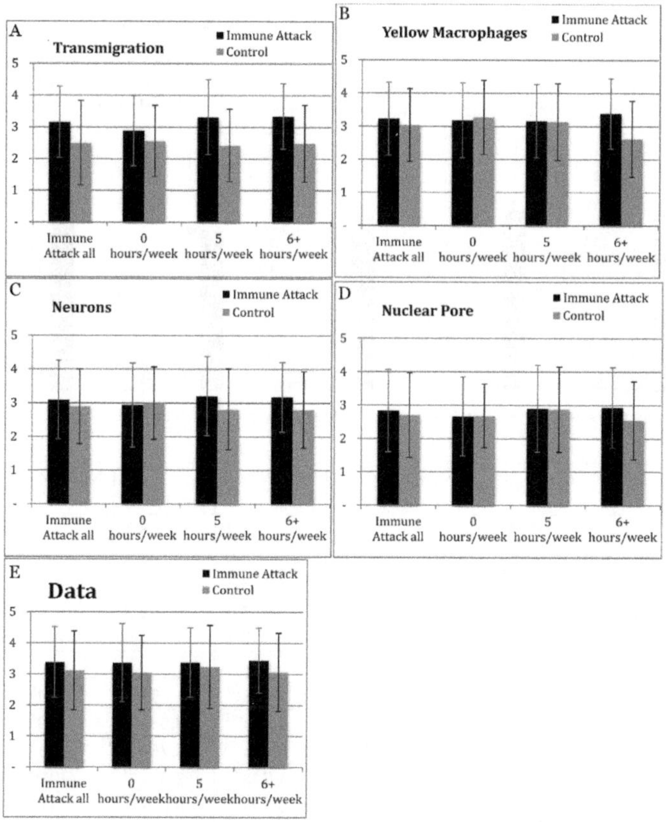

Confidence with images	n =	Trans-migration	Yellow Macro-phages	Neurons	Nuclear Pore	Data
Immune Attack all	180	3.16	3.23	3.10	2.84	3.39
0 hours/week	63	2.89	3.19	2.94	2.67	3.37
5 hours/week	66	3.32	3.17	3.21	2.91	3.38
6+ hours/week	49	3.35	3.39	3.18	2.94	3.45
Control All	161	2.50	3.04	2.90	2.71	3.12
0 hours/week	65	2.57	3.28	3.00	2.69	3.05
5 hours/week	51	2.43	3.14	2.82	2.88	3.24
6+ hours/week	42	2.49	2.63	2.81	2.55	3.07
T TESTS						
IA all vs Ctrl All		**2.3E-07**	1.2E-01	1.1E-01	3.1E-01	**3.7E-02**
IA 0 vs Ctrl 0		1.5E-01	6.7E-01	7.6E-01	9.1E-01	1.5E-01
IA 1-5 vs 1-5		**6.5E-05**	8.9E-01	6.4E-02	9.0E-01	4.9E-01
IA 6+ vs 6+		**7.1E-04**	**2.2E-03**	1.4E-01	1.1E-01	1.6E-01
Effect size						
IA All vs Ctrl Al		0.47	0.14	0.14	0.09	0.19
IA 0 vs Ctrl 0		0.22	-0.07	-0.04	-0.01	0.21
IA 1-5 vs 1-5		**0.63**	0.02	0.28	0.02	0.10
IA 6+ vs 6+		**0.66**	**0.57**	**0.28**	**0.26**	**0.27**
IA 0 hrs vs all control		0.28	0.11	0.03	-0.03	0.16

Fig. 8 (A–E) Confidence data for each of the five images, with data broken down into all students, students who report playing 0 h of video games per week, those who report playing 1–5 h, and those who report playing 6 or more hours per week.

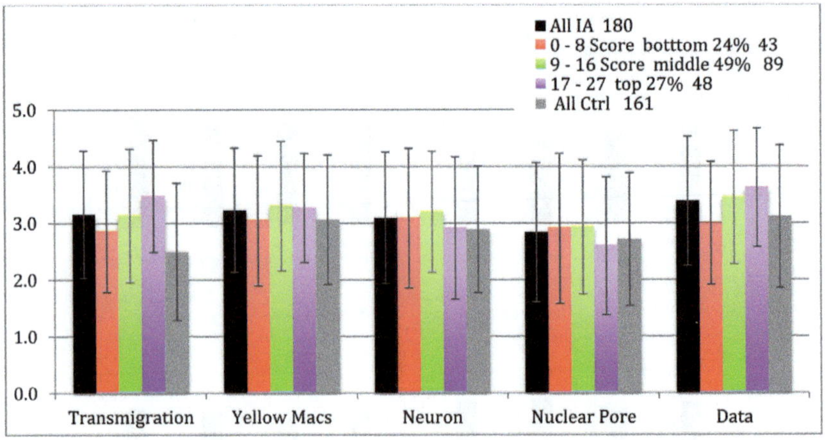

Averages	n	Trans-migration	Yellow Macs	Neuron	Nuclear Pore	Data
All IA	180	3.2	3.2	3.1	2.8	3.4
0 - 8 Score botttom 24%	43	2.9	3.0	3.1	2.9	3.0
9 - 16 Score middle 49%	89	3.1	3.3	3.2	2.9	3.4
17 - 27 top 27%	48	3.5	3.3	2.9	2.6	3.6
All Ctrl	161	2.5	3.1	2.9	2.7	3.1
TTest						
All IA vs All Ctrl		**3.62x^{-07}**	1.19x^{-01}	1.09x^{-01}	3.14^{x-01}	**3.69^{x-02}**
Bottom vs Ctrl		6.60x^{-02}	9.85x^{-01}	3.58x^{-01}	3.73^{x-01}	5.45x^{-01}
Middle vs Ctrl		**8.14x^{-05}**	8.77x^{-02}	**3.76x^{-02}**	1.51^{x-01}	**3.86x^{-02}**
Top vs Ctrl		**1.43x^{-07}**	1.74x^{-01}	9.37x^{-01}	6.01x^{-01}	**5.99x^{-03}**
Bottom vs Top		**5.55x^{-03}**	3.24x^{-01}	5.00x^{-01}	2.61x^{-01}	**6.71x^{-03}**
Effect Size						
All IA vs All Ctrl		0.47	0.13	0.15	0.08	0.19
Bottom vs Ctrl		0.26	-0.01	0.14	0.12	-0.08
Middle vs Ctrl		0.43	0.17	0.23	0.15	0.23
Top vs Ctrl		**0.75**	0.16	0.02	-0.08	0.37
Bottom vs Top		**0.50**	0.18	-0.12	-0.20	0.48

Fig. 9 Confidence data for each of the five images, with data broken down by how well students performed on the test of terms and concepts.

Conclusions

We have found that students learn molecular and cellular biology concepts and terms by playing a video game set in the molecular world. Even though the game does not explicitly teach these terms and concepts, students can answer questions on our 27 item multiple choice test significantly better than their classmates who played an unrelated video game. The higher average scores on the test of terms and concepts was true for all Immune Attack players, regardless of how many hours a week they play games and whether they are male or female. We found no factor that can predict prior to playing the game whether students will perform well on the test of terms and concepts. However, we found that students who scored in the bottom 24% on the test of terms and concepts were more likely to report that the Immune Attack is not easy to play. Additionally, understanding the game mechanism questions was the strongest indicator of whether students scored well on the test of terms and concepts. We conclude that a video game like Immune Attack – a shooter-styled, self-directed game with minimal explicit learning activities – can improve student scores on a test of terms and concepts. Importantly, a game with a more engaging introduction to the controls and game mechanism than Immune Attack may be even more effective at helping students remember terms and concepts.

Immune Attack also had a positive effect on the confidence players feel in their abilities to understand complex diagrams related to the game. However, unlike the test of terms and concepts, we found a difference between gamers and non-gamers. In particular, we found that students who reported playing zero hours of video games per week showed no difference in confidence with a molecular biology diagram as compared to the control students. So there were no confidence gains in the non-gamer group. However, Immune Attack is a complex third-person shooter, in which players need to learn to navigate a microbot in three-dimensional space, avoid many dangerous objects and shoot a ray gun at moving proteins. Yet, despite the complexity of the game, non-gamers do not show any loss of confidence compared to the control group regarding molecular biology. We did not find a difference in confidence between the gender groups, however. Boys as well as girls showed a statistically significant shift in their response to a complex scientific diagram.

Additionally, the majority of students who reported enjoying Immune Attack and who played past the third level of the game showed the greatest increase in confidence, regardless of whether they were gamers. Most interestingly, when presented with two molecular cell biology diagrams, one designed for experts and one designed for students, Immune Attack players responded to the expert diagram as favorably as if it were the simpler one.

We have shown that a large fraction of Immune Attack players gained confidence with the kinds of diagrams scientists use to communicate their data, while not causing a loss of confidence in non-gamers or students who did not perform well in the game. Immune Attack was successful at increasing the average score of all students, gamers and non-gamers alike, on our test of terms and concepts. Therefore, we conclude that a complex video game like Immune Attack that 1) requires a tutorial to teach the game mechanism similar to a hardcore video game and 2) teaches through story-based interactions with a wide array of objects but without explict text-based questions to answer can increase the scores of players on a test of terms and concepts, increase the confidence of gamers with the material and not harm the confidence levels of non gamer students. We suggest that game design improvements can positively affect the confidence levels of the non-gamer students.

Acknowledgements

Many teachers participated in our controlled evaluation, provided feedback on our assessment items, and on the game implementation and evaluation in the classroom. Caroline Pinkham at the University of Southern Maine provided expert statistical analysis. The Maine International Center for Digital Learning provided outreach and communication with participating teachers. Howard Young, Ph.D. at the National Institutes of Health has been a supporter of the Immune Attack project in many ways, including the assessment development. Immune Attack was created by the Federation of American Scientists Learning Technologies Program with funding from the National Science Foundation (USA) in collaboration with Brown University and the University of Southern California. The final game was developed by Escape Hatch Entertainment. Immune Attack was released May 2008. The Windows-only game is available for free download at ImmuneAttack.org, where there is a trailer, which previews the game action. The evaluation work presented here was funded by the National Institutes of Health, National Institute for Allergy and Infectious Diseases (NIAID) and the Entertainment Association Foundation (EASF).

References

1 K. G. Bowling, Y. Klisch, S. Wang and M. Beier, Examining an online microbiology game as an effective tool for teaching the scientific process, *Journal of Microbiol. Biol. Educ.*, 2013, **14**, 58.
2 J. Jenkinson and G. McGill, Visualizing protein interactions and dynamics: evolving a visual language for molecular animation, *CBE Life Sci. Educ.*, 2012, **11**, 103.
3 R. Joiner, J. Iacovides, M. Owen, C. Gavin, S. Clibbery, J. Darling and B. Drew, Digital games, gender and learning in engineering: do females benefit as much as males?, *J. Sci. Educ. Technol.*, 2011, **20**, 178.
4 Y. Klisch, L. M. Miller, M. E. Beier and S. Wang, Teaching the biological consequences of alcohol abuse through an online game: impacts among secondary students, *CBE Life Sci. Educ.*, 2012, **11**, 94.
5 M. W. Klymkowsky and K. Garvin-Doxas, Recognizing Student Misconceptions through Ed's Tools and the Biology Concept Inventory, *PLoS Biol.*, 2008, **6**, 1.
6 L. M. Miller, J. Moreno, V. Estrera and D. Lane, Efficacy of MedMyst: an Internet Teaching Tool for Middle School Microbiology, *Microbiol Educ.*, 2004, **5**, 13.
7 NGSS Lead States, *HS. Structure and Function, HS. Matter and Energy in Organisms and Ecosystems*, Achieve, Inc., 2013.
8 K. D. Squire, From content to context: Video games as designed experiences, *Educational Researcher*, 2006, **35**, 19.
9 K Squire, Dissertation Research, University of Wisconsin, Madison, WI, USA, 2004, http://website.education.wisc.edu/~kdsquire/research.html, accessed 2014.
10 K. Ley, C. Laudanna, M. I. Cybulsky and S. Nourshargh, *Nat. Rev. Immunol.*, 2007, 7, 678.
11 C. A. Janeway, P. Travers, M. Walport and M. Shlomshik, *Immunobiology: The Immune System in Health and Disease. 5th edition*, Garland Science, New York, USA, 5th edn, 2001.
12 C. Brunner, *A development process focused on pedagogy-centered design. Possible Worlds Technical Report and Working Paper Series*, Center for Children & Technology/Education Development Center, Inc., New York, USA, 2004, http://possibleworlds.edc.org, accessed 2014.
13 W. A. Muller, S. A. Weigl, X. Deng and D. M. Phillips, *J. Exp. Med.*, 1993, **178**, 449.

14 Sid Meier's Civilization III by Firaxis, 2001, http://www.civ3.com, accessed 2014.
15 CSI The Experience: Web Adventures, 2012, http://www.civ3.com, accessed 2014.
16 N-Squad Web Adventures, Rice University, Center for Technology in Teaching and Learning, 2011, http://n-squad.rice.edu, accessed 2014.

Faraday Discussions

PAPER

Udock, the interactive docking entertainment system

Guillaume Levieux,*[a] Guillaume Tiger,[a] Stéphanie Mader,[a] Jean-François Zagury,[b] Stéphane Natkin[a] and Matthieu Montes*[b]

Received 18th December 2013, Accepted 31st January 2014
DOI: 10.1039/c3fd00147d

Protein–protein interactions play a crucial role in biological processes. Protein docking calculations' goal is to predict, given two proteins of known structures, the associate conformation of the corresponding complex. Here, we present a new interactive protein docking system, Udock, that makes use of users' cognitive capabilities added up. In Udock, the users tackle simplified representations of protein structures and explore protein–protein interfaces' conformational space using a gamified interactive docking system with on the fly scoring. We assumed that if given appropriate tools, a naïve user's cognitive capabilities could provide relevant data for (1) the prediction of correct interfaces in binary protein complexes and (2) the identification of the experimental partner in interaction among a set of decoys. To explore this approach experimentally, we conducted a preliminary two week long playtest where the registered users could perform a cross-docking on a dataset comprising 4 binary protein complexes. The users explored almost all the surface of the proteins that were available in the dataset but favored certain regions that seemed more attractive as potential docking spots. These favored regions were located inside or nearby the experimental binding interface for 5 out of the 8 proteins in the dataset. For most of them, the best scores were obtained with the experimental partner. The alpha version of Udock is freely accessible at http://udock.fr.

Introduction

Protein–protein interactions play a crucial role in biological processes. The prediction of the geometry of protein complexes is a difficult task that has been a goal of computational chemistry. Much effort has been invested in the last few decades to develop docking methods, their performance being assessed during the CAPRI experiment.[1] Most of the available protein docking methods explore the conformational space between the proteins to be docked; this exploration

[a]Equipe Interactivité pour Lire et Jouer, Laboratoire CEDRIC, EA4626, Conservatoire National des Arts et Métiers, 292 Rue Saint Martin, 75003 Paris. E-mail: guillaume.levieux@cnam.fr
[b]Laboratoire Génomique Bioinformatique et Applications, EA4627, Conservatoire National des Arts et Métiers, 292 Rue Saint Martin, 75003 Paris. E-mail: matthieu.montes@cnam.fr

being based either on fast Fourier transform correlations,[2–5] Monte Carlo sampling[6,7] or driven by biochemical or physical information.[8] Recent developments in haptic devices allowed the emergence of interactive molecular dynamics approaches[9,10] enabling systems' simulation while receiving real-time feedback.

To date, there is no protein docking method available that allows a quick and interactive handling of the proteins to be docked to perform human driven exploration of the protein–protein interfaces. To our knowledge, the closest attempt towards this goal was the prototype of DockingShop[11] that seems to be no longer in development. The computational chemistry research field might benefit from intuitive and interactive tools[10] that would lead to a rapid gain in general knowledge on the problem, or get new ideas by trial and error exploration.

Moreover, it might be useful to have even non-experts, so-called naïve users, use these kinds of tools. Indeed, the protein docking problem can be considered as a complex 3D shapes combination problem. Humans beings are intuitively good at shape recognition and abstraction,[12] and if given appropriate tools, even naïve users could intuitively propose appropriate solutions of complex problems[13] such as protein–protein interfaces by steered trial and error exploration.

Resolving protein–protein interaction challenges can foster non-expert users' motivation because it inherently provides what is needed to create a good video game: a goal-directed task that is, according to Atari's founder Bushnell's quote "both easy to learn and very hard to master".[14]

Here, we present the first version of an interactive docking system, Udock, that would allow a quick and easy-to-handle exploration of the possible conformations of a protein complex. First, we will describe protein animation and rendering with Udock, starting from a standard molecular description file until integration into a video game physics engine. We will also present our choices with regard to binding energy calculation. Then, we explain how we simplified the protein structure representation and the docking process task, so that we allow even naïve users to perform interactive docking. As a preliminary assessment of our approach, we present the results of a two week playtest: a user-based interactive cross-docking experiment on 8 proteins, with a limited number of users that have tried to reach the best binding score for each out of the 36 possible protein complexes.

Methods

Different steps are needed in Udock before allowing users to perform interactive molecular docking: preprocessing of the coordinate files, generation of the solvent excluded surfaces (SES), and smoothed coloration of the SES according to the atomic partial charges.

Preprocessing of the coordinate files

Udock uses protonated protein mol2 files with atomic partial charges computed using AMBER12 force-field.[15] To generate such files from protein PDB files, we use the dockprep procedure as implemented in Chimera[16] using default parameters.

Generation of the SES

Once mol2 files have been generated, they can be loaded by Udock. To generate a 3D mesh out of the protein mol2 file, we use a marching cubes algorithm as described in ref. 17. For every atom in the mol2 file, we first generate the solvent accessible surface (SAS) using the sum of the atom radii and a 1.4 Å radius probe. Then, we roll the probe whose center is at the generated surface and remove all the cubes that the probe intersects, and thus obtain the solvent excluded surface (SES).

Coloration of the SES

The color of the SES surface mesh is used to represent the electrostatic potential at the surface. For every point of the surface, we calculate the mean of all the atomic partial charges within a 5 Å radius sphere. Each atom's partial charge is divided by the squared distance to a point located 1.4 Å above the surface point we calculate. We thus represent a smoothed approximation of the electrostatic potential of the protein. Fig. 1 shows the difference between the smoothed electrostatic potential displayed on the surface and the unsmoothed, basic output of atomic partial charges displayed on the surface. We use a pixel shader script to enhance the readability of the SES, mainly by using black contour lines to enhance the perception of the molecule's shape. Following the graphic chart provided by our graphical designer, neutral parts of the proteins are not white but rather a light blue color. Then, we slowly reach a strong blue or red color to indicate respectively positively and negatively charged areas.

Rendering and interaction

Once the colored SES is generated, its mesh is processed by an open source video game physics engine, Bullet,[18] to generate a collision mesh. We use the physics engine to handle the user interaction with the molecule: when the user clicks on the molecule, we apply 3D forces on the mesh based on the mouse input, and let the physic engine calculate the subsequent orientation of the molecule. To give the feeling of a molecule immersed in a solvent and facilitate interaction, we dampen angular speed and velocity so that if the user stops interacting with a protein, it takes exactly one second for it to stop moving. Moreover, the physics engine is also responsible for calculating and taking into account the collisions between the molecules. As a result, users do not have to take clashes into account when they try to dock a protein on the other one.

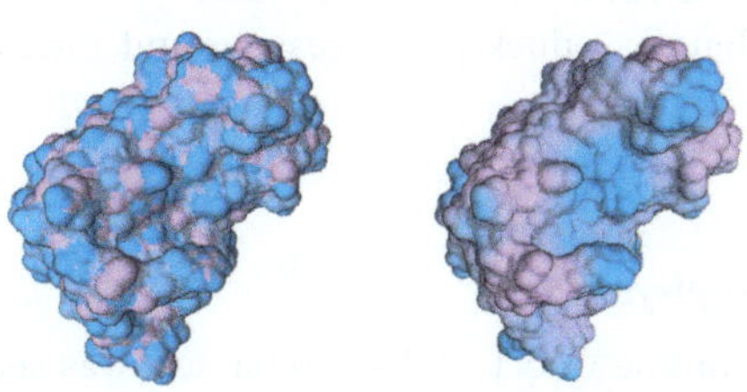

Fig. 1 Illustration of the smoothed approximation of the electrostatic potential on barnase (PDB id: 1brs, chain a). Left: atomic partial charges as computed by amber12 displayed on the solvent excluded surface. Right: our smoothed approximation of the electrostatic potential displayed on the SES.

A grapnel-based interface to perform user-steerable interactive docking

To make interactive docking a naïve-user-steerable task, we decided to use a grapnel-based representation. The users interactively select protein SES locations on which they will attach grapnels. Any number of grapnels can be attached between the SES of the proteins that will be docked. At any moment, the user can apply an attractive force on the grapnels to reduce their length gradually and put the proteins in contact. The sum of torque values applied by the grapnels on the proteins is monitored and adjusted on the fly in order to let the user orient the proteins as he wishes before collision occurs. Thanks to the physics engine, when proteins collide, no clashes between the atoms of the ligand and the atoms of the receptor are possible.

During the whole procedure, a force-field based intermolecular energy score is computed and displayed on the fly in the interface. The scoring function includes a soft van der Waals term for contacts and a Coulombic term for electrostatics. We used a distance dependent dielectric constant of $\varepsilon_0 = 20$ that resulted in balanced contributions of the different terms of the scoring function. The detailed form of the scoring function for the interaction energy of the atom pair i, j at distance r_{ij} is detailed below:

$$\text{Score} = -\left(-\frac{A_{ij}}{r_{ij}{}^{6}} + \frac{B}{r_{ij}{}^{8}} + f\frac{q_i q_j}{\varepsilon_0 r_{ij}}\right)$$

with q_i and q_j atomic partial charges of atoms i and j computed using AMBER12[15] as implemented in Chimera's dockprep.[16] A_{ij} and B_{ij} are, respectively, repulsive and attractive Lennard-Jones type parameters. f is a conversion factor for converting the electrostatics term to kcal mol^{-1}. We used $f = 332.0522$ according to the AMBER12 documentation.[15]

At any time, the user can launch a five second Monte Carlo rigid body optimization of the complex. The number of Monte Carlo steps that will be performed during the optimization procedure will depend on the number of atoms interacting in the system and the power of the CPU of the users' computer. For example, evaluation of the scoring function on a pair of atoms takes 19.5 ns on an Intel Core i7 3930 K (3.2G Hz), which corresponds to 300 Monte Carlo optimization steps of a contact between barnase and barstar within five seconds. During this Monte Carlo optimization procedure, the physics engine is switched off, allowing closer and more accurate contacts if favored by the scoring function. Indeed, the use of a soft repulsive term in the van der Waals part of the scoring function will allow (but still penalize) the existence of small clashes to simulate a pseudo-plasticity of the residues in the interface.

A detailed flowchart of Udock preprocessing and interactive docking is presented in Fig. 2.

Udock alpha version playtest

The duration of the online Udock alpha version test was set to two weeks during which the users could explore freely a test set of 4 binary complexes detailed below. The users were mostly computer science students and co-workers. Statistics on the users' age, play frequency and previous knowledge on structural biology were performed based on surveys upon registration.

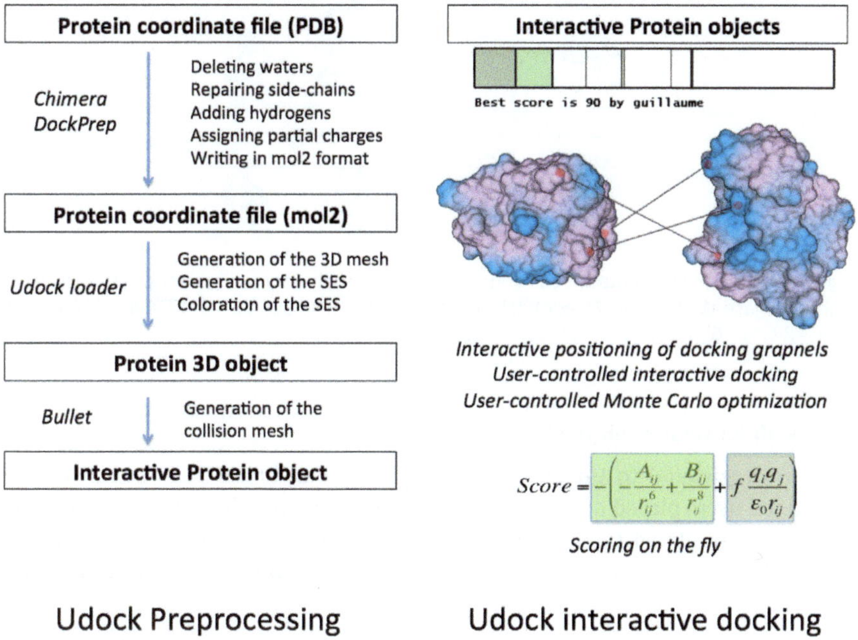

Fig. 2 Flowchart of Udock preprocessing and interactive docking.

Construction of the test set

The users explored four binary enzyme–inhibitor complexes used in the pioneer cross-docking experiment of Sacquin-Mora *et al.*,[19] namely barnase/barstar (PDB ID: 1BRS), acetylcholinesterase/fasciculin II (PDB ID: 1FSS), thermitase/eglin c (PDB ID: 2TEC) and CDC42 GTPase/CDC42 GAP (PDB ID: 1GRN) which led to 36 possible complexes to be explored by the users. In order to prevent the users from using external information about the proteins or their geometry, we anonymized the proteins in the dataset by giving them random names as detailed in Table 1. The proteins of the dataset vary in size and complexity, as can be seen in Fig. 3.

Table 1 Summary of the protein complexes investigated in the study. #residues is the number of residues of the corresponding protein. Throughout this work, we refer to the proteins by their index, given in the last column

PDB ID_Chain	Protein	Random name	#residues	Index
1BRS_A	Barnase	Dwaylith	110	**1**
1BRS_D	Barstar	Cilan	89	**2**
1FSS_A	Acetylcholinesterase	Eralg	535	**3**
1FSS_B	Fasciculin II	Taurith	61	**4**
1GRN_A	CDC42 GTPase	Bisil	200	**5**
1GRN_B	CDC42 GAP	Prok	199	**6**
2TEC_E	Thermitase	Etinna	279	**7**
2TEC_I	Eglin c	Bloc	63	**8**

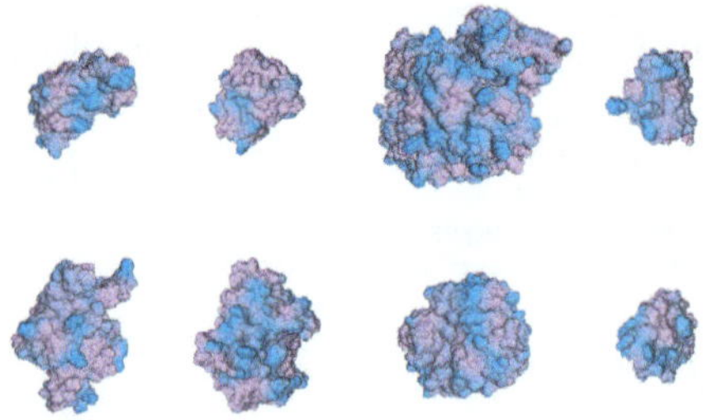

Fig. 3 Proteins of the dataset, as rendered by the Udock engine. From top left to bottom right: barnase (**1**), barstar (**2**), acetylcholinesterase (**3**), fasciculin II (**4**), CDC42 GTPase (**5**), CDC42 GAP (**6**), thermitase (**7**), eglin c (**8**).

Udock alpha version playtest users' statistics

42 users registered to Udock and played for a total of 25 h. 27 out of the 42 registered users played at least 5 min and among them only 12 played at least 30 min. The cumulated time spent by the users exploring the geometry of the 36 different complexes varied from 10 to 87 min (see Table 2). The users were on

Table 2 Cumulated time (in minutes) spent by the users on the exploration of the geometry of the 36 different possible complexes in the cross-docking dataset. A gradient of color has been applied from the less explored complexes (yellow) to the most explored complexes (dark green)

Index	*1*	*2*	*3*	*4*	*5*	*6*	*7*	*8*
1	*47*	*66*	*28*	*31*	*24*	*14*	*27*	*17*
2	*66*	*13*	*24*	*37*	*40*	*12*	*36*	*16*
3	*28*	*24*	*12*	*21*	*10*	*54*	*31*	*14*
4	*31*	*37*	*21*	*25*	*33*	*27*	*46*	*37*
5	*24*	*40*	*10*	*33*	*27*	*43*	*26*	*29*
6	*14*	*12*	*54*	*27*	*43*	*19*	*31*	*16*
7	*27*	*36*	*31*	*46*	*26*	*31*	*40*	*87*
8	*17*	*16*	*14*	*37*	*29*	*16*	*87*	*25*

average 31 years old ($\sigma = 6.7$). 26 out of the 42 registered users were frequent players. Most of the users (37 out of 42) were naïve in structural biology.

Determination of the solvent accessible surface (SAS)

To determine the SAS value of each atom, we used the marching cubes algorithm as described in ref 17 using a 1.4 Å radius probe added to each atom radius and 0.4 Å wide cubes. For each atom, we recorded the number of polygons generated as an approximation of the SAS value.

Determination of the interface atoms

All the atoms of a given protein within 4 Å of any atom of the interacting protein were considered as interface atoms.

Generation of the exploration maps

To describe the users' exploration of each possible complex of the dataset, we generated exploration maps for each of the 8 proteins of the cross-docking dataset. We logged all user-explored interface atoms every time a user called for the Monte Carlo optimization process at a specific position. Exploration maps were generated using the user-explored interface atoms' polar coordinates, θ and ϕ as follows:

$x = \frac{\theta w}{\pi}$ and $y = \frac{\phi h}{2\pi}$ with w and h being the width and height of the image space.

Definition of the experimental interface covering ratio

Experimental interface covering ratio (CR) was logged every time a user called for the Monte Carlo optimization procedure and was defined as follows:

$$\mathrm{CR} = \frac{|\mathrm{cia} \cap \mathrm{cei}|}{|\mathrm{cei}|}$$

where cia is the set of atoms in the current interface and cei, the set of atoms in the experimental interface.

Results

Udock alpha version playtest

To establish the proof of concept of Udock, we proceeded to a two week duration alpha playtest consisting of a small cross-docking experiment on 4 binary complexes, which led to 36 possible complexes to be explored by the users.

Users exploration of the dataset

After analyzing the data provided by the users during the Udock alpha playtest, we generated exploration maps for each protein of the dataset presented in Fig. 4. According to the exploration maps, we observe that the users did explore, at least one time, each atom of the experimental interface. Very few atoms of the experimental interfaces of proteins **2**, **3** and **5** have never been explored by the users during the playtest (displayed in blue in the exploration maps).

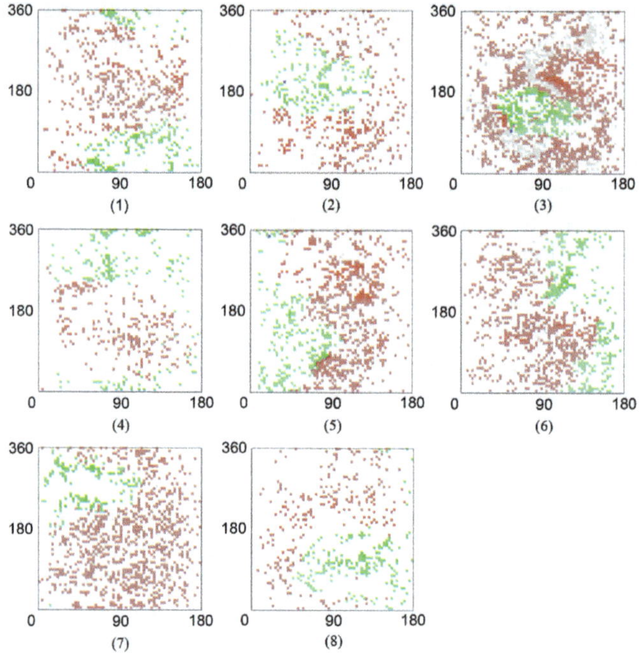

Fig. 4 Exploration maps generated for the 8 proteins of the dataset (index **1** to **8**) with the polar coordinates θ and ϕ along the horizontal and vertical axis, respectively. User-explored interface atoms within the experimental interface are displayed in green. User-explored interface atoms not within the experimental interface are displayed in red. Atoms within the experimental interface that have not been explored by the users are displayed in blue. Atoms not within the experimental interface that have not been explored by the users are displayed in grey. A gradient of grey (from light grey to dark grey) has been applied to these atoms depending on their corresponding normalized atomic SAS value. A gradient of darkness was applied to the user-explored interface atom corresponding color depending on the frequency of their exploration (lightest color: least explored interface atom; darkest color: most explored interface atom).

To enrich the information given by the exploration maps, we detailed, for each protein of the dataset, the frequency of the amount of exploration of a given atom (see Fig. 5). We wanted to highlight whether the users explored more intensely specific parts of the surface as dark red-colored surface areas of proteins **3** or **5** tend to show in their exploration maps. As expected, the frequency of the explored atoms all along the surface is not uniform, since some regions were more intensively explored. The difference of exploration in the surface points along proteins **3**, **6** and **7** was particularly striking, with a very small number of highly explored atoms and a very high number of not-highly explored atoms. For the other proteins in the dataset, the exploration of the surface atoms was more uniform.

We decided to plot the frequency of the experimental interface covering ratio (CR) to quantify how much the users explored within the experimental interfaces of each protein during Udock alpha playtests. The frequency of CR for each protein towards the entire dataset is presented in Fig. 6. The frequency of CR for each protein towards its corresponding experimental partner in the dataset is

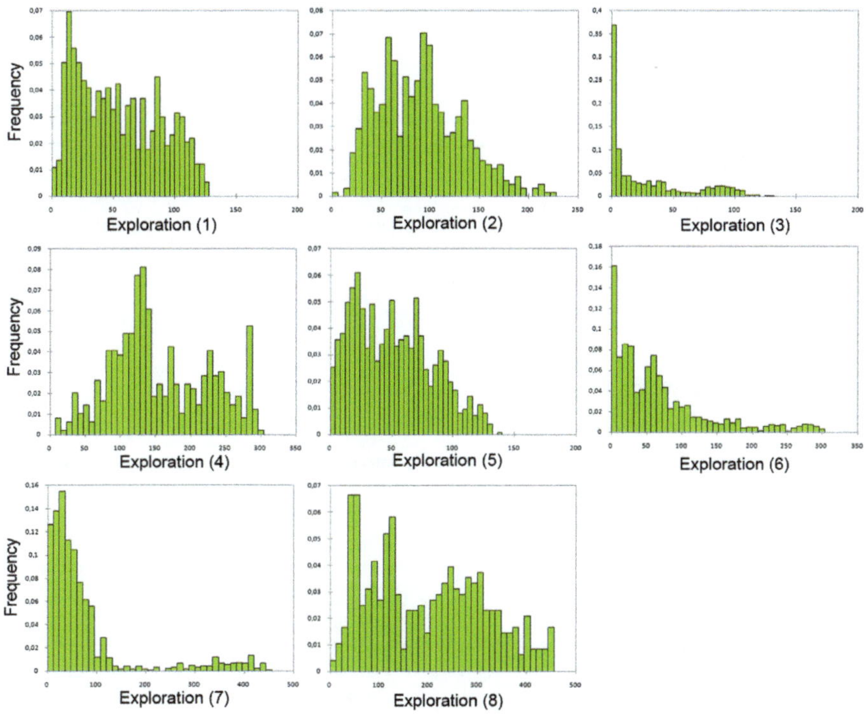

Fig. 5 Frequency of the amount of exploration for a given atom for each protein of the dataset (index **1** to **8**). An interface atom is considered as explored every time the user calls the Monte Carlo optimization procedure. The atoms with normalized SAS = 0 were not included.

presented in Fig. 7. Proteins **4** and **8** were particularly explored in the experimental binding interface whatever the protein that was involved in the complex (either the experimental partner or a decoy). The users could successively identify the experimental binding interface for the complexes acetylcholinesterase/fasciculin II (**3**,**4**) and thermitase/eglin c (**7**,**8**). They could identify the experimental interface for CDC42 GAP (**6**) but not for CDC42 GTPase (**5**).

High-scores obtained by the users during the playtest

During the Udock alpha playtest, we recorded all the best scores obtained by the users for every possible complex of the cross-docking dataset. The mean of the 3 best high scores obtained by the users for each protein with every other protein of the dataset is presented in Fig. 8. The score resulting from the rescoring with the Udock engine of the experimental geometry observed in the original PDB is also provided as an indication of a high score that could have been attained by the users in these particular cases. Except for proteins **7** and **8**, the best scores obtained by the users were far from the score they could have attained if they could reproduce the exact geometry observed in the PDB file. They found the highest score for the experimental partner for half of the proteins in the dataset, namely acetylcholinesterase/fascivulin II (**3**,**4**) and thermitase/eglin c (**7**,**8**). For barnase/barstar (**1**,**2**) and CDC42 GTPase/CDC42 GAP (**5**,**6**) the score obtained

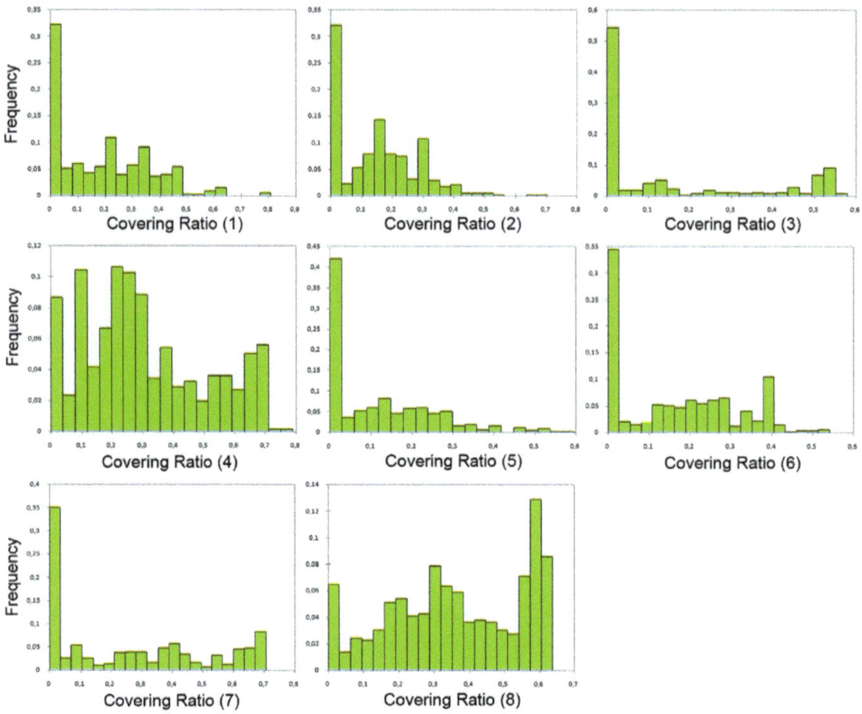

Fig. 6 Frequency of experimental interface covering ratio for the 8 proteins of the dataset (index **1** to **8**) towards the entire dataset.

with the experimental partner didn't stand out compared to the score obtained with the decoys. These observations seem consistent with the previous results, as the covering ratios show that users hardly found the correct interface for proteins **1**, **2** and **5**, leading to average docking scores for couples (**1**,**2**) and (**5**,**6**).

Discussion

Users' exploration of the dataset

Different reasons could explain the differences in the surface of the protein explored by the users. Intuitively, if proteins are explored equally, larger proteins should display more atoms that are less explored relative to smaller proteins. This is clearly the case with the 4 largest proteins (**3**, **5**, **6** and **7**), which display the highest number of less explored atoms, as illustrated in Fig. 5. It should be noted that we randomized the list of the proteins presented to the users at the start of the 2 weeks-playtest and not at each login. This randomized list exhibiting the largest proteins in the middle (**1**, **2**, **4**, **5**, **3**, **6**, **7**, **8**) could have impacted the choice of the complexes to explore selected by the users. An additional reason for this variability in the protein surface exploration is that the users could have identified at the surface of a given protein a very attractive potential docking spot that they tended to use more than other points of the surface. For all proteins, the users seemed to identify attractive spots that could be mostly in the experimental interface (proteins **1**, **4**, **6**, **7** and **8**) or out of the experimental interface (proteins **2**

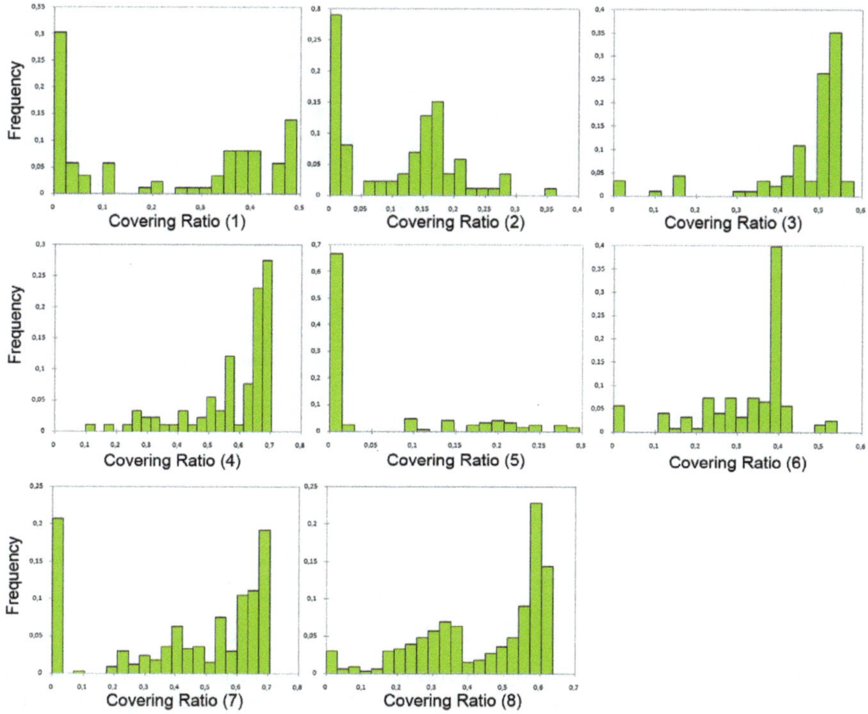

Fig. 7 Frequency of experimental interface covering ratio for the 8 proteins of the dataset (index **1** to **8**) towards their corresponding experimental partner in the dataset.

and **5**). For protein **3**, the users explored mostly in the experimental interface and its surroundings (see Fig. 4). This information from the exploration maps is confirmed by the analysis of the experimental interface covering ratios (Fig. 6 and 7), notably for proteins **4** and **8** that were particularly explored in the experimental binding interface. This points out that, even with decoy partners (Fig. 6), the experimental interface seemed particularly obvious for these proteins to the users of the alpha playtest. It is interesting to note that these proteins were the smallest of the dataset, which could result in a region of the surface that stands out in terms of geometry or charge, particularly because we used bound experimental structures to constitute the dataset. Then, when small proteins displayed particularities on the surface, as in proteins **4** and **8**, the users tended to dock them in a similar manner to the decoys or to the experimental partner since it seemed to be intuitively a good docking spot for the users. Interesting extreme profiles were also obtained for proteins **3** and **7** that were explored highly in and out of the interface. In both proteins, there is a particularly large cavity on the surface (the experimental binding site) that intuitively seemed mandatory to explore for the users, particularly for protein **3** (acetylcholinesterase).

The users successfully identified the experimental binding site for proteins **3**, **4**, **6**, **7** and **8**. For proteins **1**, **2** and **5**, they explored a lot out of the experimental interface. This could be due to less available striking particularities in terms of charge or shape that could be identified by the users as favored docking spots at the surface of these proteins.

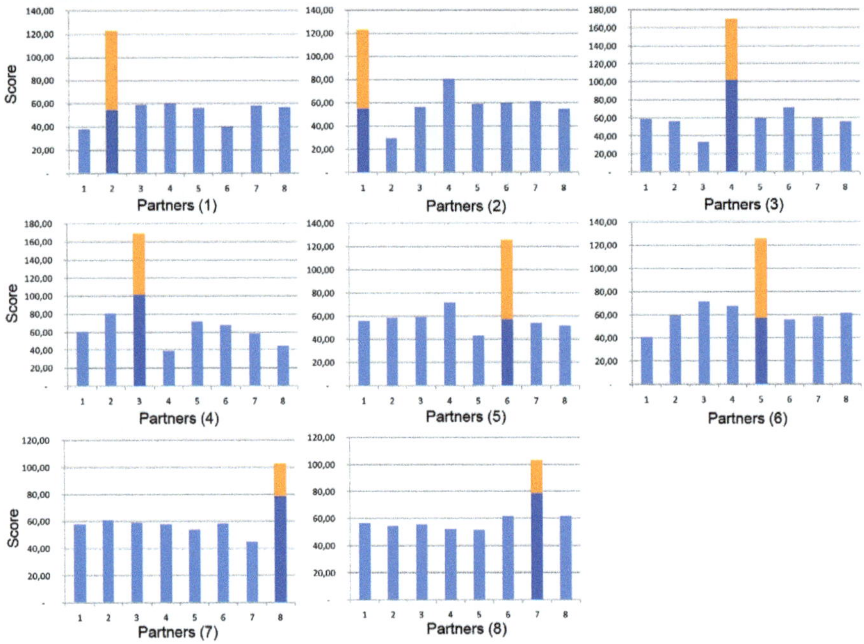

Fig. 8 Mean of the 3 best users' high scores for each protein against the whole dataset. The experimental partner is in dark blue, while the orange bar corresponds to the rescoring of the experimental geometry observed in the original PDB.

For acetylcholinesterase, fasciculin II, thermitase and eglin c (proteins **3**, **4**, **7** and **8**), the highest scores were obtained with the experimental partner. In these cases, the users seemed to be able to identify the right partner among the decoys. These complexes were the ones where the users got the closest scores to the score that could be obtained using the experimental geometry of the complex. In the other complexes, the users couldn't get higher scores with the experimental partner compared to the decoys. These results were in good correlation with the ability of the users to successfully identify the experimental interface.

The challenge of representation

Contrarily to classical protein docking approaches intended to be used by scientists who are ultimately experts in protein docking, Udock is also intended for naïve users that have not necessarily been sensitized to structural biology and protein energetics. Therefore, one of the challenges of our work was to tackle protein docking critical features in an accessible manner in order to also be performed by naïve users.

Making the representation of complex protein structures accessible to naïve users was the first challenge to overcome. Proteins are composed of several thousand atoms that render an explicit all-atom representation very confusing for non-experts. Since the problem in protein docking is to optimize the geometry of the complex by optimizing the binding energy between the two interacting partners, we chose to focus the representation of the system in Udock on these critical features, namely the protein's shape and electrostatics. Thus, we did not chose classical displays used by structural biologists to represent protein structures like wireframe, van der Waals volumes or balls-and-sticks, but focused on a global representation of the shape by displaying the solvent excluded surface (SES) of the protein. When using a van der Waals surface representation, the users' general view of the shape can be perturbed by the numerous invaginations occurring on the protein surfaces. SES carried the advantage of hiding these non-critical details about the protein shape.

We chose to use a standard 1.4 Å probe size for the SES generation, but our approach allowed different sizes of probes to be used. Indeed, we only used the generated surface for display and early collision detection. Ultimately, protein docking was performed using an all-atom rigid-body Monte Carlo optimization procedure that does take the SES into account. In the next version of Udock, we plan to use a different mesh for early collision detection and visualization, so that we would be using much more simplified representations, while still allowing proteins to be docked together. For example, the SES of barnase is displayed in Fig. 9 with 3 different probe sizes for the SES generation (1.4 Å, 2.4 Å, 3.4 Å), leading to a less and less detailed shape.

To color the SES, we decided not to use the classical CPK coloration guide[20] but to represent a simplified smoothened electrostatic potential derived from the atomic partial charges as computed by AMBER12.[15] We first represented the atomic partial charge at the surface, which resulted in precise information that was much too detailed and complex to be used during the docking process by a naïve user. When using our smoothing algorithm, we found it much easier to understand the global electrostatic configuration of the protein using larger stains of colour. It resulted in less precise but more concise information to drive the docking process, as the user tries to match globally positive areas to globally negative areas, and use the local automated optimization procedure to do the fine tuning. Moreover, this representation of the electrostatics on the surface can be helpful to the user to remember the global shape of the protein. Large and coloured stains can be used as landmarks by the user: for instance on Fig. 1, barnase can be described as featuring a positive concave region surrounded by two positive salient shapes.

We chose to generate SES surface and electrostatic potentials on the fly, every time a user loads a pair of proteins. This choice allowed us to have software that only relied on mol2 molecular description files, which are relatively small and commonly used in computational chemistry. This choice carries two advantages. First, it becomes very easy for anyone to modify the models loaded in Udock as we used a standard mol2 format. Second, these files can be easily transferred *via* the internet from our servers allowing updates of the datasets explored by the users

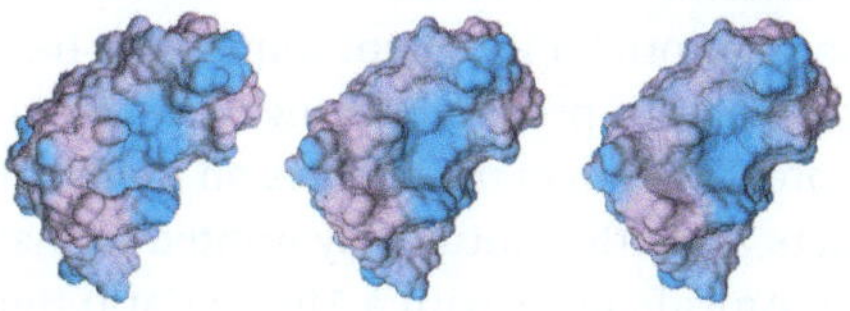

Fig. 9 Solvent excluded surface of barnase generated using probes of different size (from left to right: 1.4 Å, 2.4 Å, 3.4 Å).

without requiring large amounts of data to be downloaded every time. Yet, this choice comes with its drawbacks. Surface generation requires computational resources, and even if we optimized this step, the users reported waiting too long when loading a pair of molecules. We thus plan to optimize further our surface generation algorithms and to cache the generated models when possible to reduce the waiting time between docking runs.

The challenge of scoring on the fly during interactive docking

One of the objectives with the development of Udock was to perform interactive docking and scoring on the fly. The scoring is indeed a very difficult task to address during interactive protein docking on the fly since it takes a lot of computational resources that will also be needed by the physics and rendering engine to maintain a good fluidity in the animation of the objects and their interactions.

For instance, the calculation of the interaction score for a barnase/barstar pose takes around 50 ms, depending on the number of atoms in contact. To maximize the resource that will be used by the physics and rendering engine, we compute the interaction score on the fly every 250 ms to maintain a correct frame rate. Still, when running on computers with very limited resources, animation and manipulation of the proteins can become less fluid as the calculation of the interaction score takes too much time.

To limit the impact of this problem, we plan in the next version to let the user deactivate this feature as long as he is manipulating the molecules. Then, we would only compute the interaction score when the molecules become still, letting the user always manipulate them in a fluid fashion while still being informed on the quality of a specific pose.

Gamification of the protein docking challenge

To foster the user's motivation, we made a first step towards gamifying the protein docking process. The docking process is indeed a very interesting task to gamify: it can be viewed as a very simple pattern-matching task and as one of the most complex tasks to perform since it is not even mastered by the experts (CAPRI[1] is still considered as a very challenging experiment). The challenge was thus to make this task accessible to naïve users as a pattern-matching toy, while slowly guiding users into the realms of protein docking.

The simplification of rendering and manipulation we discussed in the previous section can be seen as the first step of the gamification process. We needed the users to be able to interact with the system very quickly, so that they could start learning by practice as soon as possible. But even if we simplified the docking process, naïve users still needed to be guided at the beginning and since we could not rely on classical documentation that would be too complex for naïve users, we created a basic tutorial. Even then, some users hardly understood some of the basic principles of the protein-docking task. For example, we did not anticipate that the representation of the charges on the surface would be counter-intuitive for naïve users since they intuitively wanted to match similar colors.

We also decided to provide users with a Monte Carlo rigid body optimization procedure in order to gamify what we felt to be the right part of the protein docking process. Indeed, we felt that users were inherently good at understanding the global shape of proteins and could be very efficient at identifying promising binding sites. Once the potential binding site was found, we felt that a local automatic optimization procedure would be much more efficient than the user at finding the precise geometry that would optimize the interaction score. One further step could be to design a gamified task that could replace or assist the optimization performed by the Monte Carlo procedure. This would be an entirely new activity for the user that would require new tools and visualization techniques to maintain Udock as accessible and motivating to naïve users. Yet, we feel that Udock is at a sweet spot between human and machines, using in an optimal manner the power of the brain and the power of the computer. As a consequence, perturbing this balance could be both destructive to the user's motivation and to the quality of the collected data.

To foster the motivation of the user, we needed to provide clear goals. Hopefully, the interaction score could be used as a game score, giving the user a clear goal of beating his own score. Also, since the users' behavior is logged on a web server, we could compute a global ranking among the users, and thus create a competitive element. When a user starts to dock two proteins together, the best interaction scores obtained by the other users with this particular complex are displayed on the interface. This provides users with a clear set of goals: beat his own high score and every high score displayed in the score bar.

We also tried to add feedback to inform the user about the quality of his performance. Feedback is fundamental to gamification as it guides the user and helps sustain his motivation.[21] Every time a user beats his own high score, we immediately inform and reward him with graphical and sound feedback. If the user beats one of the other users' score, he is also informed by the corresponding feedback, and given the name of the beaten user.

Finally, we created a compelling and immersive atmosphere by developing a dedicated soundtrack and using advice on color schemes from a graphical designer.

Still, the gamification process in Udock is far from being complete. In this first version, we could identify two major issues in the gameplay. First, even if we tried to make the docking process accessible to naïve users, the tutorial seemed clearly too short for the users to fully understand the challenge of protein docking. In the next version, we will need to make it much richer. Second, users that understood the docking process did not play for a long time because the game did not foster long term motivation. Beating scores can be seen as fun, but it's clearly not enough since we need to keep the users learning. We will need to provide tools that will assist them to perform even better docking, and provide them with new opportunities to learn and try different docking strategies. For instance, we could provide different displays of the proteins (not only the shape), give more information about the individual proteins to dock (protein sequence, for example) and thus give them the opportunity to perform protein docking, not only with regard to the shape and electrostatics, but by using other information.

Conclusion

In summary, we developed an interactive docking system, Udock, that allows a quick and easy-to-handle human driven exploration of protein–protein interfaces. We simplified the representation of protein structures and gamified the protein docking task to make it accessible to even naïve users. To validate our approach, we designed an open alpha cross-docking playtest during two weeks on 4 experimentally resolved protein complexes, leading to 36 possible complexes to explore.

Despite the small amount of time allowed for the Udock open alpha playtest and the relatively small number of active users (12 that played at least 30 min), different observations could be derived. The users explored almost all of the surfaces of the proteins that were available in the dataset but favored certain regions that seemed more attractive as potential docking spots. These favored regions were inside or close to the experimental binding interface and, for 5 out of the 8 proteins, the most explored regions covered the majority of the binding interface. For half of the proteins in the dataset (acetylcholinesterase, fasciculin II, thermitase and eglin c), the highest scores were obtained with the experimental partner.

This work could give preliminary insight on (1) the power of crowd sourcing on challenging tasks, *i.e* protein–protein docking; (2) protein–protein interfaces and interactions, as the users could identify experimental interfaces and sometimes the partners in interaction, and (3) a better way to craft games for science.

Acknowledgements

The authors are grateful to Dr Patrick Fuchs, Dr Anne Lopes, Clément Pillias and Hélène Manche for fruitful discussions. We also would like to thank all Udock alpha-testers for their time and investment.

References

1 J. Janin, K. Henrick, J. Moult, L. T. Eyck, M. J. Sternberg, S. Vajda, I. Vakser and S. J. Wodak, Critical Assessment of, P. I., CAPRI: a Critical Assessment of PRedicted Interactions, *Proteins: Struct., Funct., Genet.*, 2003, **52**, 2–9.
2 D. W. Ritchie and G. J. Kemp, Protein docking using spherical polar Fourier correlations, *Proteins: Struct., Funct., Genet.*, 2000, **39**, 178–94.
3 R. Chen, L. Li and Z. Weng, ZDOCK: an initial-stage protein-docking algorithm, *Proteins: Struct., Funct., Genet.*, 2003, **52**, 80–7.
4 D. Kozakov, R. Brenke, S. R. Comeau and S. Vajda, PIPER: an FFT-based protein docking program with pairwise potentials, *Proteins: Struct., Funct., Bioinf.*, 2006, **65**, 392–406.
5 I. A. Vakser, Low-resolution docking: prediction of complexes for underdetermined structures, *Biopolymers*, 1996, **39**, 455–64.
6 J. J. Gray, S. E. Moughon, T. Kortemme, O. Schueler-Furman, K. M. Misura, A. V. Morozov and D. Baker, Protein-protein docking predictions for the CAPRI experiment, *Proteins: Struct., Funct., Genet.*, 2003, **52**, 118–22.
7 R. Abagyan, M. Totrov and D. Kusnetsov, ICM - a new method for protein modelling and design. Applications to docking and structure prediction from the distorted native conformation, *J. Comput. Chem.*, 1994, **15**, 488–506.
8 C. Dominguez, R. Boelens and A. M. Bonvin, HADDOCK: a protein-protein docking approach based on biochemical or biophysical information, *J. Am. Chem. Soc.*, 2003, **125**, 1731–7.

9 A. M. Wollacott and K. M. Merz, Jr, Haptic applications for molecular structure manipulation, *J. Mol. Graphics Modell.*, 2007, **25**, 801–5.

10 O. Delalande, N. Ferey, G. Grasseau and M. Baaden, Complex molecular assemblies at hand *via* interactive simulations, *J. Comput. Chem.*, 2009, **30**, 2375–87.

11 T. C. Lu, J. Ding and S. N. Crivelli, DockingShop, a Tool for interactive protein docking, *IEEE Visualization 2005*, 2005.

12 A. Kawrykow, G. Roumanis, A. Kam, D. Kwak, C. Leung, C. Wu, E. Zarour, p. Phylo, L. Sarmenta, M. Blanchette and J. Waldispuhl, Phylo: a citizen science approach for improving multiple sequence alignment, *PLoS One*, 2012, **7**, e31362.

13 S. Cooper, F. Khatib, A. Treuille, J. Barbero, J. Lee, M. Beenen, A. Leaver-Fay, D. Baker, Z. Popovic and F. Players, Predicting protein structures with a multiplayer online game, *Nature*, 2010, **466**, 756–60.

14 C. Crawford, *The Art of Computer Game Design*, 1982, ISSN: B0052QA5WU.

15 D. A. Case; T. A. Darden; T. E. Cheatham; C. L. Simmerling; J. Wang; R. E. Duke; R. Luo; R. C. Walker; W. Zhang; K. M. Merz; B. Roberts; S. Hayik; A. Roitberg; G. Seabra; J. Swails; A. W. Goetz; I. Kolossvai; K. F. Wong; F. Paesani; J. Vanicek; R. M. Wolf; J. Liu; X. Wu; S. R. Brozell; T. Steinbrecher; H. Gohlke; Q. Cai; X. Ye; J. Wang; M.-J. Hsieh; G. Cui; D. R. Roe; D. H. Mathews; M. G. Seetin; R. Salomon-Ferrer; C. Sagui; V. Babin; T. Luchko; S. Gusarov; A. Kovalenko; P. A. Kollman, *AMBER 12*. University of California, San Francisco, 2012.

16 E. F. Pettersen, T. D. Goddard, C. C. Huang, G. S. Couch, D. M. Greenblatt, E. C. Meng and T. E. Ferrin, UCSF Chimera–a visualization system for exploratory research and analysis, *J. Comput. Chem.*, 2004, **25**, 1605–12.

17 W. E. Lorensen and H. E. Cline, Marching Cubes: A High Resolution 3D Surface Construction Algorithm, *ACM SIGGRAPH Comput. Graphics*, 1987, **21**, 163–169.

18 Bullet physics library, real-time physics simulation. http://bulletphysics.org.

19 S. Sacquin-Mora, A. Carbone and R. Lavery, Identification of protein interaction partners and protein-protein interaction sites, *J. Mol. Biol.*, 2008, **382**, 1276–89.

20 R. Corey and L. Pauling, Molecular Models of Amino Acids, Peptides, and Proteins, *Rev. Sci. Instrum.*, 1953, **24**, 621–627.

21 K. Salen; E. Zimmerman, *Rules of Play, Game Design Fundamentals*, MIT Press 2003, ISBN-13: 978–0262240451.

Faraday Discussions

PAPER

Exploring genomes with a game engine

Jeremiah J. Shepherd,[a] Lingxi Zhou,[a] William Arndt,[b] Yan Zhang,[a] W. Jim Zheng*[c] and Jijun Tang*[ad]

Received 20th December 2013, Accepted 20th March 2014
DOI: 10.1039/c3fd00152k

More and more evidence indicates that the 3D conformation of eukaryotic genomes is a critical part of genome function. However, due to the lack of accurate and reliable 3D genome structural data, this information is largely ignored and most of these studies have to use information systems that view the DNA in a linear structure. Visualizing genomes in real time 3D can give researchers more insight, but this is fraught with hardware limitations since each element contains vast amounts of information that cannot be processed on the fly. Using a game engine and sophisticated video game visualization techniques enables us to construct a multi-platform real-time 3D genome viewer. The game engine-based viewer achieves much better rendering speed and can handle much larger amounts of data compared to our previous implementation using OpenGL. Combining this viewer with 3D genome models from experimental data could provide unprecedented opportunities to gain insight into the conformation–function relationships of a genome.

Introduction

A significant fraction of important questions in biology research today require analysis of genomic data sources. Regulatory functions of the cell operate in three dimensions such that currently available software, which displays genetic data as linear sequences, offers insufficient insight into some problems. New experimental techniques are revealing information that impacts the 3D structure of genetic molecules such as nucleosome position distribution,[1] histone methylation,[2] and transcription factory complexes.[3] As displayed by current software, the 3D context of these information sources is not shown, which potentially obscures important spatial relationships. To address this problem, we have developed a system to visualize genomes in a way that incorporates sources of 3D information while preserving responsive viewing and interaction.

[a]*Department of Computer Science and Engineering, University of South Carolina, SC 29205, USA*
[b]*Howard Hughes Medical Institute, Janelia Farm Research Campus, Ashburn, VA 20147, USA*
[c]*School of Biomedical Informatics, U. Texas Health Science Centre at Houston, TX 77030, USA*
[d]*Key Laboratory of Systems Bioengineering of the Ministry of Education, Tianjin University, Tianjin, 300072, PR China*

A primary benefit to software that can display genomic data in three dimensions is it facilitates advances in the creation of 3D models. The chromosome conformation capture family of experiments generates data directly relevant to the 3D structure of chromatin. This data can be used to create and display small models of a few hundred thousand base pairs arranged in 3D space such as for the interesting alpha-globin gene domain on human chromosome 16,[4] all the way up to possible full human genome models using information derived from Hi-C experiments.[5] Even though data generated from current techniques are sparse and not yet sufficient to create accurate high-resolution 3D models of eukaryotic genomes, it is expected that more believable and detailed models may be derived in the future. A software platform that can visualize such high-resolution models would prepare us for the future needs of studying a 3D genome model when available.

A system that successfully displays genetic molecules in three dimensions must have the following abilities: it must display full size mammalian chromosomes on the order of billions of base pairs, display the details of important segments on the scale of individual atoms, and allow real time manipulation of the view at different levels of detail. Our software behavior mimics that of the popular Google Earth service. When using Google Earth, a user can zoom out to view the entire planet as a whole but with little detail, or seamlessly zoom in to show a progressively smaller section of the surface but with greater detail. In our case, the broadest view would display all chromosomes contained in a nucleus while zooming in would reveal the progression of more detailed features such as chromatin fiber arrangement, histone positions, base pair locations, and individual atoms. Creating such a system is not a trivial task. The volume of data needed to specify an arbitrary mammalian genome in 3D space well exceeds the limit of available memory necessitating careful loading and pruning. Google Earth only has one model to display and thus can take advantage of advanced indexing and caching optimizations, while in contrast, our software must display a wide variety of possible genomes and structures; this removes our ability to construct data structures in advance.

Most modern video games must visualize vast amounts of information at once while being interactive in real time. Most video game developers save resources by using game engines, which are off the shelf systems that implement common display and interaction features. These same engines have been shown to be useful for other applications such as simulation research.[6] Biology is no stranger to using game concepts to assist in new discoveries. Games like Fold It have been developed to help solve difficult questions about protein folding.[7] Our software is the first attempt to use game engine technology for the purpose of displaying chromatin in three dimensions. We show that these techniques developed for the video game industry can be successfully used to make useful and responsive displays of genomes in 3D.

Researchers use a variety of tools to place their experimental data in the context of reference genomes, but among these tools the ability to view spatial relationships in three dimensions is missing. Genome browsing tools have been developed by variety of institutions, such as UCSC, Ensemble, and NCBI. These genome browsers not only integrate vast amount of genome information from varied sources, but also allows users to develop custom tracks which easily integrate their own data with existing genome information.[8] Such feature-rich genome browsers have become very popular and played essential roles in several important, large-scale genome projects, including Encyclopedia of DNA Elements (ENCODE)[9] and 1000 Genome Project.[10] Existing viewers display genomes and annotations along a single dimension, the sequence coordinates, making them unsuitable for displaying some kinds of epigenetic and structural information generated by recent technologies such as chromosome conformation capture style experiments.

Genome3D was the first model-view framework developed to work with current genome browsers to address these challenges, and to facilitate multi-scale integration and visualization of large genomic and epigenomic datasets in three dimensions.[11] This model-view framework enabled researchers to infer new knowledge about the structure/function of genomes that would have been difficult to accomplish by primary sequence-based browsers. For example, the phosphate groups of different base pairs in DNA strands can be either exposed to the outside or confined between the histone proteins and DNA backbone. This information has important functional implications, as exposed phosphate groups can be easily accessible by DNA binding proteins.[12] While new sequencing technologies allow researchers to map individual nucleosomes across the whole genome,[13] it is a significant challenge for current genome browsers to display this epigenetic information in an intuitive manner. Visualizing such information using a 3D genome model can facilitate new inferences about potential regulatory behavior such as functional *versus* non-functional Single Nucleotide Polymorphism (SNP) in non-coding regions.

Unfortunately, in the years since its release, a number of limitations with the Genome3D software package were revealed, warranting the development of a new version. Genome3D had no internal throttle to manage the loading or view of very large data sets. A careless user could easily load more chromosomes than are able be displayed in real time which would slow responsiveness to unusable levels. The former software was developed in C++ using libraries that have since become obsolete. These dependencies severely restricted the portability and maintainability of the code base.

Using game engine technologies and some sophisticated visualization techniques, we intend to solve the two critical flaws of Genome3D. Game engines are designed to aid in the development of games by providing interaction and complex rendering techniques as standard features. This frees attention for the development of logic and content. Also, because game development is a large consumer industry whose products need to be highly marketable, most modern engines have multi-platform support built-in, which allows the programs developed with one to easily be ported to a number of other devices. Furthermore, game engines have previously been used for large simulation research where they are proven to work well in several different disciplines.[14] For these reasons, we found it most appropriate to use a game engine to display genomes in three dimensions.

The system

The greatest dilemma facing this system was how to both manage the large amount of information while still maintaining real time responsiveness. Since modern games engines provide open world management as a standard feature, this was chosen as a good starting point. We have used the Unity game engine, which has become highly popular and widely used by independent game developers and researchers. While the interaction features are standard to most game engines, our need to visualize data on the order of billions of base pairs is unique enough to still be challenging. The largest portion of our attention has been directed towards efficiently rendering the scene. One aspect of our approach is the level of detail (LOD) system, which is traditionally used to cut down the amount of processing power needed to generate the elements in the scene. Different parts of the scene are presented at different LODs based on criteria such as the distance between the camera and the object. An object in the far distance can be rendered with a lower LOD in a way that conserves computational resources but is indistinguishable to the viewer. An object very close to the observer will be displayed with a higher LOD and use more resources, but geometry limits the number of objects which can occupy this space and any further objects that have been occluded need not be drawn at all. To best solve the problem of determining placement and distances of the genome, we used a data structure commonly used in games, and then further processed multiple data sources in different LODs.

The system runs on multiple platforms with a consistent interface as shown in Fig. 1:

The octree

Many modern games have vast open worlds, which need to be carefully processed in order to keep the action fluid. To reduce the amount of data processed at a given time, most of these applications use a data structure called an octree. This structure subdivides 3D space into eight sections that are represented by nodes in the tree as seen in Fig. 2. These subsections can be recursively subdivided making more nodes in the tree with a finer level of detail. When a node is very far away from the camera, it is displayed directly with a lower LOD and any children it has are ignored. When the camera is moved closer to a node, then it is examined more carefully such that each of its children are fully processed and displayed.

Most commercial implementations use static octrees with a set number of subdivisions that are optimized for a single given model. However, our system is

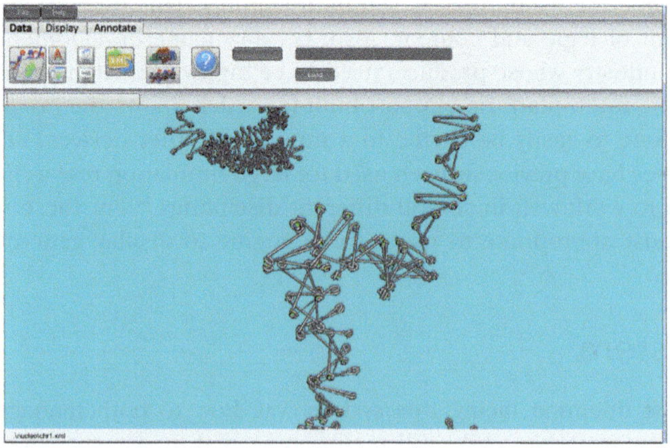

Fig. 1 System interface with a chromosome loaded at the histone level.

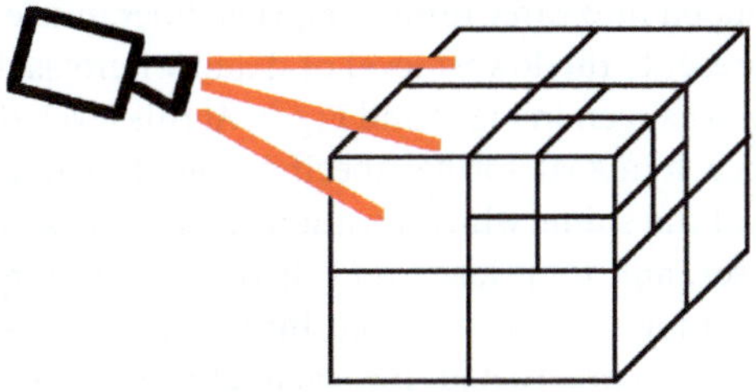

Fig. 2 Octree example. This shows how it will calculate the ranges from each cube to the camera.

distinct because our 3D models must be generated on the fly, and can only be partially built at any one time given the large amounts of data that needs to be processed and stored. This has led us to choose a dynamic octree implementation.[15] The dynamic variation of the data structure automatically subdivides into eight parts when the data contained in a node exceeds a given tolerance. In this way, we avoided having an unnecessarily big tree(s) in the memory. Also, since we are dealing with a large object, it makes sense to stay away from loose octrees.[16]

Given the difference between the input, formatted as a spline representing the chromosome fiber, and the internal storage as an octree, faithfully preserving the structure of the data was a key priority. In traditional implementations, the mesh for the objects in the scene are provided, and each polygon in the mesh is placed in a node in the tree corresponding to its spatial location. However, in our case, the mesh is generated given a series of files that contain an ordered set of control points on the curve. Instead of dividing the problem by polygons, we do so using the spline control points. Wherever the curve intersects an octree node boundary, a new control point is placed at the point of intersection and inserted into both octree nodes. While this increases memory usage slightly, it preserves information content perfectly.

Further levels of detail

In addition to the octree, a level of detail system has been implemented to manage display complexity given the context of the camera and model.

Fig. 3 Fiber level far with 22 chromosomes loaded.

Information is grouped into three different levels: fiber, nucleosome, and atomic. The fiber level represents the lowest level of detail where each chromosome's 30 nm fiber is displayed as seen in Fig. 3 and Fig. 4. At this level, the data is formatted as a series of files, each in XML format specifying an individual chromosome and the control points of the spline which positions it. The system scans the directory provided by the user, and then loads and displays each chromosome. Files are first loaded into memory and then placed into an octree. As the data proceeds, additional octrees are constructed for each chromosome to guarantee they remain segregated from one another.

Proceeding deeper, the next level of detail is the nucleosome stage. At this level, nucleosome base-pair positions plus the DNA that wraps around and links to them are displayed as seen in Fig. 5. Instead of loading all the chromosomes like the previous level, this view only displays one at a time. The files for this level of detail are significantly different from the previous section as they are much more detailed. Once again, XML file formats are used to hold the structure information for each chromosome. These files not only include the base pair number and positional information but also the orientation of each histone. Linker DNA can either be automatically generated for display or specified in the corresponding data files. An additional feature of this view, because most researchers are interested in specific interesting regions and not an entire chromosome, we implemented an option to reduce the viewing size to a user specified range as seen in Fig. 6.

Finally, the highest level of detail is at the atomic scale. Similar to the nucleosome stage, the user chooses a range from a start base pair to an ending base pair, and the system displays that atomic structure as seen in Fig. 7. This level faces an additional challenge, as two sets of information must be merged into one display. Information provided at the nucleosome level determines positional information, which is then combined with a fasta file containing the DNA sequence. The user specified position is matched to both the fasta file and the nucleosome positions, which are read in tandem and used to place the appropriate base pair atoms at their correct positions.

Other benefits

In addition to the robust interactivity and the superior display management provided by a game engine, there are other benefits to using it for this application.

Fig. 4 Fiber level with close up view of chromosome 1.

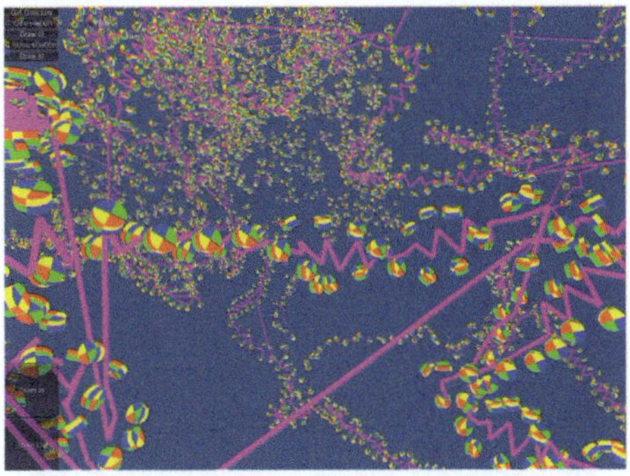

Fig. 5 Histone level of chromosome 1 in mid range. This shows 100 000 histones rendered in real time with procedurally generated linker data which is highlighted in pink.

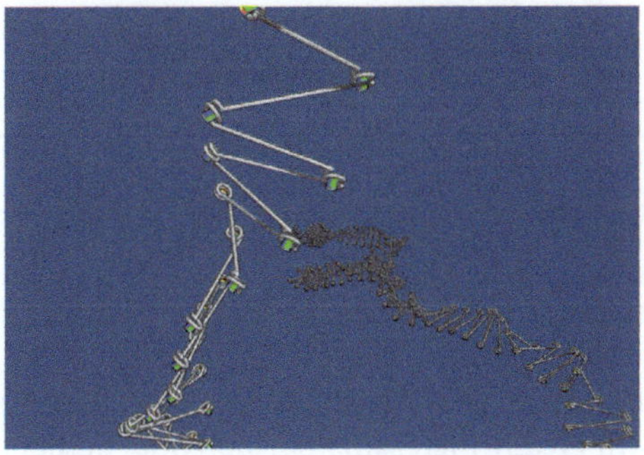

Fig. 6 Histone level of chromosome 1 at a reduced scale close up (linkers are not highlighted).

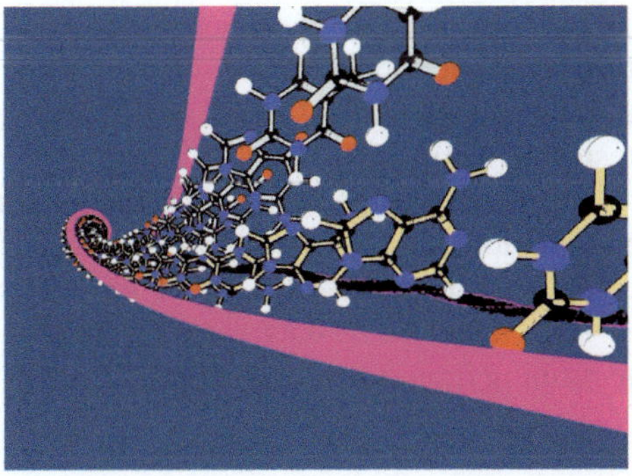

Fig. 7 Atomic level of chromosome 1 from base pairs 10 400 to 10 900.

A game with a larger marketplace is more likely to succeed, making portability to multiple platforms a desirable quality. The system we created has been deployed to Windows, Mac, and some mobile devices, including tablets as seen in Fig. 8. Each version required some minor changes but overall the system still works the

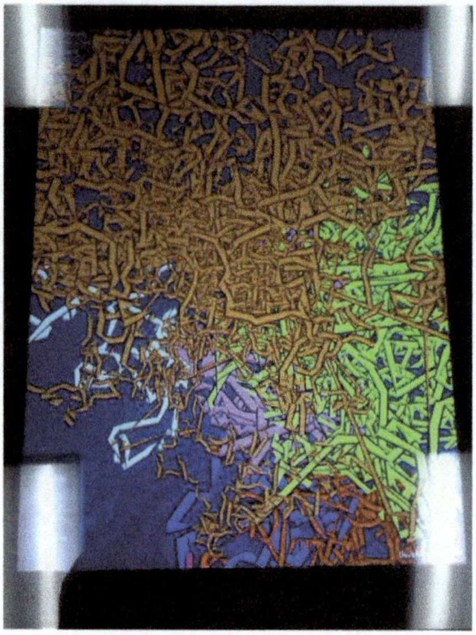

Fig. 8 iPad Example. This is an example demonstrating the octree-based LOD working on an iPad.

same with minimal added development time. For example, in order to function on mobile devices, we must limit the rendering capabilities and read the files from an external server instead of from the device itself. The simplicity of multi-platform support will greatly increasing our potential audience.

Results

In order to test the performance and functionality of this application, we compared it to the previous version of Genome3D, which is the only other implementation of this concept.[11] Each test ran on a standard desktop with 4 GB of RAM, an Intel® Xeon ® CPU at 3.07 GHz, and running 64-bit Windows 7 Professional. Our tests and results follow.

The main criteria we chose to judge success was the rendering capacity, since that is the best approximation of its functional performance from the perspective of a user. To accomplish this, we loaded a set amount of data, and then progressively added large amounts of data for each individual test, while also keeping track of the running frames per second (FPS). Each of these smaller tests

Table 1 Improved performance of game engine based viewer at the chromatin fiber level (the original viewer cannot handle data with more than one chromosome)

Program	1 chromosome	10 chromosomes	15 chromosome	23 chromosomes
Original genome viewer FPS	120	0	0	0
Game engine genome viewer's FPS range	60	60	60	30

was also run for an isolated LOD: fiber, nucleosome, and atomic levels. The first set of tests was at the fiber level with results shown in Table 1. There is a hardware refresh limit that is 120 Hz in our system. Also game engines typically set an upper limit on FPS to avoid any side effects; this limit is 60 in Unity. In Unity, we can set a targeted frame rate, which is the FPS that Unity will try to reach if the system is heavily loaded. As our genome viewer is not fast-paced, we choose 30 FPS as the target. The original genome viewer outperforms the new one with only one chromosome file loaded, but it cannot load additional chromosomes without crashing. Our new system has maximum FPS until 15 chromosomes are loaded; the FPS is still acceptable when all 23 chromosomes of human genomes are loaded.

Next, a similar test was conducted at the nucleosome level. In order to gain an accurate insight into the performance, the test varied the number of histones loaded and rendered while measuring the sustained frames per second. Results appear in Table 2. When only a few nucleosomes are loaded, the original genome viewer performs faster, but once more data is added the older system becomes unresponsive. In contrast, the performance of the new system did not begin to lose performance noticeably until the number of histones loaded exceeded 10 000.

Finally, the last set of tests was run in a similar vein as the others. Much like the nucleosome scale, the purpose was to test and evaluate each program's performance as an increasing amount of atoms were loaded and rendered, measuring the frames per second. The results, as seen in Table 3, show a more comparable trend. Both systems simply rendered the given data without any further LOD data analysis, and it first appears that the original genome viewer outperforms the game engine version. However, the original system only seems to allow twenty-five atoms to be rendered at once, so it cannot make a comparison after the initial twenty-five. On the contrary, our new system can easily handle 1000 atoms at once, without a noticeable drop in performance, achieving a 40-times increase of information loaded.

From these tests, it is obvious to see that using a game engine with sophisticated rendering techniques commonly used in games greatly expands the capacities of the former Genome3D viewer. The key to the performance enhancement was breaking apart and grouping the large amount of data in a more sophisticated way, which was facilitated by using an octree for spatial information. Also, modern game engines typically offload most rendering tasks to GPUs, which are becoming computationally more powerful and can process multiple rendering requests at once. In this way, more data could be loaded, managed, and displayed at any given time yielding significantly improved utility.

Table 2 Improved performance of game engine based viewer at the nucleosome level (the original viewer cannot handle more than 10 nucleosomes)

Program	1 nucleosome	1000 nucleosomes	5000 nucleosomes	10 000 nucleosomes
Original genome viewer FPS range	60–120	0	0	0
Game engine genome viewer's FPS range	60	60	30	30

Table 3 Improved performance of game engine based viewer at the atomic level (the original viewer cannot handle more than 25 atoms)

Program	25 atoms	500 atoms	1000 atoms	1500 atoms
Original genome viewer FPS range	60–120	0	0	0
Game engine genome viewer's FPS range	60	60	60	30

Conclusions

In this paper, we addressed the problem of visualizing genomes in three-dimensional space, and the potential benefits to such an approach. We solve this problem by applying techniques and tools that are commonly used in video games, which include using a commodity game engine, partitioning data with an octree, and grouping the files to perform multi-tier level of detail analysis (LOD). The game engine has fast rendering systems and also allows for simple and fast multi-platform support. Organizing the data with an octree reduced resource consumption allowing larger data sets to be viewed with visual fidelity and responsive performance. Finally, recognizing that the data was given in three different formats provided another level of detail, such that the system was able to segregate displayed content based on context. Finally, we showed that using this application compared with the older non-game application outperforms it in situations where the amount of data being viewed is non-trivial, and is able to load much more data before losing responsiveness. With our application, we have provided researchers with a tool that could be used for valuable genomic work.

Acknowledgements

This work is supported by NSF award #1161586 to JJS, LZ, YZ and JT, as well as #1339470 to WJZ.

Notes and references

1 R. Schopflin, V. B. Teif, O. Muller, C. Weinberg, K. Rippe and G. Wedemann, Modeling nucleosome position distributions from experimental nucleosome positioning maps, *Bioinformatics*, 2013, **29**, 2380–2386.

2 A. Barski, S. Cuddapah, K. Cui, T.-Y. Roh, D. E. Schones, Z. Wang, G. Wei, I. Chepelev and K. Zhao, High-resolution profiling of histone methylations in the human genome, *Cell*, 2007, **129**(4), 823–837.

3 P. Unneberg and J.-M. Claverie, Tentative mapping of transcription- induced interchromosomal interaction using chimeric EST and mRNA data, *PLoS One*, 2007, **2**(2), e254.

4 D. Baù, A. Sanyal, B. R. Lajoie, E. Capriotti, M. Byron, J. B. Lawrence, J. Dekker and M. A. Marti-Renom, The three-dimensional folding of the α-globin gene domain reveals formation of chromatin globules, *Nat. Struct. Mol. Biol.*, 2011, **18**(1), 107–114.

5 E. Lieberman, N. L. van Berkum, L. Williams, M. Imakaev, T. Ragoczy, A. Telling, I. Amit, B. R. Lajoie, P. J. Sabo, M. O. Dorschner, R. Sandstrom, B. Bernstein, M. A. Bender, M. Groudine, A. Gnirke, J. Stamatoyannopoulos,

L. A. Mirny, E. S. Lander and J. Dekker, Comprehensive mapping of long-range interactions reveals folding principles of the human genome, *Science*, 2009, **326**, 289–293.
6 J. Shepherd, R. Dougal, and J. Tang, Visualization and simulation potential of Microsoft's XNA, *Grand Challenge in Simulation and Modeling (GCSM)*, 2010, pp. 306–310.
7 F. Khatib, F. Dimaio, S. Cooper, M. Kazmierczyk, M. Gilski, *et al.*, Crystal structure of a monomeric retroviral protease solved by protein folding game players, *Nat. Struct. Mol. Biol.*, 2011, **18**(10), 1175.
8 W. Kent, C. Sugnet, T. Furey, K. Roskin and T. Pringle, The human genome browser, *Genome Res.*, 2002, **12**, 996–1006.
9 The ENCODE Project Consortium, The Encode (Encyclopedia of DNA Elements) project, *Science*, 2004, **306**, 636–640.
10 N. Siva, 1000 genomes project, *Nat. Biotechnol.*, 2008, **26**, 256.
11 T. Asbury, M. Mitman, J. Tang and W. Zheng, Genome3D: a viewer- model framework for integrating and visualizing multi-scale epigenomic information within a three-dimensional genome, *BMC Bioinformatics*, 2010, **11**, 444.
12 K. Eisfeld, R. Candau, M. Truss and M. Beato, Binding of NF1 to the MMTV promoter in nucleosomes: influence of rotational phasing, translational positioning and histone H1, *Nucleic Acids Res.*, 1997, **25**, 3733–3742.
13 D. Schones, K. Cui, T. Roh and A. Barski, Dynamic regulation of nucleosome positioning in the human genome, *Cell*, 2008, **132**, 887–898.
14 T. Zhang, J. Shepherd, J. Tang, and R. A. Dougal, The simulation tool for mission-optimized system design, in *Proceedings of the 2011, Grand Challenges on Modeling and Simulation Conference. Society for Modeling & Simulation International*, 2011, pp. 348–355.
15 D. Ginsburg, Octree construction, *Game Programming Gems*, 2000, pp. 439–443.
16 T. Ulrich, Loose octrees, *Game Programming Gems*, 2000, pp. 444–453.

Faraday Discussions

PAPER

Methodologies for the analysis of instantaneous lipid diffusion in md simulations of large membrane systems

Matthieu Chavent,†*[a] Tyler Reddy,†[a] Joseph Goose,[a] Anna Caroline E. Dahl,[a] John E. Stone,[b] Bruno Jobard[c] and Mark S. P. Sansom[a]

Received 16th December 2013, Accepted 20th February 2014
DOI: 10.1039/c3fd00145h

Interactions between lipids and membrane proteins play a key role in determining the nanoscale dynamic and structural properties of biological membranes. Molecular dynamics (MD) simulations provide a valuable tool for studying membrane models, complementing experimental approaches. It is now possible to simulate large membrane systems, such as simplified models of bacterial and viral envelope membranes. Consequently, there is a pressing need to develop tools to visualize and quantify the dynamics of these immense systems, which typically comprise millions of particles. To tackle this issue, we have developed visual and quantitative analyses of molecular positions and their velocity field using path line, vector field and streamline techniques. This allows us to highlight large, transient flow-like movements of lipids and to better understand crowding within the lipid bilayer. The current study focuses on visualization and analysis of lipid dynamics. However, the methods are flexible and can be readily applied to *e.g.* proteins and nanoparticles within large complex membranes. The protocols developed here are readily accessible both as a plugin for the molecular visualization program VMD and as a module for the MDAnalysis library.

1 Introduction

Our concept of biological membranes has evolved over the last 40 years, from the "fluid mosaic model"[1] to a more nuanced view of a laterally heterogeneous patchwork of proteins and lipids.[2,3] Membrane protein-lipid interactions are an area of intense research[4,5] with a number of membrane characteristics currently under scrutiny, including lateral diffusion of lipids, formation of nano-domains,

[a]*Department of Biochemistry, University of Oxford, South Parks Road, Oxford OX1 3QU, United Kingdom. E-mail: matthieu.chavent@bioch.ox.ac.uk*
[b]*Beckman Institute University of Illinois, 405 N. Mathews Avenue Urbana, IL 61801, USA*
[c]*LIUPPA laboratory, University of Pau, 1, av. de l'Université, 64013 Pau Cedex, FRANCE*
† These authors contributed equally to this work

and local membrane curvature.[6,7] The nature of lipid diffusion within membranes remains a topic of active discussion, with two major models being proposed:

– The *rattle in a cage* diffusion model describes short-range lipid movement: a lipid jumps from one location to another when it is able to escape from its immediate neighbours; and

– The *flow-like* diffusion model describes lipid movement at longer ranges: lipids can form loosely-packed clusters that move in concert, analogous to the movement of current.

While it remains difficult to experimentally assess the former theory, recent computational[8,9] and experimental[10] results tend to support the flow-like diffusion model. There has been a gap between the properties observable *via* computational approaches compared to those accessible *via* experiments.[11] However, this gap is now closing as increasing computational power provides access to simulations of biologically relevant times and length scales. To model large membrane systems, it is possible to use different levels of granularity in molecular dynamics (MD) simulations, ranging from all-atom[12] (AT) simulations to coarse-grained[13] (CG) models in which small numbers of atoms are grouped into representative particles, to Dissipative Particle Dynamics (DPD) simulations in which lipids may be modelled by just three[14] to eight particles.[15] As a result, the size and the complexity of computationally-accessible membrane systems are also increasing.[16] The membrane simulation field has thus progressed from *e.g.* a single α-helix embedded in a small patch of membrane for a duration of 1–2 ns[17] to the modelling of *e.g.* photosynthetic membrane vesicles[18] and HIV virion membranes.[19,20] At this level, visual analysis becomes challenging.[21,22] Despite using the latest graphical hardware for efficient system display,[23–26] interpretation of very large membrane simulations by visual inspection is becoming unproductive due to their intrinsic complexity. There is therefore a substantial motivation to provide a simple and visually clear methodology to understand the complex dynamics of large-scale membrane systems. The challenge of computational visualization of very large systems has already been addressed in a number of areas of physics.[27,28] To some extent, visualization of very complex biomembrane systems, especially in the context of results from MD simulations, is less well developed. For example, the display of lipids is often limited to the representation of static states (*i.e.* simulation snapshots) and does not effectively render the complex dynamics of such systems. Progress has been made *e.g.* using arrows to depict discretised movements[8] but approaches are needed to more fully capture complex dynamical processes.

Here, we describe methods for informative visualization of lipid motion in large and complex systems using approaches derived from biomolecular simulations, physics, and computer visualization. We have developed visual and quantitative analysis methodologies for both molecular positions and their velocity field using vector field and streamline techniques. This allows us to highlight substantial and transient flow-like movements of lipids and to potentially assess the effects of crowding within models of bacterial cell membranes.[29] We have evaluated and applied our methodology using two examples of large biomembrane systems recently simulated in our research group: First, a model of a crowded, planar bacterial membrane with lateral dimensions ~1200 Å and containing 256 copies of a simple bacterial outer membrane protein (OmpA); second, a 300 Å radius vesicle the lipid composition of which matches the influenza virion membrane as determined in recent lipidomics studies[30] (see Fig. 1). With

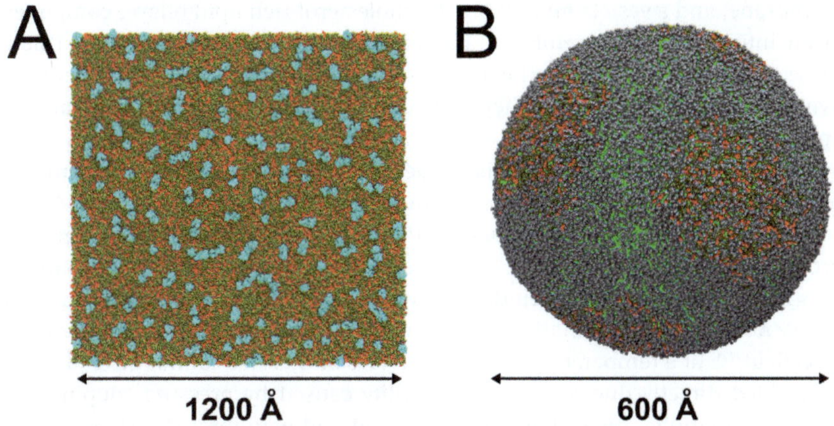

Fig. 1 Membrane simulation systems discussed in this study. (A) A planar membrane containing 256 OmpA proteins (cyan) in a bilayer of 28 260 molecules of POPE (brown) plus 9420 molecules of POPG (red). (B) A spherical lipid vesicle comprising 45 349 lipids (see Table 1 for details) including sphingolipids (grey), zwitterionic (brown) and anionic (red) phospholipids and cholesterol (green).

Table 1 Lateral diffusion coefficients for the membrane constituents of both systems. The values of D_{xy} were derived from the MSD curves shown in the ESI, Fig. S1,† by fitting these curves between 100 and 900 ns for each system. The error estimates are included in brackets

Components	Number of molecules	D_{xy} (10^{-7} cm^2 s^{-1})
Planar membrane		
OmpA	256	2.04 (± 0.26)
POPE	28 260	5.16 (± 0.14)
POPG	9420	5.14 (± 0.14)
Vesicle		
POPS	6802	0.23 (± 0.01)
DOPX	4081	0.24 (± 0.01)
DOPE	2268	0.24 (± 0.01)
CHOL	23 804	0.17 (± 0.01)
SM	8389	0.08 (± 0.01)

membrane surfaces of more than 10 000 Å^2 (~14 400 Å^2 and ~11 300 Å^2 for the planar membrane and the vesicle, respectively), these systems are representative of the largest currently simulated by CG approaches.[31–33] Furthermore, these systems provide two distinct test cases, namely a planar membrane with a simple lipid composition but with crowded membrane proteins compared with a vesicle system containing a more complex lipid mixture with a high concentration of cholesterol (see Methodology for full details). This allows us to test the robustness of our approach with rather different biomembrane model systems.

2 Methods

2.1 CG MD simulations

We have used two systems that differ in terms of shape and composition: a planar lipid bilayer with embedded OmpA proteins that mimics a crowded bacterial membrane, and a vesicle mimicking the cholesterol-rich lipid bilayer component of an influenza virion membrane but without any embedded proteins. Due to these intrinsic differences, the two systems are expected to exhibit distinct dynamic behaviour. Hence, they provide realistic test cases to evaluate our approach.

2.1.1 Planar membrane. This model mimics a large patch of a crowded bacterial membrane. The initial system size was 1180 × 1180 × 104 Å^3 and contained 37 680 lipids (1 : 3 POPG to POPE) and 256 OmpA proteins, giving a protein : lipid ratio of roughly 1 : 150 (see Table 1). The total number of particles is ~1 202 500 (~577 900 excluding solvent). The system was modelled using GROMACS 4.5.3 in combination with a modified in-house version of the MARTINI forcefield[34,35] at a temperature of 313K. Bilayers of this size cannot be readily self-assembled directly due to system instability caused by pressure coupling and initial fluctuations. Instead we self-assembled and equilibrated a smaller patch containing approximately 150 lipids which was then replicated. POPE and POPG do not phase separate and are randomly distributed within the bilayer during this initial equilibration. The small lipid patch was then analyzed for leaflet asymmetry and corrected such that the numbers of lipids in the two leaflets were balanced. The patch was then tiled to form a 4 × 4 patch containing ~2500 lipids followed by a second equilibration and another round of tiling, resulting in a final patch containing ~38 000 lipids. The final equilibration was performed for 100 ns. During each round of equilibration after membrane tiling, the system was observed to shrink modestly in the *x*–*y* plane and expand slightly in the *z* plane. The final system was simulated for 1 μs, during which frames were saved every 0.2 ns giving a 5000-frame trajectory for subsequent analysis.

2.1.2 Vesicle. In the course of constructing a full-scale computational model of a human influenza A virion, a vesicle with lipid composition matching the experimentally-determined influenza A lipidome[30] was produced. The details of this construction process will be described in more detail elsewhere (Reddy *et al.*, ms. in preparation). In brief, the vesicle construction process starts off with random placement of lipids within two leaflets of a spherical system with an outer diameter of ~740 Å. The random seeding of lipid positions within a set of spherical geometric constraints was handled by employing the Packmol program[36] for addition of lipids interleaved with energy-minimization in GROMACS. The asymmetric lipid distribution in the vesicle was based on the data of Gerl *et al.*[30] Thus the inner leaflet contains POPS, DOPX (ether-linked DOPE) and DOPE with an approximate 3 : 2 : 1 ratio, while the outer leaflet contains sphingomyelin (SM) molecules (with saturated C16 chains). Cholesterol (CHOL) molecules were distributed between the two leaflets and comprise >50% of all lipid molecules in the vesicle. Some mixing (*i.e.*, flip–flop) of leaflet components occurred during the equilibration process. An inner core of restrained CG particles was employed to mimic the interior of the virion which was not modelled explicitly. The system contained ~3 006 000 CG particles in total (~500 000 without solvent). Lipids were initially somewhat sparsely packed, forming a well-packed lipid vesicle (~600 Å outer diameter) after a 300 ns equilibration in water at 323 K using GROMACS 4.5.5[37] and the MARTINI 2.1 forcefield.[38,39] The resultant lipid vesicle model was simulated for 5 μs at 295 K, with the final microsecond being analysed in the current study.

2.2 System pre-processing

It is necessary to perform a series of pre-processing operations on the MD simulation data prior to employing our analysis methodologies. First, for each system, the motion of the centre of mass was removed to avoid possible artefacts. Subsequently, we performed low-pass filtering to remove high frequency noise using the GROMACS g_filter function. The filter shape is

$$\cos\left(\frac{\pi t}{N}\right) + 1$$

between $-N$ and $+N$, where N is the number of frames, as detailed in Affentranger *et al.*[40] In order to study the effect of such filtering on visual results, we have filtered the two systems over a series of time scales: 10 frames, 20 frames and 40 frames corresponding respectively to 2 ns, 4 ns and 8 ns. The largest window used corresponds to 0.8% of the simulation. To study the behaviour of each lipid membrane leaflet independently they were separated using in-house python code with the MDAnalysis library.[41]

2.3 Diffusion analysis and clustering

The mean square deviation (MSD) was calculated using the g_msd module in the GROMACS 4.5.4 package.[37,42] This module was previously employed to calculate lipid diffusion coefficient[43,44] using the Einstein relation. "Restart" parameters of 2 ns and 5 ns were compared to measure their influence on the MSD curve and diffusion coefficient. Protein clustering was quantified using an in-house tcl script. To do so, the distance between the centre of mass of each protein was calculated and if this distance was smaller than the radius of the protein plus 5 Å we assumed that the two proteins were in contact. Each protein neighbour at time t was stored and the clusters were identified by a connectivity test. The calculations were performed every 10 ns on the 1 μs trajectories.

2.4 Algorithm implementation and availability

The algorithms presented here are primarily implemented in both Tcl and Python. The Tcl code will be incorporated as a plugin for the molecular visualization program VMD[45] while the Python code is designed to be a module for the MDAnalysis library.[41] Scripts as well as documentation are available at the address: http://sbcb.bioch.ox.ac.uk/flows/

2.4.1 Path line visualization. The path line visualization is currently only implemented in VMD. We have used VMD 1.9 to display and calculate the performance of such rendering. In the case of the planar membrane we used a tail of 5 frames (1 ns) and a mean displacement value of 1.2 Å. To obtain an equivalent rendering for the vesicle system, we chose a window size of 20 frames (4 ns) and a mean displacement value of 0.5 Å.

2.4.2 Vector field visualization. The vector field visualization is executable from within either VMD or a Python module depending on MDAnalysis. For VMD, it is based on Tcl code launchable from the VMD plugin. For this representation, two pieces of code are available: a 2D version and a 3D version. The 2D version flattens all the z values on one layer while the 3D version uses the complete 3D coordinates. Plotting in the MDAnalysis module (written in Python) leverages the matplotlib library in 2D[46] and the MayaVi[47] package in 3D.

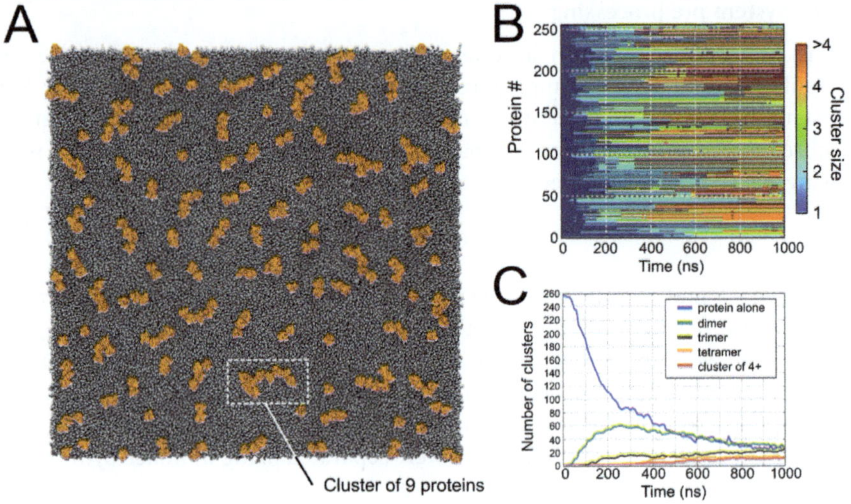

Fig. 2 Protein clustering analysis for the planar bacterial membrane model. (A) VMD snapshot of the planar membrane at the end of the 1 μs, with the proteins shown in orange. (B) Aggregation of proteins as a function of time. Each line represents a unique protein and the colours illustrate each protein's association to a cluster of a certain size. A cluster of size one corresponds to a single, unclustered protein. (C) The frequency of occurrence of clusters of different sizes as a function of time.

2.4.3 Streamline visualization. Streamline visualization is executable from within either VMD or a Python module depending on MDAnalysis. Our in-house Python code used the matplotlib[46] streamplot‡ function to produce 2D streamlines and, for the 3D streamlines visualization, we used the MayaVi package.[47]

3 Results and discussion

In parts 3.2 to 3.4 we focus on the movement of lipid phosphate headgroup particles for one leaflet to demonstrate our method on a simple case. However, the method can be applied to other cases such as water molecule diffusion.

3.1 Global analysis of dynamic behaviour of lipids and proteins

Prior to applying our flow analysis tools, we will briefly describe the dynamic and clustering behaviour of lipids and proteins in our simulation systems.

3.1.1 Clustering of proteins. The planar membrane model (1 : 150 protein–lipid ratio or *ca.* 15% of the surface occupied by proteins) provides an example of a relatively crowded membrane system (see Fig. 2). Our final snapshot at 1 μs depicts a crowded membrane environment with numerous clusters of OmpA proteins, qualitatively consistent with recent results from high-speed AFM[48] studies of protein clustering in bacterial outer membranes. Visual inspection reveals approximately linear clusters within the bilayer plane. Comparable *chains* of interacting proteins were recently highlighted by computational work for smaller systems.[29] At the end of

‡ http://matplotlib.org/examples/images_contours_and_fields/streamplot_demo_features.html

the simulation, the system consists of 24 single proteins, 27 dimers, 25 trimers and 9 tetramers (see Fig. 2-C). There are also 11 clusters formed by more than four proteins. The largest cluster (of 9 proteins) is highlighted in Fig. 2-A.

We also calculated the clustering of these proteins as a function of time (see Fig. 2B, C). Analysis reveals that the number of monomeric OmpA proteins decreased rapidly (also see Sup. Movie 1†), with conversion of monomers to dimers reaching a maximum after ~250 ns. Dimers recruit additional monomeric proteins to form trimers, and may also aggregate to form larger clusters. After 200 ns, we observed the formation of tetramers followed by the formation of higher order clusters at ~300 ns. The clusters continued to evolve reaching a nearly steady state by the end of the simulation.

3.1.2 Diffusion of lipids and proteins. We have estimated the diffusion coefficients for proteins and lipids in the two systems (see Table 1). As anticipated, the OmpA proteins diffuse more slowly than lipids in the planar membrane, with the two lipid species, POPE and POPG, showing similar diffusion characteristics. We have also calculated the diffusion coefficient of these lipids in each leaflet separately and the results were equivalent to the ones in Table 1. The diffusion coefficient for OmpA proteins is in agreement with the value of $\sim 1.5 \times 10^{-7}$ cm^2 s^{-1} obtained recently for a matching system composition of smaller membrane size.[29]

The vesicular lipid diffusion coefficients are clearly lower by a factor of 20 compared to the planar membrane. Although this may be explained in part by the lower temperature (295 K for vesicle; 313 K for planar membrane) it is mainly due to the high content (>50% molecular species) of cholesterol.[49] These diffusion coefficients are not far from the value of 0.7×10^{-7} cm^2 s^{-1} obtained using magic angle spinning NMR spectra experiments on an influenza lipid vesicle at 290 K.[50] If we compare between lipids, the differences in diffusion coefficients relate to the asymmetry in the vesicle. The inner leaflet is mainly composed of POPS, DOPE and DOPX lipids while the outer leaflet is largely composed of SM, with CHOL present in both leaflets but at a higher level in the outer (62%) relative to the inner (38%) leaflet. So, the diffusion coefficients reveal an inner leaflet which is *ca.* 2–3 times more diffusionally mobile than the outer leaflet.

Whilst the calculation of average diffusion coefficients can provide insights into the overall lateral fluidity of a membrane it does not reveal more complex spatial and temporal variations in behaviour for a given species in different regions of the membrane (*e.g.*, proximal *versus* distal to proteins). Furthermore, analysis of *e.g.* lipid clustering is complex due to their loose, dynamic packing.[8,9,51] For these reasons, we have applied specialized visualization methodologies to our MD simulation systems to provide a more detailed insight into the dynamic properties of lipids in membranes. Thus the dynamic dissimilarities of the two membrane systems we have simulated provide test cases to assess the robustness of our approach.

3.2 Following each lipid by path line visualization

Current published methodologies to visualise the displacement of lipids fall into two categories:

(1) Using arrows to represent individual lipid displacement between a time t and a time $t + dt$.[8,9,52,53]

(2) Using lines to connect the different positions of individual lipids over time.[8,54,55]

The latter is generally used to present an overview of the global region explored by a lipid while the former is used to depict lateral displacements of lipids in a dedicated area between time t and $t + \mathrm{d}t$.

We have combined these two representations to a single depiction that can evolve as a function of time to enhance visual interpretation of molecular mobility in a membrane. In practice, lines (or cylinders) depict the path of lipids over time like a tail. For a defined time window, positions of a lipid at different times (t_i through t_n) are linked to the current time, t (see Fig. 3A). At time $t + \mathrm{d}t$ the earliest element is removed while a new line is added at the front of the tail segment (see Fig. 3B). We have coloured segments and set the thickness of the tail based on the displacement between two time steps (see Fig. 3C). This highlights changes to lipid motion over time and removes the noise of quasi-static lipids.

This first approach allows visualizing large and transient correlated lipid movements for the planar system (see Fig. 4A and Sup. Movie 1†). These transient movements can be linear as well as rotational – *i.e.* vortices. This behaviour has

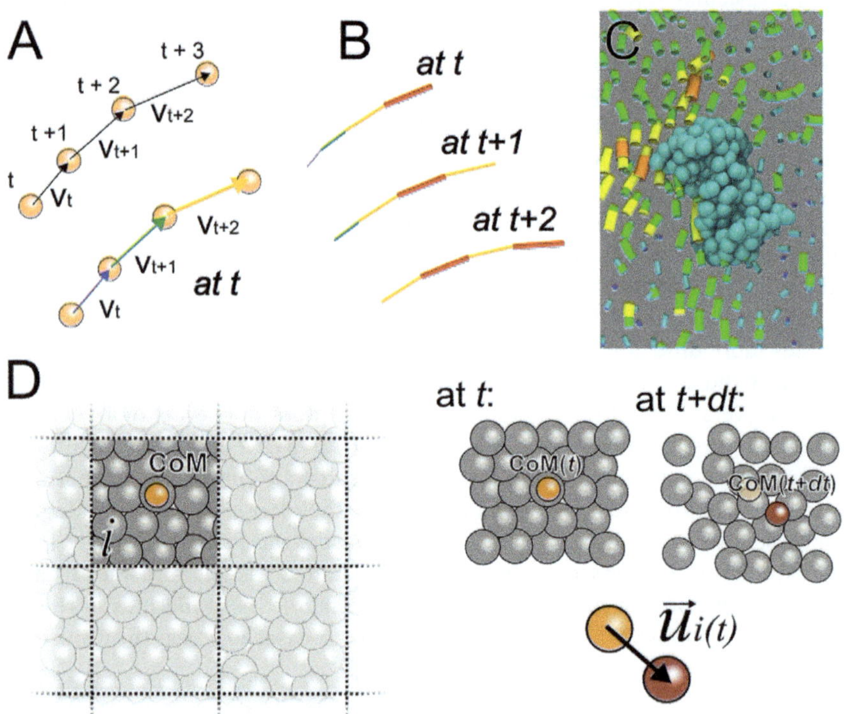

Fig. 3 Path lines and vector field construction. (A) Construction of path lines. (B) Evolution of a path line in function of the time. (C) Lipid path lines around proteins. For (B) and (C), colour and thickness of the tail are function of lipid displacement between two time steps. (D) Simple 2D schematic of the creation of the vector field. After defining a grid, the centre of mass of each lipid selection (in each grid square) is calculated at time t. Then, the centre of mass of the same selection is calculated at time $t + \mathrm{d}t$, and the two centres are used to define the vector.

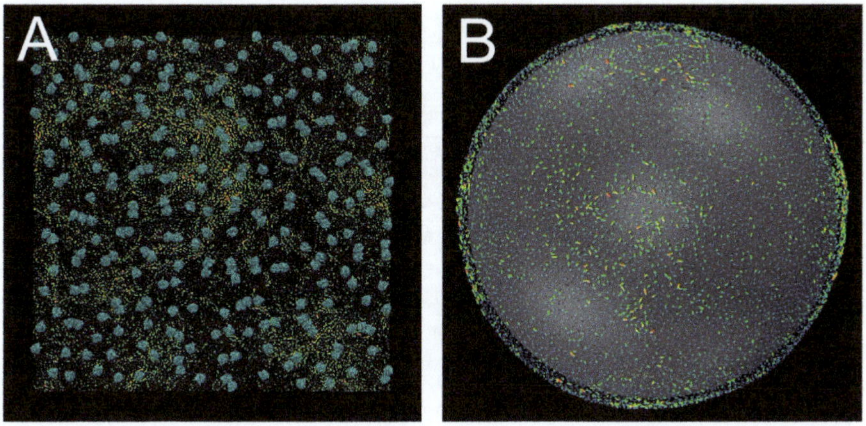

Fig. 4 Path line visualization of a filtered (8 ns time window) dataset. (A) Path line rendering of the upper leaflet of the large planar membranes containing 256 OmpA proteins. (B) Path line rendering of the outer leaflet of the large vesicle membrane. Movies of (A) and (B) are available in the ESI.† Snapshots were captured with the tcl plugin in VMD.

been reported in previous computational[56,57] and experimental[58] studies. Furthermore, Sup. Movie 1† highlights transient correlation between the lipid movements and the displacement of proteins. The translational and rotational motions of the lipids appear to be correlated with comparable motions of the proteins. A more detailed analysis of such phenomena will be presented elsewhere (Goose, Chavent & Sansom, in preparation). In contrast, the vesicle system presents only small changes and did not exhibit large well-defined coherent lipid motions (Fig. 4B and Sup. Movie 2†). Instead, this system appeared to present large patches of motionally stable lipids.

In some respects, analysing the membranes in these systems may be likened to analysing macroscopic systems such as oceans or atmosphere in which transient flows appear§. To study such flow formation, it is common to use evenly-spaced probes depicting the underlying vector field.

3.3 Approaches for following a group of lipids

3.3.1 Vector field visualization. In material physics, vector field visualization has been used to detect phase transitions in large ferroelectric materials.[27] In computational biology, vector fields can be used to depict evolution of regulatory networks[59] or to study the role of water molecules in protein folding processes.[60,61] Here, we use a vector field approach to follow a group of lipids and thereby simplify the representation of their dynamics.

Fig. 3D presents the simple steps used to create a vector field:

(1) The systems were divided into grids. For each cell i (square in 2D or cube in 3D), we calculate the centre of mass, $CoM_{i(t)}$, of the constituent lipids at a time t.

(2) Then, we calculate $CoM_{i(t+\mathrm{d}t)}$,the centre of mass of the same selection at time $t + \mathrm{d}t$,

§ For a comparison see http://www.nasa.gov/topics/earth/features/perpetual-ocean.html or http://hint.fm/wind/

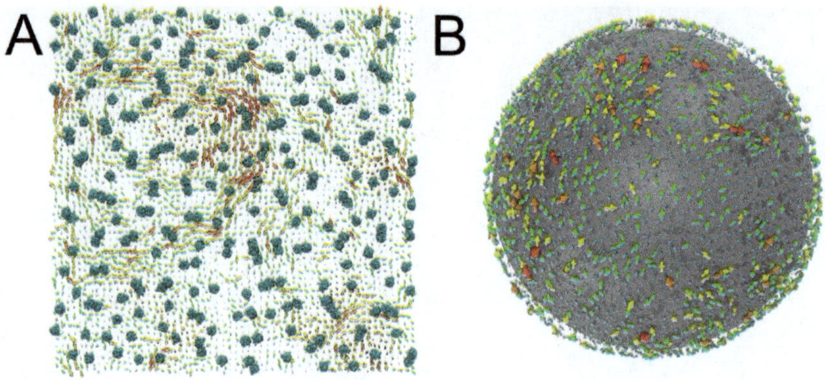

Fig. 5 Vector field visualization using a 20 Å grid resolution. (A) 2D results on the planar membrane. (B) 3D results on the vesicle system, for which a transparent sphere was added to clarify the visualization. Snapshots were captured with the tcl plugin in VMD.

(3) These two centres of mass are used to define the vector at time t following the equation:

$$\vec{u}_{i(t)} = CoM_{i(t+\mathrm{d}t)} - CoM_{i(t)} \tag{1}$$

The vector origin is the centre of mass of the lipids at time t. In this paper, we have chosen a $\mathrm{d}t = 0.2$ ns (*i.e.* 1 frame). This value is quite small as we wish to avoid potentially large spread of the lipid selection at $t + \mathrm{d}t$. For the examples presented below we have used a grid with a resolution of 20 Å.

This visualization has some advantages over the path line visualization. Reducing the number of graphical objects (*i.e.* vectors representing a group of lipids here instead of lines corresponding to each lipid in the previous representation) evidently clarify the representation (Fig. 5 and Sup. Fig. 2†). It is especially valuable for the creation of a static picture. Furthermore, this representation provides more insights because it is possible to track the direction of the lipid movements using arrows. This vector field can be then used as input to perform more advanced rendering.

3.3.2 Streamlines visualization. Mathematically, streamlines are curves instantaneously tangent to the motion vectors. Streamline curves have been used for several decades to represent flow patterns.[62] A streamline is a curve that describes the trajectory of a particle in a stationary vector field at a given time. More formally, given a 2D time-dependent vector field $\boldsymbol{v}(x,t) = (v_x,v_y)$, a streamline S is a parametric curve $S(\tau)$ defined at time t and initiated from a seed point p. $S(\tau)$ is given by the equation:

$$\frac{\mathrm{d}s}{\mathrm{d}\tau} = \boldsymbol{v}(S(\tau), t) \text{ with } S(0) = p$$

Any sample point of the streamline S is given by:

$$S(l) = S(0) + \int_0^l \boldsymbol{v}(S(\tau), t)\mathrm{d}\tau$$

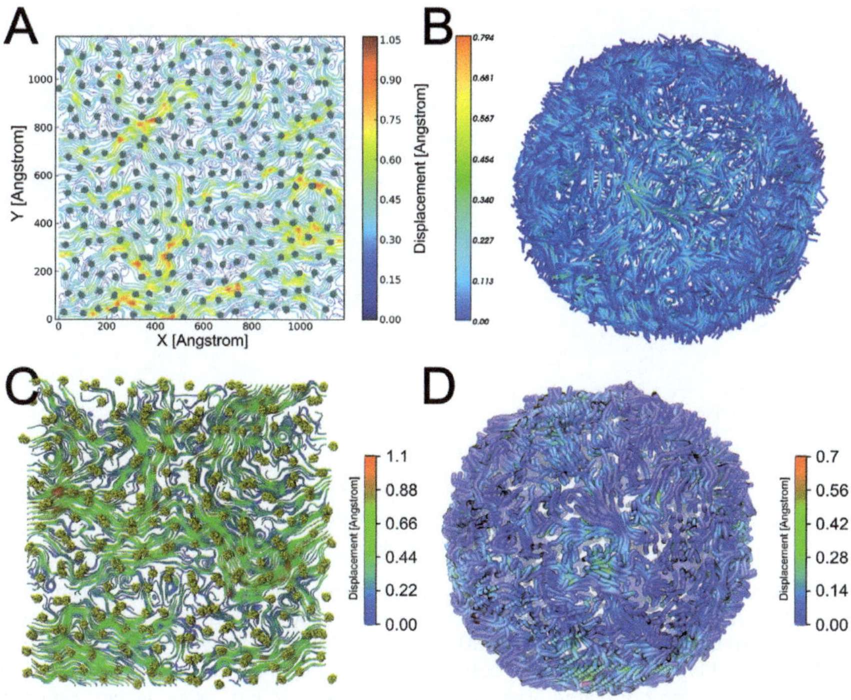

Fig. 6 Streamline visualizations for the planar membrane and the vesicle system. (A) and (C) 2D visualization of the upper leaflet of the planar membrane system. (B) and (D) 3D visualization of the vesicle outer leaflet. (A) and (B) are rendered using in-house Python code employing the MDAnalysis module and matplotlib/MayaVi while (C) and (D) are displayed using VMD program. For (A) and (D) associated movies are available in the ESI.†

with the curvilinear coordinate $l \in [-l_1, l_2]$ and $l_1, l_2 \geq 0$. A standard Runge–Kutta integration scheme¶ is used to sample a streamline backward ($l \in [-l_1, 0]$) and forward ($l \in [0, l_2]$) from its seed point. We can notice that in function of the algorithm used the streamlines can be integrated in one direction or in both directions. Then, the placement of the streamlines constitutes a key step and has been studied for a long time.[63]

An analogous approach, display of field lines, is used in computational biology to depict electrostatic fields calculated with algorithms such as the Adaptive Poisson–Boltzman Solver (APBS).[64] Streamlines were also recently used to describe water flows revealed by MD simulations of photosystem II.[65]

It is possible to display streamlines in 2D or 3D. This streamline representation can combine the advantages of the two previous representations. Like the vector field, it follows a group of lipids but not all the lipids while, like the path lines, the streamlines help to better delineate the "flow" pattern of the lipids creating continuous lines between discretised vectors (Fig. 6).

For the planar membrane, this visualization enables display of both linear and circular nanometer-scale movements of lipids (Fig. 7A and B) as well as small protein-proximal vortices (see Fig. 7C). Sup. Movie 3† shows that the correlated

¶ http://en.wikipedia.org/wiki/Runge-Kutta_method

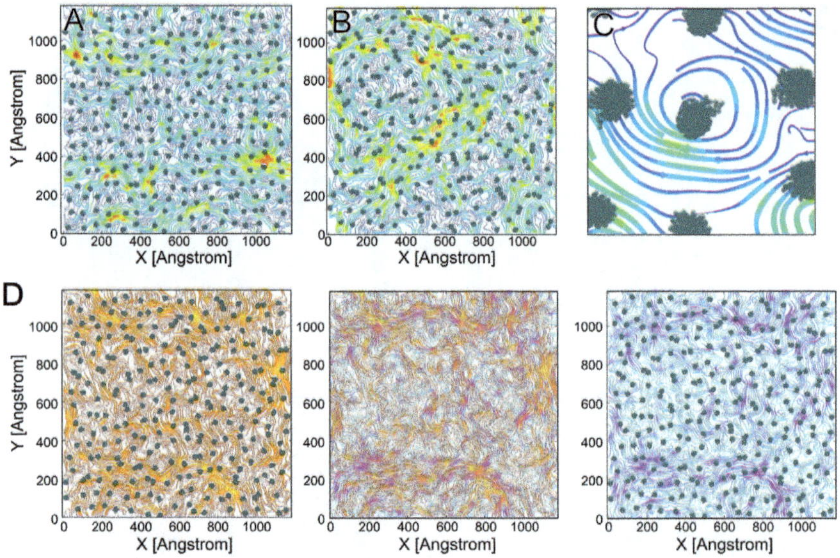

Fig. 7 Lipid motions patterns highlighted by the streamline visualization. (A) Large linear correlated motions crossing the entire system (for $y \sim 300$ Å and $y \sim 900$ Å). (B) Large circular correlated motions (from 0 to 800 Å in x and 400 to ~1100 Å in y). Same snapshot as Fig. 4A and Fig. 5A. (C) Vortex around a protein. (D) Comparison of upper and lower leaflets highlighting correlated areas across the bilayer. On left, upper leaflet; on right, lower leaflet; in the middle superimposition of the lower and the upper leaflet. Snapshots made with the MDAnalysis Python plugin.

lipid patterns are transient (nanosecond scale lifetime) and constantly changing. Furthermore, visual inspection of the results tends to show that these flow patterns are correlated with areas of substantial lipid displacement. The streamlines rendering also highlight how the lipid flows can bridge several proteins far away from each other in large correlated motions (see Fig. 6A and C and Sup. Movie 3†) extending the model of protein and lipids dynamic complexes[66] to transient networks of protein and lipids.

The streamline visualization did not depict very large nanometer-scale flow for the vesicle system. It was only possible to display little vortices (see Fig. 7D and Sup. Movie 4†). This result is in agreement with the path line visualization and the calculation of the diffusion coefficient. It is also possible to use streamlines to compare correlated lipid motions between the two leaflets of a membrane (see Fig. 7D).

3.5 Analysis tool to complement the visualizations: leaflet correlations

Streamline visualization revealed interesting flow lipid patterns. However, the representations may contain considerable information and it is therefore important to be able to isolate selected regions for comparisons. There have been several proposals to filter streamline data to aid such analysis (for examples see ref. 67 and 68). We applied comparable methodology to our systems to simplify the streamline representation using calculations based on the vector fields. As noted above, streamline visualizations can highlight correlated lipid motions between two leaflets. Here, we filter the data to highlight only the areas where the lipid motions are correlated above a certain threshold.

One can define $C_{l(t)}$, the leaflet correlation function at time t:

$$C_{l(t)} = \frac{1}{N}\sum_{i=1}^{N}\frac{\vec{u}_{i(t)}\cdot\vec{v}_{i(t)}}{|\vec{u}_{i(t)}||\vec{v}_{i(t)}|} \tag{2}$$

where · denotes the scalar product of vectors $\vec{u}i(t)$ and $\vec{v}i(t)$ as defined by equation (1). $\vec{u}i(t)$ and $\vec{v}i(t)$ are lower and upper leaflet vectors, respectively. N is the number of non-empty cells in the grid. This type of correlation was recently used to identify correlated intermolecular dynamics of short polymer chains.[69] Note that $-1 \leq C_{l(t)} \leq +1$, where -1 represents movements in opposite directions and +1 represents movements in the same direction. A $C_{l(t)}$ value near 0 depicts no correlation between the leaflets.

For the planar membrane we used the previously-defined Cartesian grid. Due to the shape of the vesicle, we used spherical coordinates (see ESI†). To select the same areas of interest, we chose the same θ and φ between the two leaflets while changing the radial distance r to correspond to the average radius of phosphate headgroup particles in each leaflet. After obtaining these coordinates, we created cubes of dimensions 20 Å to select the lipids for vector calculations (as described in section 3.3).

Fig. 8 presents the inter-leaflet lipid flow correlation results for each system using data filtered over an 8 ns time window. There is a high correlation between the two leaflets in the case of the planar membrane (with $C_{l(t)} > 0.6$) while, for the vesicle, the correlation values are lower, with $C_{l(t)} \sim 0.4$. These inter-leaflet correlations are rather more pronounced than those seen in simulations of a simple DOPC bilayer system.[70] The difference in the magnitude of the correlations between these two systems may be linked to the differences in their compositions. As noted above, the planar membrane has a symmetric, relatively simple lipid composition. In contrast the lipid composition for the vesicle system is highly

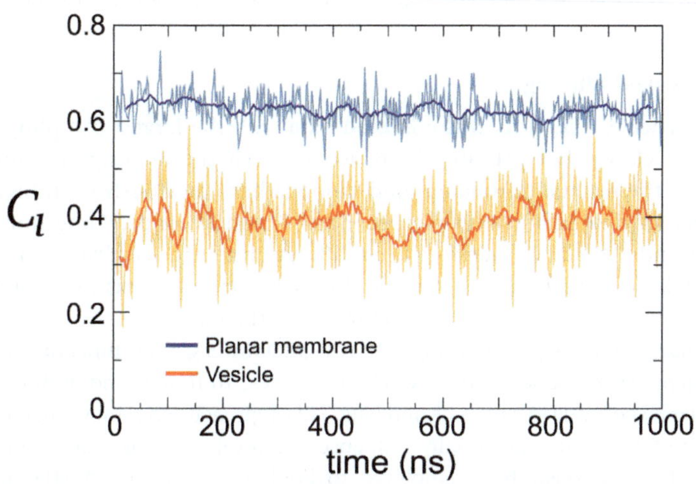

Fig. 8 Inter-leaflet lipid flow correlation assessment for the planar bilayer (blue) and the vesicle (red) systems. The bold lines depict the averaged values on 10 steps.

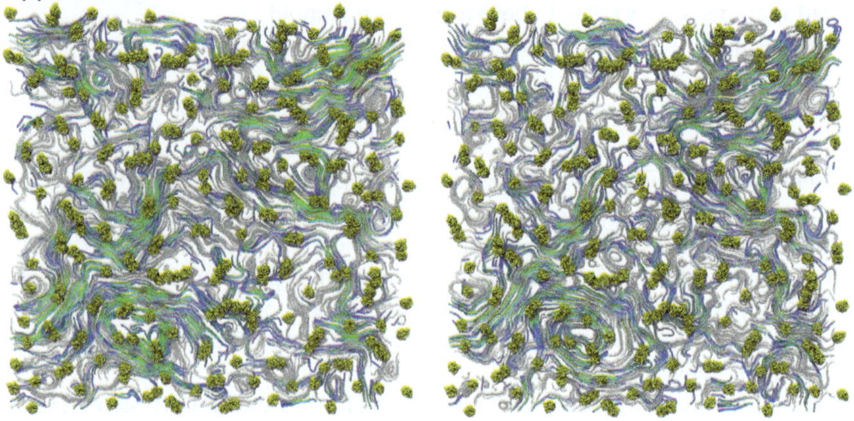

Fig. 9 Examples of correlated dynamics for the planar membrane visualized using VMD. Coloured streamlines depict the correlated parts between the two leaflets (with $C_{l(t)} > 0.8$) while the remaining streamlines are coloured in gray.

asymmetric (see above), with a more slowly diffusing outer leaflet which is nearly twice as rich in cholesterol as the inner leaflet. We also note that the planar membrane correlation may be slightly overestimated due to the flattened 3D coordinates in the x–y plane.

The analysis of correlations may be used to filter the streamlines by using a cut-off threshold for these inter-leaflet lipid correlation values. For example, for the *E. coli* bilayer system, we can display the streamlines with a $C_{l(t)}$ value superior to 0.8 (see Fig. 9). This represents around 50% of vectors used to define the streamlines. This highlights large inter-leaflets patterns visually correlated with the areas of high velocity. We postulate that the slower diffusion of the proteins can be, in part, explained by the fact that sometimes the flows between leaflets are not correlated which can have an influence on trans-membrane proteins. So, using this filtering can help to select only high correlated parts of the membrane for further analysis.

3.7 Low pass filtering

In this article we have used the filtered data with the largest sampling window (8 ns) for clarity, but it is important to note that the filtering granularity can influence the results. Correlated lipid motions have previously been highlighted for different time scales: over picoseconds,[10,69] nanoseconds,[8,9] and hundreds of nanoseconds.[9] We have therefore tried different filters to see if this can highlight a particular flow pattern on a specific time scale. Here, we examine the effect of the filter window size on the representation of the data.

We have used low pass filtering with the GROMACS g_filter function to filter the two systems over a series of time scales: 10 frames, 20 frames and 40 frames corresponding respectively to 2 ns, 4 ns and 8 ns. This approach can remove high frequency motions of lipids and potentially influence the results of this study. This type of method has previously been employed to lipid membrane-based MD simulation systems[71] and for polypeptide folding.[40] Recently, Gypsa and colleagues have also used a low pass filter to highlight different characteristics of membrane curvature.[72]

Filtering has no effect on the diffusion coefficients calculated for our simulations, with any discrepancies between filtered and unfiltered values falling within the bounds of the estimated error (data not shown). We calculated the leaflet lipid motion correlation values for different filter window sizes (see Sup. Fig. 3†). The results for the planar membrane indicate that the leaflet correlation is affected by the filtering. The raw data presents a leaflet correlation value greater than 0 ($C_{l(t)} \sim 0.3$), consistent with some correlated lipid movement between leaflets in the absence of a filter. This value drastically increases with the addition of even a short time window filter ($C_{l(t)} \sim 0.55$ for 2 ns filter). Increasing the time window size of the filter further does not lead to a substantial increase in calculated leaflet correlation, consistent with potential convergence to a consensus correlation value ($C_{l(t)} \sim 0.6$ for 4 ns filter and $C_{l(t)} \sim 0.63$ for 8 ns filter).

We reused equation (2) to compare the data obtained with different filter time window sizes within a single leaflet. To do this, instead of comparing vectors from two different leaflets, we compare vectors for the same leaflet but with different filter windows. We called this function filter correlation: $C_{f(t)}$. The raw data has a small correlation with 8 ns filtering ($C_{f(t)} \sim 0.1$) but this value drastically increases even for a small filter over 2 ns ($C_{f(t)} \sim 0.5$) and is even higher for data filtered on a 4 ns timescale ($C_{f(t)} \sim 0.75$) (see Sup. Fig. 4†). Indeed, visual inspection of the streamline plots indicates conserved patterns for all the filtered data (see Sup. Fig. 5†). The effect of this filtering is to remove high frequency noise, which can potentially be related to jiggling motions of lipids around their positions. Thus, the filtering can allow focusing of the data representation over longer time scale motions rather than residual noise from local fluctuations.

3.8 Algorithm comparison and efficiency

We have tested our implementations on an Intel 6 cores (12 threads) i7-3930K with a Geforce GTX 680. The tcl code was used on 1 thread while the Python code, implemented to handle multithreads calculation, was tested on 1 and 12 threads (see Sup. Table 1†). The calculation of vectors are clearly faster for the Python code than for the tcl code, this can be in part explained by the use of Numpy[73] to efficiently deal with large arrays in the Python code. So, we believe that there is room for improvement for the tcl code in translating some parts of the calculation in C/C++ or CUDA language. At the inverse the visualization of streamlines is clearly faster using VMD: this is due to an implementation in C in the core of the program which is well optimized in comparison to the Matplotlib/MayaVi code. We can also notice that for our systems the use of multithread calculation with Python only improves the results by a small amount. The VMD visualization takes a little bit longer when high quality pictures (as presented in this article) are rendered using the VMD ray-tracing engine called Tachyon.[74]

3.9 Grid size and time scale parameters

We tried different grid sizes for the presented CG models (data not shown). A resolution of 20 Å seems appropriate. This value is coherent with the work of Roark *et al.* defining correlated lipid motions with a length of 25 Å.[70] Furthermore, this resolution value can be linked to the average number of lipid headgroups in each grid cell. This value is 5.7 for the planar membrane. The use of a higher grid resolution of 10 Å resulted in very few lipid headgroups in a given cell, which can defeat the purpose of analysing lipids in unison. Conversely, a lower grid resolution of 40 Å produced reasonable results, albeit less accurately than at 20 Å resolution. So, a minimum 2D system size for use of the streamline visualization with lipids is approximately 200 $Å^2$ – giving a maximum grid of 100 cells. As we have seen with the planar membrane example system, far greater dimensions are well suited for such a visualization. For the vesicle system, the number of lipid headgroups per cell is 21.5 on average (18 for the outer leaflet and 25 for the inner leaflet). So, we could have diminished the cube size to increase the resolution but we preferred to keep the same grid (20 Å side) to compare the two systems with the same conditions. Furthermore, at this resolution, the MayaVi program has difficulty rendering the scene.

In terms of the time scale over which visualizations are produced, in this paper we have observed that two consecutive frames (0.2 ns) produce an acceptable result. We have not tested our analyses over much longer time scales but we will investigate this in a future study.

3.10 Combining the streamline visualization with other analyses: example with membrane undulations

Correlating diffusional streamlines with visualization of other membrane parameters can be used to highlight subtle dynamic behaviours. In Fig. 10, we have colour-mapped the vertical displacement (z coordinate values along bilayer normal) of the membrane in the presence of streamlines. By inspection, the diffusional

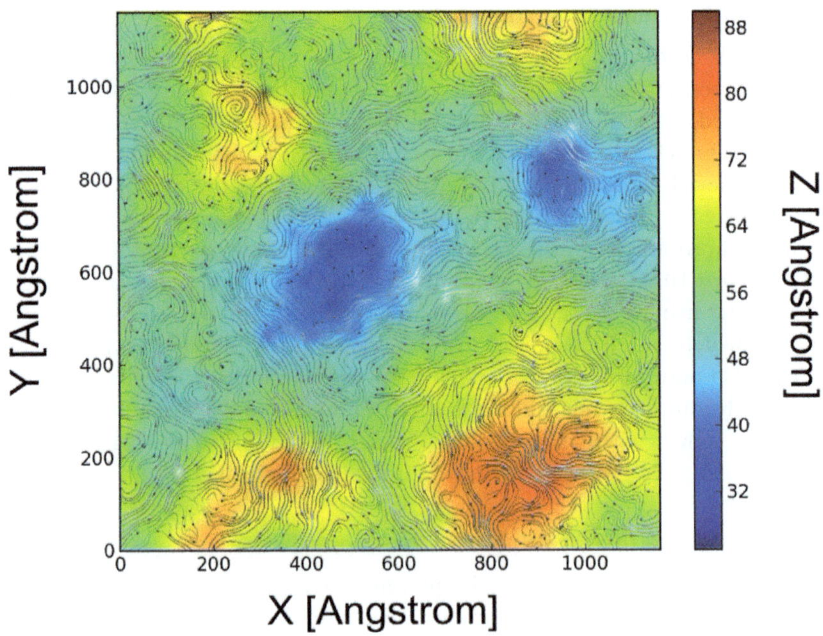

Fig. 10 Streamline visualization coupled with height map for the planar membrane. The colours represent the vertical z values of the membrane. This representation highlights the undulations of the membrane in the z direction. High displacement parts of the streamlines are coloured in white. An associated movie is available in ESI.†

streamlines seem to evolve in a manner which correlates with the membrane undulations (See Sup. Movie 5†) and can circulate around vertical undulations of the membrane (see Fig. 10). The influence of the surface shape on lipid dynamics has been highlighted by experimental results.[75,76] This could be related to membrane tension and pressure with the less curved parts of the membrane presenting fewer constraints and allowing the lipids to circulate more easily. It will therefore be of interest to examine complex membranes by combining visualization of streamlines with analyses such as the calculation of pressure fields and tension lines,[77] area per lipid visualization[78] or lipid phase analysis.[79]

3.11 Future directions

We have focused this study on the collective behaviour of lipids in bilayer membranes, and are extending this work to analyse the effects of such lipid flows on the diffusion and the clustering of membrane proteins (Goose *et al.*, in preparation). We suggest that these nano-flows can be a more general phenomenon. As the visualisation and analysis method we have described is flexible, it should be possible to apply this approach to other large-scale ensembles of molecules. For example, one might analyse data from simulations of water or solutes (*e.g.* ions) in the vicinity of transbilayer pores to explore possible nanoflows. In particular, as larger scale membrane simulations become more practicable (and common) it will be important to model such a systems using all-atomistic simulations in order to evaluate the robustness of our approach to different levels of granularity. We also plan to develop a new GPU-accelerated streamline integration and rendering engine for VMD using CUDA, in order to improve the interactive calculation and rendering performances.

4 Conclusion

We have presented a new visualization approach to aid our understanding of the complex dynamic properties of biological membranes. We have illustrated our analysis *via* two different systems with contrasting lipid compositions, shapes and dynamics. Our methodology has been successfully applied to render lipid motions for these two systems, and allows us to highlight large concerted lipid motions within the plane and between the two leaflets of a membrane. The lipid motions were seen to be linear as well as circular (*i.e.* vortices) in nature and may be linked to membrane deformations perpendicular to the bilayer plane. Finally, we have shown that the leaflet asymmetry in lipid composition can be related to an asymmetry in term of dynamics.

Our set of tools is not intended to solve by itself the complex question of transient lipid flows, but it can be used to focus attention on particular aspects of the dynamic behaviour of membrane prior to more detailed statistical analysis. As larger and more complex lipid bilayer simulations become the norm, we anticipate that these visualization techniques will help us to better define the complexities of lipid movements. This, in turn, will enable improved parameterization of more coarse grained (*e.g.* DPD) and/or mesoscopic models of biological membranes. If they can accurately capture the key elements of complex membrane dynamics, such higher level models will then allow us to relate molecular properties of complex biological membranes to *e.g.* super-resolution microscopy imaging of the dynamics of membrane in cells.[80]

Scripts are available at the address: http://sbcb.bioch.ox.ac.uk/flows/

Acknowledgements

We thank Drs Philip Fowler, Sebastian Busch and Mickael Lelimousin for discussions and Dr Oliver Beckstein for his comments. This work is supported by grants from the Wellcome Trust and the BBSRC. T.R. is supported by a Canadian Institutes of Health Research (CIHR) post-doctoral fellowship and a Fulford Junior Research Fellowship from Somerville College, Oxford. A.C.E.D is supported by an Engineering and Physical Sciences Research Council Systems Biology Doctoral Training Centre studentship. Computational resources were provided in part by the Partnership for Advanced Computing in Europe (PRACE) grant number PP2012061115. VMD development is supported by NIH grant 9P41GM104601.

References

1 S. J. Singer and G. L. Nicolson, *Science*, 1972, **175**, 720–731.
2 D. M. Engelman, *Nature*, 2005, **438**, 578–580.
3 F. Spira, N. S. Mueller, G. Beck, P. von Olshausen, J. Beig and R. Wedlich-Söldner, *Nat. Cell Biol.*, 2012, **14**, 640–648.
4 H. Sprong, P. van der Sluijs and G. van Meer, *Nat. Rev. Mol. Cell Biol.*, 2001, **2**, 504–513.
5 E. J. Denning and O. Beckstein, *Chem. Phys. Lipids*, 2013, **169**, 57–71.
6 S. J. Marrink, A. H. de Vries and D. P. Tieleman, *Biochim. Biophys. Acta, Biomembr.*, 2009, **1788**, 149–168.
7 S. M. Loverde, D. A. Pantano, D. A. Christian, A. Mahmud, M. L. Klein and D. E. Discher, *Curr. Opin. Solid State Mater. Sci.*, 2011, **15**, 277–284.
8 E. Falck, T. Róg, M. Karttunen and I. Vattulainen, *J. Am. Chem. Soc.*, 2008, **130**, 44–45.
9 T. Apajalahti, P. Niemelä, P. N. Govindan, M. S. Miettinen, E. Salonen, S.-J. Marrink and I. Vattulainen, *Faraday Discuss.*, 2010, **144**, 411–30, – discussion 445–81.
10 S. Busch, C. Smuda, L. C. Pardo and T. Unruh, *J. Am. Chem. Soc.*, 2010, **132**, 3232–3233.
11 P. F. J. Fuchs, *Methods Mol. Biol.*, 2010, **654**, 403–421.
12 M. Karplus and G. A. Petsko, *Nature*, 1990, **347**, 631–639.
13 S. J. Marrink and D. P. Tieleman, *Chem. Soc. Rev.*, 2013, **42**, 6801–6822.
14 B. J. Reynwar, G. Illya, V. A. Harmandaris, M. M. Müller, K. Kremer and M. Deserno, *Nature*, 2007, **447**, 461–464.
15 F. J.-M. de Meyer, M. Venturoli and B. Smit, *Biophys. J.*, 2008, **95**, 1851–1865.
16 E. Lindahl and M. S. P. Sansom, *Curr. Opin. Struct. Biol.*, 2008, **18**, 425–431.
17 L. R. Forrest, D. P. Tieleman and M. S. Sansom, *Biophys. J.*, 1999, **76**, 1886–1896.
18 M. K. Sener, J. D. Olsen, C. N. Hunter and K. Schulten, *Proc. Natl. Acad. Sci. U. S. A.*, 2007, **104**, 15723–15728.
19 G. S. Ayton and G. A. Voth, *Biophys. J.*, 2010, **99**, 2757–2765.

20 P. L. Freddolino, A. S. Arkhipov, S. B. Larson, A. McPherson and K. Schulten, *Structure*, 2006, **14**, 437–449.

21 E. Mendez-Villuendas, S. Baoukina and D. P. Tieleman, *Journal of Physics*, 2012, **385**, 012002.

22 S. I. O'Donoghue, A.-C. Gavin, N. Gehlenborg, D. S. Goodsell, J.-K. Hériché, C. B. Nielsen, C. North, A. J. Olson, J. B. Procter, D. W. Shattuck, T. Walter and B. Wong, *Nat. Methods*, 2010, **7**, S2–S4.

23 M. Chavent, B. Lévy, M. Krone, K. Bidmon, J.-P. Nominé, T. Ertl and M. Baaden, *Briefings Bioinf.*, 2011, **12**, 689–701.

24 M. Krone, J. E. Stone, T. Ertl and K. Schulten, *EuroVis-Short Papers*, 2012, 67–71.

25 J. E. Stone, K. L. Vandivort and K. Schulten, *Lect. Notes Comput. Sci.*, 2011, **6939**, 1–12.

26 J. E. Stone, K. L. Vandivort, and K. Schulten, *UltraVis '13 Proceedings of the 8th International on Ultrascale Visualization*, 2013.

27 2012 IEEE Visualization Contest Winner: K. Scharnowski, M. Krone, F. Sadlo, P. Beck, J. Roth, H. R. Trebin and T. Ertl, Visualizing Polarization Domains in Barium Titanate, *IEEE Comput. Graphics Appl.*, 2013, **33**, 9–17.

28 H. Yu, C. Wang, R. W. Grout, J. H. Chen and K.-L. Ma, *IEEE Comput. Graphics Appl.*, 2010, **30**, 45–57.

29 J. E. Goose and M. S. P. Sansom, *PLoS Comput. Biol.*, 2013, **9**, e1003033.

30 M. J. Gerl, J. L. Sampaio, S. Urban, L. Kalvodova, J.-M. Verbavatz, B. Binnington, D. Lindemann, C. A. Lingwood, A. Shevchenko, C. Schroeder and K. Simons, *J. Cell Biol.*, 2012, **196**, 213–221.

31 S. Baoukina, S. J. Marrink and D. P. Tieleman, *Biophys. J.*, 2012, **102**, 1866–1871.

32 K. M. Davies, C. Anselmi, I. Wittig, J. D. Faraldo-Gómez and W. Kühlbrandt, *Proc. Natl. Acad. Sci. U. S. A.*, 2012, **109**, 13602–13607.

33 M. Louhivuori, H. J. Risselada, E. van der Giessen and S. J. Marrink, *Proc. Natl. Acad. Sci. U. S. A.*, 2010, **107**, 19856–19860.

34 P. J. Bond, C. L. Wee and M. S. P. Sansom, *Biochemistry*, 2008, **47**, 11321–11331.

35 P. J. Bond and M. S. P. Sansom, *J. Am. Chem. Soc.*, 2006, **128**, 2697–2704.

36 L. Martínez, R. Andrade, E. G. Birgin and J. M. Martínez, *J. Comput. Chem.*, 2009, **30**, 2157–2164.

37 B. Hess, C. Kutzner, D. Van Der Spoel and E. Lindahl, *J. Chem. Theory Comput.*, 2008, **4**, 435–447.

38 S. J. Marrink, H. J. Risselada, S. Yefimov, D. P. Tieleman and A. H. de Vries, *J. Phys. Chem. B*, 2007, **111**, 7812–7824.

39 L. Monticelli, S. K. Kandasamy, X. Periole, R. G. Larson, D. P. Tieleman and S.-J. Marrink, *J. Chem. Theory Comput.*, 2008, **4**, 819–834.

40 R. Affentranger and X. Daura, *J. Comput. Chem.*, 2010, **31**, 1889–1903.

41 N. Michaud-Agrawal, E. J. Denning, T. B. Woolf and O. Beckstein, *J. Comput. Chem.*, 2011, **32**, 2319–2327.

42 D. Van Der Spoel, E. Lindahl, B. Hess, G. Groenhof, A. E. Mark and H. J. C. Berendsen, *J. Comput. Chem.*, 2005, **26**, 1701–1718.

43 A. Raudino, S. J. Marrink and M. Pannuzzo, *J. Chem. Phys.*, 2013, **138**, 234901.

44 D. Hakobyan and A. Heuer, *J. Phys. Chem. B*, 2013, **117**, 3841–3851.

45 W. Humphrey, A. Dalke and K. Schulten, *J. Mol. Graphics*, 1996, **14**, 33–8, 27–8.

46 J. D. Hunter, *Comput. Sci. Eng.*, 2007, **9**, 90–95.

47 P. Ramachandran and G. Varoquaux, *Comput. Sci. Eng.*, 2011, **13**, 40–51.
48 I. Casuso, J. Khao, M. Chami, P. Paul-Gilloteaux, M. Husain, J.-P. Duneau, H. Stahlberg, J. N. Sturgis and S. Scheuring, *Nat. Nanotechnol.*, 2012, **7**, 525–529.
49 T. Róg, M. Pasenkiewicz-Gierula, I. Vattulainen and M. Karttunen, *Biochim. Biophys. Acta, Biomembr.*, 2009, **1788**, 97–121.
50 I. V. Polozov, L. Bezrukov, K. Gawrisch and J. Zimmerberg, *Nat. Chem. Biol.*, 2008, **4**, 248–255.
51 G. S. Ayton and G. A. Voth, *Biophys. J.*, 2004, **87**, 3299–3311.
52 S. Busch and T. Unruh, *Biochim. Biophys. Acta, Biomembr.*, 2011, **1808**, 199–208.
53 M. Javanainen, L. Monticelli, J. Bernardino de la Serna and I. Vattulainen, *Langmuir*, 2010, **26**, 15436–15444.
54 Y. Andoh, S. Okazaki and R. Ueoka, *Biochim. Biophys. Acta, Biomembr.*, 2013, **1828**, 1259–1270.
55 M. Javanainen, H. Hammaren, L. Monticelli, J.-H. Jeon, M. S. Miettinen, H. Martinez-Seara, R. Metzler and I. Vattulainen, *Faraday Discuss.*, 2013, **161**, 397–417.
56 I. Vattulainen and T. Róg, *Cold Spring Harbor Perspect. Biol.*, 2011, **3**(4), DOI: 10.1101/cshperspect.a004655.
57 Yaohong Wang, Jon Karl Sigurdsson, Erik Brandt and Paul J. Atzberger, *Phys. Rev. E: Stat., Nonlinear, Soft Matter Phys.*, 2013, **88**, 023301.
58 A. R. Honerkamp-Smith, F. G. Woodhouse and V. Kantsler, *Phys. Rev. Lett.*, 2013, **111**, 038103.
59 J. J. Tyson, K. Chen and B. Novak, *Nat. Rev. Mol. Cell Biol.*, 2001, **2**, 908–916.
60 J. K. Weber and V. S. Pande, *J. Chem. Phys.*, 2013, **138**, 085103.
61 J. K. Weber and V. S. Pande, *J. Chem. Phys.*, 2013, **138**, 165101.
62 T. McLoughlin, R. S. Laramee, R. Peikert, F. H. Post, and M. Chen, 2010, **29**, 1807–1829.
63 B. Jobard and W. Lefer, *Visualization in Scientific Computing'97*, 1997, 43–55.
64 N. A. Baker, *Proceedings of the National Academy of Sciences*, 2001, 98, 10037–10041.
65 S. Vassiliev, P. Comte, A. Mahboob and D. Bruce, *Biochemistry*, 2010, **49**, 1873–1881.
66 P. S. Niemelä, M. S. Miettinen, L. Monticelli, H. Hammaren, P. Bjelkmar, T. Murtola, E. Lindahl and I. Vattulainen, *J. Am. Chem. Soc.*, 2010, **132**, 7574–7575.
67 H. Yu, C. Wang, C.-K. Shene and J. H. Chen, *IEEE Trans. Visualization Comput. Graphics*, 2012, **18**, 1115–1124.
68 M. Schlemmer, I. Hotz, B. Hamann, F. Morr, and H. Hagen, *VGTC Symposium on Visualization*, 2007, 227–234.
69 H. Morhenn, S. Busch, H. Meyer, D. Richter, W. Petry and T. Unruh, *Phys. Rev. Lett.*, 2013, **111**, 173003.
70 M. Roark and S. E. Feller, *J. Phys. Chem. B*, 2009, **113**, 13229–13234.
71 S. J. Marrink, O. Berger, P. Tieleman and F. Jähnig, *Biophys. J.*, 1998, **74**, 931–943.
72 V. Gapsys, B. L. De Groot and R. Briones, *J. Comput.-Aided Mol. Des.*, 2013, **27**, 845–858.
73 T. E. Oliphant, *Comput. Sci. Eng.*, 2007, **9**, 10–20.
74 J. E. Stone, University of Missouri-Rolla.

75 A. Roux, D. Cuvelier, P. Nassoy, J. Prost, P. Bassereau and B. Goud, *EMBO J.*, 2005, **24**, 1537–1545.
76 R. Parthasarathy, C.-H. Yu and J. T. Groves, *Langmuir*, 2006, **22**, 5095–5099.
77 P. M. Kasson, B. Hess and E. Lindahl, *Chem. Phys. Lipids*, 2013, **169**, 106–112.
78 G. Lukat, J. Krüger and B. Sommer, *J. Chem. Inf. Model.*, 2013, **53**, 2908–2925.
79 E. Jefferys, M. S. P. Sansom and P. W. Fowler, *Faraday Discuss.*, 2014, DOI: 10.1039/c3fd00131h.
80 C. Eggeling, C. Ringemann, R. Medda, G. Schwarzmann, K. Sandhoff, S. Polyakova, V. N. Belov, B. Hein, C. von Middendorff, A. Schönle and S. W. Hell, *Nature*, 2009, **457**, 1159–1162.

Faraday Discussions

PAPER

Rapid decomposition and visualisation of protein–ligand binding free energies by residue and by water†

Christopher J. Woods,*[a] Maturos Malaisree,[b] Julien Michel,[c] Ben Long,[d] Simon McIntosh-Smith[d] and Adrian J. Mulholland[b]

Received 6th December 2013, Accepted 31st January 2014
DOI: 10.1039/c3fd00125c

Recent advances in computational hardware, software and algorithms enable simulations of protein–ligand complexes to achieve timescales during which complete ligand binding and unbinding pathways can be observed. While observation of such events can promote understanding of binding and unbinding pathways, it does not alone provide information about the molecular drivers for protein–ligand association, nor guidance on how a ligand could be optimised to better bind to the protein. We have developed the waterswap (C. J. Woods *et al.*, *J. Chem. Phys.*, 2011, **134**, 054114) absolute binding free energy method that calculates binding affinities by exchanging the ligand with an equivalent volume of water. A significant advantage of this method is that the binding free energy is calculated using a single reaction coordinate from a single simulation. This has enabled the development of new visualisations of binding affinities based on free energy decompositions to per-residue and per-water molecule components. These provide a clear picture of which protein–ligand interactions are strong, and which active site water molecules are stabilised or destabilised upon binding. Optimisation of the algorithms underlying the decomposition enables near-real-time visualisation, allowing these calculations to be used either to provide interactive feedback to a ligand designer, or to provide run-time analysis of protein–ligand molecular dynamics simulations.

1 Introduction

Recent advances in hardware and software enable the running of long timescale (microsecond plus) atomistic, condensed phase molecular dynamics simulations of protein and protein–ligand systems.[2–5] This is sufficient to allow biochemical

[a]School of Chemistry, University of Bristol, Bristol, BS8 1TS, UK. E-mail: Christopher.Woods@bristol.ac.uk
[b]School of Chemistry, University of Bristol, Bristol, BS8 1TS, UK
[c]EastCHEM School of Chemistry, Joseph Black Building, The King's Buildings, Edinburgh, EH9 3JJ, UK
[d]Department of Computer Science, University of Bristol, Bristol, BS8 1TR, UK

† Electronic Supplementary Information (ESI) available: Graph comparing binding of *m*-chlorophenyl and benzamidine ligands, graph showing the convergence of the residue components of the waterswap free energy and an animated version of Fig. 8. See DOI: 10.1039/c3fd00125c/

events such as protein folding[4] and ligand binding and unbinding[5,6] to be observed directly. This is a great advance, but watching events as they occur during a dynamics trajectory provides limited quantitative information on the molecular driving forces of these events. In particular, examining a trajectory where a ligand binds or unbinds once or twice from a protein provides little insight into the binding affinity of the ligand, nor much guidance on how the ligand could be modified to increase binding and so design a better drug.

Binding free energy methods calculate the free energy change associated with ligand binding. There is a direct relationship between binding free energies and equilibrium binding constants. Negative (favourable) binding free energies show that the equilibrium is tilted in favour of the bound complex, while positive values indicate that the equilibrium is tilted towards the unbound protein and ligand. While knowledge of binding free energies is useful, it is difficult to relate changes in structure to changes in binding affinity. One reason is that protein–ligand binding can be viewed as a competition between the ligand and water for the protein active site. This means that, in some cases, differences in binding may be caused by differences in solvation of the ligand in the protein active site.[7,8] New methods need to be developed that can link molecular dynamics and binding free energy calculations. This would allow changes in structure and solvation observed during a dynamics trajectory to be related directly to their contribution to the binding affinity of the ligand. In particular, new ways of visualising this data also need to be developed, so that the link between structure, solvation and binding can be easily recognised and understood by drug designers. Such visualisations would ensure that changes in structure and solvation upon ligand binding are readily available and relatable to binding affinities, thereby providing inspiration to drug designers to suggest modifications to the ligand that could improve binding. Here, we derive and then demonstrate how the waterswap[1] absolute binding free energy method can be used to analyse snapshots from molecular dynamics trajectories. We show how these binding free energies can be decomposed into components that can be used to annotate the output of a molecular dynamics trajectory. These annotations provide detailed graphical visualisations that reveal the molecular drivers for binding, providing clear suggestions for how stronger-binding ligands could be developed.

2 Background

Many methods have been developed to estimate binding free energies of protein–ligand complexes. Many rely on calculating differences in binding free energy between different ligands to the same protein.[9] These relative binding methods, in which one ligand is alchemically morphed onto another, have been used successfully to investigate structure–activity relationships (SARs).[10,11] However, these methods struggle to accurately predict differences in binding free energy when the binding modes of the two ligands are different, or when the two ligands vary significantly in size. An alternative is to use an absolute binding method. These methods attempt to calculate the binding free energy of a ligand to a protein directly. Double-decoupling[12–14] (or double annihilation[15,16]) uses two simulations; one that calculates the free energy of decoupling the ligand from the protein binding site and transfers it to vacuum, and another that calculates the free energy of similarly decoupling the ligand from water. The decoupling free energies must be calculated with a high accuracy. This is because, for large or polar ligands, they are typically on the order of 5–20 times the magnitude of the binding affinity. This can lead to large errors because the binding free energy is calculated as the difference between the free energy to decouple fully the ligand from bulk water and the free energy to decouple fully the ligand from the protein. Another problem is that, as the ligand is transferred to vacuum, it leaves behind a cavity in the solvent. This cavity has to be filled by bulk solvent, which is a slow process. One way to solve the cavity problem is to use implicit solvent methods, such as Poisson–Boltzmann Surface Area (PBSA)[17,18] or Generalized Born Surface Area (GBSA).[19,20] These use an implicit, continuum model of water, with the binding free energy calculated as a combination of the interaction energy between the protein and ligand, and the difference in hydration free energies of the protein–ligand complex, and the individual protein and ligand molecules. PBSA or GBSA methods can be used to analyse the binding observed in molecular dynamics simulations, *e.g. via* MM-PBSA or MM-GBSA.[19–23] Snapshots are taken from the trajectory, and PBSA or GBSA used to calculate the binding free energy for each. The average from all snapshots is taken as the absolute binding free energy, with individual snapshot values giving insight into how changes in binding mode during the trajectory affect the binding. While popular, these techniques have drawbacks. The hydration free energies of the protein and protein–ligand complex are large, and since the difference between them needs to be evaluated, errors or small variations in their value can greatly affect the calculated binding free energy. In addition, the entropy of binding is neglected (or included *via* an approximation[24]). Most significantly, the use of an implicit model of water means that the molecular detail of protein–water, and ligand–water interactions is lost. This is particularly important where individual water molecules may form important bridging interactions between the ligand and the protein.[25]

2.1 Waterswap

Waterswap[1] is an absolute binding free energy method that we developed to solve the cavitation and difference-of-large-values problems of double-decoupling. The method uses an explicit model of water, so includes the molecular detail of protein–water, ligand–water and protein–water–ligand interactions that are missing in continuum solvent methods. Waterswap is a comparatively new method, and has seen successful application in the prediction of mutants of influenza neuraminidase that show reduced drug binding.[26] The method has been described in detail elsewhere,[1] but in summary, it works by using a single simulation that swaps the ligand bound to the protein with an equivalent shape and volume of bulk water molecules. The effect is to push the ligand into bulk water, while filling the resulting cavity in the protein with a cluster of water molecules. This removes the cavitation problem of double-decoupling. In addition, the use of a single reaction coordinate means that the free energy is not calculated as the difference of two large numbers, and so highly accurate values are not required.

The method works by using a pair of coupled simulation boxes: (a) the protein box, containing the protein–ligand complex solvated in a periodic-boundaries box of explicit water molecules, and (b) the water box, containing just a periodic box of water. Both boxes are coupled to the same thermostat, enabling energy transfer between boxes through the heat bath in which both are theoretically placed.

Waterswap uses an identity constraint[1,27] to identify a cluster of water molecules in the water box that has approximately the same shape and size as the ligand. It works by changing how water molecules are labelled during a simulation. Rather than labelling water molecules based on their index in an input coordinate file, water molecules are instead labelled by their location in space. This is achieved by placing identity points in space. For example, let us consider a blue and a green identity point (Fig. 1(a)). These are used to label the blue and green water molecules, with the labels assigned so as to minimise the sum of the distances between the centre of the blue water molecule and the blue point, and the centre of the green water molecule and the green point (Fig. 1(b)). As the simulation progresses, and the water molecules move, this constraint is continually re-evaluated. If another water molecule moves such that, for example, it becomes closer to the green point (Fig. 1(c)), then the identities of the water molecules are swapped, and it then becomes the green water molecule (Fig. 1(d)). In this way, the identity of the cluster is maintained, as water molecules that diffuse away from the cluster are replaced by water molecules that diffuse into it. Note that the identity constraint only affects the labelling of the water molecules. The molecular coordinates are unchanged, no forces or restraints are required, and the thermodynamics of the system is unaffected.

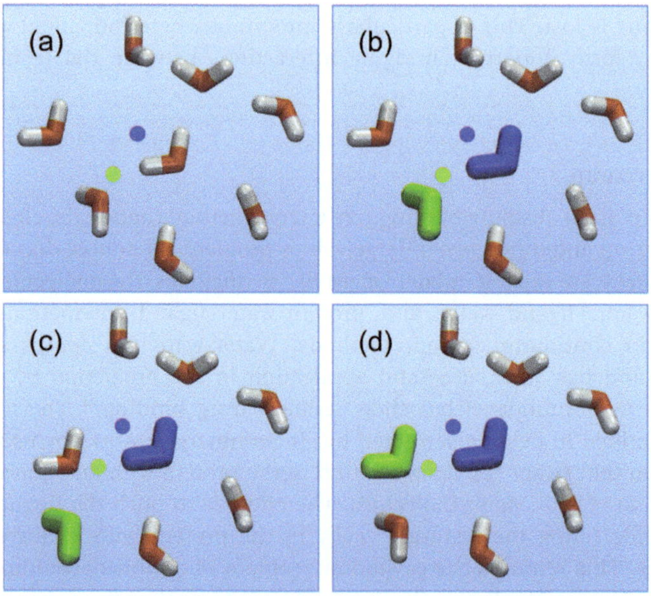

Fig. 1 Illustration of the identity constraint. Water molecules are labelled by their location in space by using identity points. A green and blue identity point (a) identify a green and blue water molecule (b). This assignment is based on minimising the sum of the distances between each identified water molecule and its corresponding point. As the water molecules diffuse during simulation (c), this constraint is updated to switch the identity to a better matching water molecule (d).

Waterswap uses the identity constraint to pick out a cluster of water molecules from the water box that are in the same position, and are approximately the same shape and volume as the ligand in the protein box. Identity points are placed onto selected atoms of the ligand, and these points are copied into the water box to identify a cluster of water molecules. With the cluster identified, the total energy of the system is evaluated using

$$\begin{aligned} E(\lambda) = {} & E_{\text{proteinbox}} + E_{\text{waterbox}} + E_{\text{ligand}} + E_{\text{cluster}} \\ & + (1-\lambda)\left(E_{\text{ligand:proteinbox}} + E_{\text{cluster:waterbox}}\right) \\ & + (\lambda)\left(E_{\text{cluster:proteinbox}} + E_{\text{ligand:waterbox}}\right) \end{aligned} \tag{1}$$

where $E_{\text{proteinbox}}$ is the energy of all molecules except the ligand in the protein box, E_{waterbox} is the energy of all molecules except the identified water cluster in the water box, E_{ligand} is the intramolecular energy of the ligand, and E_{cluster} is the intermolecular energy between all of the water molecules in the water cluster. The interaction energy between the ligand and all other atoms in the protein box, $E_{\text{ligand:proteinbox}}$, and the interaction energy between the water cluster and all other water molecules in the water box, $E_{\text{cluster:waterbox}}$, is scaled by $1 - \lambda$, where λ is an alchemical reaction coordinate. The effect is to decouple the ligand from the protein box while simultaneously decoupling the water cluster from the water box. At the same time, the ligand is coupled to the water box, and the water cluster is coupled to the protein box. This is achieved by calculating the energy between the ligand and all water molecules in the water box, $E_{\text{ligand:waterbox}}$ and the water cluster and the molecules in the protein box, $E_{\text{cluster:proteinbox}}$, and scaling this by λ. The result is that λ is a single reaction coordinate that alchemically morphs the system from $\lambda = 0$, where the ligand is bound to the protein in the protein box, to $\lambda = 1$, where the ligand has unbound and is in bulk water, and a corresponding cluster of water has been transferred to the protein box to fill the resulting cavity.

The absolute binding free energy can be calculated using thermodynamic integration.[28–30] This involves calculating the gradient of the energy with respect to λ,

$$\begin{aligned} \frac{\mathrm{d}E}{\mathrm{d}\lambda} = {} & \left(E_{\text{cluster:proteinbox}} + E_{\text{ligand:waterbox}}\right) \\ & -\left(E_{\text{ligand:proteinbox}} + E_{\text{cluster:waterbox}}\right), \end{aligned} \tag{2}$$

and then calculating the ensemble average of this at different values of λ to obtain the free energy gradient across λ,

$$\left(\frac{\mathrm{d}G}{\mathrm{d}\lambda}\right)_\lambda = \left\langle\frac{\mathrm{d}E}{\mathrm{d}\lambda}\right\rangle_\lambda. \tag{3}$$

Because the identity constraint introduces discontinuities in the energy surface (associated with identity reassignment events), the free energy gradients must be averaged using Monte Carlo sampling[31] at each value of λ. The binding free energy is then obtained by integrating the gradients across λ,

$$G_{\text{bind}} = -\int_0^1 \left(\frac{\mathrm{d}G}{\mathrm{d}\lambda}\right)_\lambda \mathrm{d}\lambda. \tag{4}$$

Note that the negative of the integral is used as the waterswap reaction coordinate models unbinding (pulling the ligand out of the protein). The quality of the result depends on the number of Monte Carlo simulations used to obtain free energy gradients across λ. We have found that sixteen Monte Carlo simulations, generating sixteen free energy gradients spaced across λ to be sufficient. The gradients can be integrated either numerically (*e.g. via* quadrature) or analytically by fitting the gradients to a polynomial expansion and integrating the resulting expression.[32]

2.2 Free energy decomposition

Waterswap calculates absolute binding free energies using a single reaction coordinate. The free energy is calculated by averaging the gradient of the total waterswap energy with respect to λ to form the free energy gradient (eqn (3)), and then integrating this over the entire waterswap reaction coordinate to obtain the absolute binding free energy (eqn (4)). An advantage of waterswap that will be explored and developed in this paper is that it is equally valid to average the gradient of components of the total waterswap energy with respect to λ. This would result in an approximation of the gradient of the corresponding free energy component. For example, eqn (5)–(8) show how the free energy change associated with only the protein box, $G_{\text{proteinbox}}$, could be obtained by integrating the gradient of the energy components that involve only interactions between the ligand and water cluster and the molecules of the protein box ($E_{\text{proteinbox}}(\lambda)$).

$$E_{\text{proteinbox}}(\lambda) = (1-\lambda)E_{\text{ligand:proteinbox}} + \lambda E_{\text{cluster:proteinbox}} \tag{5}$$

$$\frac{\mathrm{d}E_{\text{proteinbox}}}{\mathrm{d}\lambda} = E_{\text{cluster:proteinbox}} - E_{\text{ligand:proteinbox}} \tag{6}$$

$$\left(\frac{\mathrm{d}G_{\text{proteinbox}}}{\mathrm{d}\lambda}\right)_{\lambda} \approx \left\langle\frac{\mathrm{d}E_{\text{proteinbox}}}{\mathrm{d}\lambda}\right\rangle_{\lambda_{\text{WSRC}}} \tag{7}$$

$$G_{\text{proteinbox}} = -\int_0^1 \left(\frac{\mathrm{d}G_{\text{proteinbox}}}{\mathrm{d}\lambda}\right)_{\lambda} \mathrm{d}\lambda \tag{8}$$

The result, $G_{\text{proteinbox}}$, provides an estimate of the contribution to the total binding free energy resulting from only the molecules in the protein box. A similar decomposition can be used to estimate the contribution to the total binding free energy from only the molecules in the water box, G_{waterbox}.

$$E_{\text{waterbox}}(\lambda) = (1-\lambda)E_{\text{cluster:waterbox}} + \lambda E_{\text{ligand:waterbox}} \tag{9}$$

$$\frac{\mathrm{d}E_{\text{waterbox}}}{\mathrm{d}\lambda} = E_{\text{ligand:waterbox}} - E_{\text{cluster:waterbox}} \tag{10}$$

$$\left(\frac{\mathrm{d}G_{\text{waterbox}}}{\mathrm{d}\lambda}\right)_{\lambda} \approx \left\langle\frac{\mathrm{d}E_{\text{waterbox}}}{\mathrm{d}\lambda}\right\rangle_{\lambda_{\text{WSRC}}} \tag{11}$$

$$G_{\text{waterbox}} = -\int_0^1 \left(\frac{\mathrm{d}G_{\text{waterbox}}}{\mathrm{d}\lambda}\right)_{\lambda} \mathrm{d}\lambda \tag{12}$$

Note that these decompositions are approximate because the free energy is formed as an average over configurations sampled using the total waterswap energy function (as indicated by λ_{WSRC} in the equations). Because of this, the sum of $G_{proteinbox}$ and $G_{waterbox}$ will not equal G_{bind} exactly. In this work, we will explore whether or not this error is small, and will examine if the resulting decomposition provides useful information. Our expectation is that this decomposition will reveal whether a favourable binding free energy is a result of the ligand having a strong affinity for the protein ($G_{proteinbox}$ is large and favourable), or because the ligand has a low affinity for water (*i.e.* has a low solubility, with $G_{waterbox}$ being the dominant term).

2.3 Thrombin

The binding of ten ligands to thrombin was chosen as the test system for this work.[33,34] The ligands are shown in Fig. 2, and represent a congeneric series with increasing size and increasing experimentally measured binding affinity. These ligands were chosen because it is known that their binding affinity to thrombin increases in proportion to the hydrophobic contact area with the protein, suggesting that selectivity is driven by hydrophobic (van der Waals) interactions.[34] This would provide a strong difference to the binding of the water cluster, where binding should be driven by hydrophilic (hydrogen bonding) interactions.

3 Results

3.1 Waterswap binding free energies

The first step was to discover whether waterswap could correctly rank the binding affinities of the ten ligands to thrombin. Models of each of the ten thrombin–ligand complexes were built using the Amber FF99SB[35] and GAFF[36] forcefields. These were solvated in a periodic box of TIP3P[37] water, and minimised and equilibrated for 1 nanosecond (ns). Each solvated complex was subjected to a further 50 ns of molecular dynamics using the GPU-accelerated version of the

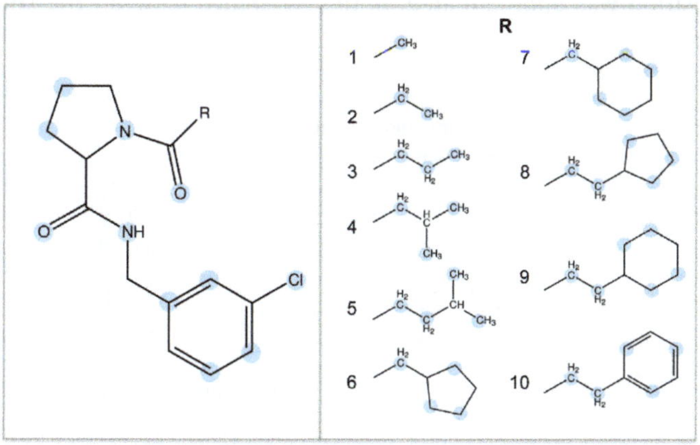

Fig. 2 The ten thrombin ligands studied in this work. They share a common core, and differ only in the R-group. The atoms used to anchor the identity points during the waterswap calculations are highlighted in blue.

PMEMD.cuda module from Amber 12.[2,38,39] Snapshots were taken every 10 ns from each of the 50 ns molecular dynamics trajectories for analysis *via* waterswap. Full technical details for all of these simulations are given in the Simulation protocols section. The result of all this work was five waterswap binding free energies per ligand, which were combined together to yield a time-averaged value. Fig. 3 shows the resulting time-averaged waterswap *versus* experimental binding free energies for each of the 10 ligands. It is satisfying that there is a strong degree of correlation between the simulation and the expected values (R^2 of 0.8). There is a large difference between the absolute values predicted by waterswap and as measured by experiment. This can in part be explained by waterswap not measuring important contributions to binding, *e.g.* the effect of protein conformational change, or the loss of translational or rotational entropy for the ligand. This is because waterswap calculates the free energy change for swapping the ligand and water cluster for the conformation and binding mode of the complex used as input to the calculation. While the ligand and protein are flexible, Monte Carlo sampling allows only limited sampling of protein backbone motion. This means that the conformational change between the bound and unbound form of the protein is unlikely to be sampled. Waterswap also does not model concentration effects, as the protein and ligand are both effectively alone in an infinitely-sized heat bath (*i.e.* they both have an effective concentration of zero).[1] Waterswap also overestimates the absolute binding affinity because of the large difference in ionic strength between the protein box (high, due to charged protein residues) and water box (zero). However, these missing contributions should all be roughly equal for the thrombin ligands. The observation of strong correlation, together with a correct prediction of the strongest and weakest binding ligand, suggests that time-averaged waterswap calculations capture the important drivers for binding in this system.

3.2 Free energy decomposition

The protein box and water box energy components were separately averaged and integrated to obtain estimates for $G_{\text{proteinbox}}$ and G_{waterbox} for each protein–ligand complex, for each snapshot from the dynamics simulations. The error in the decomposition was approximated as the difference between the sum of these

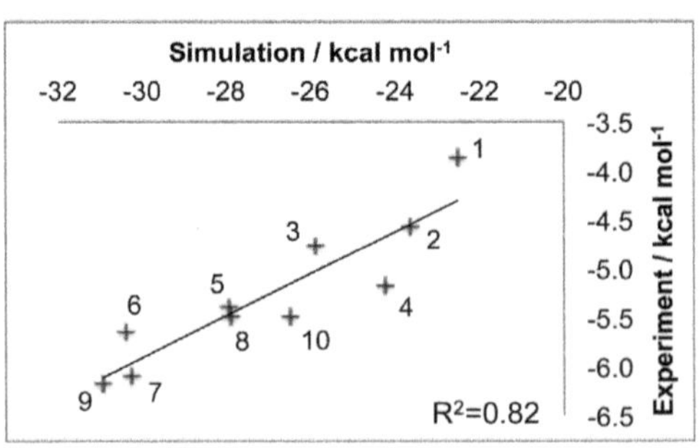

Fig. 3 Absolute binding free energies of ten ligands to thrombin as calculated using time-averaged waterswap calculations *versus* the experimental value. Points are labelled with the ligand number, and the line of best fit with corresponding R^2 correlation coefficient are shown.

components ($G_{proteinbox}$ + $G_{waterbox}$) and the total predicted binding free energy (G_{bind}) for each snapshot. The maximum error was 1.0 kcal mol^{-1} with a mean unsigned error of 0.9 kcal mol^{-1}. The low error, of less than 5% of the binding free energy, suggests that the decomposition should be meaningful. Fig. 4 compares the time averaged, decomposed free energies against the experimental and waterswap binding affinities of each ligand. It is clear from this figure that the waterswap binding free energy is dominated by the contribution from the water box. This accounts for between 60–85% of the total binding free energy. This is unsurprising as the ligands are predominantly hydrophobic, and so it would be expected that there would be a penalty for swapping the ligand into bulk water. In contrast, the protein has several charged and hydrophilic residues in the active site, so it is reassuring that there is little penalty in exchanging the hydrophobic ligand with (obviously) hydrophilic water. What is interesting is the correlation between the water box and protein box components with the experimental binding affinity. This is shown in Fig. 5. It was hoped that there would be a correlation for $G_{proteinbox}$, indicating that selectivity of the ligands was driven by specific interactions with the protein. Instead, little to no correlation is seen, and instead strong correlation is found between $G_{waterbox}$ and experiment ($R^2 > 0.8$). This suggests that increases in binding affinity of the ligands are not driven by increased affinity with the protein, but are instead driven by decreased affinity with bulk water, *i.e.* lower solubility.

3.3 Residue-based decomposition

The method used to decompose G_{bind} to $G_{proteinbox}$ can be used to decompose the waterswap binding free energy into even smaller components. One possibility is to decompose the free energy into per-residue components ($G_{residue}$). To do this, the interaction energy between each residue and the ligand ($E_{ligand:residue}$) and each residue in the water cluster ($E_{cluster:residue}$) can be averaged and integrated according to

$$E_{residue}(\lambda) = (1 - \lambda)E_{ligand:residue} + \lambda E_{cluster:residue} \tag{13}$$

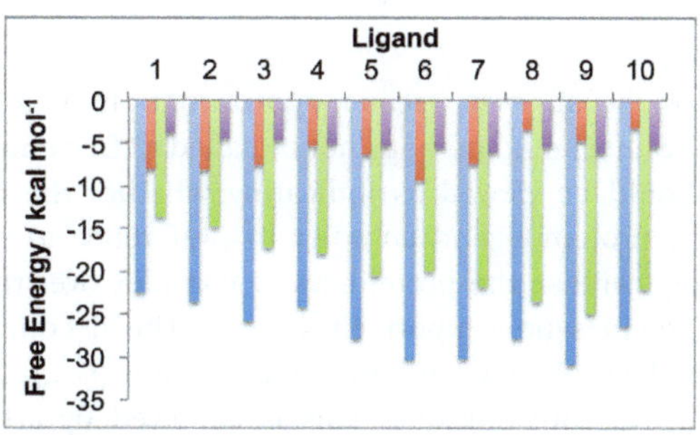

Fig. 4 Comparison of the time-averaged waterswap (ΔG_{bind}, blue) binding free energy of the ten ligands to thrombin against the protein box ($\Delta G_{proteinbox}$, red) and water box ($\Delta G_{waterbox}$, green) free energy components. The experimental binding affinities are shown in purple.

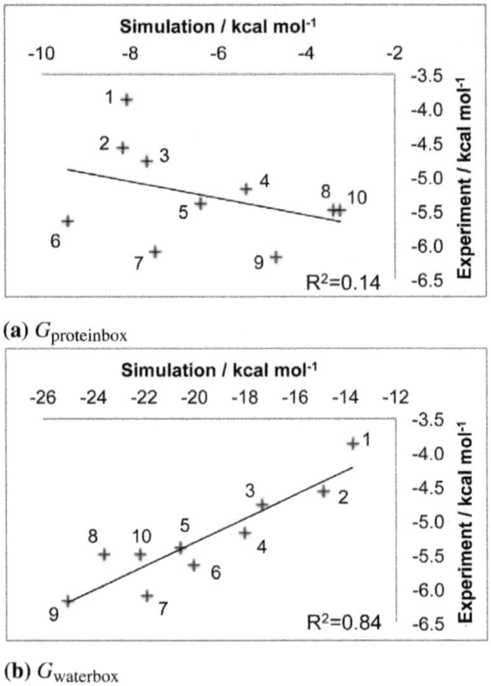

Fig. 5 (a) Protein box and (b) water box contributions to the binding free energy of ten ligands to thrombin as calculated using a decomposition of the waterswap free energy *versus* the experimental binding affinity. Points are labelled with the ligand number, and the line of best fit and corresponding R^2 correlation coefficients are shown.

$$\frac{\mathrm{d}E_{\text{residue}}}{\mathrm{d}\lambda} = E_{\text{cluster:residue}} - E_{\text{ligand:residue}} \tag{14}$$

$$\left(\frac{\mathrm{d}G_{\text{residue}}}{\mathrm{d}\lambda}\right)_{\lambda} \approx \left\langle\frac{\mathrm{d}E_{\text{residue}}}{\mathrm{d}\lambda}\right\rangle_{\lambda_{\text{WSRC}}} \tag{15}$$

$$G_{\text{residue}} = -\int_0^1 \left(\frac{\mathrm{d}G_{\text{residue}}}{\mathrm{d}\lambda}\right)_{\lambda} \mathrm{d}\lambda. \tag{16}$$

This is likely to be an even greater approximation than for $G_{\text{proteinbox}}$ or G_{waterbox}, and the result should perhaps not be called a "free energy". Instead, this decomposition should be viewed as indicating whether the residue either (a) helps stabilise the protein–ligand complex in preference to the protein–water complex, (b) helps stabilise the protein–water complex over the protein–ligand complex, or (c) provides no support either way. The decomposition for each residue can in turn be broken down into its electrostatic ($G_{\text{residue[coulomb]}}$) and van der Waals ($G_{\text{residue[LJ]}}$) contributions. This is achieved by averaging and integrating only the coulombic or Lennard Jones (LJ) parts of $E_{\text{ligand:residue}}$ and $E_{\text{cluster:residue}}$, respectively. The values for all residues within 15 Å of each ligand were calculated and time-averaged. As the sum of $G_{\text{residue[coulomb]}}$ and $G_{\text{residue[LJ]}}$

should equal $G_{residue}$, the error on the approximation was estimated by examining the difference. Across all snapshots of all ligands and all residues, the maximum error was 0.016 kcal mol^{-1} with a mean unsigned error of 0.0008 kcal mol^{-1}.

The time-averaged values of the total ($G_{residue}$) components for the twelve residues that show the strongest preferences for the any of the ten ligands or their corresponding swapped water clusters are shown in Fig. 6. This shows that charged residues (Lys60, Arg173 and Asp189) stabilise the water cluster (indicated by positive values), while the ligand is stabilised by its large hydrophobic contact with the neighbouring residues Ser214, Trp215 and Gly216. Ile174 sits at the bottom of the binding site, and can only come into direct contact with the large ligands. This is seen in its increasing contribution to the binding free energy with ligand size, although it is worth noting that this contribution alternates between supporting the water cluster and supporting the ligand.

To gain further insight, Fig. 7 shows the further decomposition of these values into their corresponding electrostatic and van der Waals terms. This breakdown reveals that the protein residues that are in direct contact with the ligand (His57, Leu99, Ser195, Ser214, Trp215 and Gly216) show a strong van der Waals preference for the ligand. For residues that have partial contact (Val213 and Ile174 for larger ligands), there is some van der Waals stabilisation, but it is weaker and occasionally in favour of the water cluster. This is unsurprising as these ligands are hydrophobic, with binding affinities that increase with increasing hydrophobic contact area with the protein.[34] This is a good result, showing that this decomposition is revealing what is known experimentally about the ligands. What is more interesting is that this decomposition shows that the increased van der Waals stabilisation carries an electrostatic penalty. The contact residues almost all show strong electrostatic stabilisation in favour of the swapped water cluster.

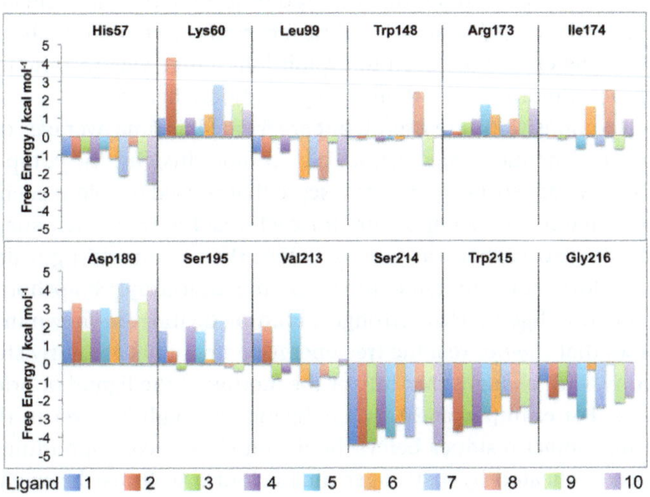

Fig. 6 Time-averaged decomposition of the waterswap binding free energy into selected per-residue components. The values for each of the ten ligands are shown, with positive values indicating that the residue stabilises the complex with the water cluster, while negative values indicate stabilisation of the protein–ligand complex. Residues are numbered to match their identification in the PDB (2ZC9[33]).

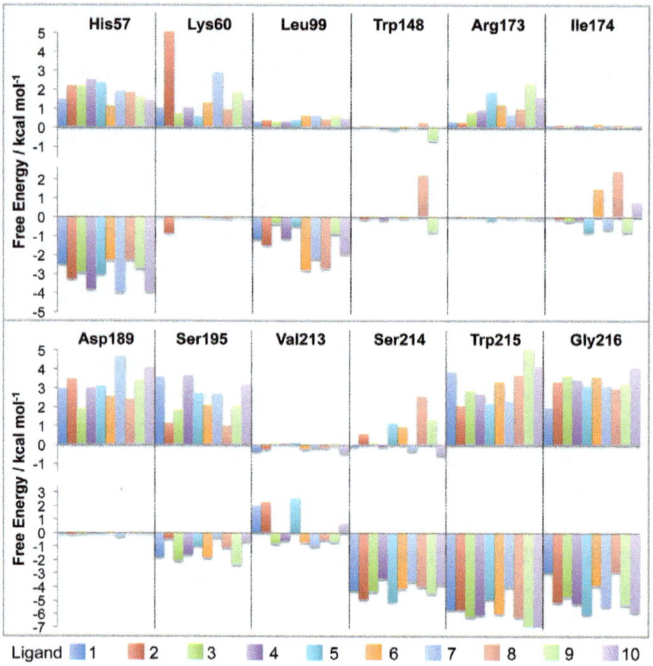

Fig. 7 Time-averaged decomposition of the waterswap binding free energy into selected per-residue electrostatic (top) and van der Waals (bottom) components for each of the ten thrombin ligands. The twelve residues selected are the same as shown in Fig. 6.

Examination of the structures shows that this is because the water cluster is able to form hydrogen bonds with the backbone of many of these residues. In addition, the decomposition shows that the stabilisation of the water cluster from the non-contact and charged residues is almost entirely electrostatic. The polar water molecules in the cluster are seen to exploit long range electrostatic interactions with the protein more readily than the hydrophobic ligands.

These decompositions are useful, but are formed as time averages over each of the molecular dynamics trajectories. To gain more insight, the components can be visualised by converting them into a score that is used to colour each residue in a molecular viewer. The components for each residue from each snapshot were converted into a score that was used to colour the corresponding residue in that snapshot to form an annotated movie of the trajectory. Snapshots from the resulting movie of ligand 9 (the strongest binding ligand) are shown in Fig. 8. The movie shows that the per-residue free energy components are fairly stable across the trajectory, with changes that mirror the motion of the ligand or orientation of the residues. For example, at 20 ns, the ligand sits slightly lower in the binding site, and the contact residues below the ligand show a correspondingly stronger stabilisation (indicated by a stronger blue colour). At 30 ns the *m*-chlorobenzyl substituent on the right of the ligand rotates up, increasing its contact and stabilisation with Trp148 above. As the ligand moves back down at 50 ns, this stabilisation is weakened. The motion of the residues also appears to be important, with Lys60 showing greater stabilisation for the water cluster (brighter red colour)

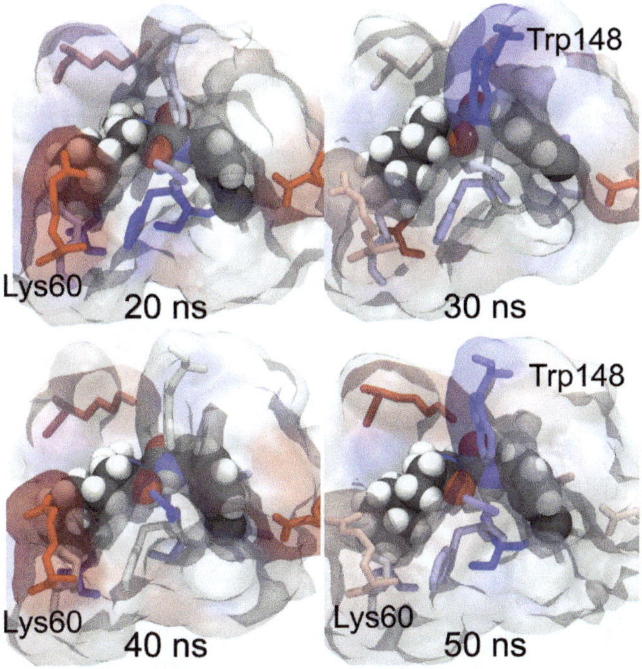

Fig. 8 Snapshots taken at selected times from the trajectory of ligand 9 complexed with thrombin. The residues are coloured by their component of the waterswap binding free energy calculated for that timestep. Red values indicate that the residue stabilises the water cluster, while blue values indicate that the residue stabilises the protein–ligand complex. Residues discussed in the text are highlighted.

in snapshots when it points towards the ligand than when it points away. Fig. 6 shows that Lys60 can provide a lot of electrostatic stabilisation to the water cluster, revealing the potential to design a better ligand by adding on hydrophilic substituents that target hydrogen bonding with that residue. This visualisation provides new insight by providing a direct link between changes in structure and their impact on binding affinity.

Movies can be made looking at the total stabilisation, or at the electrostatic or van der Waals components. Of equal interest is the conformations adopted by the ligand and water cluster during the Monte Carlo trajectories used to calculate the waterswap free energy. Fig. 9 shows the annotated structure at the end of the Monte Carlo trajectories at $\lambda = 0.005$ (ligand) and $\lambda = 0.995$ (water cluster) for the 20 ns snapshot for ligand 9. The residues are coloured by their total, electrostatic and van der Waals free energy components. This shows how the swapped water cluster packs into the volume originally occupied by the ligand. Features such as the cluster-equivalent of the cyclohexyl and *m*-chlorobenzyl rings are clearly apparent. The figure shows that the water cluster is able to form hydrogen bonds with the oxygens of the contact residues along the backbone of the protein. This provides the cluster with strong electrostatic stabilisation (indicated by red) compared to the ligand. In compensation, the ligand receives strong van der Waals stabilisation (indicated by blue) compared to the cluster from those

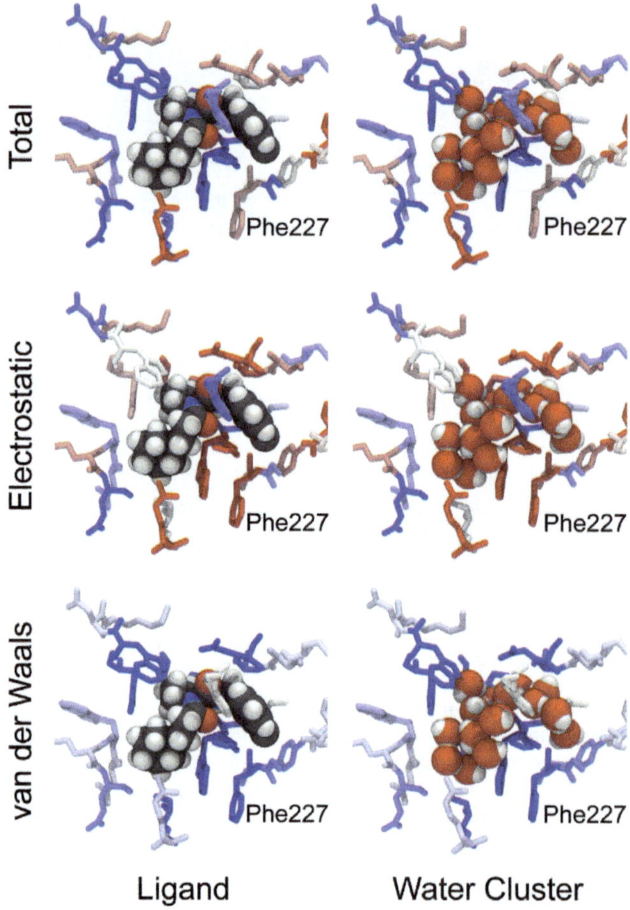

Fig. 9 Final configurations at $\lambda = 0.005$ (ligand, left) and $\lambda = 0.995$ (swapped water cluster, right) from the Monte Carlo simulations used to calculate the waterswap free energy for the 20 ns snapshot of ligand 9 bound to thrombin. The coordinates of the ligand and swapped water cluster are shown. The residues are coloured by their total (top), electrostatic (middle) and van der Waals (bottom) components of the waterswap free energy, with red values indicating stabilisation of the water cluster, and blue values indicating stabilisation of the ligand. Phe227, discussed in the text, is highlighted.

residues. This decomposition reflects the experimental observation that the ligands bind *via* hydrophobic interactions. The decomposition also shows that a path to strong-binding ligands may be to add hydrogen bond donor groups to pick up the same hydrogen bonding interactions that are observed to stabilise the swapped water cluster. In particular, it may be worth replacing the chlorine on the *m*-chlorobenzyl group with a hydroxyl or other group that can pick up the hydrogen bond to the oxygen on the backbone of Phe227. In addition to providing insight into how changes in structure affect the electrostatic and van der Waals components of the binding affinity, this visualisation also reveals the molecular detail of how water solvates the binding site. To bind, the ligand must displace the water molecules of the swapped water cluster, and make interactions with the protein that are stronger than those made by the water. The visualisation in Fig. 9 reveals the detail of the interactions between the protein and water. This should provide inspiration for the design of ligands that can replicate these interactions, thereby leading to the development of compounds with improved binding affinity.

3.4 Water-based decomposition

In addition to decomposing the waterswap binding free energy per-residue, it is also possible to decompose it to per-water molecule components. To do this, the identity of water molecules in the binding site has to be constrained. This is to ensure that the free energy gradient is averaged over energy components calculated using a consistently identified water molecule, and so that free energy gradients calculated across λ can be correctly matched up and integrated. The identity constraint was used to hold onto the identity of each of the water molecules in the binding site around the ligand. Identity points were placed on all of the oxygen atoms of water molecules in the protein box within 5 Å of the ligand. These points were fixed in space, with a water molecule identified by each point. As each Monte Carlo trajectory across λ started from the same configuration, it was possible to identify a single water molecule across λ and to match and integrate its corresponding free energy gradients. Gradients of the component of the waterswap free energy for each water molecule were calculated in a similar manner to those for each residue. These components were themselves decomposed into electrostatic and van der Waals terms. These were all collected and integrated for each snapshot for each ligand to obtain G_{water}, $G_{\text{water[coulomb]}}$ and $G_{\text{water[LJ]}}$ for each identified water molecule. Fig. 10 shows the annotated structure at the end of the Monte Carlo trajectories at $\lambda = 0.005$ (ligand) and $\lambda = 0.995$ (water cluster) for the same 20 ns snapshot for ligand 9 as seen in Fig. 9. The residues and water molecules are coloured by their total, electrostatic and van der Waals free energy components. This shows that water molecules near the cyclohexyl substituent on the solvent-exposed face of the ligand provide net stabilisation to the water cluster. This is to be expected, as the cyclohexyl substituent is hydrophobic and will likely disrupt the bulk water that is in contact with the surface of the protein–ligand complex. This suggests that one way to improve the binding affinity of the ligand would be to add hydrophilic groups that point out towards the surface of the complex. These would stabilise the water molecules, or at least reduce the preferential stabilisation provided to the water cluster. More interestingly, this figure also reveals a bridging water molecule that aids in the electrostatic stabilisation of the water cluster by residue Asp189. When the ligand is bound, this water molecule is trapped in the binding pocket in contact with the *m*-chlorobenzyl group. Here it provides van der Waals stabilisation to the ligand. However, when the water cluster is bound, the water molecule forms a hydrogen bonding bridge between the water cluster and Asp189. This allows both the water molecule and Asp189 to provide electrostatic stabilisation to the water cluster. The electrostatic stabilisation from this water molecule to the cluster is cancelled out by the van der Waals stabilisation provided to the ligand, so the net stabilisation is near zero. Despite this, the role of this water molecule in bridging to Asp189 is clear. A similar pattern is seen with another water molecule that bridges from the water cluster to Glu192. This helps both it and Glu192 to electrostatically

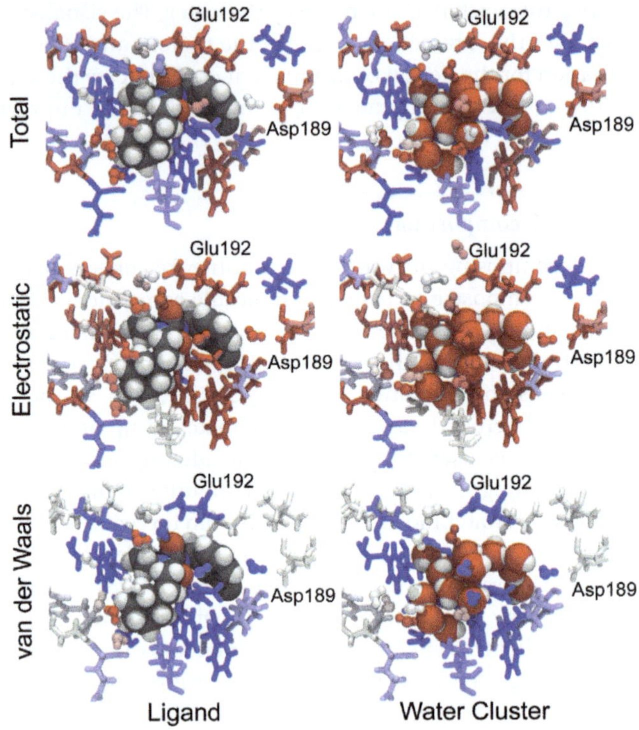

Fig. 10 Final configurations at $\lambda = 0.005$ (ligand, left) and $\lambda = 0.995$ (swapped water cluster, right) from the Monte Carlo simulations used to calculate the waterswap free energy for the 20 ns snapshot of ligand 9 bound to thrombin. The coordinates of the ligand and swapped water cluster are shown. The residues and selected binding site water molecules are coloured by their total (top), electrostatic (middle) and van der Waals (bottom) components of the waterswap free energy, with red values indicating stabilisation of the water cluster, and blue values indicating stabilisation of the ligand. Residues discussed in the text are highlighted.

stabilise the cluster over the ligand. This suggests that ligands that are designed to displace these water molecules and that make direct hydrogen bond or salt bridge interactions with Asp189 or Glu192 may have stronger binding affinities. Fortunately, the original experimental study of these ligands included a series where the *m*-chlorobenzyl group was replaced with benzamidine. Benzamidine analogs of all ten ligands were synthesised and their binding to thrombin measured.[34] It was seen that these analogs did indeed displace water and form a direct interaction with Asp189, with each having a binding affinity that is about 2 kcal mol^{-1} stronger than that for the corresponding *m*-chlorobenzyl ligand (see ESI†).

These visualisations show directly how the ligand disrupts solvation of the protein. The swapped water cluster is seen to provide a seamless hydrogen-bonding network between all of the water molecules in the active site. In contrast, the ligand breaks this network, with lone water molecules trapped in what are now revealed to be electrostatically unfavourable conformations between the protein and the ligand. This new way of visualising ligand binding, as a competition between the ligand and water, should ensure that the impact of the ligand on solvation in the active site is readily visible and comprehendible throughout the drug design process. This should aid drug designers in their quest for better-binding ligands that do not unwittingly destabilise active site water molecules. In addition, these visualisations should help when rationalising the structure–activity relationships (SARs) of ligands where changes in structure cause differences in solvation. One such example of this is the different specificity of ligands binding to Itk, Src, cKit and Flt-3 kinases.[7,8] In this case, the different SARs to different kinases were rationalised by using WATERMAP[40] calculations to work out how water would be distributed in the active site of the apoproteins, and then consider how different ligands would displace that water on binding. We believe that waterswap decomposition and visualisation will provide a deeper and more detailed understanding of this system. We are now performing these calculations, the results of which we plan to discuss in a future paper.

4 Discussion

Decompositions of the waterswap binding free energy have been shown to provide useful insight and meaningful annotation of trajectories generated by molecular dynamics simulations of protein–ligand complexes. Performing a waterswap calculation is relatively straightforward. A waterswap executable is available as part of Sire[41] and performs the calculations directly from the Amber-format topology and restart files that are produced by PMEMD.cuda during a molecular dynamics simulation. The only user intervention that is suggested is to specify the atoms of the ligand on which the identity points used to identify the swapped water cluster are placed. We have developed an algorithm to do this automatically, but it is not yet sufficiently robust to rely on blindly. The calculations are relatively quick, taking four minutes per iteration (50 thousand Monte Carlo moves per λ value) on a 2 × 8 core Intel Xeon E5-2670 compute node. We performed these calculations using 1000 iterations (50 million Monte Carlo moves per λ value), which took nearly 3 days. This is comparable to the 4 days needed to generate each 50 ns molecular dynamics trajectory (accelerated using nVidia Tesla M2090 GPUs). While the calculations were performed in series, it is not hard to imagine that a compute cluster could run heterogenous jobs involving waterswap calculations running on the CPUs and dynamics calculations on the GPUs. This would allow analysis to be performed on snapshots from a dynamics simulation as it was being generated. In addition, for this work we over-specified the number of iterations so that we could fully converge the free energy averages. Acceptable binding free energies, albeit with larger errors, could be generated with 400–600 iterations. Because the phase space is smaller, the residue- and water-free energy components converge extremely quickly, with initial estimates of their value available within the first 2–5 iterations, and acceptable values available after about 200 iterations (see the ESI†). This means that the colour-coding of the residues and water could be performed in near-real time (10–20 min). Waterswap is currently implemented in a prototyping code, and optimisation is on-going to create a custom application. Our goal is for this optimisation to accelerate the code by at least a factor of 10, which would allow estimates of components in 1–2 min, with converged free energies available within 5 h. This would allow near-interactive applications, *e.g.* allowing a drug designer to quickly see the effect on residue and water stabilisation of modifications made to the ligand in an interactive viewer. Non-interactive jobs could then be submitted to refine the components and binding free energies of promising designs.

One major challenge posed by these decompositions is that they significantly increase the amount of information that accompanies each molecular dynamics trajectory. Annotation adds several extra dimensions to the data. In addition to viewing a trajectory by "time", waterswap adds in a λ dimension for swapping in the water cluster, a Monte Carlo "timestep" dimension for each simulation at each value of λ, and a "type" dimension that allows either total, electrostatic or van der Waals components to be visualised. Analysis and easy visualisation of all of this data is a significant challenge. We have written and rely on custom Python and Tcl scripts in VMD[42] to manage this data deluge, but this is not ideal and not intuitive for non-experts. We feel that there is an urgent need for molecular visualisation tools that can readily present complex, multidimensional data in a manner that would be easy to navigate and manipulate by drug designers.

5 Conclusions

The utility of free energy decompositions of the waterswap binding free energy for annotating and visualising molecular dynamics trajectories of thrombin–ligand complexes has been demonstrated. These visualisations provide detailed molecular-level information that should be useful to drug designers aiming to create better-binding ligands. While we tested these decompositions on thrombin, we believe that they should be just as useful when applied to other protein–ligand systems. The calculations are robust, computationally inexpensive, almost fully automated, and implemented in software[41] that is available for download under an open source license. These results show that waterswap, and its decomposition into free energy components, provides a promising new direction in the search for better tools for analysing and visualising the molecular driving forces in simulations of protein–ligand complexes.

6 Simulation protocols

A crystallographic structure of human thrombin in a complex with a thrombin ligand structurally related to the ligands simulated in this work was downloaded from the PDB databank (PDB code 2ZC9[33]). All subsequent structure modifications were done with the software Maestro.[43] The hirugen chain was removed from the structure. The side-chain of Arg75 in chain H was completed in a solvent exposed conformation. The incomplete light chain was capped before Glu1C with an ACE residue and after Ile14L with an NME residue. The incomplete heavy chain was capped after Gly246 with an NME residue. Missing residues Trp148, Thr149, Ala149A, Asn149B, Val149C, Gly149D, Lys149E in chain H were modelled in the structure using the FALC-Loop web server.[44] Standard protonation states were assumed for protein side-chains. On the basis of visual inspection of hydrogen bonding patterns, His57 and His71 were modelled in their uncharged, δ-tautomer. His91, His119 and His230 were modelled in the δ-tautomer. Disulfide bridges were modelled between Cys42–Cys58, Cys1–Cys122, Cys168–Cys182 and Cys191–Cys220. Models of the ligands were prepared by editing the structure of the ligand present in the crystal structure. Substituents were positioned in the S3 pocket in plausible starting conformations, on the basis of the available experimental data.[33,34] The protein was modelled using the Amber FF99SB forcefield,[35] with GAFF[36] used for the ligands. Ligand parameters and AM1BCC partial charges were generated using Antechamber.[45] Each protein–ligand complex was solvated in a periodic box of TIP3P[37] water molecules using the LEaP module of Amber, with the water molecules added to create a buffer of at least 10 Å around the protein. The resulting box had dimensions of approximately $70 \times 71 \times 74$ Å^3. Seven chloride ions were added using LEaP to neutralise the system. This provided a starting point for minimisation, equilibration and production using the GPU-accelerated PMEMD.cuda module of Amber 12 (version 12.1, bugfix 9, released August 2012).[2,38,39] The mixed single-precision/fixed precision (SPFP) version of pmemd.CUDA was employed. The Particle Mesh Ewald (PME) method was used to account for long-range electrostatics, with a 10 Å real space cutoff, with the same cutoff used for the Lennard Jones potential. SHAKE was used to constrain bonds involving hydrogen. Each complex was subjected to two cycles of 3000 steps of minimisation, after which molecular dynamics (MD) simulations were carried out, also using pmemd.CUDA. A timestep of 2 fs was used, with the simulation divided into thermalisation, equilibration and production phases. In the thermalisation step, the temperature was increased linearly to 300 K over a period of 100 ps using canonical (NVT) dynamics. A Langevin thermostat was used to maintain temperature, using a collision frequency of 5 ps^{-1}. Next, the system was equilibrated for a further 100 ps of NVT dynamics, then a further 800 ps of isothermal–isobaric (NPT) dynamics, with a Langevin thermostat used to maintain a temperature of 300 K, and the isotropic pressure scaling algorithm implemented in pmemd.CUDA used to maintain the pressure at 1 bar, using a pressure relaxation time of 1 ps. Visual inspection of each system together with monitoring of RMSD confirmed that the starting structures closely matched the original crystal structures. Production dynamics was performed for 50 ns at 300 K and 1 bar. Thermalisation, equilibration and production was carried out using a single M2090 nVidia Tesla GPU per protein–ligand complex, with production dynamics of each 33 thousand atom system running at a rate of approximately 14 ns of sampling per day. Waterswap calculations were performed using the waterswap executable in Sire (devel branch version 2169, obtainable *via* Google Code,[46] and released as Sire 2014.1),[41] with calculations using the same force field parameters and solvent model as the dynamics simulations. Identity points were placed manually onto the ligands, anchored to the atoms highlighted in Fig. 2. These points were set to move as the ligand moved, thereby ensuring maximum overlap between the ligand and swapped water cluster. The points were spread evenly over the ligand, with care taken to ensure that the identified water clusters had approximately the same shape and size as the ligands. The waterswap binding free energy was calculated using thermodynamic integration[28–30] over 16 λ windows (values 0.005, 0.071, 0.137, 0.203, 0.269, 0.335, 0.401, 0.467, 0.533, 0.599, 0.665, 0.731, 0.797, 0.863, 0.929 and 0.995). The free energy gradients were fitted to a polynomial of order 14 and integrated analytically.[32] 50 million Monte Carlo (MC) moves were performed for each window, with free energy averages collected over the last 30 million steps. The location and orientation of the ligand was fixed, but all internal degrees of freedom were sampled using Z-matrices that were automatically generated by Sire.[41] Non-bonded interactions were subjected to a 15 Å coulomb and Lennard Jones cutoff. PME is too expensive to use with Monte

Carlo, so instead a shifted-force[47] electrostatic cutoff was used. This has been shown to show give good agreement with PME.[47] The free energy components of all residues within 15 Å of the ligand, and for all water molecules within 5 Å were evaluated, with interactions evaluated and averaged every 1000 MC steps. To aid convergence of the free energy averages, a soft-core potential was used to evaluate the interaction energies between the ligand, water cluster and the protein and water boxes, using the "Set A" parameters already developed for use with waterswap (shift delta of 1.2, water softening parameter of 1.1 and coulomb power of 0).[1] This meant that the interaction energies involving the ligand or swapped water cluster had a dependency on λ, and so the simple differentiation used to obtain the free energy gradients in eqn (3), (7), (11) and (15) was not possible. Instead, a finite-difference approximation was used.[48,49] The free energy gradient, $\left(\frac{\mathrm{d}G}{\mathrm{d}\lambda}\right)_\lambda$, was approximated using the Zwanzig equation[50] to calculate the free energy difference between λ and $\lambda + \Delta\lambda$,

$$\left(\frac{\mathrm{d}G}{\mathrm{d}\lambda}\right)_\lambda \approx \frac{\Delta G_\lambda}{\Delta\lambda} \tag{17}$$

$$\Delta G_\lambda = G_{\lambda+\Delta\lambda} - G_\lambda \tag{18}$$

$$= -kT\ln\left\langle \exp\left(-\frac{E(\lambda+\Delta\lambda)-E(\lambda)}{kT}\right)\right\rangle_\lambda \tag{19}$$

This finite difference approximation was used to calculate all of the free energy gradients, using a value of $\Delta\lambda$ of 0.001. To check the approximation, both forwards gradients (using $E(\lambda + \Delta\lambda) - E(\lambda)$) and backwards gradients (using $E(\lambda) - E(\lambda - \Delta\lambda)$) were evaluated and compared.

To improve the efficiency of the calculation, all residues and all water molecules wholly beyond 15 Å of the ligand were fixed. All residues and water molecules within 15 Å were free to move. Each mobile residue had a full range of motion, with sampling of both the intra-residue degrees of freedom, and a backbone-anchored translation and rotation move that enabled limited backbone sampling. Each mobile water molecule was kept rigid, with sampling only of its translational and rotational degrees of freedom. To prevent water molecules from moving beyond 15 Å of the ligand, a reflection sphere boundary condition was employed. This condition ensured that any Monte Carlo moves that would take the centre of mass of the water molecule out of a sphere of radius 15 Å centred on the ligand, were reflected back into the sphere. This boundary condition was developed and implemented in Sire,[41] and has the advantage that no restraints are needed to keep the mobile water molecules inside the sphere. In addition, as all fixed water molecules and protein residues outside the reflection sphere were still included in the energy calculation, it meant that there were no artefacts associated with mobile water molecules encountering a vacuum boundary (as is the case in "droplet" boundary conditions, whereby fixed atoms are deleted and mobile atoms kept in the sphere through use of a half-harmonic boundary potential). A further benefit of the reflection sphere was that as the molecules/residues outside the sphere were fixed, evaluating their contribution to the long range non-bonded interactions could be optimised, *e.g.* by using pre-allocated arrays and pre-calculated values. While not used for this work, it would also be possible to use an electrostatic grid to pre-compute the electrostatic interactions between the fixed and mobile region, thereby further improving the efficiency of the calculation.

In total, this work generated 510 ns of dynamics trajectories (requiring access to 10 nVidia Tesla M2040 GPUs for four days, and generating 14 GB of data) and required 8 billion Monte Carlo moves (requiring access to 10 16-core Intel Xeon E5-2670 compute nodes for 3 days, and generating 57 GB of data). We are grateful to e-Infrastructure South for providing access to the Emerald GPU cluster to run the dynamics simulations, and to the ACRC at Bristol for access to the newly-opened bluecrystal3 cluster for the waterswap calculations. We are also grateful to the Bristol Research Data Storage Facility for providing backed-up disk space to hold the data for this work.

Acknowledgements

The work was carried out with the kind support and the computational facilities of e-Infrastructure South (http://einfrastructuresouth.ac.uk) and the University of Bristol Advanced Computing Research Centre (http://www.bris.ac.uk/acrc). We thank the EPSRC (EP/I030395/1) for funding the development of Sire and the development of the fundamental research underpinning waterswap and its decomposition, and the BBSRC (BB/K016601/1) for funding this work. AJM is an EPSRC leadership fellow (EP/G007705/1).

References

1 C. J. Woods, M. Malaisree, S. Hannongbua and A. J. Mulholland, *J. Chem. Phys.*, 2011, **134**, 054114.

2 A. W. Goetz, M. J. Williamson, D. Xu, D. Poole, S. L. Grand and R. C. Walker, *J. Chem. Theory Comput.*, 2012, **8**, 1542–1555.

3 R. O. Dror, R. M. Dirks, J. P. Grossman, H. Xu and D. E. Shaw, *Annu. Rev. Biophys.*, 2012, **41**, 429–452.

4 D. E. Shaw, R. O. Dror, J. K. Salmon, J. P. Grossman, K. M. Mackenzie, J. A. Bank, C. Young, M. M. Deneroff, B. Batson, K. J. Bowers, E. Chow, M. P. Eastwood, D. J. Ierardi, J. L. Klepeis, J. S. Kuskin, R. H. Larson, K. L. Larsen, P. Maragakis, M. A. Moraes, S. Piana, Y. Shan and B. Towles, *Proceedings of the Conference on High Performance Computing Networking, Storage and Analysis*, New York, NY, USA, 2009, pp. 1–11.

5 Y. Wang and E. Tajkhorshid, *Proc. Natl. Acad. Sci. U. S. A.*, 2008, **105**, 9598–9603.

6 C. J. Woods, M. Malaisree, B. Long, S. McIntosh-Smith and A. J. Mulholland, *Biochemistry*, 2013, **52**, 8150–8164.

7 J.-D. Charrier, A. Miller, D. P. Kay, G. Brenchley, H. C. Twin, P. N. Collier, S. Ramaya, S. B. Keily, S. J. Durrant, R. M. Knegtel, A. J. Tanner, K. Brown, A. P. Curnock and J.-M. Jimenez, *J. Med. Chem.*, 2011, **54**, 2341–2350.

8 K. R. M. A. and D. Robinson, *Mol. Inf.*, 2011, **30**, 950–959.

9 B. L. Tembe and J. A. McCammon, *Comput. Chem.*, 1984, **8**, 281–283.

10 J. Michel, N. Foloppe and J. W. Essex, *Mol. Inf.*, 2010, **29**, 570–578.

11 J. Michel and J. W. Essex, *J. Comput.-Aided Mol. Des.*, 2010, **24**, 639–658.

12 D. Hamelburg and J. A. McCammon, *J. Am. Chem. Soc.*, 2004, **126**, 7683–7689.

13 J. Hermans and L. Wang, *J. Am. Chem. Soc.*, 1997, **119**, 2707–2714.

14 C. Barillari, J. Taylor, R. Viner and J. W. Essex, *J. Am. Chem. Soc.*, 2007, **129**, 2577–2587.

15 W. L. Jorgensen, J. K. Buckner, S. Boudon and J. Tirado-Rives, *J. Chem. Phys.*, 1988, **89**, 3742–3764.

16 S. B. Singh and P. A. Kollman, *J. Am. Chem. Soc.*, 1999, **121**, 3267–3271.

17 M. A. L. Eriksson, J. Pitera and P. A. Kollman, *J. Med. Chem.*, 1999, **42**, 868–881.

18 B. Honig and A. Nicholls, *Science*, 1995, **268**, 1144–1149.

19 F. Godschalk, S. Genheden, P. Soderhjelma and U. Ryde, *Phys. Chem. Chem. Phys.*, 2013, **15**, 7731–7739.

20 J. Michel, M. L. Verdonk and J. W. Essex, *J. Med. Chem.*, 2006, **49**, 7427–7439.

21 I. Massova and P. A. Kollman, *Perspect. Drug Discovery Des.*, 2000, **18**, 113–135.

22 I. Massova and P. A. Kollman, *J. Am. Chem. Soc.*, 1999, **121**, 8133–8143.

23 L. T. Chong, Y. Duan, L. Wang, I. Massova and P. A. Kollman, *Proc. Natl. Acad. Sci. U. S. A.*, 1999, **96**, 14330–14335.

24 S. Genheden, O. Kuhn, P. Mikulskis, D. Hoffmann and U. Ryde, *J. Chem. Inf. Model.*, 2012, **52**, 2079–2088.

25 Z. Li and T. Lazaridis, *Phys. Chem. Chem. Phys.*, 2007, **9**, 573–581.

26 C. J. Woods, M. Malaisree, B. Long, S. McIntosh-Smith and A. J. Mulholland, *Sci. Reports*, 2013, **3**, 3561.

27 M. D. Tyka, R. B. Sessions and A. R. Clarke, *J. Phys. Chem. B*, 2007, **111**, 9571–9580.

28 D. A. Pearlman and P. S. Charifson, *J. Med. Chem.*, 2001, **44**, 3417–3423.

29 B. C. Oostenbrink, J. W. Pitera, M. M. H. Lipzig, J. H. N. Meerman and W. F. van Gunsteren, *J. Med. Chem.*, 2000, **43**, 4594–4605.

30 X. Barril, M. Orozco and F. J. Luque, *J. Med. Chem.*, 1999, **42**, 5110–5119.

31 N. Metropolis, A. W. Rosenbluth, M. N. Rosenbluth, A. H. Teller and E. Teller, *J. Chem. Phys.*, 1953, **21**, 1087–1092.

32 F. M. Y. C. Shyu, *J. Comp. Chem.*, 2009, **30**, 2297–2304.

33 B. Baum, L. Muley, A. Heine, S. M., D. Hangauer and G. Klebe, *J. Mol. Biol.*, 2009, **391**, 552–564.

34 L. Muley, B. Baum, M. Smolinski, M. Freindorf, A. Heine, G. Klebe and D. G. Hangauer, *J. Med. Chem.*, 2010, **53**, 2126–2135.

35 V. Hornak, R. Abel, A. Okur, B. Strockbine, A. Roitberg and C. Simmerling, *Proteins: Struct., Funct., Bioinf.*, 2006, **65**, 712–725.

36 J. Wang, R. M. Wolf, J. W. Caldwell, P. A. Kollman and D. A. Case, *J. Comput. Chem.*, 2004, **25**, 1157–1174.

37 W. L. Jorgensen, J. Chandrasekhar, J. D. Madura, R. W. Impey and M. L. Klein, *J. Chem. Phys.*, 1983, **79**, 926–935.

38 R. Salomon-Ferrer, A. W. Goetz, D. Poole, S. L. Grand and R. C. Walker, *J. Chem. Theory Comput.*, 2013, **9**, 3878–3888.

39 D. Case, T. Darden, T. Cheatham, C. S. III, J. Wang, R. Duke, R. Luo, R. Walker, W. Zhang, K. Merz, B. Roberts, S. Hayik, A. Roitberg, G. Seabra, J. Swails, A. Goetz, I. Kolossvry, K. Wong, F. Paesani, J. Vanicek, R. Wolf, J. Liu, X. Wu, S. Brozell, T. Steinbrecher, H. Gohlke, Q. Cai, X. Ye, J. Wang, M. J. Hsieh, G. Cui, D. Roe, D. Mathews, M. Seetin, R. Salomon-Ferrer, C. Sagui, V. Babin, T. Luchko, S. Gusarov, A. Kovalenko and P. Kollman, *AMBER 12*, University of California, San Francisco, 2012.

40 R. Abel, T. Young, R. Farid, B. J. Berne and R. A. Friesner, *J. Am. Chem. Soc.*, 2008, **130**, 2817–2831.
41 C. J. Woods and J. Michel, *Sire 2014.1, a complete molecular simulation framework*, 2014, http://siremol.org.
42 W. Humphrey, A. Dalke and K. Schulten, *J. Mol. Graphics*, 1996, **14**, 33–38.
43 *Suite 2011: Maestro, version 9.2, Schrödinger*, LLC, New York, NY, 2011.
44 J. Ko, D. Lee, H. Park, E. A. Coutsias, J. Lee and C. Seok, *Nucleic Acids Res.*, 2011, **39**, W210–W214.
45 J. Wang, W. Wang, P. A. Kollman and D. A. Case, *J. Mol. Graphics Modell.*, 2006, **25**, 247–260.
46 C. J. Woods and J. Michel, *Sire, Public Source Code Repository*, 2013, http://sire.googlecode.com.
47 C. J. Fennell and J. D. Gezelter, *J. Chem. Phys.*, 2006, **124**, 234104.
48 M. Mezei, *J. Chem. Phys.*, 1987, **86**, 7084–7088.
49 C. R. W. Guimaraes and R. B. Alencastro, *Int. J. Quantum Chem.*, 2001, **85**, 713–726.
50 R. W. Zwanzig, *J. Chem. Phys.*, 1954, **22**, 1420–1426.

Faraday Discussions

DISCUSSIONS

Applications and serious games: from docking to protein folding: general discussion

DOI: 10.1039/c4fd90022g

Dr O'Donoghue opened the discussion of the paper by Dr Stegman by remarking: Currently there is an enthusiasm in the academic community for developing computer games that help players learn science. Your work – and that of others – has demonstrated that such games can indeed be effective learning tools. However, creating a successful computer game often requires considerable time and effort, both in developing the concepts, as well as in implementation. Do you believe that the current funding mechanisms for education and research are adequate to support educational game development? If not, which mechanisms could be introduced to improve the situation?

Dr Stegman replied: To create a game that teaches a significant amount of science, contains a large amount of engaging game play, and provides clear and immediate feedback to the player requires significant development time. After the game mechanism and content are established, after three iteration cycles, then two or three more cycles are required to work out the timing, balance, feedback and difficulty level of the game. Most funding in the US is for research on games: the games are created for use in testing not for release in the commercial market. Games that show great promise to teach but that require a year or two of polish currently have no funding source. However, If a game is fun and well designed, it will be paid for by players. And more and more, game developers are realizing that setting their game in reality is a road to financial success, and this is largely due to the success of the game Plague Inc. This game is an excellent example of a well polished game. The game was developed by an independent game developer, not a scientist or educator, and the game has sold millions of copies, see http://www.gamasutra.com/view/news/172161/Top_iPhone_game_apps_Plague_Inc_Catapult_King_head_up_paid_charts.php

Dr Chavent remarked: Nowadays, sophisticated rendering plays a central role in "blockbuster" games and even in independent games with much smaller budgets. The design of characters, environments, *etc.*, are early steps in the overall process. Conversely, I have the impression that in science the rendering can be a little bit "dry" and that scientists are even afraid/sceptical in front of a pretty picture. Don't you think that the serious/scientific games are susceptible to this pitfall? In your case, have you thought about the character and environment design as a central part of your project?

Dr Stegman replied: I agree with you that science games could be more fantastic. We can make the molecular world look like a fantasy landscape, while still portraying it accurately. We can also make the molecular world look like it is interactive, with hints drawn onto the molecules to indicate where they interact or where their allosteric inhibitor binding sites are. Using games and modern game art, we can reveal to the non scientists the wonderful intricate puzzle that we are able to imagine in our minds. See MolecularJig.com for the final release of Immune Defense soon.

Dr Chavent asked Dr Stegman: Would it be possible to add more fancy effects or simplified representations as presented in the paper by Cipriano *et al*?[1] This rendering could have several advantages: (1) attract people due to the fancy colour and shape; (2) simplify the problem due to the smoothed surface; and (3) help people select the correct answer by adding "stickers" *i.e.* texture onto the molecular surface. This feature can then be removed for more advanced players.

1 1.G. Cipriano and M. Gleicher, *IEEE Trans. Vis. Comput. Graph.*, 2007, **13**, 1608–1615.

Dr Stegman responded: I agree with you completely. The molecular world is not just vast and beautiful, but every molecule has its function and its own role to play, so game scenarios are endless. Allowing players to choose their own method of displaying molecules – letting them change the "stickers" on the models of molecules, would be a powerful way to let players take advantage in the game of the many ways we present molecules. If the player needs to mark a binding site or use a particular beta sheet for a purpose in the game, the player is more likely to remember what binding sites and beta sheets are.

Professor Olson opened the discussion of the paper by Dr Montes by addressing both Dr Montes and Dr Stegman: Could you both tell me how you recruited testers for your games, how you publicized the games, and how much time and effort went in to analyzing the results of the tests?

Dr Montes replied: Since it was a closed alpha version, we recruited the testers among our colleagues and students in our institute specialized in disciplines not related to structural biology (mostly computer science).

For the closed alpha, we created a website (http://udock.fr) to gather produced data and make the software available for download. During the playtests, the software sends data to the webserver in real time. There is an export function in the software that downloads and processes all data from the server. Most of the work of analysis was performed before the alpha playtests since it was critical to choose which data produced by the players was relevant to save on the webserver since it was out of reach to save everything. Since it is very difficult and time consuming to gather people to perform the playtests, it is of major importance to save as much data as possible.

Dr Stegman answered: The game Immune Attack was developed by the Federation of American Scientists (FASLearningTech.org) and we collaborated with the Maine International Center for Digital Learning (MICDL.org) to evaluate the learning and attitude changes in players. In the US, we need an "Internal Revue Board" to approve our research protocols if we want to publish our data. MICDL.org helped us arrange for that approval and also helped us contact teachers. Many teachers contacted FASLearningTech.org directly, and most found our site by searching for "Biology Game" on the internet. Our data was collected by an online survey. Once a teacher registered as interested (Over 6000 registered) we contacted them by e-mail to see if they could complete our three-day controlled protocol. Over 50 teachers have successfully completed our protocol over the 4 years of test development and data collection. If however, we only want to do product testing, and not publish our data, then we do not need IRB approval, we just need to follow rules and let players know data is being collected. So I show my games to students aged 13 and older and ask them what they think as often as possible. In Washington, DC I worked with a local high school, McKinley Technology High School. These students are learning themselves how to design and program video games. So they consistently give me very useful critiques.

We spent about three years on test development, overlapped with two years of data collection and one year on data analysis. We created our own test of knowledge gains. Because we were asking specific questions about molecular biology as they pertain to our game, there are no standardized tests that we could use. Additionally, asking teachers for any questions they knew to be tried and true was fruitless. No one is asking students aged 13–19 any questions about the specific proteins involved in white blood cell behavior. Questions written for their age level are too general and do not address the behaviors of molecules, while questions written for college students require too much background knowledge to answer. Addressing molecular behavior is, as you know, a small field. Creating a test of knowledge gained is going to require time, a large numbers of students, and an education research expert who can calculate several measures of reliability for your test.

Dr Zoppè stated: I have several comments: on the difficulties of creating games, on funding and on the different aims of game development.

Developing a game, and especially a scientific, discovery or teaching game, is a multidisciplinary task, that involves many skills. From the basic science (which we have) to the design of the game logic, with mechanisms of rewards, stages, and gestual engagements, and the visuals associated, to the computational design, that must be capable of elaborating sometimes very complex tasks in real time. As such, the production of a game is also a very expensive task, since very few research laboratories can have all the required experts in house. Furthermore, after a game has been developed, if it is going to have some impact and be played by many people, it must be supported by a community, it should be updated, which, again, requires dedicated people working on this, and taking care of the life of the game outside the lab. An example of a game that manages to do all this is Angry Birds™ (http://freeangrybirdsgame.org/), which has become so popular that, despite being distributed for free, has generated huge revenues for the developers thanks to merchandise. But we are not supposed to make games for profit, and we normally look for grant funding in our activities.

This combination of inter- or multidisciplinary, and of novelty, makes it difficult to obtain significant money, since funding agencies rely on experts of single disciplines to evaluate any project. So this is the first problem.

Dr Zoppè continued: The use of games for teaching, both school students and the general public, is an idea that has only become common in recent times. In this group we are in a sense among the openers of this possibility. It takes for sure more time to get the politicians to put (public) money on it. So in this respect I think that things may change in the next few years, already some signals are there, for example in the Horizon 2020 program. It is up to us, with the few examples that we already have, to demonstrate that they work better (if they do) than traditional teaching instruments.

However, once games as teaching tools become common, there will be games to learn history, geography, literature and other subjects. In this respect, competition will be more relevant, but here we may have an advantage since we have started earlier.

Game developers can have several different scopes: until now, most games have been produced as commercial, then one can use them to teach, and again they can be used as research tools. Do you think that the three can be all optimized?

Dr Zoppè addressed Dr Montes: Very nice work. I also tried it during the coffee breaks. I was wondering about showing the electrostatic potential using field lines, possibly calculated on the fly, instead of simply using colors on the surface which might make the task simpler. Such a solution would also make it possible to use surface colors (or other visual clues, like degrees of grey, or texture features) to show another property that is important in protein–protein interaction, that is the lipophilic potential.

Finally, I found it difficult to get the depth information while using the game. Maybe you could use some clues to suggest depth, such as having a floor that receives shadows, or introducing a fade effect on the depth axis.

Dr Montes replied: Many different modes of representation can be used to visualize additional properties. Even more than field lines, we are working on dynamic visualization of different properties using smoke-like particle systems for instance.

We didn't use shadows since we feel it burdens the view by adding visual edges that are not part of the surface mesh. Depth of field is a good suggestion indeed to increase the depth perception.

Dr Reiher asked Dr Stegman: How much of game development can be formalized before starting to work on the actual implementation of a game?

Say, you have a number of messages that you want the player to understand in the course of a game. Can you plan how much of this information you may get across in each level of the game? Or is it necessarily an iterative process, in which playing the game during development is the only way to find out whether the scientific information the game provides has been chopped up into small enough pieces? In the end, the games you envision provide a level by level understanding of complex scientific knowledge. Knowing *a posteriori* (once (iterative) game development has been finished) how the scientific information had to be dissected into pieces appears to me to be important information on the whole learning process. Is such analysis done and exploited for other didactical purposes?

Dr Stegman replied: Game development, any kind of game development, is an iterative process. Fig. 1 shows the process of science game design that I follow. In the beginning, we decide on the learning objectives and we choose a game mechanism. To teach most effectively, we choose a game mechanism that will require students to use the facts and concepts to be learned in order to win the game. The meaning of the game is in the mechanism, see my essay on how this relates to learning games, with a reference to Soren Johnson's original talk on the topic. http://www.FASLearningTech.org/2011/02/theme-is-not-meaning. Once the learning objectives and game mechanism are chosen, paper prototypes are created and tested first in our group and then with members of the target audience. Observations and follow up questions help us see whether the game mechanism is clear to the player. If the player cannot identify the objects in the game, cannot explain the goal of the game, or what the win condition would be, then the mechanism is too complex or too poorly executed. Then we iterate on the paper prototype. We consult with many scientists and educators during each round of iteration.

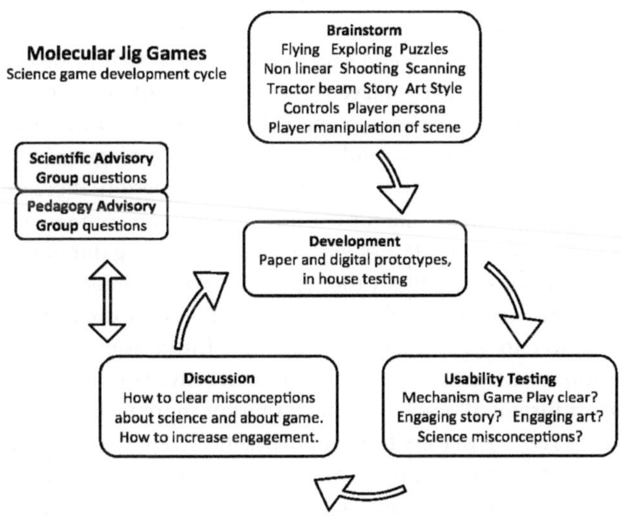

Fig. 1 This is the Molecular Jig Games production schedule template. It is optimized for involving scientists and teachers throughout our entire game development process. See www.molecularjig.com/blog for more information

In this room, where many of you have already created digital simulations of molecular behavior, I would suggest that you consider what you could teach to various types of audiences using your simulation, and then choose the game mechanism to match your learning objectives. Remember that players will only remember what they need to win the levels of your game. We found this is true with Immune Attack players (see the slides from my presentation here: http://www.molecularjig.com/research/). If you make a chess game using molecules for chess pieces your players will still learn chess. If you make chess but add on the requirement that your proteins need to face their opponents with their binding sites pointed towards each other, your players will know that molecules have binding sites and will learn whatever symbol you draw on them to represent the binding site. Once we have prototyped a game mechanism we can start play testing with our target audience. At this point, leaning game development and entertainment game development are very similar. We all want our players to keep playing. We all need players to maintain interest. To do this we must make the player feel competent, capable and challenged. Game designers know that information needs to be delivered in steps, and that small learning steps with trial and error being required by the player will put the player in a state called "Flow." Coined by a psychologist, Mihaly Csikszentmihalyi,[1] who studies what makes people happy. Working on a problem we think we can solve keeps us focused and if the problem is too easy or too hard we are more likely to put it down. So, in answer to this question, after the learning objectives are well matched to the mechanism, the goal is the same for all games: Create game play that makes the target audience keep playing. When we create game play experience that is complex enough to involve the player but not too much to overwhelm the player we will by definition find that point at which the science "has been chopped up into small enough pieces." This is why my paper is significant: because Immune Attack is a complex game that requires flying a ship in three dimensional space and shooting small targets. It could very well have been that these complex game dynamics could have reduced player's confidence with the science material. However, we showed that non-gamers and those who found Immune Attack not "easy to play" did not have less confidence in their cell biology abilities compared to control students. Therefore, this is a step in the direction of creating in depth and engaging game mechanisms for learning games: the game mechanic can be complex enough to engage players without scaring non-gamers away from the science. An important point about the types of learning objectives: if memorization of many facts is intended, then a simpler game that presents the facts and requires the player to guess, such as Medmyst (http://webadventures.rice.edu) may be appropriate. Medmyst is more of an interactive comic book than a game, but this simpler format allows the game to present more facts and provides more opportunities to require players to recall the facts. These games are very effective at teaching facts. However, Immune Attack and its sequel, Immune Defense, are designed to teach broad concepts. Immune Defense is designed to require the player to repeatedly make decisions about which protein to use to solve puzzles. Our assessment at each round is to learn whether players learn and deeply understand that proteins mediate interactions between cells. I hope that others will use the data in this paper to advance their own research and also reach out to me to collaborate. I think our next few video games at Molecular Jigs Games will address the question of how to impart not just knowledge but a deep enough understanding that non-intuitive concepts such as the random diffusion of molecules pervades players' thoughts and affects how they solve related but new problems.

1 M. Csikszentmihalyi, Flow: The Psychology of Optimal Experience, Harper and Row, New York, 1990.

Dr Zheng commented to Dr Stegman: To be convinced that the gaming approach to education works, I think we would need to go to schools with the poorest performance in the toughest neighborhoods, and do an experiment by letting some students use the game based education. In the end, if these students performed better than students who had no access to the game based education, then it would be a very convincing story that the game based education really works.

Dr Stegman replied: The evidence that games teach and affect us in many positive ways is piling up. See this review of the field, http://www.sri.com/work/projects/glasslab-research.

Mr Hall asked Dr Stegman by asking: It would appear that the control group for your experiment did not get exposed to the terminology of cellular or molecular biology since the videogame that they played lacked this information and since they did not receive a lecture on this information in class. Is this the case? If so, then does that not mean that you were comparing your videogame as a learning technique to the control case of not receiving the information in anyway at all? Do you plan to compare your videogame to other learning techniques, *e.g.* readings or lectures, in the future?

Dr Stegman answered: You are correct. In the paper I just presented, our students only learned molecular biology if they played Immune Attack. The names of the white blood cells and the names of proteins required to mediate their behaviors are all certainly new to our students. Our test does cover more general questions, however, such as what makes a T Cell different from a Macrophage and the answer is their proteins, not their DNA. When Immune Attack was developed, in 2008, many people thought that students might not remember vocabulary learned in a video game and that new concepts introduced without prior teacher instruction might not be understood. Our data clearly showed that this is not the case: Immune Attack players can remember vocabulary and concepts, while the fact that the material in the game was new did not affect their confidence levels. Now that we have a test, we can use it in several ways. One way is to allow teachers to use the game first and then follow up with lectures and compare with teachers who use videos of game play only with their lectures. This test would require many more teachers and students, to control for the variance introduced by the teachers. This would test for whether playing the game is the key, not just seeing the molecules and cells moving. We believe that giving the player agency, that is, allowing them to affect the outcome of the scenario and also giving them a goal to reach will engage the students better than a video and if the game is well designed, the goals will require the player to understand the principles in order to succeed. In order to facilitate this kind of testing, we created a new game called Immune Defense. Immune Defense will allow us to recruit more teachers, because the game is playable on PC, Mac, Web browsers, Linux and tablets. Immune Defense requires players to make use of molecular and cellular behaviors to win the levels. The molecular behaviors are closely designed into the game mechanism. Immune Defense is a Real Time Strategy game, with some puzzle aspects. We are recruiting teachers at MolecularJig.com.

Dr Johnson opened the discussion of the paper by Dr Zheng by asking: Even though 6 m of DNA are packed into the relatively small volume of the nucleus, I'm stunned by how much free space appears to remain. Do you know what the expected/theoretical volume difference is *i.e.* typical nuclear compartment volume minus DNA volume for your model cell type?

Does the input data capture "longer range" interactions, *i.e.*, rather than DNA–protein–DNA, can Hi-C provide interaction distances that might span massive protein complexes, *e.g.* tens of nm, such as DNA–enhanceosome–DNA?

Dr Zheng replied: Yes, there is still a lot of free space in the nucleus, you probably can see it from the movie we have. My guess is that this free space also allows protein machinery to pass through for regulation purposes. I am not sure I know the answer about the "longer range" as we don't have enough data to answer the question. However, I suspect this will have something to do with the degree of cross-linking. If the cross-linking is extensive, I can imagine that large complexes can be cross-linked together as the cross-linker used is a small molecular and can diffuse quite efficiently. So there is a chance the "longer range" could happen, but I don't think we have data to support it now.

Dr O'Donoghue addressed Dr Zheng: Your 3D genome viewer is built on the assumption that within the foreseeable future we will have structural models of whole genomes at something approaching atomic resolution. However, it should be pointed out that the current experimental efforts towards this goal are very much at the beginning stage. The main approaches currently being tried are based on the HiC method pioneered by Lieberman Aidan *et al.*[1], as well as related methods. However, these methods have not yet been widely reproduced, and there is considerable disagreement in the community about the accuracy of the resulting data. There is some evidence that these data may have a high false positive rate. Even assuming the data are accurate, the current approaches produce only a very sparse fraction on all molecular contacts that occur – these data are vastly inadequate for producing atomic-resolution models. There are also more fundamental issues, such as accounting for the large structural rearrangements occurring throughout the cell cycle, as well as the issue that experiments to date use immortalised cell lines in which genome structure is significantly altered compared to healthy cells.

1 Nynke L. van Berkum, Erez Lieberman-Aiden, Louise Williams, Maxim Imakaev, Andreas Gnirke, Leonid A. Mirny, Job Dekker, Eric S. Lander, Hi-C: A Method to Study the Three-dimensional Architecture of Genomes, DOI:10.3791/1869

Dr Zheng replied: We definitely agree and share the same view about the current status of the technology on the HiC method. However, the field is moving very fast, and there maybe some other new technologies that can overcome the current hurdle. In addition, the knowledge we have about chromatin and other genomic aspects through years of basic research could also help with model construction. That said, what we focus on is a tool that can accommodate the data and technology at different stages, and can be used to visualize genome structure at different scales. We hope that as the technology moves along, the tool is always there to integrate and visualize the data and the model. We want visualization and integration to be movers and facilitators, rather than a bottle neck.

Dr O'Donoghue continued the discussion of the paper by Dr Zheng by asking: Now that you have implemented 3D genome visualisation using a game engine, have you considered the possibility of creating an educational computer game, along the lines of the Immune Attack game presented by Melanie Stegman at this Discussion.[1]

1 Melanie Stegman, Immune Attack players perform better on a test of cellular immunology and self confidence than their classmates who play a control video game, *Faraday Discuss.*, 2014, **169**, DOI: 10.1039/c4fd00014e

Dr Zoppè addressed Dr Zheng: Very nice program, it will be useful in many studies.

I was wondering if it could also be used for dynamic elaboration. As you know, data derived from 3C experiments are referred to millions of cells, and not all of their DNA is necessarily in the same conformation. So on one side we have very scattered information and missing contact data (as all the repetitive sequences are discarded from original reads), on the other we may have contact information apparently contradictory, when we consider some DNA fragment that is apparently in contact with more than one other DNA fragment.

I really think that the best condition to obtain useful data, at this time is the comparison of a cell in different states, like before/after activation, or in response to different stimuli. In this case it might even be possible to see how the different conformations are obtained, that is, obtaining a dynamic representation of DNA. Also, there are very important conformational changes during the chromosome condensation at mitosis.

The system that you built is based on a game engine, and it should be possible to introduce some time activities into it: do you have any plans to do it?

Dr Zheng replied: Yes, using a game engine is exactly our intention to expand the capacity of the current software to capture dynamic information. As for the technology, there are already attempts to perform the high-C method on single cells. So when that technology is mature, it will be possible to perform the analysis you suggested to capture dynamic changes and population difference in DNA conformation among different cells, and we hope our software will be ready for that stage when the data becomes available.

Dr Procter asked Dr Zheng: Have you considered how DNA conformational changes might be explored with the 3D genome viewer? (*i.e.* could the system be easily adapted to display multiple conformations, and would the rendering model effectively portray conformational change?)

Dr Zheng responded: At this stage, we can store different DNA conformations as state files, and load individual state files to visualize the changes in conformation. In the future, we plan to implement dynamic visualization where conformation changes can be captured and rendered dynamically. The move to use game engines is preparing for that stage.

Dr O'Donoghue continued the discussion of the paper by Dr Zheng stating: You have focused your current work on representing the human genome in 3D. However, I believe that significant experimental efforts are underway that focus on determining 3D genome information for model organisms. These efforts are more likely to lead to accurate genome 3D structures long before such structures will be determined for the human genome. Do you know which model organisms are being most actively investigated in this way, *via* HiC and other related methods? Also, do you have plans to incorporate these data into the system you have developed?

Dr Zheng replied: There are works ranging from bacteria to yeast, to fruit fly and mouse. Due to the size of the genome, the bacteria may have the highest resolution by Hi-C. However, there are dramatic differences between prokaryotic and eukaryotic genomes. As for eukaryotes, I think yeast would be the easiest one to do as they are haploid with a relatively small genome size. We do plan to incorporate these data in the system we are developing.

Dr Zheng replied: Yes, we do have an interest in doing that in the near future.

Dr Mouawad opened the discussion of the paper by Dr Woods by asking: Your presentation was very interesting, however my question concerns your assumption that the cavity adopts the same conformation whether the ligand is bound or not. We know that this doesn't correspond to the majority of the cases, so how do you deal with the proteins whose cavity presents structural modifications due to ligand binding?

Dr Woods replied: We admit that we are not calculating the "true" absolute binding free energy, you are right, we do not account for the free energy of the protein to change conformation from the free to bound forms. While we sample protein motion during the simulation, Monte Carlo is unable to sample the (potentially) large backbone conformational changes between bound and free forms of some proteins. However, Waterswap swaps the ligand with water, and so will calculate the difference in free energy between the bound ligand/protein complex, and the unbound protein and unbound ligand, where the protein is in the ligand-bound conformation. Since the free energy to change from the unbound protein in the ligand-bound conformation to the unbound protein in the unbound conformation will be (roughly) the same for all ligands, we believe that Waterswap will allow for accurate ranking of binding affinities of a series of ligands. Where the "true" absolute binding free energy is desired, then we would recommend, *e.g.* a single umbrella sampling simulation to be performed to calculate the conformational free energy difference between the bound and unbound conformation of the protein.

Dr Zheng addressed Dr Woods: The evolution may have worked out how the water molecular will be excluded when a typical enzyme–substrate binding occurs, maybe through some conformation change when the substrate goes in. Is it possible to look into those bindings, such as heat shock response protein, that need a quick response, and see how water molecules may be excluded? Obviously for these kind of fast response bindings, the cell cannot tolerate a slow exclusion of water molecules during the binding.

Dr Woods replied: Waterswap only looks at the thermodynamics of binding, not at the kinetics. It would be difficult to use Waterswap to investigate binding where it is believed that the water structure is kinetically trapped in a thermodynamically unfavourable state. Fortunately, for small-ligand binding (Waterswap cannot model protein–protein binding) these situations are rare (or indeed non-existent – I am unaware of any cases). Binding of water from MD simulation is in the range of 10s to 100s of nanoseconds, so does not seem to cause any difficulty to small ligands.

Dr Brancale asked Dr Woods: What happens to water molecules that could be involved in ligand binding? Can you treat them differently?

Also, how does your system perform when the bound ligand itself is extensively exposed to the solvent?

Dr Woods replied: Waterswap selects waters from a free water box and swaps them into the protein box in place of the ligand. This means that water molecules in the protein box that are involved in binding with the ligand remain in the protein box, and will form hydrogen bonds *etc.* with the water molecules swapped in place of the ligand. We don't treat these waters any differently than any other water molecules in the protein box as we don't want to disrupt them during the simulation. The method performs well even if the ligand is extensively exposed to solvent, as the solvent is free to interact with the ligand, and then with the swapped-in water molecules during the simulation. From the perspective of the solvent, it sees a ligand molecule slowly perturb into a cluster of water molecules, and the solvent will thus adapt its conformation accordingly (*e.g.*, we have seen the solvent change from forming a hydrophobic cage around a hydrophobic ligand to forming nice hydrogen bonds with the water that is swapped in in-place of the ligand).

Professor Hirst asked Dr Woods: Does the (re-)labelling of water molecules based on their location in space influence the entropy of the system?

Dr Woods replied: Possibly. It is something that we are thinking about. The relabelling shouldn't affect the thermodynamics if the labels were used only as labels (as labelling a molecule doesn't affect how it moves or the phase space that it can explore). However, we definitely know that Waterswap doesn't account for the entropy of confining the ligand in the binding site, and we suspect that the identity constraint also doesn't account for the entropy of confining the water in the binding site either. It is something we are currently investigating and thinking about. One thing we do know is that changing the number of confined waters doesn't affect the binding free energy (varying from, *e.g.* 12–18 water molecules didn't affect the calculated result). This means that if there is an entropy contribution we are missing, it either cancels out (as water is confined in the binding site of the protein, and is also confined within the water box in an identical manner), or is independent of the number of waters (as otherwise the binding free energy would have a linear relation to the number of confined waters).

Dr Turpin asked Dr Woods: The Waterswap algorithm is a very nice method for calculating ligand–protein binding energies. Do you think it would be possible to use a similar approach to calculate the binding energies of larger molecules, such as in protein–protein binding?

Dr Woods replied: Yes, we agree that Waterswap is potentially a very nice method for calculating protein–ligand binding free energies. However, we don't think it would be suitable for calculating protein–protein binding free energies. This is because Waterswap swaps the "ligand" with an equivalent volume of water, so a protein–protein calculation would involve swapping one protein with a potentially large volume of water. It is unlikely that the free energy average in this calculation would converge, and also, limitations to Waterswap (only limited protein sampling, ligand constrained to sit in a small volume) would have a big impact on the swapped protein.

Professor Olson commented: I agree that consideration of water is important in determining the binding free energy of ligand protein interaction. A comprehensive analysis of 392 high-resolution crystal structures of protein interacting with ligands showed that in almost all the complexes (95%) one or more waters are present and interact with the ligand upon binding, and 72% of them are actively mediating interactions by bridging protein–ligand contacts. This presents a challenge for rapid docking and virtual screening of large ligand libraries. We published a paper in 2012 that describes a method to include displaceable waters in AutoDock dockings[1] We found that this is especially important in docking fragment libraries to get the correct pose of the ligand.

1 S. Forli, and A. J. Olson, A forcefield with discrete displaceable waters and desolvation entropy for ligand docking., *J. Med. Chem.*, 2012, **55**(2), 623–638, PMC3319101.

Dr Procter asked Dr Woods: The Waterswap approach appears to be extremely useful for medicinal chemists, but the data it generates is not easy to comprehend for scientists not familiar with the approach. Have you considered the effectiveness of different ways to present these data to chemists actively engaged in lead optimisation?

Dr Woods replied: I agree that the method can be potentially very useful, but that it produces a lot of data that can be difficult to understand. We are actively working on developing visualisation and tools that can help medicinal chemists understand this data and have partners working in different pharmaceutical companies who are providing us with advice. We are always open to suggestions and welcome input from people who have more experience in this area than ourselves. I have a strong belief that simulations such as Waterswap are most useful if they provide good visualisations that can be used to inspire medicinal chemists. The actual "value" or "result" they generate is less important than the clear picture that lets a medicinal chemist immediately understand how a drug binds well, and provides inspiration as to how the drug could be improved.

Dr Baaden asked Dr Woods: The Waterswap method focuses on water molecules in a binding site as a basis for assessing protein–ligand binding energies. How are ions present in the aqueous medium taken into account? Locally, ions may enter a binding pocket and affect the water structure as well as the free energy convergence. Globally, ionic strength and electrostatic fields induced by the moving ions may impact on free energy results and pose a sampling issue. To what extent do these issues affect the accuracy of the Waterswap approach or can they be accounted for?

Dr Woods replied: This is a very good question. Ions are included in the protein box but we currently don't have any ions in the water box (it could be argued that perhaps we should, so that the water box has the same ionic strength as the protein box). Ions are free to diffuse during the Waterswap simulation, but you are right that the timescales in a single simulation are insufficient to fully sample all ion conformations. To overcome this (and also to allow better sampling of the protein backbone) we ran Waterswap calculations from snapshots taken from a long timescale MD simulation (100+ nanoseconds). We assume that, as the ion conformations are sampled well during the MD simulation, then snapshots taken from the resulting trajectory will have a statistically correct distribution of ion conformations, and so the average of the Waterswap free energies calculated from those snapshots will include the average influence of the ions.

Dr Procter opened the discussion of the paper by Dr Chavent: Your use of streamline analysis seems to emphasise long range correlations in membrane behaviour. Are you investigating higher order behaviour in these models, and is there the possibility of creating a multiscale membrane model from this work?

Dr Chavent replied: Indeed, this approach would be really interesting to use on lower resolution models such as Dissipative Particles Dynamics, but unfortunately we have not yet tested such models. We are currently trying to analyze and display movements of molecules using the OPEP force-field coupled with hydrodynamics.[1] We also envisage that our streamline representations may be compared with the same rendering using lower resolution methodologies in order to compare equivalent systems with different granularities. Identification of any discrepancies will allow tuning the methodologies to improve the convergence of the different approaches.

1. Fabio Sterpone *et al.*, The OPEP Protein Model: From Single Molecules, Amyloid Formation, Crowding and Hydrodynamics to DNA/RNA Systems, *Chem. Soc. Rev*, 2014, **43**, 4871–4893, DOI:10.1039/c4cs00048j.

Professor Olson asked Dr Chavent: In your paper you refer to the method used to populate a lipid bi-layer vesicle, but say that the details are in a forthcoming paper. Could you give us an idea of the general approach that was taken, and how much computational time the method took?

Dr Chavent replied: The lipid bilayer vesicle is constructed by alternating Packmol[1] packing steps with GROMACS[2] energy minimization steps, with a starting outer diameter that exceeds the final target diameter to reduce the likelihood of major steric conflicts. The equilibrating lipid vesicle proceeds through stages of overall shape deformation and develops large bilayer perforations before eventually attaining a well-packed spherical shape in approximately 300 ns. This method can require a few days to complete. However, we now have far superior methods of building lipid vesicles and enveloped viruses by exploiting the GROMACS free energy machinery, and these methods allow for vesicle/virion system construction workflows on the order of a few hours.[3]

1 L. Martinez, R. Andrade, E. G. Birgin, and J. M. Martinez, PACKMOL: a package for building initial configurations for molecular dynamics simulations. *J. Comput. Chem.*, 2009, **30**, 2157–2164.
2 B. Hess, C. Kutzner, D. van der Spoel and E. Lindahl, Gromacs 4: Algorithms for highly efficient, load-balanced, and scalable molecular simulation. *J. Chem. Theory Comput.*, 2008, **4**, 435–447. http://pubs.acs.org/doi/abs/10.1021/ct700301q.
3 E. Jefferys, P. W. Fowler and M. S. P. Sansom, Alchembed: A New Method for Incorporating Molecules into Equilibrated Lipid Bilayers, (in preparation).

Professor Olson asked Dr Chavent: In your lipid bi-layer simulations do you see flipping of phospolipids from one leaflet to the other – a phenomenon observed experimentally?

Dr Chavent responded: In conventional CG-MD simulations we do observe rare flip-flop events for certain phospholipids, and more frequent flip-flop events for cholesterol. In the case of the vision simulations (which is the next modeling step after the vesicle creation), we have found evidence that the presence of immobilized membrane proteins (*i.e.*, immobilized by the underlying matrix) may drive a modest mixing of lipids between leaflets. The largest transfers of lipids between leaflets actually appears to occur during equilibration of the lipid vesicle, when certain lipid species are able to involute over openings in the sparsely-packed envelope.

Dr Sommer asked Dr Chavent: During the discussion of a previous question you presented an image showing the flow visualization of a lipid double layer vesicle. This is a perfect application case for 3D stereoscopic visualization with scientific relevance (which I have discussed during the poster presentation at the VizBi 2014 in Heidelberg). With a simple 2D projection onto the screen it is difficult to differentiate between the foreground and background of the image, especially in the case of overlapping double lipid layers.

Question 1: VMD has supported for years stereoscopic visualization, but is it possible to interactively visualize the flow with VMD in combination with your approach?

Question 2: And in case this is not possible by now, is it possible to export just a 3D stereoscopic movie with VMD of your flow visualization?

Mr Stone remarked: It is possible to create both interactive and pre-rendered movies of stereoscopic streamline visualizations produced within VMD. Production of stereoscopic movies is a bit of a specialised topic beyond the scope of the current discussion, but this can be accomplished with a small amount of scripting to render the left and right eye views and by subsequently using a stereo-capable movie encoder or video editing package to generate the final movie bitstream.

Dr Chavent replied: Answer to question 1: It will be possible to visualize streamlines with stereoscopic visualization because the methodology will be embedded in VMD and will benefit from the visualization features already implemented in VMD – from stereoscopic visualization but also advanced rendering such as GLSL visualization or Tachyon rendering. Nevertheless, the animation of streamlines is, for the moment, computationally demanding and can be a limitation for interactive rendering? see also John Stone's answer to this question. Answer to question 2: It is possible to export data to create streamlines using Paraview (http://www.paraview.org). This program runs in immersive environments and so can perform 3D stereoscopic rendering. It is also notable that the streamlines visualization in VMD can be exported to many file formats including Wavefront OBJ, VRML, and the new X3D file format used by a variety of free model viewers.

Mr Stone commented: It is possible to animate a stereoscopic display of the streamline visualization in the existing versions of VMD, but it is not as straightforward as it could be. At present, there is not a "trajectory" file format for velocity grids, so each timestep would have to be loaded into a separate molecule. The computation of the streamlines is the most computationally demanding step for VMD, but once they have been computed, they can be animated rapidly using the "Multiple Molecule Animation" plugin found in the visualization extensions menu. This mechanism, while not an elegant way of accomplishing the visualization, should be fast enough to display a time-varying streamline visualization, even on a stereoscopic display. In the future it should be possible to greatly accelerate the calculation of the streamlines by porting this algorithm to the GPU. I should point out that other groups have developed fast GPU-accelerated streamline visualization algorithms that would also be candidates for a future implementation.

Dr O'Donoghue continued the discussion of the paper by Dr Chavent with: At this conference, many presenters have used rainbow colour scales to communicate qualitative data. For many of us who studied physics, using this scale seems like a natural and good choice, since the order of colours in the rainbow is something we are very familiar with, and is determined by the physics of light. However, in almost all cases, using rainbow colour scales for quantitative data is widely regarded as a bad idea by data visualisation experts (*e.g.*, see "Dear NASA: No More Rainbow Color Scales, Please" at http://blog.visual.ly/rainbow-color-scales/). Using the rainbow scale creates many problems: the spacing of colours is very uneven, which leads to arbitrarily overemphasising some regions of the scale. Worse, for people with colour impaired vision, rainbow scales can render a figure unintelligible. In short, using a rainbow scale typically results in a beautiful figure, but it obscures the scientific meaning you are trying to convey. Much better is to use either greyscale, or, if your data contains positive and negative numbers, use two colours with a gradual increase in lightness and/or decrease in saturation, so that zero becomes white. See "Mapping quantitative data to colour" – http://

www.nature.com/nmeth/journal/v9/n8/full/nmeth.2134.html. It would be great if as a community we could stop using rainbow colour scales for quantitative data.

Dr Chavent replied: We agree with Sean and indeed, we used a rainbow colour scale because it was familiar, but it is possible to change this in all the figures of our paper.[1] The different programs used to depict the streamlines have numerous other schemes in order to tackle the issue described by Sean in his question. For more information on the different colour scales available see also the answers to the question on colourblindness.

1 Matthieu Chavent, Tyler Reddy, Joseph Goose, Anna Caroline E. Dahl, John E. Stone, Bruno Jobard and Mark S. P. Sansom, Methodologies for the analysis of instantaneous lipid diffusion in md simulations of large membrane systems, *Faraday Discuss.*, 2014, **169**, DOI:10.1039/c3fd00145h

Dr Baaden addressed Dr Chavent: We saw many visualization examples where the specific use of colours enhanced the information feedback to the reader. The videos accompanying your paper[1] are a good example. The colours used in this case give the impression that the fast diffusing areas light up and are thus more easily visible than the slow or immobile regions. How was the color scale for this particular example chosen? More generally (now addressing all delegates), what considerations may be used to decide on colours, for instance related to perceptual cues, physiological markers (*e.g.* color blindness), experience or convention?

1 Matthieu Chavent, Tyler Reddy, Joseph Goose, Anna Caroline E. Dahl, John E. Stone, Bruno Jobardc and Mark S. P. Sansom, Methodologies for the analysis of instantaneous lipid diffusion in md simulations of large membrane systems, *Faraday Discuss.*, 2014, **169**, DOI:10.1039/c3fd00145h

Dr Chavent replied: For Fig. 4, we have chosen the colour scale and the background to especially highlight such movement. We chose a dark background and a very simple and common colour scale for the lipid displacement: a sort of reduced "rainbow" colour scale that passes from blue to red *via* cyan, green, yellow and orange colours. This colour scale will be changed in the near future to use the colour range implemented in VMD. Ultimately, the user can select an appropriate colour scale. Indeed, the streamline implementation in VMD – see Fig. 6 C and D – will use the colour scale embedded in VMD. It is possible to change the colour scale in VMD by clicking on Graphics > Colors > Color Scale.

For the MDAnalysis 2D implementation of streamlines visible in Fig. 6-A, we use the matplotlib library which gives access to a varied colour palette – see: http://wiki.scipy.org/Cookbook/Matplotlib/Show_colormaps.

For the MDAnalysis 3D implementation of streamlines visible in Fig. 6-A, we use MayaVi, which also features a varied colour palette.

Mr Krone continued the discussion of the paper by Dr Matthieu Chavent by asking: Have you considered using Streaklines or Pathlines instead of Streamlines? These methods could be useful in your case, since they might show different phenomena in your dynamic vector fields.

Dr Chavent replied: The first representation presented in the paper – see paragraph 3.2 – can be related to the pathlines representation. We have not yet tested the streaklines approach but it is indeed an interesting methodology to try and compare with pathlines in order to investigate the dynamic behavior of the system. These streaklines would also be useful to study the lipid clustering.

Mr Stone remarked: In regards to highlighting areas of interest or data values on a surface, there are many published computer graphics algorithms that use texturing techniques such as spots, line integral convolution, and other techniques, beyond the more commonly used coloring approaches. Some of these techniques are very computationally demanding, but GPU acceleration and parallel algorithms could make some of these approaches practical to use for molecular visualization.

Mr Quareshy asked: Having seen all the work presented over the 3 days and one of the main agendas of the *Faraday Discussion 169* being visualisation; I wanted to ask a general question to all the delegates. When depicting your visualisation, be it image or video, do you take into consideration the colours used? And do you have in mind an audience like myself who are colourblind?

Dr Zoppè replied: Thank you for the question (I'm sorry I did not get it entirely, or I would have certainly replied during the Conference).

At the Scientific Visualization Unit (www.scivis.it) we have dedicated some thoughts about the use of colors in bio-molecular representations.

Our final decision was to build images and movies only in Black and White, as in Fig. 2, and see our movies on vimeo (www.vimeo.com/scivis). The world at nanoscale has no intrinsic color; indeed the color is in the eye of the observer, not in the object itself, as first understood by the great astronomer, scientist and artist Galileo Galilei about 400 years ago. A physical description of colors can be made with the use of wavelength: in this case, we see that the size of most molecular

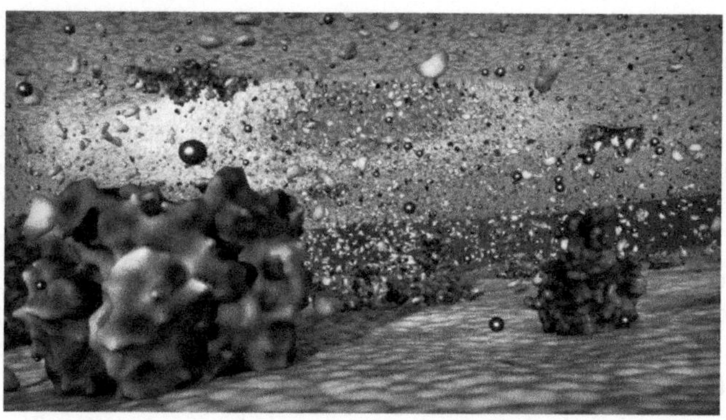

Fig. 2 Representation of a portion of the synaptic space: the presynaptic button is releasing a load of neurotransmitter (serotonin) form the vescicle; in the postsynaptic membrane the most prominent protein is the ionotropic serotonin receptor (5HT-R3); also in the synaptic space are present various macromolecules (mostly glucose) and the spherical ions. The image, produced by our group, is extracted from the Film 'The Dark Gene' (FilmTank, 2014).

objects is too small to get such a feature (smaller than wavelength). Proteins are indeed invisible to our eyes; what we look at is a representation of their shape (molecular surface), and those of us that use colors should do it with some meaning. We think that applying colors to molecules might lead people (both expert and naive) to dismiss other visual features, which are important determinants in the physical, chemical and biological behavior of proteins. The use of BW induces viewers to pay more attention to shape and other features, which are those that they should focus on if they want to understand how proteins work. Besides the problem of colorblind people, which is obvious and must always be taken in account, the use of colors has profound cultural meanings, different in different cultures, and those should also be considered when using them in a scientific setting. In our work, we use color to transmit some information which might be important in the social context (like in the short movie "The Challenge of HIV Research"), but we never use it in a scientific context, in which we show the shape of proteins, with the lipophilic potential 'painted' on the surface using a grey-scaled texture, and the electrostatic potential in the form of field lines, with particles flowing along the lines, showing both intensity and direction at the same time.

In future work, we intend to consider the use of colors to convey other information, possibly related to environmental aspects, such as pH, or osmotic pressure, or temperature, *etc.*

Dr O'Donoghue commented: The majority of data visualisations created by scientists are made as part of the initial process of exploring and understanding data, while only a small minority of visualisations are created for the later processes of publication or presentation. Since the initial stage visualisations are intended to be viewed only by a small group of close collaborators, any choice of colours that works for that team is fine. However, when creating visualisations intended for publication or presentations, it is all too easy to forget that many of our intended audience may have a very different perception of colour. Failing to take this into account can result in figures that are unintelligible to some of our audience, thus limiting the impact of our work. Many journals have specific instructions to authors designed to ensure that published illustrations are understandable by readers with colour impaired vision. Hopefully more journals will adopt and enforce such policies. However, when preparing presentations no such guidelines are generally available. A good rule of thumb is to preview your slides and illustrations in black and white (*e.g.*, import image into Photoshop and select the Image menu > Adjustments > Destaturate). If it still makes sense in black and white, all is good. If not, consider making greater use of lightness and saturation levels in addition to hue to distinguish different parts of your illustration. For more fine-grained tests, select the View menu > Proof Setup > Colour Blindness – this enables you to preview how your illustration will be perceived by people with different types of colour blindness.

Ms Laustsen continued the discussion by commenting: Try using Vischeck.com, which simulates how colorblind people would see an image.

Dr Sommer remarked: Ms Laustsen mentioned the following website during the meeting:

http://www.vischeck.com This tool can be used to check how your images will appear to color-blind people or those with a red/green as well as blue/yellow color deficit. This website was a really good hint, thank you!

Dr O'Donoghue opened the discussion of the paper by Professor Frederick Brooks by asking: You mentioned that controlled user studies are the gold standard in software engineering. Isn't the number of users a better standard? Ultimately, we aim to create tools that will be used; user studies can be helpful in achieving that, but they have some fundamental limitations. In particular, when we design user studies we often unintentionally build in hidden assumptions about our tools and users. One example of this would be the CAVE work you mentioned, based on the concept that scientists would be greatly empowered if they could walk around a molecule. Any user metrics designed in the early stage of that work would have missed the critical point that was initially unknown: that people prefer systems that require less effort to use. You mentioned the example of the QWERTY keyboard layout, which is famously significantly less efficient than other layouts (*e.g.*, Dvorak). However, the continued success of the QWERTY layout shows that efficiency isn't the most important factor. Personally, I used the Dvorak layout for several years, but gave up because of the constant battle required against such a ubiquitous standard. In this case, there may even be more subtle issues at play – perhaps the inefficiency of the QWERTY layout actually has some hidden benefits. For example the slowed down typing speed could potentially be better matched to the rate at which most people can formulate their thoughts into language. Thus, I feel that usage may be a better gold standard than controlled user studies.

Professor Brooks replied: I believe I claimed that controlled user studies were the gold standard in evaluating human–machine interaction techniques and devices, not in software engineering.

But in many fields we know the most productive technique or device is not the one that attracts the most users. Hence, user adoption cannot be used as a measure of productivity excellence. The Dvorak *vs.* QWERTY keyboard layouts is a well-known excellence.

As to the occasional benefit of lower-productivity techniques that better match thinking speed, you are certainly right. I type at the computer all my correspondence and routine documents, since I am a trained typist. But I permanently switched some 60 years ago to using first a dip pen and inkwell, and then a felt-tipped pen, for writing first drafts of papers and books. I can type faster than I think, and what comes out has too many unnecessary words that have to be edited out again.

In general, constraints can indeed help creativity, as you argue and Bach demonstrated. See Chapter 11, "Constraints Are Friends" in my 2010 book.[1]

1 Frederick P. Brooks, Jr., The Design of Design: Essays from a Computer Scientist, Pearson Education, Inc., Boston, MA, USA, 2010, ch. 11, p. 128.

Faraday Discussions

PAPER

Impressions by a dinosaur – summary of Faraday discussion 169: molecular simulations and visualization

Frederick P. Brooks, Jr*

Received 12th June 2014, Accepted 15th July 2014
DOI: 10.1039/c4fd00130c

A computer scientist was selected to prepare summary comments on the Faraday discussion. Much fine work was reported at the discussion. The tools and techniques reported at the conference have made radical improvements over past decades, and those are severally celebrated. Concerns and caveats about the papers are discussed.

1. Who's this, and why is he here?

Surprise and thankfulness met your invitation for me to close this Faraday discussion on visualization and simulation in chemistry. Surprise, because I am not a chemist but a computational toolsmith. Thankfulness, because it was a welcome opportunity to catch up on work I followed from 1965–2000. I am an admirer of chemists. The work reported is impressive and very pretty.

My training was in physics and then computer science and engineering (1953–56). I worked as an architect of the IBM 7950 Harvest cryptanalytic engine and its host, the IBM 7030 STRETCH supercomputer. The nine copies of the STRETCH reigned as the world's fastest supercomputers from 1961–1965. Then I had in 1961–65 the privilege of leading first the hardware development of the IBM System/360 (mainframe) product line and then the software development of its Operating System/360. Both the hardware product line and the operating system family are still actively updated today.

When in 1964 I went to the University of North Carolina to start the Computer Science Department, the environment was not suitable for continuing either my earlier research in computer architecture or that in software engineering. We picked computer graphics as a promising nascent field, and then molecular structure as an application and collaboration area. Our first collaborators were Jan Hermans, Jr, David and Jane Richardson, and Sung Hou Kim. As far as we can determine, the Richardsons were the first to solve a protein, Bovine Cu,Zn superoxide dismutase, by computer graphics without needing to build a brass Kendrew model.[1,2]

Brooks Computer Science Building, University of North Carolina at Chapel Hill, CB# 3175, Chapel Hill, NC 27599-3175, USA. E-mail: brooks@cs.unc.edu

We operated at UNC an NIH Research Resource in Molecular Graphics from 1974 until 1999, when such tools became generally available to chemists. Our computer science research included not only visual and aural human–computer interaction, but also pioneering testing of haptic displays, starting in 1968.[3–5] We had the delight of looking over the shoulders of some 40 teams of chemists during those 25 years. Hence my pleasure at seeing the progress during these 15 recent years.

2. Tools change science

Art Olson has remarked here that new tools change the science that people undertake. Sung Hou Kim put it dramatically to the NIH site visit team when our UNC molecular graphic resource became funded: "With this molecular graphics system [the UNC GRIP 75 system] I can solve a molecule in a week that formerly took me three months. I now do different chemistry. I can now dope my crystal [yeast phenylalanine tRNA] with a different carcinogen every week, and see how the structures differ."

Since I am a toolsmith, I shall remark on how these papers illustrate how tool power has evolved. Your papers show how you as scientists have harnessed that power for new science.

Progress in computational chemistry tool power

Obviously, the biggest change in the tools of computational chemistry is in raw computing power. Any order of magnitude change in a tool's power or cost becomes a qualitative change in how one thinks about it and how one uses it. Just since transistorized computers, we've seen 12 orders of magnitude change in raw computational power! The STRETCH supercomputer (1961) computed at about one megaflop. The Cray XK7 Titan at Oak Ridge nominally peaks at 27 petaflops, more than one teraflop for each of the ~19 000 nodes.

Nevertheless, some things seem unchanged. When I started collaborating with Los Alamos physicists in 1956, most long runs turned out to be about a week. I get hints that is still true in molecular dynamics. Is this user behavior indeed a constant of computational science?

These papers reflect two evolutionary changes in tool use in computational chemistry.

• Most papers reported the use of multiple different tools to achieve new and rich results.

• Most papers exploited community tools as well as home-grown tools, substantially enhancing productivity. Hooray!

Molecular graphics progress

When we started doing molecular graphics in 1965, we visited and learned from the handful of pioneers: Robert Diamond at MRC in Cambridge, Robert Langridge at UCSF, Charles Molnar at Washington University at St. Louis.

Building such hardware and software systems was difficult and expensive then. Now, the development of consumer products aimed at other markets has made molecular graphics readily available for chemists everywhere.

• Television and frame buffers made displays cheap.

- Consumer personal computers made computation cheap.
- Game cards made graphic rendering cheap.
- Off-the-shelf software systems made programming cheap(er).

The power molecular graphicists now have, inherited as others' by-products, is today remarkable!

Scientific visualization progress

The biggest progress in scientific visualization seen in this conference is in the use of integrated multiscale models, showing both overview and detail with consistency.

In the field at large, we now see integrated multimedia models: visual, aural, haptic. Here I heard no sound.

The set of actual visualization techniques for data has not changed much, and there still seems to be no coherent theory for data visualization, the perception thereof, and the comprehension thereof.

Molecular dynamics simulation progress

The greatest recent computational chemistry progress has been in the simulation of molecular dynamics. This, too, is a consequence of supercomputing generally—progress funded principally by military applications:

- Supercomputer memory capacity, radically reducing the slow input and output traffic.
- Circuit miniaturization, hence speed.
- Massive parallelism of cheap processors.
- The good fit of the embarrassingly parallel molecular dynamics problem to the only kind of supercomputers we now know how to scale.
- Algorithmic invention.

Cray XK7 "Titan" Supercomputer at Oak Ridge National Laboratories

• Result analysis techniques.

• Advances in experimental data quantity and quality to validate computer models.

Dr Grubmüller emphasized a crucial point:

Don't measure and then model—[one can always fudge models]. Instead, model and first predict some quantity that can be measured precisely and has not yet been. Then check how faithfully and precisely the measurement matches the prediction. If then the model is verified in this particular, it encourages belief in the model's predictions of properties and phenomena far more detailed than can be measured.

A sixth computer technology revolution – the GPU

I have watched six revolutions in computer technology and architecture since I fell in love with computers in 1944. The sixth is the development of the graphics rendering card, a single-instruction-stream-parallel-data-stream computer. Devised as a plug-in supplementary computer for the graphics rendering process, it has outgrown its host, now incorporating a billion transistors. Architecturally, the parallel floating-point pipelines are not surprising. The streaming table-look-up capability is. First seen in cryptanalytic engines in the 1960s, it is an extremely powerful operation, one whose exploitation will enable many new algorithms.

Now incredibly cheap because of the mass game market, the GPU has been adopted, but hardly adapted, for scientific computing. Floating-point word-lengths are painfully short; memories painfully small. I doubt if scientific computing will ever make a big enough relative market to get those things fixed.

Human interaction devices

The market now brings us a new shower of economical devices for human–computer interaction. We saw a rich variety used in papers here.

• The PHANTOM haptic device, 6 degrees of freedom in; 3 out.

• The KINECT motion and depth detection device.

• The LEEP motion detector.

• The smartphone, with accelerometers and gyros.

Many papers here have reported early use of these devices for ligand docking, and also as a viewpoint positioner. The smartphone serves as both viewpoint specification and as a wayfinder display for a large-scale display.

I am continually re-amazed at the iPhone and its contrast with the 1961 STRETCH Supercomputer. The iPhone has its 64 bit processor, iOS operating system, 10 I/O devices, a display, a touch screen and up to 64 gigabytes of memory. One can choose among ~775 000 apps (in 2012). Fifty million iPhones were sold in 1Q14 alone, 450M total so far!

3. Concerns and caveats

Evaluation of human–machine techniques

If this Faraday discussion is representative, the computational chemistry community is still primitive in this art. Researcher experience and opinion is real. Any user's anecdotal report ("Mikey likes it.") is real evidence and hence not to be ignored.[6] To carry much credibility, evaluations need to go beyond that, at least to standard *ex post facto* Likert-scale quantitative questionnaires by a sample of disinterested users. Next in power come comparative task performance measures over a sample of tasks, all performed by a sample of users.

Widespread adoption and prevalence of users are not reliable evidences of superiority. Too many other factors play into popularity. Consider COBOL and the QWERTY keyboard.

The gold standard for HCI technique evaluation is the formal controlled user study yielding significant results, costly and effortful though such evaluation is. Psychologists know well how to do these studies; we always include one on Ph.D. committees advising HCI technique dissertations.

WARNING re: tool and techniques papers

The development of tools and techniques is an engineering discipline, not a scientific one, concerned with making, not discovering. In engineering, as opposed to science, novelty is not *per se* a contribution. Any fool can make something that is just different.

The challenge for Ph.D. students in the synthetic disciplines is therefore to show that their creations are either.

- Better in some measurement space or;
- Made by a new process that at least promises to be both better and generalizable to other products.

Games for education

The most precious and costly commodity in education is *student time*. If you doubt that assertion, ask yourself what educator debates are most fervent. A central and perennial question is broad *vs.* deep. Intense exposure to a sample of concepts and applications, or completeness of coverage?

Computer games typically come out in the deep family of techniques—a relatively small sample of concepts are intensely exposed. Fine; that's an important part of all education. But the trade-off in coverage must continually be assessed as part of a balanced education in any subfield.

Even more challenging is evaluating cost-effectiveness *vs.* the best alternative educational technique. Many papers on computer-based education omit the rigor of such a comparative evaluation. Good games are not only narrow, they are costly to build, prototype, refine, and test.

Such neglect of evaluating cost-effectiveness is understandable. Such controlled comparative studies take much effort. In our experience, variance among student learners is high, hence to show significant superiority takes lots of experimental participants.[7]

Vision papers

Tool and technique papers fall into two quite distinct categories: "What we did," *vs.* "What we are going to do," or "could do." We saw both kinds at this conference.

Good vision papers are very useful. Ivan Sutherland's great 1965 "The Ultimate Display" founded and has challenged the whole research field of virtual environments for a half-century.[8]

Burks, Goldstine, and von Neumann's 1946 "Preliminary Discussion... of an Electronic Computing instrument" sparked into existence today's computer field.[9] Sir Maurice Wilkes said, "I recognized this at once as the real thing," and he went on to build at Cambridge, the first useful implementation of the stored-program computer.[10]

Vision papers must, however, be clearly labeled as such.

4. Net impressions of Faraday discussion 169

First, I must say this dinosaur was delighted! What a pleasure to see so much progress in fields I've been quite unable to follow for fifteen years.

Second, the format is very effective indeed. Although the Faraday discussions are over a century old, I never encountered a similar format before. It really is a discussion, totally unlike the normal handful of questions about each compressed paper in the usual conference. The depth and substance of questions and comments was rich, participation broad, back-and-forth exchange common.

Last, you have deep knowledge and an abundance of ideas. I frequently tell high school students that in 1944 I picked computers as my career. It proved for me the ideal career field for the last half of the twentieth century. Today, for the first half of the twenty-first century, I commend to them the intersection of computation and biology. You are there. The intellectual breakthroughs at this intersection during your lifetimes will be marvellous indeed!

Best wishes to you as you share in this great service, this great adventure!

References

1 J. S. Richardson, K. A. Thomas, B. H. Rubin and D. C. Richardson, Crystal Structure of Bovine Cu,Zn Superoxide Dismutase at 3 Å Resolution: Chain Tracing and Metal Ligands, *Proc. Natl. Acad. Sci. U. S. A.*, 1975, **72**, 1349–1353, DOI: 10.1073/pnas.72.4.1349, PMC 432531.

2 J. S. Richardson and D. C. Richardson, Doing Molecular Biophysics: Finding, Naming, and Picturing Signal Within Complexity, *Annu. Rev. Biophys.*, 2013, **42**, 1–28, DOI: 10.1146/annurev-biophys-083012-130353, PMCID: PMC3695750, NIHMSID: NIHMS478919.

3 J. J. Batter and F. P. Brooks, Jr, GROPE-1: A Computer Display to the Sense of Feel, *Information Processing 71*, North-Holland Publishing Co., Amsterdam, 1971, pp. 759–763, IFIP Congress '71, Ljubljana, Yugoslavia.

4 M. Ouh-young, D. V. Beard and F. P. Brooks, Jr, Force Display Performs Better Than Visual Display in a Simple 3-D Docking Task, *Proceeding of IEEE Robotics and Automation Conference*, Scottsdale, AZ, May 14–18, 1991, pp. 1462–1466.

5 F. P. Brooks, Jr, M. Ouh-Young, J. J. Batter, P. J. Kilpatrick, "Project GROPE—Haptic Displays for Scientific Visualization, *Proceedings of SIGGRAPH '90*, (Dallas, Texas, August 6–10, 1990). In *Computer Graphics*, 24, 4 (August 1990), ACM SIGGRAPH, New York, 1990, pp. 177–185.

6 F. P. Brooks, Jr, Grasping Reality Through Illusion—Interactive Graphics Serving Science, *CHI '88 Conference Proceedings*, ed. E. Soloway, 1988, pp. 1–11.

7 J. C. Mudge, Human Factors in the Design of a Computer-Assisted Instruction System, Ph.D. dissertation, 1973, Univ. of NC at Chapel Hill.

8 I. E. Sutherland, The Ultimate Display. *Proceedings of IFIP 65*, 1965, vol. 2, pp. 506–508.
9 A. W. Burks, H. H. Goldstine and J. von Neumann, Preliminary discussion of the logical design of an electronic computing instrument, 1946, *Collected Works of John von Neumann*, ed. A. H. Taub, 1961, vol. 5, pp. 34–79, also at http://research.microsoft.com/people/gbell/Computer_Structures_Readings_and_Examples/00000112.html.
10 M. V. Wilkes, Personal communication, 2007.

Poster titles

Activation by closure: exploring the relationship between agonist efficacy and c-loop closure in cys-loop receptors, **R. Yu, T. Greiner, E. Hurdiss, R. Lape, L. G. Sivilotti and P. C. Biggin**, *University of Oxford, UK*

Advanced features for the exploration of the solvent accessible surface using a haptic feedback device, **S. D. Laycock, R. Easdon and S. Hayward**, *University of East Anglia, UK*

Beware of the routine use of structure-based virtual screening: identification of significant discrepancies in four docking programs, **L. Chaput, J. M. Sanz, E. Quiniou, N. Saettel and L. Mouawad,** *INSERM, Institut Curie, France*

BioBlender 2. visualization of biological molecules, their interactions and the forces guiding them, **M. Zoppe, T. Loni and F. Milanese,** *Scientific Visualization Unit, Italy*

Capturing spontaneous binding of human VECTOR islet to anionic membranes using a highly mobile membrane mimetic model, **K. Skeby, E. Tajkhorshid and B. Schiøtt,** *iNANO and Aarhus University, Denmark*

Computer modelling of the radiolysis of adsorbed water on oxide surfaces in nuclear systems, **M. Smith, N. Buton and S. M. Pimblott,** *University Manchester, UK*

Computational study of phosphonium ionic liquids, **R. V. Bansode, N. Karodia, A. Paradkar, R. C. Dash and M. E. S. Soliman,** *University of Bradford, UK*

Cooperative swarms of simulation replicas for optimisation and exploration in biomolecular phase space: application to free energy calculations, **K. K. Burusco, N. J. Bruce, I. Alibay, A. Atzori and R. Bryce,** *University of Manchester, UK*

Crossing the scales of computational chemistry with high performance computing: from QM/MM to polarizable to coarse-grain, **S. Riahi, J. Besaw and C. N. Rowley,** *Memorial University, Canada*

Dynamics and spectroscopy of peptides, **R. E. Hill, N. T. Hunt, A. W. Parker and J. D. Hirst,** *University of Nottingham, UK*

Exploring chemical reactivity interactively, **M. P. Haag, A. C. Vander, M. Bosson, S. Redon and M. Reiher,** *ETH Zurich, Switzerland*

Extension of MONETA to analysis of allosteric communication in protein–protein and protein–DNA complexes, **Y. Guarracino, F. Langenfeld, N. Panel and L. Tchertanov,** *LBPA- ENS Cachan, France*

Holographic molecules in the classroom, **O. Gillespie,** *Royal Society of Chemistry, UK*

BUDE: High performance *in silico* drug screening on many-core processors, **S. McIntosh-Smith, J. Price, R. Sessions and A. A. Ibarra,** *University of Bristol, UK*

Interactive visual analysis of cavities in molecular simulations, **J. Parule, C. Turkay, N. Reuter, I. Viola and H. Hauser,** *University of Bergen, Norway*

Membrane rafts affecting GPCR signalling? **D. Robinson,** *University of Nottingham, UK*

Modelling the influence of solvent and conformational dynamics on protein spectroscopy, **F. S. Husseini and J. D. Hirst,** *University of Nottingham, UK*

Poking holes in bacteria; understanding the mechanism of trans-membrane pore formation by antibiotic peptides, **E. R. Turpin, B. Boner, J. D. Hirst,** *University of Nottingham, UK*

Potential directions and features for future molecular visualization software, **K. W. Hall, P. G. Kusalik and S. Carpendale,** *University of Calgary, Canada*

Sweet UnityMol, molecular graphics for Glycoscience, **T. Tubiana, M. Baaden, A. Imberty and Serge Perez,** *Laboratoire de Biochimie Théorique, France*

The PKD1L1 mutant and the left-right asymmetry disorder: characterisation by molecular dynamics simulation, **M. T. Buenavista, L. J. McGuffin, A. M. Mallon, R. Hussain and D. R. Nutt,** *University of Reading/Diamond Light Source/Medical Research Council Harwell, UK*

Understanding picking of moving targets: towards new paradigms for interactive and immersive molecular simulations, **A. Kouyoumdjian, P. Bourdot, S. Huot and N. Férey,** *LIMSI CNRS, France*

Unraveling novel binding sites on human neutrophil elastase, **A. K. Laustsen, M. Christensen and B. Schiøtt,** *Aarhus University, Denmark*

CELLmicrocosmos 2.2.2 membrane editor and APL@Voro: interactively generate and analyse PDB-based membrane models, **B. Sommer and G. Lukat,** *Bielefeld University, Germany*

RiboVision: a macromolecular journey through dimensions, **C. R. Bernier, A. S. Petrov, Z. J. Wartell and L. D. Williams,** *Georgia Institute of Technology, USA*

Visualizing executable models of biological networks using the BioModelAnalyzer: technique and challenges, **D. Benque, B. Cook, J. Fisher, B. A. Hall, S. Ishtiaq, N. Piterman and A. Taylor,** *Microsoft Research Cambridge, UK*

Avogadro: open source libraries and application for chemistry, **M. D. Hanwell and J. Thomas,** *Liverpool University, UK*

Comparative analysis of sequences, structures and simulations, **J. B. Procter and U. Zachariae,** *University of Dundee, UK*

Aquaria: integrating sequence and structure, **S. T. O'Donoghue, K. Sabir, C. Stolte, M. Kalemanov, B. Wellmann, V. Ho, F. Buske, M. Roos, N. Perdigão, J. Heinrich, B. Rost and A. Schafferhans,** *CSIRO & Garvan, Australia*

Visualisation of cyclic peptide energy landscapes, **L. Smeeton, M. Oakley and R. Johnston,** *University of Birmingham, UK*

The Skinner Prize for the best poster was awarded to Thibault Tubiana of Laboratoire de Biochimie Theorique, France, for his poster on Sweet UnityMol, molecular graphics for glycoscience.

There were two prizes awarded for Best Poster Overall at FD169. These prizes were kindly sponsored by Yasara and were awarded to Yann Guarracino for the poster on Extension of MONETA to analysis of allosteric communication in protein–protein and protein–DNA complexes, and to Björn Sommer for the poster on CELLmicrocosmos 2.2.2 membrane editor and APL@Voro: interactively generate and analyse PDB-based membrane models.

List of participants

Mr Saleh Alarfaji, *King Khalid University, Saudi Arabia*
Dr Andrew Almond, *University of Manchester, United Kingdom*
Dr Rossen Apostolov, *KTH Royal Institute of Technology, Sweden*
Dr Marc Baaden, *Institut de Biologie Physico-Chimique, France*
Mr John Baker, *University of Nottingham, United Kingdom*
Mr Peter Balshaw, *NVIDIA, United Kingdom*
Mr Ratnadeep Bansode, *University of Bradford, United Kingdom*
Dr Alisa Becker, *Royal Society of Chemistry, United Kingdom*
Mr Chad Bernier, *Georgia Institute of Technology, U.S.A.*
Dr Andrea Brancale, *Cardiff University, United Kingdom*
Professor Frederick Brooks, *Department of Computer Science, University of North Carolina at Chapel Hill, U.S.A.*
Dr Richard Bryce, *University of Manchester, United Kingdom*
Ms Maria Teresa Buenavista, *University of Reading/Diamond Light Source/Medical Research Council Harwell, United Kingdom*
Dr Kepa Burusco-Goni, *University of Manchester, United Kingdom*
Professor Silvana Claudia Caglieri, *Universidad Tecnológica Nacional, Argentina*
Mr Allan Chappell, *None, United Kingdom*
Dr Ludovic Chaput, *Institut Curie, France*
Dr Matthieu Chavent, *SBCB, Biochemistry Department, United Kingdom*
Dr Olivier Delalande, *University of Rennes 1, France*
Professor Uri Dinur, *Ben-Gurion University, Israel*
Mr Matthieu Dreher, *INRIA, France*
Professor Dr Thomas Ertl, *University of Stuttgart, Germany*
Dr Nicolas Férey, *LIMSI CNRS - Université Paris-Sud, France*
Dr Philip Fowler, *Department of Biochemistry, University of Oxford, United Kingdom*
Mr Oscar Gillespie, *Royal Society of Chemistry, United Kingdom*
Dr David Glowacki, *University of Bristol, United Kingdom*
Mr Mike Goodstadt, *CNAG, Spain*
Professor Dr Helmut Grubmüller, *Max Planck Institute for Biophysical Chemistry, Germany*
Mr Yann Guarracino, *LBPA- ENS Cachan, France*
Mr Moritz Haag, *ETH Zurich, Switzerland*
Mr Kyle Hall, *University of Calgary, Canada*
Dr Ben Hall, *Microsoft Research Cambridge, United Kingdom*
Dr Steven Hayward, *University of East Anglia, United Kingdom*
Mr Jonathan Higham, *The University of Manchester, United Kingdom*
Ms Rachel Hill, *University of Nottingham, United Kingdom*
Professor Jonathan Hirst, *University of Nottingham, United Kingdom*
Mr Fouad Husseini, *University of Nottingham, United Kingdom*
Mr Georgios Iakovou, *University of East Anglia, United Kingdom*

Dr Graham Johnson, *University of California, San Francisco, U.S.A.*
Dr Mikhail Kabeshov, *University of Cambridge, United Kingdom*
Miss Caroline Karanu, *Kenyatta University, Kenya*
Dr Colin King, *Royal Society of Chemistry, United Kingdom*
Mr Alexandre Kouyoumdjian, *LIMSI CNRS - Université Paris Sud, France*
Mr Michael, *Krone, VISUS, University of Stuttgart, Germany*
Ms Anne Laustsen, *Aarhus University, Denmark*
Dr Stephen Laycock, *University of East Anglia, United Kingdom*
Dr Guillaume Levieux, *CNAM, France*
Dr Helen Lunn, *Royal Society of Chemistry, United Kingdom*
Mr Robbie Mackenzie, *University of Nottingham, United Kingdom*
Dr Juan Martinez-Sanz, *Institut Curie, France*
Mr Simon McIntosh-Smith, *University of Bristol, United Kingdom*
Miss Anne-Elisabeth Molza, *University of Rennes 1, IGDR CNRS UMR 6290, France*
Dr Matthieu Montes, *CNAM, France*
Dr Liliane Mouawad, *Institut Curie, France*
Mr Sam Mulholland, *University of Nottingham, United Kingdom*
Dr Dmitry Nerukh, *Aston University, United Kingdom*
Professor Arthur Olson, *The Scripps Research Institute, U.S.A.*
Dr Seán O'Donoghue, *CSIRO & Garvan, Australia*
Mr Nicolas Panel, *LPBA- ENS de Cachan, France*
Dr Julius Parulek, *Department of Informatics, University of Bergen, Norway*
Dr Anton S. Petrov, *Georgia Institute of Technology, U.S.A.*
Dr James Procter, *University of Dundee, United Kingdom*
Mr Mussa Quareshy, *University of Warwick, United Kingdom*
Dr Bruno Raffin, *INRIA, France*
Professor Dr Markus Reiher, *ETH Zurich, Switzerland*
Dr David Robinson, *University of Nottingham, United Kingdom*
Professor Christopher Rowley, *Memorial University of Newfoundland, Canada*
Dr Dmitri Rozmanov, *The University of Calgary, Canada*
Professor Mark Sansom, *Department of Biochemistry, University of Oxford,, United Kingdom*
Ms Katrine Skeby, *iNANO and Department of Chemistry, Aarhus University, Denmark*
Mr Stephen Skowron, *University of Nottingham, United Kingdom*
Dr Eric Sliwinski, *University of Cambridge, United Kingdom*
Mr Lewis Smeeton, *University of Birmingham, United Kingdom*
Ms Marisa Smith, *University Manchester, United Kingdom*
Dr Björn Sommer *Bio-/Medical Informatics Department, Bielefeld University, Germany*
Dr Melanie Stegman, *FAS Learning Technologies Program, U.S.A.*
Mr John Stone, *University of Illinois, U.S.A.*
Ms Ioanna Danai Styliari, *Nottingham University, United Kingdom*
Professor Dr Luba Tchertanov, *ENS Cachan, France*
Mr Thibault Tubiana, *Laboratoire de Biochimie Théorique, France*
Mr Jens Thomas, *Liverpool University, United Kingdom*
Mr Mikael Trellet, *LIMSI CNRS - Université Paris Sud, France*
Dr Abbie Trewin, *Lancaster University, United Kingdom*
Dr Eleanor Turpin, *University of Nottingham, United Kingdom*
Dr Jamie Twycross, *University of Nottingham, United Kingdom*
Miss Xun Wang, *University of Pittsburgh, U.S.A.*
Miss Gemma Wilkins, *Royal Society of Chemistry, United Kingdom*

Dr Christopher Woods, *University of Bristol, United Kingdom*
Dr Rilei Yu, *University of Oxford, United Kingdom*
Dr Jim Zheng, *School of Biomedical Informatics, The University of Texas Health Science Center at Houston, U.S.A.*
Dr Monica Zoppè, *Scientific Visualization Unit, Italy*